ADP-RIBOSYLATION: METABOLIC EFFECTS AND REGULATORY FUNCTIONS

Developments in Molecular and Cellular Biochemistry

Series Editor: Naranjan S. Dhalla, Ph.D., FACC

1. V.A. Najjar (ed.): *Biological Effects of Glutamic Acid and Its Derivatives.* 1981 ISBN 90-6193-841-4
2. V.A. Najjar (ed.): *Immunologically Active Peptides.* 1981 ISBN 90-6193-842-2
3. V.A. Najjar (ed.): *Enzyme Induction and Modulation.* 1983 ISBN 0-89838-583-0
4. V.A. Najjar and L. Lorand (eds.): *Transglutaminase.* 1984 ISBN 0-89838-593-8
5. G.J. van der Vusse (ed.): *Lipid Metabolism in Normoxic and Ischemic Heart.* 1989 ISBN 0-7923-0479-9
6. J.F.C. Glatz and G.J. van der Vusse (eds.): *Cellular Fatty Acid-Binding Proteins.* 1990 ISBN 0-7923-0896-4
7. H.E. Morgan (ed.): *Molecular Mechanisms of Cellular Growth.* 1991 ISBN 0-7923-1183-3
8. G.J. van der Vusse and H. Stam (eds.): *Lipid Metabolism in the Healthy and Diseased Heart.* 1992 ISBN 0-7923-1850-1
9. Y. Yazaki and S. Mochizuki (eds.): *Cellular Function and Metabolism.* 1993 ISBN 0-7923-2158-8
10. J.F.C. Glatz and G.J. van der Vusse (eds.): *Cellular Fatty-Acid-Binding Proteins, II.* 1993 ISBN 0-7923-2395-5
11. R.L. Khandelwal and J.H. Wang (eds.): *Reversible Protein Phosphorylation in Cell Regulation.* 1993 ISBN 0-7923-2637-7
12. J. Moss and P. Zahradka (eds.): *ADP-Ribosylation: Metabolic Effects and Regulatory Functions.* 1994 ISBN 0-7923-2951-1
13. V.A. Saks and R. Ventura-Clapier (eds.): *Cellular Bioenergetics: Role of Coupled Creatine Kinases.* 1994 ISBN 0-7923-2952-X

KLUWER ACADEMIC PUBLISHERS – DORDRECHT / BOSTON / LONDON

ADP-Ribosylation: Metabolic Effects and Regulatory Functions

Edited by

JOEL MOSS

Laboratory of Cellular Metabolism,
National Heart, Lung and Blood Institute, NIH,
Building 10, Room 5N-307
Bethesda, MD 20892, USA

and

PETER ZAHRADKA

Division of Cardiovascular Sciences,
St. Boniface General Hospital Research Center,
351 Tache Avenue,
Winnipeg, MB R2H 2A6, Canada

Reprinted from *Molecular and Cellular Biochemistry*, Volume 138 (1994)

Kluwer Academic Publishers
Dordrecht / Boston / London

Library of Congress Cataloging-in-Publication Data

ADP-ribosylation: metabolic effects and regulatory functions / edited
by Joel Moss and Peter Zahradka.
p. cm. -- (Developments in molecular and cellular biochemistry: 12)
ISBN 0-7923-2951-1
1. ADP-ribosylation. I. Moss, Joel. II. Zahradka, Peter,.
III. Series.
QP625, A29A347 1994
574, 87'6--dc20 94-21849
CIP

ISBN 0-7923-2951-1

Published by Kluwer Academic Publishers,
P.O. Box 17, 3300 AA Dordrecht, The Netherlands.

Kluwer Academic Publishers incorporates
the publishing programmes of
D. Reidel, Martinus Nijhoff, Dr W. Junk and MTP Press.

Sold and distributed in the U.S.A. and Canada
by Kluwer Academic Publishers,
101 Philip Drive, Norwell, MA 02061, U.S.A.

In all other countries, sold and distributed
by Kluwer Academic Publishers Group,
P.O. Box 322, 3300 AH Dordrecht, The Netherlands.

Printed on acid-free paper

Printed in the Netherlands

Molecular and Cellular Biochemistry:

An International Journal for Chemical Biology in Health and Disease

CONTENTS VOLUME 138, No. 1–2, 1994

ADP-RIBOSYLATION: METABOLIC EFFECTS AND REGULATORY FUNCTIONS
Joel Moss and Peter Zahradka, guest editors

Molecular and Cellular Biochemistry **138**: 1–2, 1994.

Preface

Over a quarter of a century has passed since Paul Mandel and his associates first established that ADP-ribose is synthesized by eucaryotic cells. During these 25 years, much has been accomplished in defining the basic biochemical and molecular characteristics of ADP-ribosylation reactions and in identifying the cellular functions of this modification system. It is now obvious that two classes of post-translational modification, mono(ADP-ribosyl)ation and poly(ADP-ribosyl)ation, are distinguishable in eucaryotes. The different metabolic and functional parameters associated with each reaction has led to a dichotomy within the field, with poly(ADP-ribose) being the domain of those predominantly interested in nuclear processes and investigators of mono(ADP-ribose) generally examining cytoplasmic events. The recent discoveries of ADP-ribosylation factors and cyclic ADP-ribose, however, have expanded the scope of this discipline in ways previously unforeseen and demonstrated that ADP-ribose can influence cellular events at any point between the plasma membrane and the nucleus. Within this context, ADP-ribose can be viewed as a participant in the receptor-mediated activation of second messenger systems, ion flux mechanisms, membrane transport processes, chromatin organization and nucleic acid metabolism.

Considering the current interest in cellular regulation and intracellular signalling systems, it is surprising that the contribution of ADP-ribosylation reactions to the modulation of a variety of specific cell processes, in parallel with other post-translational modifications such as phosphorylation, has not been generally recognized. While it was not feasible to cover all aspects of ADP-ribosylation, the thirty-one articles contained in this volume should provide a valuable overview of recent progress in the field within the context of cell control mechanisms. For the convenience of the reader, the various topics have been grouped into several sections: (a) poly(ADP-ribosyl)ation, (b) mono-ADP-ribosylation, (c) toxin mono-ADP-ribosylation, (d) inhibitors and activators, (e) protein modification with ADP-ribose and its analogues, and (f) non-modification forms of ADP-ribose. The contents of the individual chapters reflect the ideas of the contributors, many of whom have spent their careers attempting to resolve the biological functions of ADP-ribosylation. We hope that this publication will serve as a useful reference for those investigators that are new to the area as well as those who are actively studying ADP-ribosylation.

Dedication

This book is dedicated to Paul Mandel (1908-1992), a dedicated scientist and pioneer, whose laboratory was among the first to identify ADP-ribosylation as an essential cell process and who continued to be a leader in the field.

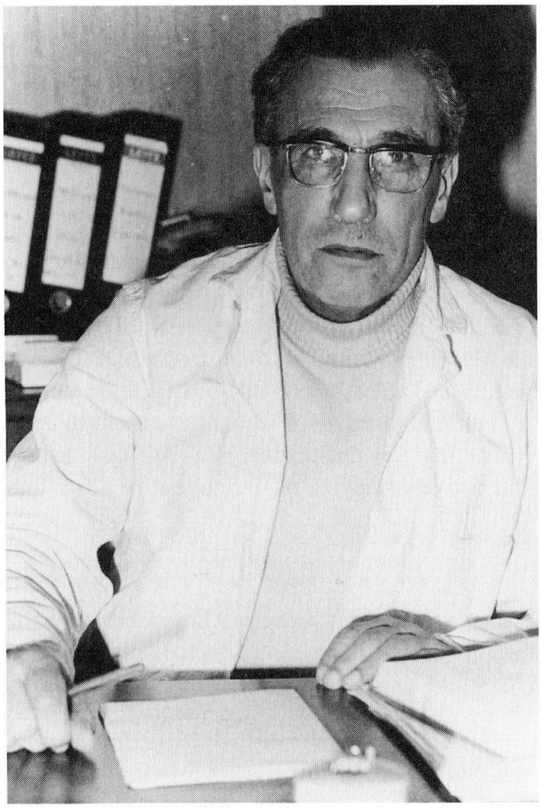

Dr. Paul Mandel
(1908–1992)
who pioneered the studies of ADP-Ribolysation

PART I

HISTORICAL PERSPECTIVE

Molecular and Cellular Biochemistry 138: 5–12, 1994.

Poly(ADP-ribose): Historical perspective

Takashi Sugimura[1] and Masanao Miwa[2]

[1]National Cancer Center, Tsukiji 5-chome, Chuo-ku, Tokyo 104 Japan,[2] Department of Biochemistry, Institute of Medical Sciences, The University of Tsukuba, Tsukuba, Ibaraki 305, Japan

Abstract

The early historical background of the discovery of poly(ADP-ribose) and the following development of science on poly(ADP-ribose) are reviewed. Fundamental knowledge on the natures of poly(ADP-ribose), poly(ADP-ribose) polymerase and enzymes degrading poly(ADP-ribose) are summarized with brief description on the methodology for their purification and characterization. Future prospect of research on biological significance of poly(ADP-ribose) has also been discussed briefly. (Mol Cell Biochem **138:** 5–12, 1994)

Key words: discovery of poly(ADP-ribose), phosphoribosyl-AMP(φ-ADP-ribose), poly(ADP-ribose) polymerase, phosphodiesterases, poly(ADP-ribose) glycohydrolase, branched structure of poly(ADP-ribose)

Initiation of studies on poly(ADP-ribose) in the laboratory of Dr. Paul Mandel

Studies on the ADP-ribosylation reaction have two different origins. One was the finding by the late Dr. Paul Mandel in Strassbourg in 1963 that poly(ADP-ribose) is formed in mammalian cell nuclei [1]. The other was the finding by Dr. R.J. Collier and Dr. A.M. Pappenheimer in 1964 [1] that NAD was required for inhibiting protein synthesis in a cell-free system by diphtheria toxin which led later to the discoveries of mono(ADP-ribosyl)ation reactions of cytoplasmic and membrane proteins of mammalian cells by bacterial toxins.

In this chapter, the historical background of the discovery and development of the early stage of science on poly(ADP-ribose) is described. The prospect for the future of science on poly (ADP-ribose) will be briefly mentioned, but avoidance of any overlapped descriptions with those in other chapters will be intended.

The first report on poly(ADP-ribose) formation by Drs. P. Chambon, J. Weil and P. Mandel was published in 1963 [1]. They found that the incorporation of (^{14}C-adenine)-labeled ATP into the acid-insoluble fraction by a nuclear preparation from chicken liver was enhanced 1,000 times by nicotinamide mononucleotide (NMN). At first, they suspected that the acid-insoluble product was poly(A). But Dr. Doly, who joined Dr. Mandel's group in the early phase of the study, told me that they soon realized that the product might not be poly(A), and they reported data on the nature of the product in 1966 [3].

Studies on poly(ADP-ribose) were also initiated in the National Cancer Center, Tokyo, by our group [4]. Using a nuclear preparation from rat liver, we confirmed the observation by Dr. Mandel's group that NMN greatly enhanced the formation of acid-insoluble material from [^{14}C-ade]ATP. Our studies were motivated by the idea that the biosynthesis of high-molecular weight compounds, such as RNA and poly(A) could be regulated by allosteric effectors of low molecular weight compounds, such as a nucleotide, NMN. We suspected that abnormality of a control regulatory system such as this might be related to autonomous growth of cancer cells.

We confirmed that the product was hydrolysed by snake venom phosphodiesterase, as reported by Dr. Mandel's group. However, we found that the product was resistant in alkaline conditions under which poly(A) should be hydrolyzed to adenosine 3'-monophosphate [4–6]. Furthermore, on paper chromatography, the mobility of the product, with an [^{14}C]adenine moiety, hydrolized, by the snake venom phosphodiesterase was different from that of 5'[^{14}C-

Address for offprints: T. Sugimura, National Cancer Canter, Tsukiji 5 Chome, Chuo-ku, Tokyo 104, Japan

ade]AMP but close to that of 5'[^{14}C-ade]ADP. The major hydrolysis product contained one molecule of adenine, two molecules of ribose and two molecules of phosphate. These phosphates were removed by alkaline phosphomonoesterase and were not in a phosphodiester bond or a pyrophosphate bond. Since only one mole of periodate was consumed per mole of this major hydrolysis product, the 2'-OH or the 3'-OH of one of the two riboses was not free, and was substituted by other residues. The major hydrolysis product was deduced to be 2'(or 3')-(5"-phosphoribosyl)-5'AMP, which was named "phosphoribosyl AMP (PR-AMP)" [7]. Information on this new poly(ADP-ribose) and PR-AMP were presented at the Annual Meeting of the Japanese Biochemical Society in Kyoto in 1966 and also communicated personally with Dr. Osamu Hayaishi's group in Kyoto University, who were investigating the biosynthesis of nicotinamide from L-tryptophan. Variously labeled NMN and NAD compounds were used for determination of the structure of the acid-insoluble product. The Kyoto group started also the research on poly(ADP-ribose) [8], and the competition between the Tokyo and Kyoto laboratories resulted in very rapid progress in the study on poly(ADP-ribose) in Japan.

At the end of Autumn, 1966, we were almost ready to submit a paper on the structure of PR-AMP and poly(ADP-ribose). At almost the same time the Kyoto group came to the same conclusion [7,8]. At the end of December, 1966, we were also excited by a report by Dr. Mandel's group, describing the structures of poly(ADP-ribose) and the major hydrolysis product, named φ-ADP-ribose corresponding to PR-AMP [3]. From these findings, it became obvious that NAD was formed from NMN and ATP by NAD pyrophosphorylase in nuclei, and that the ADP-ribose moiety of NAD was then converted to poly(ADP-ribose) with concomitant release of nicotinamide. Thus the main stream of research on poly(ADP-ribose) originated from the observation of NMN enhancement of acid-insoluble product formation from [^{14}C-ade]ATP by Dr. Mandel in 1963 and two streams, from Tokyo and Kyoto have joined. Meanwhile Dr. Petek and Ms. Doly in Dr. Mandel's group made much effort to determine the substitution of the 2'-OH group of one of the two ribose molecules [9].

I first met Dr. Mandel when he visited Tokyo on the occasion of International Congress of Biochemistry in 1967. It was a very pleasant experience for me to find out how broad minded Dr. Mandel was as a scientist. I realized that the fair competition, with mutual regard for the work of other groups and adequate citations of their papers is very important for maintaining joy in research and friendship among scientists. This experience was probably more important for me than the discovery of poly(ADP-ribose).

The review articles are chronologically available by T. Sugimura [10], by M.R. Purnell, P.R. Stone and W.J.D. Whish [11], by P. Mandel, H. Okazaki and C. Niedergang [12], by T. Sugimura and M. Miwa [13], by S. Shall [14], by P.A. Cerutti [15], by K. Ueda and O. Hayaishi [16], by J.C. Gall and C.K. Pearson [17], by K. Ueda [18], by F.R. Althaus and C. Richter [19], by G. de Murcia, A. Huletsky and G.G. Poirier [20] and by J. Moss and M. Vaughan [21], by M. Miwa and T. Sugimura [22], and by T. Boulikas [23].

Enzymes related to poly(ADP-ribose)

Poly(ADP-ribose) polymerase [EC 2.4.2.30]

Poly(ADP-ribose) polymerase is bound to chromatin but can be solubilized with high salt and purified to homogeneity. As described in detail in other chapters of this volume, the enzyme requires DNA with nicks or double strand breaks and NAD as substrate. Poly(ADP-ribose) polymerase was later demonstrated in the slime mold, *Physarum polycephalum* by Shall *et al.* [24] and it is thought to be present in all eucaryotes but not in procaryotes like *Escherichia coli*. However, there is as yet no conclusive evidence for its existence in the yeasts *Saccharomyces* and *Torulae*, although ribosyl-adenosine was reported as a contaminant in tRNA fractions from yeast by Hall [25], suggesting the existence of poly(ADP-ribose) polymerase and poly(ADP-ribose) in yeast.

As first reported by Shizuta's group [26], the purified enzyme is composed of three domains: DNA binding, automodification and NAD binding domains. Cloning of cDNA of human poly(ADP-ribose) polymerase and determination of its full length nucleotide sequence were reported in 1987. Later, the full length cDNAs were isolated from bovine, mouse, chicken, and *Drosophila melanogaster*. Several conserved sequences have bee demonstrated, for instance two zinc finger structures [27]. The zinc fingers of the enzyme have longer loops consisting of 28-30 amino acid residues than those of most other DNA-binding proteins. Differential functions of two zinc-finger structures are demonstrated as described in other Chapter. Recently, leucine zipper-like structures were found in the enzymes of *Drosophila* and other eukaryotes [28]. It would be interesting to see if poly(ADP-ribose) polymerase can bind to nuclear transcription factors or oncogene and tumor-suppressor gene products through the leucine zipper motif or in some other way.

Phosphodiesterases

Snake venom phosphodiesterase (EC.3.1.4.1). Snake venom phosphodiesterase can split the pyrophosphate bond of poly(ADP-ribose), yielding PR-AMP as described before, and 5'AMP should be produced from a terminus. The molar ratio of PR-AMP to 5'AMP reflects the chain length of poly(ADP-ribose). The hydrolysis proceeds "endonucleolytically" [10], in contrast to the reaction of rat liver phosphodiesterase, the mode of which proceeds "exonucleolytically" [10].

Rat liver phosphodiesterase (EC.3.1.3.1). Futai *et al.* purified a phosphodiesterase from the particulate fraction of a rat liver homogenate and showed that it splits poly(ADP-ribose) at pyrophosphate bonds [10]. This enzyme splits poly(ADP-ribose) exonucleolytically, but not endonucleolytically, and so does not produce oligomers of PR-AMP. The direction of hydrolysis is from the AMP terminus to the other terminus, at which ADP-ribosylation of protein is initiated to form poly(ADP-ribose) [10].

Plant phosphodiesterase. The phosphodiesterase isolated from cultured tobacco cells also hydrolyzes poly(ADP-ribose) at the pyrophosphate bonds. This enzyme also hydrolyzes the pyrophosphate bonds in the cap structure of the 5'-terminus of mRNA without hydrolyzing the phosphodiester bonds of RNA [29].

Poly(ADP-ribose) glycohydrolase

During studies on the breakdown of poly(ADP-ribose), the formation of ADP-ribose but not of PR-AMP was noticed, leading to the discovery of a new enzyme. This enzyme splitting ribose-ribose bonds was reported by Miwa *et al*, [30,31] and Ueda *et al.* [32]. This enzyme is also bound to chromatin. Its mode of hydrolysis is reported to be like exoglycosidase. However, there is as yet no information on whether the direction of hydrolytic processing is from the free terminus or from the terminus bound to protein. The chain elongation of poly(ADP-ribose) by addition of ADP-ribose at the free terminus, and removal of ADP-ribose from the free terminus, would better explain the regulatory mechanism through reversible elongation and shortening of poly (ADP-ribose).

ADP-ribosyl protein lyase

Initiation of the formation of poly(ADP-ribose) starts with ADP-ribosylation of protein. Proteins such as histone H1 serve as acceptors of ADP-ribosylation. The amino acid involved is glutamic acid and its carboxy group is esterified through the OH-group of C-1 of ADP-ribose [16]. Okayama *et al.* found a novel enzyme responsible for removal of ADP-ribose from protein and named it ADP-ribose-histone lyase [33]. The bonds between ADP-ribose and glutamic acid residues at amino acid positions 2, 14 and 116 in histone H1 and at position 2 of histone H2B are relatively labile and break down spontaneously, making enzyme assay difficult. Later, Williams *et al.* reported a hereditary disorder with neurological symptoms, in which 5'-phosphoribosyl glutamic acid is deposited in lysosomes [34]. A possible source of the 5'-phosphoribosyl glutamic acid is the ultimate degradation product of ADP-ribosylated histones. This disease may be caused by a defect in the activity of ADP-ribosyl protein lyase.

ADP-ribosylation of proteins occurs, in addition to histone H1, on topoisomerase I, calcium-magnesium dependent endonuclease and DNA polymerase β [19]. However, ADP-ribosylation of those enzymes in intact cells have not been proven.

Structure of poly(ADP-ribose)

Linear structure of poly(ADP-ribose)

Fig. 1 shows the linear structure of poly(ADP-ribose) and the enzymes involved in its synthesis and degradation. The chain length of poly(ADP-ribose) synthesized *in vitro* is estimated to range from a few to about 30 ADP-ribose units. The terminus, which is attached to protein should yield 5'-phosphoribose or 5'-phosphoribosyl glutamic acid or peptides. But as this terminal structure is not well preserved, during preparation of poly(ADP-ribose), there is not many papers demonstrating the presence of 5'-phosphoribose, 5'-phosphoribosyl glutamic acid or 5'-phosphoribosyl peptide. Steric configuration between nicotinamide and ribose on NMN is in β-form but that between ribose-ribose bond or PR-AMP is in α-form [19].

Branched structure of poly(ADP-ribose)

Miwa et al. first noticed that the molecular weight calculated from the ratio of PR-AMP to 5'AMP is not consistent with the S value determined by centrifugation. This suggested aggregation of poly(ADP-ribose) chains themselves or a branched structure, as in the case of polysaccharides such as starch and glycogen. Poly(ADP-ribose) prepared enzymatically *in vitro* and purified by column chromatography is heterogeneous in chain length, although the average chain length is about 20 ADP-ribose units as calculated from the molar ratio of PR-AMP to 5'AMP. This heterogeneity was demonstrated as a ladder on electrophoresis in polyacrylamide gel [35], shorter chains moving faster. It was suspected that the molecules, remaining at the top of the gel had chains of more than 100 ADP-ribose units. Electromicroscopy showed that the molecules had Christmas-tree like figures [36,37].

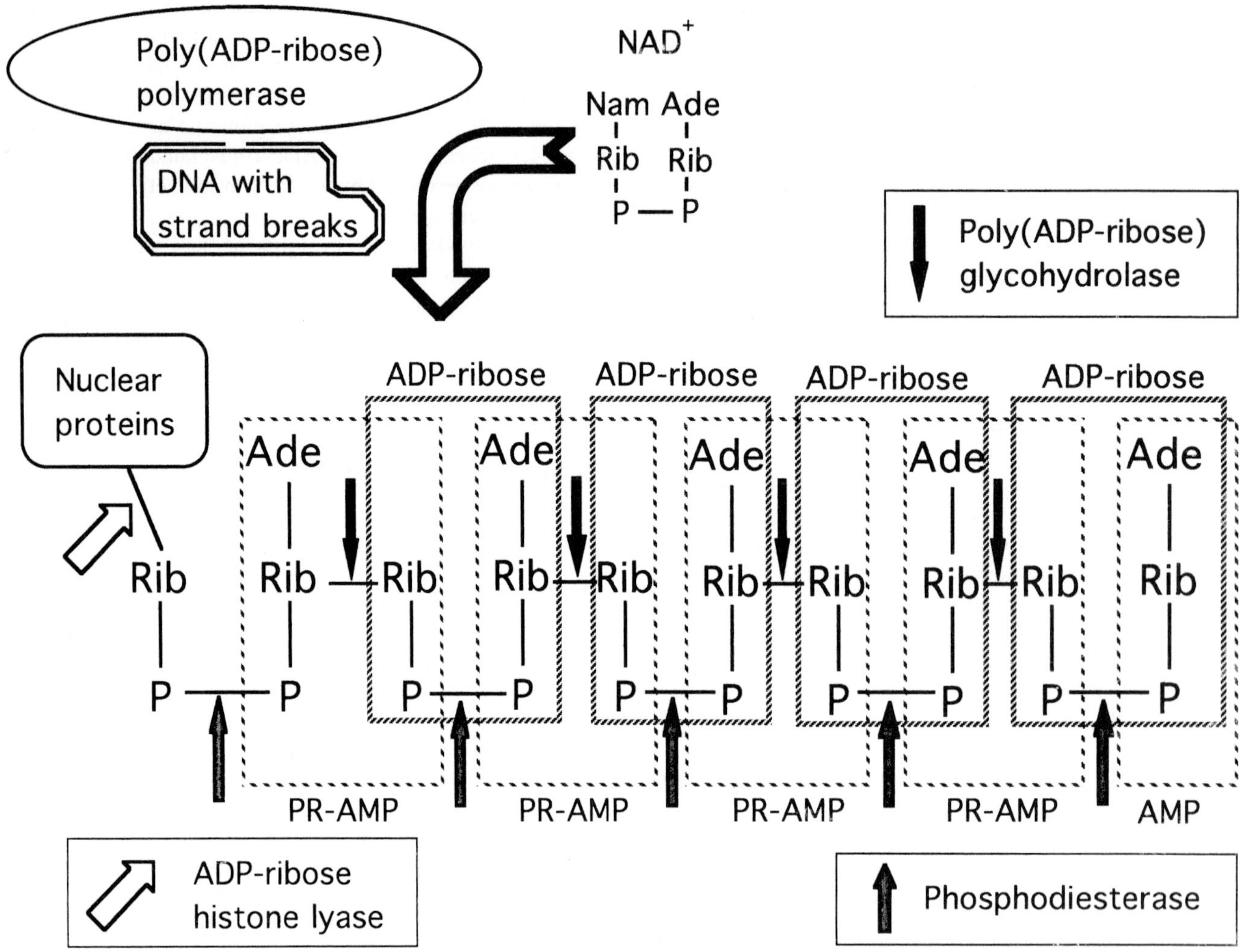

Fig. 1. Enzymes involved in synthesis and degradation of poly (ADP-ribose).

A minor peak in addition to those of PR-AMP and 5′–AMP was observed on DEAE sephadex column chromatography after complete hydrolysis of poly(ADP-ribose) by snake venom phosphodiesterase. The material in this peak was collected and subjected to chemical and physiochemical analyses. The material consisted of one molecule of adenine, three molecules of ribose and three molecules of phosphomonoesterase sensitive phosphates at the 5′ position of each ribose. Final structure was determined to be *O*-α-D-ribofuranosyl-(1‴ → 2″)-*O*-α-D-ribofuranosyl-(1″ → 2′)-adenosine-5′,5″,5‴-tris(phosphate), abbreviated as $(PR)_2$-AMP (38), as shown in Fig. 2. Poly(ADP-ribose) synthesized enzymatically *in vitro* and eluted with 0.22-0-40 M phosphate buffer from a hydroxyapatite column showed a molar ratio of 5′AMP, $(PR)_2$-AMP and PR-AMP of about 1: 0.5: 24, and is consistent of 2 linear chains of poly(ADP-ribose) with an average chain length of 25 ADP-ribose units and one branched structure. Snake venom phosphodiesterase hydrolyses branched poly(ADP-ribose) more slowly than linear poly(ADP-ribose), indicating the difference of conformation [19].

Occurrence of poly(ADP-ribose) and other compounds with related structure

The natural occurrence of poly(ADP-ribose) is supported by much circumstantial evidence, including the ubiquitous presence of enzyme synthesizing and degrading poly(ADP-ribose), which are associated with chromatin. Antibody against poly(ADP-ribose) was first prepared by Kanai *et al.* in 1974 [39]. Later, antibody against poly(ADP-ribose) was found in the sera of patients with systemic lupus erythematosus, but not in those with rheumatoid arthritis [40]. The monoclonal antibody has been available [41]. The

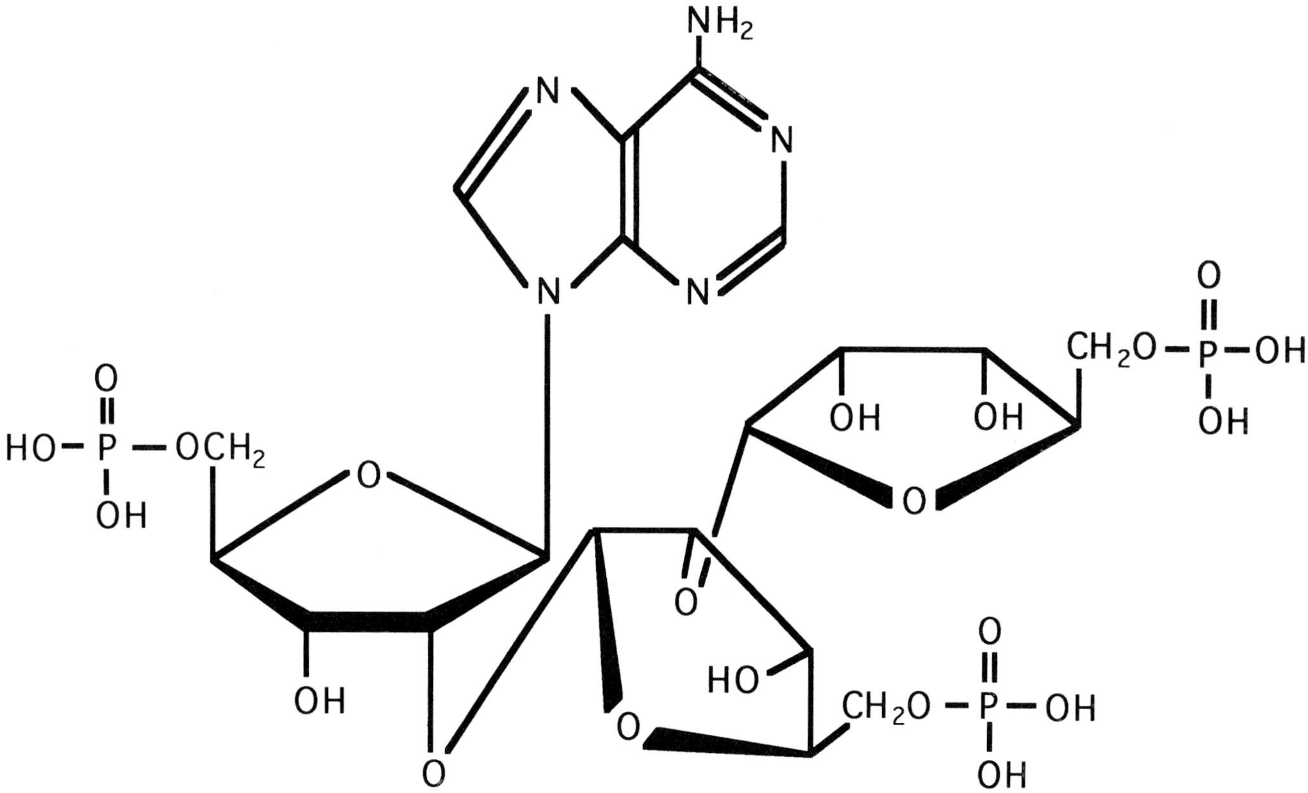

Fig. 2. Structure of isolated branched portion of poly(ADP-ribose), namely *O*-α-D-ribofuranosyl (1‴ → 2″)-*O*-α-ribofuranosyl (1″ → 2′) adenosine 5′,5″5‴-tris(phosphate).

antibody has been widely used for immunohistochemical demonstration of the presence of poly(ADP-ribose). All types of mammalian nucleated cells, except granulocytes with segmented nucleus, stain immunohistochemically for poly(ADP-ribose) [42]. The specificity of the antibody against poly(ADP-ribose) needs careful consideration because poly (A).poly (U) can induce antibody reacting with poly(ADP-ribose) [43]. Several chemical methods have been reported to demonstrate poly(ADP-ribose) *in vivo* [19].

Other unique structures related to poly(ADP-ribose) which have ribose-ribose bonds include polyribophosphate of *Hemophilus influenzae* type b, in which the ribose moieties of polyribophosphate chains have β(1-1) bonds with the ribose moieties of other polyribophosphate chains [44]. ADP-ribosyl NAD was isolated from *Azotobacter vinelandii* [45]. The prosthetic group of citrate lyase form *Klebsiella aerogenes* has (1 → 2) ribose-ribose bonds in 2′-(5″-phosphoribosyl) dephospho-CoA [46]. It would be interesting to determine whether poly(ADP-ribose) glycohydrolase can hydrolyze these ribose-ribose bonds.

Future prospect

The most urgent matter to be elucidated is the biological function of poly(ADP-ribosyl)ation. The observation made by Shall's group in Sussex that 3-aminobenzamide (3AB), an inhibitor of poly(ADP-ribose) polymerase, causes a delay in rejoining of DNA strands after dimethyl sulfate treatment of cultured cells was a landmark in DNA repair studies [47]. Recent findings using cellular mutants with decreased poly(ADP-ribose) polymerase activity [48.49] and using antisense poly(ADP-ribose) polymerase RNA [50] are consistent with a role in DNA repair and/or cell growth functions, as previously suggested by experiments using inhibitors of poly(ADP-ribose) polymerase. However it is still unknown how poly(ADP-ribose) polymerase and its inhibitors are really involved in DNA repair. Satoh and Lindahl showed poly(ADP-ribose) polymerase to bind to DNA strand breaks and DNA excision repair to occur after poly(ADP-ribose) polymerase was removed by auto-poly(ADP-ribosyl)ation [51]. The interaction of poly(ADP-ribose) polymerase with nicks or double strand breaks is possible through its unique zinc finger structures. Interactions of poly(ADP-ribose) polymerase with other proteins might also occur [28].

Lectin-stimulated blast formation of lymphocytes is accompanied by marked elevation of the transcript from the gene of poly(ADP-ribose) polymerase, suggesting a biological relationship between DNA replication and ADP-ribose) polymerase [22]. The promoter structure of DNA polymerase β gene has similarity with that of poly(ADP-ribose) polymerase gene [52]. 3AB and benzamide, inhibitors of poly(ADP-ribose) polymerase, caused sister chromatid exchanges [22], and that benzamide caused deletion of exogenously transfected oncogenes, resulting in reversion of the transformed phenotype of NIH3T3 cells to flat cell phenotype [53].

Thus, poly(ADP-ribose) synthesis has been suggested to be involved in various physiological events, including DNA excision repair, DNA recombination, DNA replication, cell differentiation, cell growth and transformation [22]. Its involvement in carcinogeneses *in vivo* and in longevity make it particularly interesting [22,23,54]. Potentiation of the antitumor effect of DNA damaging drugs by inhibitors of poly(ADP-ribose) polymerase induces endonuclease-mediated apoptosis in tumor cells [55].

Recently cyclic ADP-ribose (i.e. cyclized ADP-ribose having an *N*-glycosyl linkage between the anomeric carbon of the terminal ribose unit and the amino group at the 6-position of the adenine ring [56]) was found to be another derivative of ADP-ribose produced enzymatically from NAD. Cyclic ADP-ribose can increase calcium induced calcium release, apparently through a different mechanism from that of inositol 1,4,5-tris(phosphate) (IP_3). Observation of dramatic change of pyridine nucleotides after fertilization of sea-urchin eggs led investigator to discovery of cyclic ADP-ribose [57]. Okamoto *et al.* found that cADP-ribose and calcium, but not IP_3, induced insulin secretion in digitonin-permeabilized islets [58].

So far, the reports of ADP-ribosylation of proteins produced by oncogenes and tumor suppressor genes *in vivo* are not available. However, it is known that ADP-ribosylation of p21 of *v*-Ha-*ras* and *c*-Ha-*ras* proteins occurs *in vitro* catalyzed by purified endogenous ADP-ribosyltransferases [22]. In addition, p21 ARF(ADP-ribosylation factor), which is known as one of GTP binding proteins that enhances ADP-ribosylation by choleragen, is suggested to be related to the structure and function of Golgi-derived vesicles [59].

Current progress in biotechnology should be promising. Gene targeting for enzymes involved in the metabolism of poly(ADP-ribose) or isolation of mutant animals completely lacking the above enzymes should answer the key question of "biological function" of poly(ADP-ribose) polymerase and the poly(ADP-ribosyl)ation reaction.

It is reasonable to anticipate that new and important observation, elucidating the role of poly(ADP-ribosyl)ation reaction on carcinogenesis, will be made soon.

References

1. Chambon P, Weil JD, Mandel P: Nicotinamide mononucleotide activation of a new DNA-dependent polyadenylic acid synthesizing nuclear enzyme. Biochem Biophys Res Commun 11: 39-43, 1963
2. Collier RJ, Pappenheimer AM Jr: Studies on the mode of action of diphtheria toxin. II Effect of toxin on amino acid incorporation in cell-free systems. J Exp Med 120: 1019-1039, 1964
3. Chambon P, Weil JD, Doly J, Strosser MT, Mandel P: On the formation of a novel adenylic compound by enzymatic extracts of liver nuclei. Biochem Biophys Res Commun 25: 638-643, 1966
4. Fujimura S, Sugimura T, Okabe K, Yoshida T: NMN-activated poly(A) polymerase in nuclei from rat liver and hepatoma cells (In Japanese). Proc Ann Meeting of Jpn Biochem Soc 38: 691, 1965
5. Hasegawa S, Fujimura S, Shimizu Y, Okuyama H, Sugimura T: Venom phosphodiesterase hydrolysis product of ADPR polymer synthesized enzymatically from NAD (In Japanese). Proc Ann Meeting of Jpn Biochem Soc 39: 194, 1967
6. Fujimura S, Hasegawa S, Sugimura T: Nicotinamide mononucleotide-dependent incorporation of ATP into acid-insoluble material in rat liver nuclei preparation. Biochim Biophys Acta 134: 496-499, 1967
7. Fujimura S, Hasegawa S, Shimizu Y, Sugimura T: Polymerization of the adenosine 5'-diphosphate-ribose moiety of nicotinamide-adenine dinucleotide by nuclear enzyme. I. Enzymatic reactions. Biochim Biophys Acta 145: 247-259, 1967
8. Nishizuka Y, Ueda K, Nakazawa K, Hayaishi O: Studies on the polymer of adenosine diphosphate ribose. 1. Enzymatic formation from nicotinamide adenine dinucleotide in mammalian nuclei. J Biol Chem 242: 3164-3171, 1967
9. Doly J, Petek F: Étude de la structure d'un composé "poly(ADP-ribose)" synthétisé par des extraits nucléaires de foie de poulet. CR Hebd Seanc Acad Sci Ser D Sci Nat 263: 1341-1344, 1966
10. Sugimura, T: Poly(adenosine diphosphate ribose). Prog Nucleic Acid Res Mol Biol 13: 127-151, 1973
11. Purnell MR, Stone PR, Whish WJD: ADP-ribosylation of nuclear proteins: Biochem Soc Trans 8: 215-227, 1980
12. Mandel P. Okazaki H, Niedergang C: Poly(Adenosine diphosphate ribose). Prog Nucleic Acid Res Mol Biol 27: 1-51, 1982
13. Sugimura T, Miwa M: Poly(ADP-ribose) and cancer research. Carcinogenesis 4: 1503-1506, 1983
14. Shall S: ADP-ribose in DNA repair: a new component of DNA excision repair. Adv Radiat Biol 11: 1-69, 1984
15. Cerutti PA: Prooxidant states and tumor promotion. Science 227: 375-381, 1985
16. Ueda K, Hayaishi O: ADP-ribosylation Annu Rev Biochem 54: 73-100, 1985
17. Gaal JC, Pearson CK: Eukaryotic nuclear ADP-ribosylation reactions. Biochem J 230: 1-18, 1985
18. Ueda K: Nonredox reactions of pyridine nucleotides. In: D. Dolphin, P. Rozanne and O. Arramovic (eds.). Pyridine Nucleotide Coenzymes: Chemical, Biochemical and Medical Aspects. John Wiley & Sons, Inc., New York, 1987, pp 549-597
19. Althaus FR, Richter C: ADP-Ribosylation of Proteins: Enzymology and Biological Significance. Springer-Verlag, Berlin, 1987, 1-230
20. De Murcia G, Huletsky A, Poirier GG: Review: modulation of chromatin structure by poly(ADP-ribosyl)ation. Biochem Cell Biol

66: 625-635, 1988

21. Moss J, Vaughan M: ADP-ribosylation of guanyl nucleotide-binding regulatory proteins by bacterial toxins. Adv Enzymol 61: 303-379, 1988
22. Miwa M, Sugimura T: ADP Ribosylation and carcinogenesis. In J. Moss and M. Vaughan (ed.). ADP-Ribosylating Toxins and G Proteins: Insight into Signal Transduction. American Society for Microbiology, Washington DC, 1990, pp 543-560
23. Boulikas T: Relation between carcinogenesis, chromatin structure and poly(ADP-ribosylation)(Review). Anticancer Res 11: 489-528, 1991
24. Brightwell MO, Leech CE, O'Farrell MK, Wish JD, Shall S: Poly(adenosine diphosphate ribose) polymerase in *Physarum polycephalum*. Biochem J 1476: 119-129, 1975
25. Hall RH: A general procedure for the isolation of "minor" nucleosides from ribonucleic acid hydrolysates. Biochemistry 4: 661-670, 1965
26. Kameshita I, Matsuda Z, Taniguchi T, Shizuta Y: Poly(ADP-ribose) synthetase. Separation and identification of three proteolytic fragments as the substrate binding domain, the DNA binding domain, and the automodification domain. J Biol Chem 259: 4770-4776, 1984
27. Uchida K, Miwa M: Poly(ADP-ribose) polymerase: Structural conservation among different classes of animals and its implications. Moll Cell Biochem (in press)
28. Uchida K, Hanai S, Ishikawa K, Ozawa Y, Uchida M, Sugimura T, Miwa M: Cloning of cDNA encoding *Drosophila* poly(ADP-ribose) polymerase: Leucine zipper in the auto-modification domain. Proc Natl Acad Sci USA 90: 3481-3485, 1993
29. Shinshi H, Miwa M, Sugimura T, Shimotohno K, Miura K: Enzyme cleaving the 5'-terminal methylated blocked structure of messenger RNA. FEBS Lett 65: 254-257, 1976
30. Miwa M, Sugimura T: Splitting of the ribose-ribose linkage of poly(adenosine diphosphate-ribose) by a calf thymus extract. J Biol Chem 246: 6362-6364, 1971
31. Miwa M, Tanaka M, Matsushima T, Sugimura T: Purification and properties of a glycohydrolase from calf thymus splitting ribose-ribose linkages of poly(adenosine diphosphate ribose). J Biol Chem 249: 3475-3482, 1974
32. Ueda K, Oka J, Narumiya S, Miyakawa N, Hayaishi O: Poly ADP-ribose glycohydrolase from rat liver nuclei, a novel enzyme degrading the polymer. Biochem Biophys Res Commun 46: 516-523, 1972
33. Okayama H, Honda M, Hayaishi O: Novel enzyme from rat liver that cleaves an ADP-ribosyl histone linkage. Proc Natl Acad Sci USA 75: 2254-2257, 1978
34. Williams JC, Chambers JP, Liehr JG: Glutamyl ribose 5-phosphate storage disease. A hereditary defect in the degradation of poly(ADP-ribosyl)ated proteins. J Biol Chem 259: 1037–1042, 1984
35. Tanaka M, Hayashi K, Sakura H, Miwa M, Matsushima T, Sugimura T: Demonstration of high molecular weight poly-(adenosine diphosphate ribose). Nucleic Acids Res 5: 3183-3194, 1978
36. De Murcia G, Jongstra-Bilen J, Ittel M-E, Mandel P, Delain E: Poly(ADP-ribose) polymerase automodification and interaction with DNA: Electron microscopic visualization. EMBO J 2: 543-548, 1983
37. Hayashi K, Tanaka, M., Shimada T, Miwa M, Sugimura T: Size and shape of poly(ADP-ribose): Examination by gelfiltration, gel electrophoresis and electron microscopy. Biochem Biophys Res Commun 112: 102-107, 1983
38. Miwa M, Ishihara M, Takishima S, Takasuka N, Maeda M, Yamaizumi Z, Sugimura T, Yokoyama S, Miyazawa T: The branching and linear portions of poly(adenosine diphosphate ribose) have the same $\alpha(1\rightarrow 2)$ ribose-ribose linkage. J Biol Chem 256: 2916-2921, 1981
39. Kanai Y, Miwa M, Matsushima T, Sugimura T: Studies on anti-poly(adenosine diphosphate ribose) antibody. Biochem Biophys Res Commun 59: 300-306, 1974
40. Kanai Y, Kawaminami Y, Miwa M, Matsushima T, Sugimura T, Moroi Y, Yokohari R: Naturally occurring antibodies to poly(ADP-ribose) in patients with systemic lupus erythematosus. Nature 265: 175-177, 1977
41. Kawamitsu H, Hoshino H, Okada H, Miwa M, Momoi H, Sugimura T: Monoclonal antibodies to poly(adenosine diphosphate ribose) recognize different structures. Biochemistry 23: 3771-3777, 1984
42. Ikai K, Ueda K, Fukushima M, Nakamura T, Hayaishi O: Poly(ADP-ribose) synthesis, a marker of granulocyte differentiation. Proc Natl Acad Sci USA 77: 3682-3685, 1980
43. Kanai Y, Sugimura T, Matsushima T: Induction of specific antibodies to poly(ADP-ribose) in rabbits by double-stranded RNA, poly(A).poly(U). Nature 274: 809-812, 1978
44. Rosenberg E, Zamenhof S: A new naturally occurring disaccharide, β-D-ribofuranosyl β-D-ribofuranoside. J Biol Chem 237: 1040-1042, 1962.
45. Imai T, Okuda S, Suzuki S: Natural occurrence of a new pyridine nucleotide consisting of nicotinamide adenine dinucleotide and adenosine diphosphate ribose joined through a ribosidic link. J Biol Chem 244: 4547-4544, 1969
46. Oppenheimer NJ, Singh M, Sweeley CC, Sung SJ, Srere PA: The configuration and location of the ribosidic linkage in the prosthetic group of citrate lyase (*Klebsiella aerogenes*). J Biol Chem 254: 1000-1002, 1979
47. Durkacz BW, Omidiji O, Gray DA, Shall S: (ADP-ribose)n participates in DNA excision repair. Nature 283: 593-596, 1980
48. Chatterjee S, Hirschler NV, Petzold SJ, Berger SJ, Berger N: Mutant cells defective in poly(ADP-ribose) synthesis due to stable alterations in enzyme activity or substrate availability. Exptl Cell Res 184: 1-5, 1989
49. Yoshihara K, Itaya A, Hironaka T, Sakuramato S, Tanaka Y, Tsuyuki M, Inada Y, Kamiya T, Ohnishi K, Honma M, Kataoka E, Mizusawa H, Uchida M, Uchida K, Miwa M: Poly(ADP-ribose)polymerase-defective mutant cell clone of mouse L1210 cells, Exptl Cell Res 200: 126-134, 1992
50. Ding R, Pommier Y, Kang VH, Smulson M: Depletion of poly(ADP-ribose) polymerase by antisense RNA expression results in a delay in DNA strand break rejoining. J Biol Chem 267: 12804-12812, 1992
51. Satoh MS, Lindahl T: Role of poly(ADP-ribose) formation in DNA repair. Nature 356: 356-358, 1992
52. Ogura T, Nyunoya H, Takahashi-Masutani M, Miwa M, Sugimura T, Esumi H: Characterization of a putative promoter region of the human poly(ADP-ribose) polymerase gene: structural similarity to that of DNA polymerase ß gene. Biochem Biophys Res Commun 167: 701-710., 1990
53. Nakayasu M, Shima H, Aonuma S, Nakagawa H, Nagao M, Sugimura T: Deletion of transfected oncogenes from NIH3T3 transformants by inhibitors of poly(ADP-ribose) polymerase. Proc Natl Acad Sci USA 85: 9066-9070, 1988
54. Grube K, Bürkle A: Poly(ADP-ribose) polymerase activity in mononuclear leukocytes of 13 mammalian species correlates with species-specific life span. Proc Natl Acad Sci USA 89: 11759-11763, 1992
55. Rice WG, Hillyer CD, Harten B, Schaeffer CA, Dorminy M, Lackey III DA, Kirsten E, Mendeleyev J, Buki KG, Hakam A, Kun E: Induction of endonuclease-mediated apoptosis in tumor cells by C-nitroso-substituted ligands of poly(ADP-ribose) polymerase. Proc Natl Acad Sci USA 89: 7703-7707, 1992
56. Lee HC, Walseth TF, Bratt GT, Hayes RN, Clapper DL: Structural determination of a cyclic metabolite of NAD^+ with intracellular Ca^{2+}-mobilizing activity. J Biol Chem 264: 1608-1615, 1989
57. Galione A: Cyclic ADP-Ribose: A new way to control calcium. Science 259:L 325-326, 1993

58. Takasawa S, Nata K, Yonekura H, Okamoto H: Cyclic ADP-Ribose in insulin secretion from pancreatic ß cells. Science 259: 370-373, 1993
59. Serafini T, Orci L, Amherdt M, Brumer M, Kahn RA, Rothman JE: ADP-ribosylation factor is a subunit of the coat of Golgi-derived COP-coated vesicles: a novel role for a GTP-binding protein. Cell 67: 239-253, 1991

Addendum

Scientists all over the world who are working on poly(ADP-ribose) and ADP-ribosylation are very friendly and often met at the occasions of International Meetings of ADP-ribosylation taken place in Hamburg, Germany (by H. Hilz, 1972), Bethesda, USA (by O. Hayaishi, 1974), Blankenese, Germany (by H. Hilz, 1976), Bethesda, USA (by M. Smulson and T. Sugimura, 1979), Tokyo, Japan (by M. Miwa, O. Hayaishi, S. Shall, M. Smulson and T. Sugimura, 1982), Vitznau, Switzerland (by F.Althaus, H. Hilz and S. Shall, 1984), Forth Worth, USA (by M. Jacobson and E. Jacobson, 1987), Quebec, Canada (by G. Poirier and P. Moreau, 1991) and Tsukuba, Japan (by M. Miwa, H. Esumi, K. Ueda and S. Shall, 1992). I would like to mention that Dr. Paul Mandel attended all conferences except the last two. In the Quebec meeting, through international phone-call system we could hear his voice and only a few weeks after the Tsukuba meeting, we were all shocked by the news that he passed away on October 6, 1992.

PART II

POLY(ADP-RIBOSYL)ATION

Molecular and Cellular Biochemistry **138**: 15–24, 1994.

Structure and function of poly(ADP-ribose) polymerase

Gilbert de Murcia[1], Valérie Schreiber[1], Miguel Molinete[1], Bénédicte Saulier[2], Olivier Poch[3], Murielle Masson[1], Claude Niedergang[1] and Josiane Ménissier de Murcia[1]

Ecole Supérieure de Biotechnologie de Strasbourg, Unité de Cancérogénèse et de Mutagénèse Moléculaire et Structurale du Centre National de la Recherche Scientifique, Boulevard Sébastien Brant, 67400 Illkirch-Graffenstaden, France[1], Institut de Biologie Moléculaire et Cellulaire, Unité des Mécanismes de Division Cellulaire et du Développement, 15 rue René Descartes, 67084 Strasbourg Cedex, France[3], Laboratoire de Biologie et Génétique du Développement, Centre National de la Recherche Scientifique, Université de Rennes, 35042 Rennes Cedex, France[2]

Abstract

Poly(ADP-ribose) polymerase (PARP) participates in the intricate network of systems developed by the eukaryotic cell to cope with the numerous environmental and endogenous genetoxic agents. Cloning of the PARP gene has allowed the development of genetic and molecular approaches to elucidate the structure and the function of this abundant and highly conserved enzyme. This article summarizes our present knowledge in this field. (Mol Cell Biochem **138**: 15–24, 1994)

Key words: zinc–finger protein, nuclear localization signal, base excision repair, microinjection, dominant-negative mutant, NAD metabolism

Introduction

DNA-strand breaks introduced in the genome of eukaryotic cells exposed to radiation or genotoxic agents triggers the endogenous synthesis of poly(ADP-ribose) by the enzyme poly(ADP-ribose) polymerase (PARP, EC 2.4.2.30). During DNA repair, PARP catalyses the transfer of the ADP-ribose moiety from its substrate, NAD^+, to a limited number of protein acceptors involved either in chromatic architecture or in DNA metabolism (for reviews, see 1). Therefore poly(ADP-ribosyl)ation is a posttranslational modification of proteins. Although the precise biological function(s) of the enzyme and the polymer have not been fully elucidated, PARP has recently emerged as a critical regulatory component of the cellular response to DNA damage. Cloning of the PARP gene has opened new areas of investigation, and allowed the development of novel molecular biological approaches aimed at better understanding the biological function of poly(ADP-ribosyl)ation reactions.

The present article reviews the recent findings dealing with the structure-function relationship of this abundant nuclear enzyme, with a particular emphasis on the work performed in our laboratory.

Modular organization of the PARP molecule

Poly(ADP-ribose) polymerase is a multifunctional enzyme. Three main functional elements were identified in the PARP molecule by Kameshita *et al.* [2] following limited proteolysis: a 46 kDa fragment including the DNA binding domain, a central 22 kDa polypeptide fragment containing the automodification sites and a C-terminal fragment of 54 kDa bearing the NAD-binding domain (Fig. 1a). This modular aspect of the PARP molecular structure was fully confirmed by site-directed mutagenesis and deletion mapping experiments (see below).

Address for offprints: G. de Murcia, Ecole Supérieure de Biotechnologie de Strasbourg, Unité de Cancérogénèse et de Mutagénèse Moléculaire et Structurale du Centre National de la Recherche Scientifique, Boulevard Sébastien Brant, 67400 Illkirch-Graffenstaden, France.

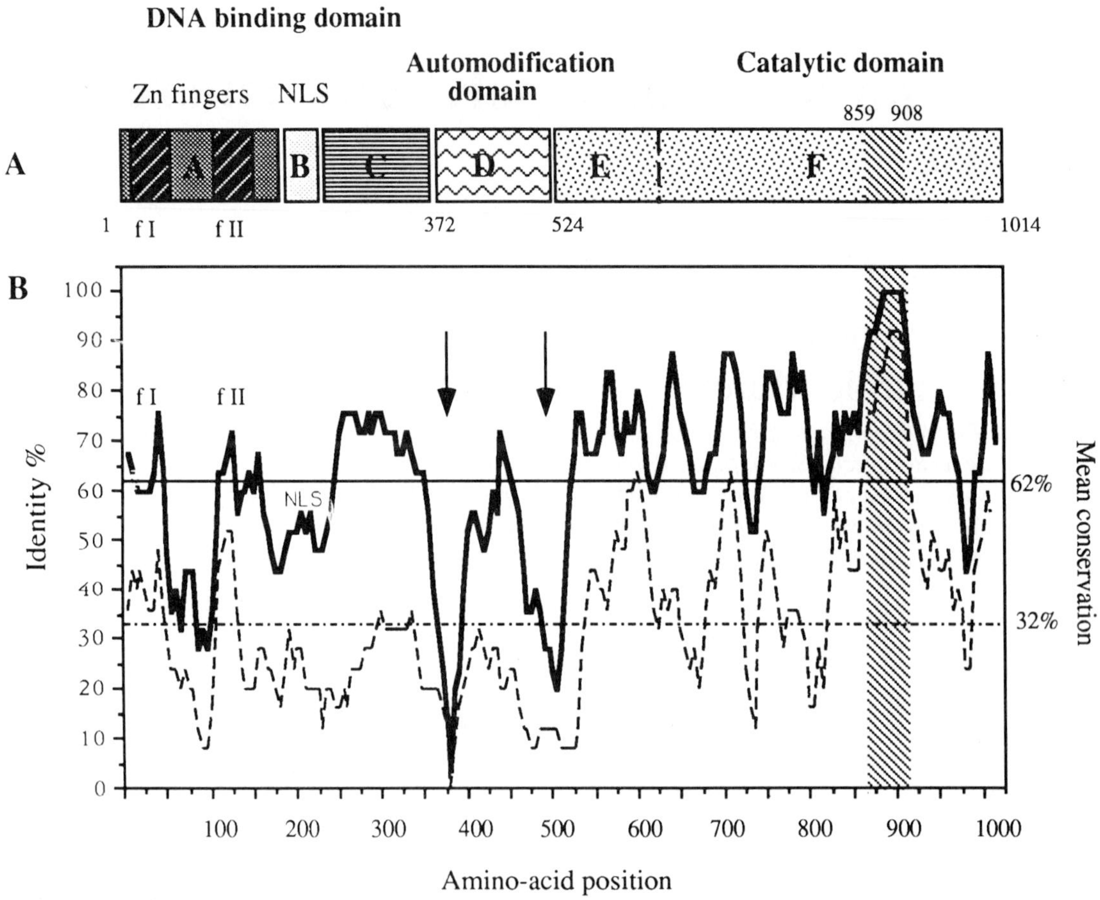

Fig. 1. **(a)** Schematic representation of the three functional domains of human PARP, obtained after mild trypsin and papain digestion, according to Kameshita *et al.* [2]. **(b)** Identity profiles obtained after alignment of the human [3], murine [4], bovine [5], chicken [6], xenopus [7] (accession number Z12139) and *Drosophila* [8] PARP sequences: the solid line corresponds to the profile calculated only with the vertebrate PARP sequences, the dashed line corresponds to that of the vertebrates and the invertebrate sequences. Each position of the alignment was scored as +1 when the amino acid was strictly conserved in all the compared sequences. The average percentage of identity was calculated within a window of 20 amino acids running with a step of 5 residues. The arrows indicate the regions of lowest conservation common to the two profiles.

This organization of concatenated functional elements is also clearly reflected by the identity profiles (Fig. 1b) obtained from the multiple alignment of the human [3], mouse [4], bovine [5], chicken [6] xenopus [7], and *Drosophila* [8] PARP sequences. Careful examination of the amino acid sequences revealed an extensive conservation at the amino acid level. The two regions of lowest conservation (indicated by arrows) coincide perfectly with the major interdomains previously determined by limited proteolysis [2]. This holds for the vertebrate sequences (solid line) or all the PARP sequences including that of *Drosophila* (dashed line). Inclusion of the *Drosophila* sequence dramatically decreases the overall sequence conservation from 62% to 32%. This divergence, particularly visible in region C (unknown function) and in region D (automodification and putative dimerization functions), may result from evolutionary distance and/or from the existence of distinct function(s) specific to the invertebrate enzyme. However, the conserved residues are clustered into blocks of strong conservation which correspond precisely to the PARP functional domains and subdomains: the zinc fingers fI and fII involved in DNA strand-break recognition [9,10] (subdomain A), the bipartite nuclear localization signal [11] (subdomain B) and the catalytic region [12–14] (E,F).

The C-terminal region of PARP (subdomains E,F) is by far the most conserved part of the protein. Region F (catalytic domain) in particular contains a block of 50 amino acids (residues 859–908) strictly conserved among the vertebrates (100%) and highly conserved (92%) among all species (see Fig. 6). This region can be considered now as the PARP signature.

The DNA binding domain (DBD)

Since the work of Zahradka and Ebisuzaki [15], we know that PARP is a metalloenzyme and needs zinc for its activity. Using the Energy Dispersive X-ray Fluorescence (EDXRF) technique, we determined that each PARP molecule contains two Zn^{2+} ions [16]. The zinc-blot assay, which was then developed [17], allowed us to locate the metal binding sites within the 29 kDa N-terminal region (domain A) of the purified calf thymus PARP [16]. Because this region is a part of the DBD, the function of metal ions in the DNA-binding properties of PARP was questioned. Indeed, from South-Western experiments we demonstrated that this domain binds nicked DNA in a zinc-dependent manner [16,9].

involved in DNA damage and repair including PARP, XPA-C [18], Fpg [19] and Kin-17 (Mazin *et al.*[1] submitted for publication).

There are two other distinctive features of domain A. First, there is a set of invariable aromatic residues (Y9, Y13, W51, H53, F57, F113, Y117, W157, H159, F163 and F184), some of which are duplicated; i.e. the cluster W y/h/f **H** X_2CF (Fig. 2) present at the C-terminal part of each finger. They may play a role either in the structure of the DBD or in the interaction with DNA by intercalation of aromatic residues into the double helix at DNA ends. Second, a number of

```
                                     Finger I
                           ●
                           ◆  ◆                    ◆  ◆
HumPARP   1 MaessdklYrveYAksGRAsCKkCsesIpKdslrmAiMVqSpmfDgKvPhWyHfsCFwKvghsirhpDvqvdgfseLRwdDQqkvkktaEagg...vtgkg 98
MouPARP   1 ---a-er----q-----------------------------------------------------q-------e-----------------------...-a--- 98
BovPARP   1 -----------------------------------i---f--e-------i-----l---------f--w----e-e-------------ti--m--t--rtd-s--- 101
ChiPARP   1 ---tg--p--a----------------g---a-----l-l---------------h-y-----rarivs-t-..i---p----e---e-i--ai-t-alqeek-gt 100
XenPARP   1 ---------------------------gdn-a-e--gl-----------------h-y-----rarvlsqg-..iy-yt----e---emi--ai-t--..aaa-a- 97
DroPARP   1 -.d.ielp-la---rt----t--g-kst-s--t--i-v-----afh-a---n-f-kt---f-nqrpssvg-..iqnign---fa---keltdlv-..niqe-isaq 95

HumPARP  99 qdgigsKa.ektLgDFaaeYAKSnRStCkgCmeKIeKgqvRlsKkmvdpekp.qlGmidrWyHpgCFvknReELgfrpeysasqlkGFslLateDkealKK 197
MouPARP  99 ---s-g--.----------------m----l--------m-------------.--------------t-----k-d----------------------sa-------- 197
BovPARP 102 ---v---t.----i--g-g-----------s------d----------vvy-d--.-----v-c-----k----qk--------------f--th-m----v-ta---q-t---- 200
ChiPARP 101 rkev.g--.--s-t--------------------eq-----qi-i-----h----.--------n------d----sr-a------l-a-g-t--l----i-ka-----t---- 199
XenPARP  98 g-sk-g-g.-m--n-----------------a-----eq--------i-i---s--v-r-.------------------d----ss-----d-l-s-----------ti-sa----ds---- 196
DroPARP  96 lgkkr---fnla-k--gi-------s-----r---eq---n-dl---r-tvy-t-vgmky-gqpl-h-le---aql-s----wfa..-gedmp---qs---dd-qakv-n 194
                              ◆  ◆  *      ●                          ◆  ◆
                              ●
                                                  Finger II
```

Fig. 2. Multiple alignment of the sequence of the amino-terminal region of human PARP (HumPARP, residues 1-197, domain A) with those of mouse (MouPARP), bovine (BovPARP), chicken (ChiPARP), *Xenopus* (XenPARP), and *Drosophila* (DroPARP) enzymes. The strictly conserved residues are *boldfaced* and *boxed*. Cys and His involved in zinc coordination are indicated by a diamond (◆). Crucial residues for DNA-binding and enzymatic activity, identified by site-directed mutagenesis, are indicated by closed circles (●). The asterisk (* at bottom line) indicates the location of the epitope recognized by the monoclonal antibody C^1_9.

The N-terminal region of the human PARP DBD contains a repeated sequence (residues 2–97 and 106–207) in which 35 amino acids are duplicated (Fig. 2). Among them, a zinc finger-like motif of the form: **C** - Xaa_2 - **C** $Xaa_{28,30}$ - **H** - Xaa_2 - **C**, is also conserved suggesting that the DBD may have arisen by sequence divergence of a primordial element following its duplication.

Crucial residues involved in metal coordination of the first (C21, C56) and second zinc fingers (C125, C162) were identified by the combined approach of site-directed mutagenesis, expression of the mutant recombinant proteins in *E.coli.* and the zinc-blot assay(17). Interestingly, these residues, as well as the two other sets involved in metal coordination (H53, C56 and H159, C162 respectively), are strictly conserved during evolution (Fig. 2), thus confirming the existence of two modules, independently folded in zinc-containing min-domains, fI and fII. Although some authors prefer to restrict the term zinc finger to the original TFIIIA-like peptide domains, when now used in a broader sense it also includes a new family of zinc finger proteins

basic residues (K22, K30, K47, K59, R78, K105, K119, R122, K131, K134, R138, K141, R167 and K196), potential candidates for DNA interaction, are strictly conserved but few of them are duplicated (i.e. K30 and K131), From mutational analysis, we know that not only are the residues involved in zinc coordination important for structure-activity but other basic amino acids i.e. K131 and R138, are also important (see below).

PARP-DNA interaction

Given that DNA interruptions are essential for enzyme activation [20], we asked if the metal-binding domain was involved in DNA interaction and especially if this binding occurs at DNA strand breaks. This question was assessed using a footprinting assay where PARP (or proteolytic fragments of it) was immobilized onto nitrocellulose and allowed to interact with a 66 base pairs (bp) DNA harbouring a single-stranded interruption at a defined position.

[1] Mazin A, Angulo J, Ménissier-de Murcia J, Schreiber V, Timchencko T, de Murcia G, Dévoret R: Kin 17, a mouse zinc finger protein, binds preferentially to curved DNA. (submitted for publication).

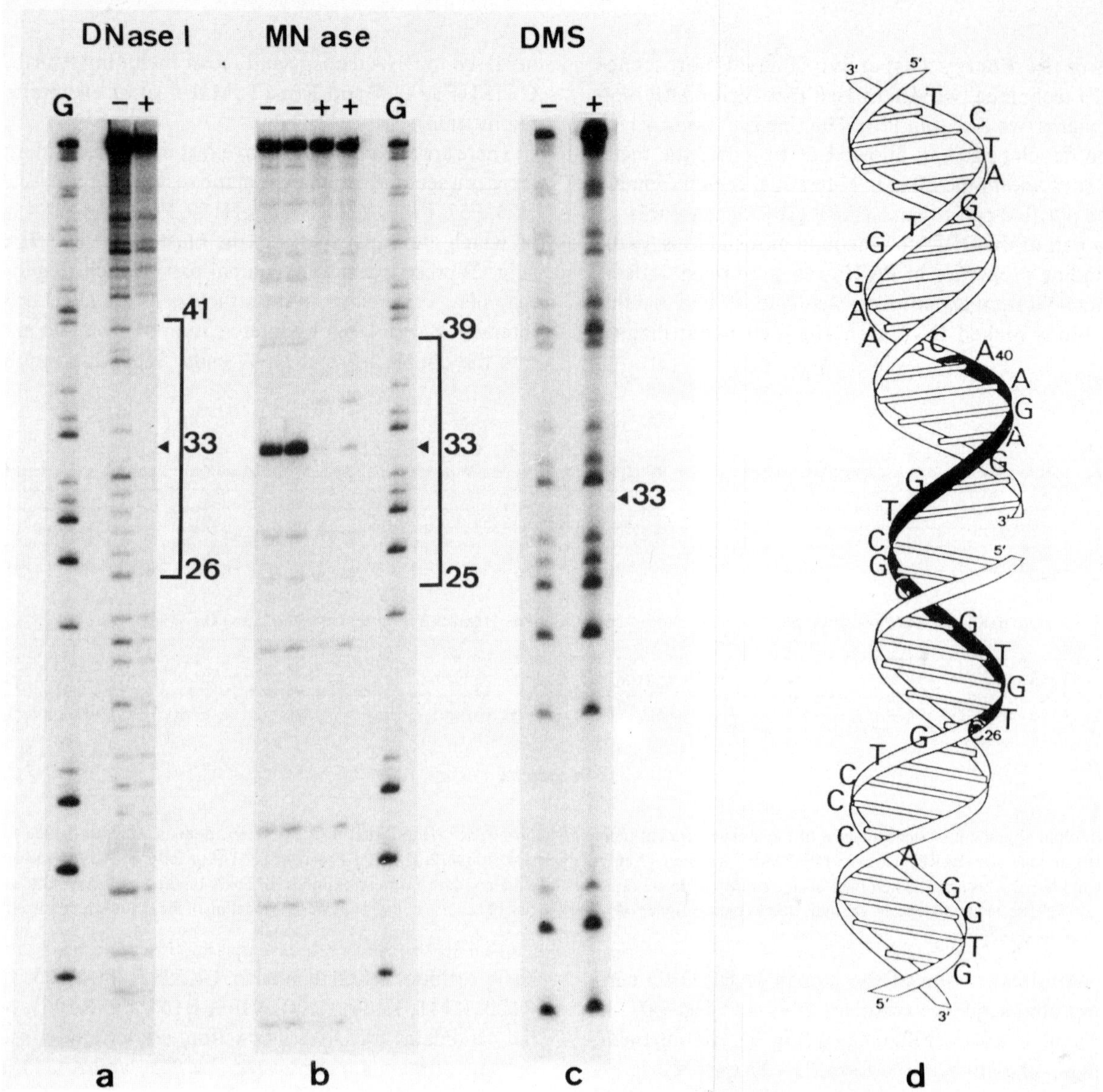

Fig. 3. Specific recognition of a single-stranded break located at position 33 in a DNA duplex by the PARP DBD. (**a**): DNase I footprinting. (**b**): micrococcal nuclease footprinting. (**c**): DMS footprinting. Lanes - : degradation products obtained in the absence of protein. Lanes + -: footprint of nicked DNA-DBD complex. G Lane: degradation products of the G sequencing reaction. (**d**): schematic representation of the protected region in the continuous strand of the nicked DNA probe by the PARP DBD.

Nucleases were used as structural probes to give a view of the outside of the double helix (the ribose–phosphate backbone). As shown in Fig. 3a, when DNase I was used, we found that PARP protects symmetrically 7(±1) nucleotides of the continuous strand, at each side of the nick located at position 33 [9]. This protection is sequence-independent since the same result was obtained when the nick was shifted to position 23. The diagram (Fig. 3d) indicates that the PARP DBD protects one and half turns of the double helix, suggesting that two DNA strands are in contact with the DBD at the nick. Alternatively the footprint symmetry could reflect the dimerization of the DBD following nick binding. In that case, each PARP monomer would protect 7 nucleotides at each side of the nick. Digestion of the PARP-DNA complex by micrococcal nuclease (Fig 3.) confirmed the extent of protection on the continuous strand which is particularly noticeable at position 33.

Dimethylsulfate (DMS) which probes the guanines inside the major groove, was also used to investigate the interaction of PARP with the same nicked DNA. No footprint was observed in that case (Fig. 3c) confirming that continuity of the sugar–phosphate backbone is recognized by PARP, independent of the sequence.

What does PARP recognize in nicked DNA? To answer

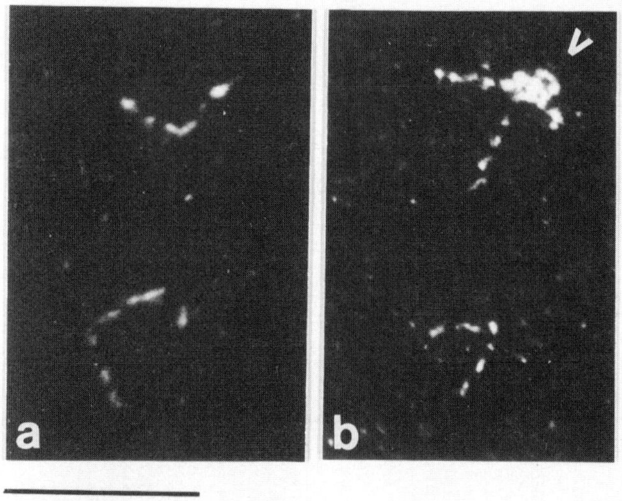

Fig. 4. Electron micrographs of 139 bp DNA fragments harbouring a single strand break at position 69 in the presence (panel **b**, arrow) or in the absence of PARP (panel **a**) [22]. The bar equals 50 nm.

this question a 139 bp DNA duplex containing a unique single-stranded break in its center was constructed and visualized by dark field electron microscopy in the presence (Fig. 4b) or in the absence (Fig. 4a) of PARP. As displayed in Fig. 4a, the introduction of a break in one strand of the DNA duplex, induces a strong flexibility in this region which is clearly reflected in a characteristic V-shape of the DNA molecule. Specific binding of PARP to the centre of the V was observed (Fig. 4b). The V-shaped conformation of DNA was slightly enhanced by PARP interaction at the nick [39]. This observation is reminiscent of the preferential binding of PARP to DNA crossovers [21] and to DNA cruciforms [22]. Both structures have in common some topological similarities with nicked DNA: the close proximity of two DNA duplexes, forming an angle which could be primarily an attractive symmetrical target for PARP. However, DNA bending and DNA flexibility are not sufficient to trigger the PARP polymerizing activity; the nick is an essential cofactor.

Although the structure of the protein interface contacting the DNA ends is not known yet, we have learned from site-directed mutagenesis experiments that the second zinc finger (f II) plays an important role in the binding of DNA containing single-stranded breaks. When the fragments AB [10] or ABC [23] are used, point mutations affecting f II (i.e. C125G, C162H, R138I) decrease its capacity to bind nicked DNA. As a consequence, the enzymatic activity is strongly reduced when a recombinant mutant enzyme (R138I, for example) is activated by DNA containing a single-stranded break (Gradwohl *et al.* unpublished results). When DNA containing double-stranded breaks is used, about 80% of residual activity is maintained in this same mutant. Mutations affecting f I (i.e. C21G, Δ f I) completely abolish enzymatic activity whatever the nature of the DNA breaks. These results together with those of Ikejima *et al.* [24], highlight the differential role of f I and f II in the recognition of breaks and in PARP activation. The two fingers are clearly individualized as two subdomains in the identity profile (Fig. 1), which is also suggestive of a specific role for each.

The bipartite nuclear localization signal of PARP

The nuclear localization signal (NLS) is a short amino acid sequence generally rich in basic residues, that targets proteins larger than 40 to 60 kDa to the cell nucleus (reviewed in ref. 25). Sequence analysis of the entire PARP revealed three basic motifs matching the criteria for a nuclear-targeting signal. By deletion analysis we found that the DNA-binding domain, fused to the N-terminus of ß-galactosidase, contains a functional NLS (residues 207–226) of the form KRK-X_{11}-KKKSKK (see Fig. 5). Extensive site-directed mutagenesis performed in this short region revealed that each of the two basic cluster (D1 and D2) was essential but not sufficient on its own for this function, while both motifs together are, thus demonstrating that the human PARP NLS is bipartite [11]. Moreover, in each cluster, crucial amino acids (K207, R208, and K222) were identified. They correspond precisely to the basic residues which are strictly conserved in the NLS motif during evolution (Fig. 5).

What do we know about the NLS structure? Recent NMR studies of a 30 amino acid peptide corresponding to the bipartite human PARP NLS have shown that the region encompassing the second basic cluster D2 has the propensity to form an α-helix (K. Holland and J. F. Lefèvre, unpublished results). To verify this hypothesis, an alanine was introduced at position 214 of both the wt and the mutant K222I. The first mutation did not influence the nuclear phenotype, whereas the same insertion before the KIKSKK sequence (of the mutant K222I) partially reverted its cytoplasmic phenotype. These results strongly suggests that a helical-type structure containing a lysine in a critical position (K222) is necessary for the nuclear-targeting activity. It is also worthwhile to point out that D2 is likely to be located at the surface of the PARP molecule since this sequence is preferentially cleaved by plasmin [26] and trypsin [27] precisely after the crucial residue K222.

Finally it was further demonstrated *in vitro* that the DNA-binding capacity and the polymerizing activity of PARP were independent of the nuclear targeting function, thus supporting the concept that human PARP NLS which is encoded by exon 5 [28], is an autonomous functional element and belongs to the class of bipartite NLSs.

```
                      ●●                ●
HumPARP 202 vKsEg....KRkgdEvdgvdevakKKskkekdkdsklekaLKa 240
MouPARP 202 i-n--....---------t--------r--t--y--------- 240
BovPARP 205 i-g-r....---------i---t----------ei-------- 243
ChiPARP 202 t-t--....----e----.nv------r---e-e--q--q--e 239
XenPARP 201 --n--....---a-----hsaat---i----e-e-----l--e 239
DroPARP 199 i---elpdt--akm-lsdtneege--qr............--d 229

            ----------------    -------------
                   D1                D2
```

Fig. 5. Multiple alignment of the sequence of the Nuclear Localization Signal (NLS, domain B) of human PARP (HumPARP, residues 202–238) with those of mouse (MouPARP), bovine (BovPARP), chicken (ChiPARP), *Xenopus* (XenPARP), and *Drosophila* (DroPARP) enzymes. The strictly conserved residues are *boldfaced* and *boxed*. Crucial residues for nuclear targeting function, identified by site-directed mutagenesis are indicated by closed circle (●). D1 and D2 constitute the two basic clusters of the human bipartite NLS which are necessary and sufficient to target PARP into the nucleus [11].

The catalytic domain

The C-terminal region of PARP was identified by Kameshita *et al.* [2] as the NAD-binding domain. With respect to the NAD binding function, PARP is of special interest since the NAD molecule is used as a substrate rather than a coenzyme for electron transfer. To study this highly conserved region of the PARP molecule, a novel assay was developed, the "activity blot", which allows the detection of transferred and renatured polypeptides bearing the poly(ADP-ribose) synthesis function [29]. Proteolytic degradation experiments using purified calf thymus PARP and deletion analysis of the human PARP cDNA expressed in *E.coli* demonstrated that the C-terminal region of the PARP molecule (domain F, 40 kDa) retains basal polymer synthesis when immobilized on nitrocellulose membranes [12,14]. Recently the cDNA encoding domain F was inserted into an expression vector and the 40 kDa polypeptide was overproduced in *E. coli* and purified to homogeneity [14]. The different catalytic activities usually associated with the full length enzyme were also preserved in this domain in solution: abortive NAD^+ hydrolysis, initiation, elongation, branching, and termination of ADP-ribose polymers synthesized either in the catalytic domain (automodification) or on histones H1 and H2B (heteromodification); no DNA dependence was observed. The K_m of domain F for NAD^+ (65μM) was comparable to that of full-length enzyme (50 μM) but the specific activity was approximately 500-fold lower than that of the whole enzyme activated by DNA strand beaks. The processive mode of polymer attachment, as well as the regulatory influence of the acceptor on the polymer size [30] were also detected in domain F. All together these results indicate that region F not only binds NAD but also constitutes the PARP catalytic domain containing the basal and autonomous capacity to synthesize branched polymers [14]. This basal activity is the same as that of the full-length PARP operating in the absence of DNA. The high specific activity of PARP is absolutely dependent on binding to DNA strand breaks and hence on the integrity of the zinc finger domain.

Based on sequence similarities between the C-terminal region of PARP and the $NAD(P)^+$ binding domains of Leu and Glu dehydrogenases (L/G DH), site-directed mutagenesis of the strictly conserved (K893, D914, D993) and the non conserved (K953) residues revealed the crucial importance of this region which encompasses the PARP signature (Fig. 6). In a first set of mutants, a non-conservative change (K893I and D993A) totally inactivated the polymerizing function, while the conservative mutants (K893R, and D993E) were still enzymatically active with a K_m close to 60 μM but with a strongly reduce k_{cat}. These results indicated that a charged residue is required both at positions 893 and 993, and that they may likely be involved in cataly-

```
                                                 O
                          ■          ■           ●
HumPARP 852 kqLHNRrLLWHGSRtTNFaGILSqGLRIAPPEAPvTGYMFGKGIYFADMVSKSANYChtsqgdp 915
MouPARP 851 r--------------------------------------------------------------- 914
BovPARP 854 ---------------------------------------------------------------- 917
ChiPARP 849 ------q------------------------------------------------------a-- 912
XenPARP 848 ------q---------------------------------------------------amp-s- 911
DroPARP 834 -k----k-------l---v----h----------p----------------------c---qns 897
```

Fig. 6. Amino acid alignment of the highly conserved region (positions 859–908) present in the catalytic domain of the human PARP. Crucial residues for enzymatic activity identified either by site-directed mutagenesis (●) (12,13) or by random mutagenesis (■) are indicated by closed symbols. The binding site of 2′-azido NAD+ is indicated by an open circle (O).

sis of the reaction. In the second set of mutants, conservative (D914E, K953R) and non-conservative mutations (D914A, K953I) led to the retention of enzymatically active PARP recombinant proteins having a K_m for NAD^+ close to the wt but, again, a strongly reduced k_{cat}. The residues D914 and K953 may participate indirectly in catalytic activity and do not seem to be involved in NAD^+ binding [13].

The recent resolution of the crystallographic structure of the NAD^+-linked glutamate dehydrogenase from *Clostridium symbiosium* (31) prompted us to superimpose the sequence containing the conserved amino acids between PARP and L/G DH, on the schematic representation of the tridimensional structure (Fig. 7). Interestingly, all these residues are located near or in the deep catalytic cleft where NAD^+ and glutamate are bound (Fig. 7), supporting the hypothesis that conservation of these residues may reflect a critical role in the structural organization of the PARP catalytic site. Among the mutated residues, two are critical for the PARP activity (K893 and D993). K893 may be part of a cluster of basic residues conservatively maintained between PARP (K852, R858, R878 and K893) and L/G DH (K89, R93, K113 and K125) in which they are supposed to stabilize glutamate binding [31]. Since PARP preferentially modifies glutamic residues within a polypeptide chain, it is attractive to speculate that these 4 basic residues may be implicated as well, in the stabilization of the negative charge present on the acceptor amino acid, thus allowing its proper positioning with respect to the NAD^+ molecule.

According to the three-dimensional structure of the GDH,

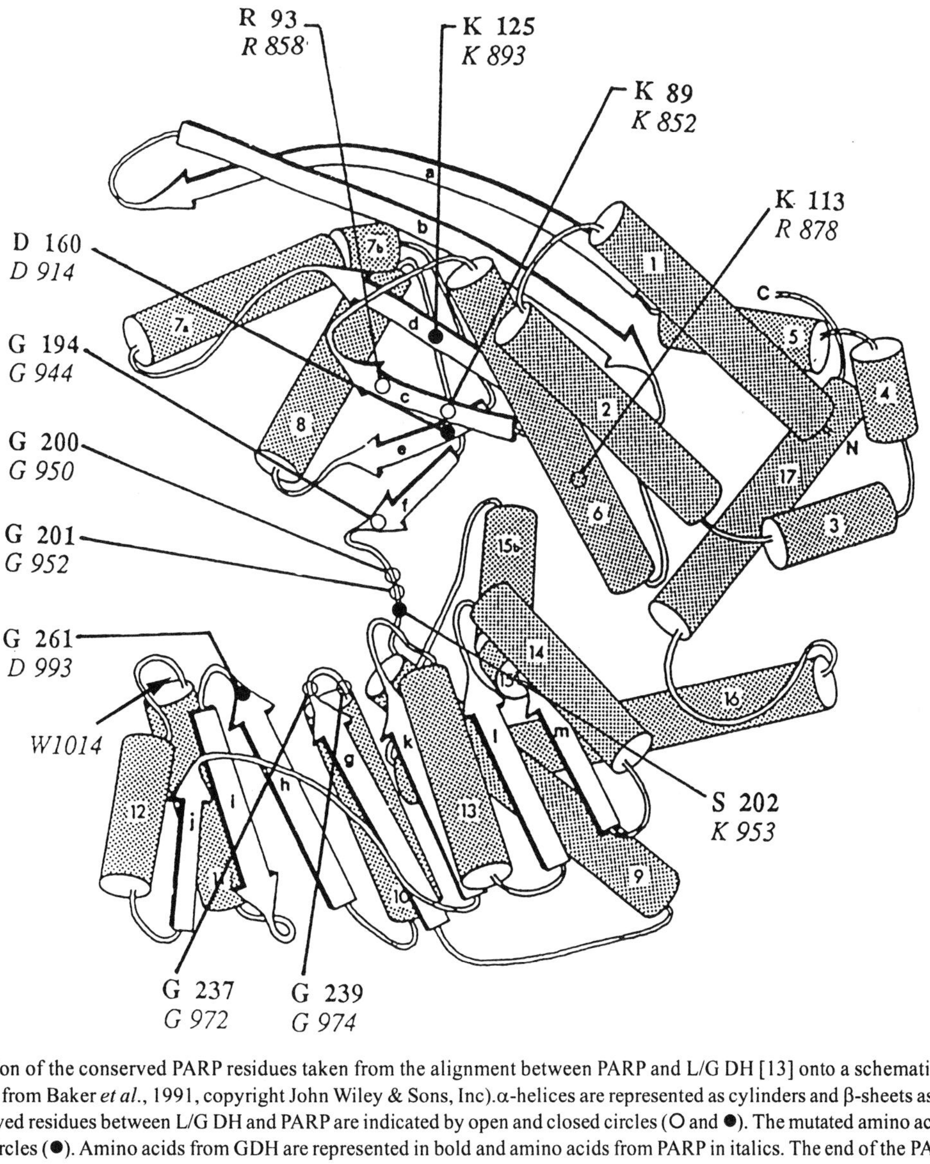

Fig. 7. Superimposition of the conserved PARP residues taken from the alignment between PARP and L/G DH [13] onto a schematic representation of the GDH structure (taken from Baker *et al.*, 1991, copyright John Wiley & Sons, Inc).α-helices are represented as cylinders and β-sheets as arrows. The structural position of the conserved residues between L/G DH and PARP are indicated by open and closed circles (O and ●). The mutated amino acids in human PARP are indicated by closed circles (●). Amino acids from GDH are represented in bold and amino acids from PARP in italics. The end of the PARP sequence (W1014) is indicated by an arrow.

the amino acid D993 may be involved in the super secondary structure βαβ termed a Rossman fold, known to bind specifically the ADP moiety of the dinucleotide NAD+ [32]. The total loss of catalytic activity in the mutant D993I suggests that this is a most likely hypothesis (see Fig. 7). This prediction turned out to be even more plausible, since we recently found from photolabelling experiments that the adenine moiety contained in 2′-azido NAD analog was specifically bound to the residue W1014 (Kim *et al.* in preparation). Thus, using the GDH structure as a model for the PARP catalytic site (Fig. 7), one can suggest that the C-terminal part of domain F is involved in the binding the ADP moiety. In this context, the residue D993 would interact with the 2′-OH of the ribose relative to the adenine, whereas W1014 would be very close to the adenine ring.

A second site for 2′-azido NAD binding (K893) was also identified in the same photolabelling experiment, suggesting that there are two possible sites for NAD and/or ADP binding: (i) the βαβ region containing the residues D993 and W1014 and (ii) the highly conserved region, corresponding to the "PARP signature" (see Fig. 1) including the amino acid K893 as well as two others, N 868 and A880 (Fig. 6) identified as crucial residues by random mutagenesis (O'Farrell, unpublished results). It is tempting to speculate that the first site ending at residue 1014, represents the polymer elongation site capable of binding an AMP extremity, whereas the second one is the extremely conserved catalytic site, bearing the specific function of NAD+ hydrolysis and initiation of poly (ADP-ribose) synthesis on a glutamate acceptor. In line with this it is also useful to remind here the work of Marsischky *et al.* [33] who found that the residue D988 plays a critical role in polymer elongation.

Modules at work: an approach to the physiological role of PARP *in vivo*

The two characteristic properties of the PARP DBD, i.e. specific binding to DNA breaks and nuclear targeting, were elegantly exploited by Küpper *et al.* [34] to inhibit the endogenous PARP activity by overproducing its DBD. Under low or moderate DNA damage conditions (10 μM of the alkylating agent N-methyl-N′-nitro-N-nitrosoguanidine, MNNG) a trans-dominant inhibition of the resident PARP activity was observed due to competition between endogenous PARP and its overproduce DBD. This ability of the DBD to interfere with endogenous PARP made it possible to assess the role of PARP in two different repair pathways. To this aim, the wt PARP DBD (domains ABC, 46 kDa) and several derivatives thereof, mutated in f I or f II zinc-fingers, were overproduced in CV-1 monkey cells or in human fibroblasts. The trans-dominant inhibition of the

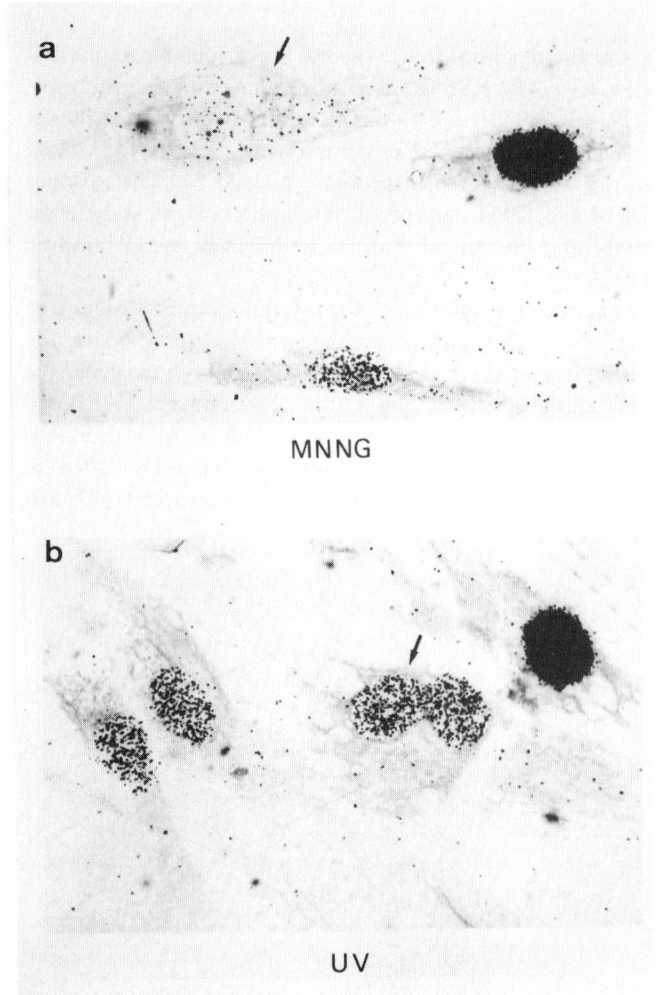

Fig. 8. Microinjection of the 46kDa PARP-DBD protein in C5RO repair-proficient human fibroblasts. UDS was assayed following MNNG treatment (100 μM; panel **a**); or UV irradiation (10 J/m^2; panel (**b**). The corresponding fields were prepared for immunofluorescence with the monoclonal C^{I}_{9} anti-DBD antibody to identify the injected homo-polykaryons (arrows). Non-injected monokaryons in S phase are also visible (upper right) [23].

endogenous PARP activity and the consequences on DNA repair induced either by MNNG treatment or by UV (254 nm) irradiation were then studied. We found a positive correlation between the *in vitro* binding capacity of the recombinant DBD polypeptides and their inhibitory effect *in vivo*, on PARP activity stimulated by MNNG. As a consequence, overproduction or micro injection of the wt DBD into human fibroblasts blocked unscheduled DNA syntheses (UDS) (Fig. 8a). In this process the absolute requirement of DNA-break binding is critical both for the trans-inhibition and the UDS block to occur [23].

When UV (5 J/m^2, 254 nm) was used as a DNA damaging agent of CV-1 cells or human fibroblasts, no stimulation of PARP activity could be detected by immunofluorescence using a monoclonal antibody against poly(ADP-ribose). Neither trans-inhibition of PARP activity was observed nor UDS inhibition was detected (Fig. 8b) pointing out a specific role for PARP in base excision repair (BER) but not in nucleotide excision repair (NER). A similar conclusion was independently reached by Satoh *et al.* [35] who clearly demonstrated that in a cell-free system, the stimulation of PARP activity by UV irradiation of DNA at 254 nm is due to the repair of minor pyrimidine hydrate lesions processed by the BER machinery rather than by the repair of cyclobutane pyrimidine dimers and 6-4 photoproducts which are mainly produced at this wavelength and processed by the NER pathway.

Conclusions and prospects

The convergence of biochemical, genetic and molecular biology strategies in the study of the PARP molecule stresses the importance of two major functional modules: a N-terminal domain acting as a real molecular nick-sensor and a C-terminal region which synthesizes poly (ADP-ribose) following recognition and binding of PARP to a DNA interruption by the DBD module. Regarding the interdependence of these two domains, ours results suggest that the DBD does not repress continuously the catalytic domain; the activation mechanism may include an allosteric step by which the DBD informs domain F that a nick is present on DNA.

These two modules have contradictory effects as far as the binding of PARP to DNA is concerned, but both may contribute positively and sequentially to the protection of the genome. The scanning function of the DBD drives the PARP molecule to DNA breaks induced either directly or indirectly by the repair of DNA lesions, primarily to prevent genomic instability. An example of this protective function is illustrated *in vivo*, by the inhibition of UDS induced by MNNG (Fig. 8a). This block may also have negative consequences on the progression of the replicative fork, leading DNA polymerases to stall on DNA in the vicinity of a non-repaired break occupied by PARP. This could explain the co-purification of PARP with a high number of these DNA replicative forks, as we observed long ago under the microscope [36].

The polymerizing function associated with domain F has an opposite effect: to decrease the DNA binding affinity of PARP and the chromosomal acceptors, leaving the DNA interruption available for the repair enzymes. An example of this effect is the relaxation of chromatin superstructure that we observed *in vitro* [37, 38] inviting the speculation that, in this second phase of the poly (ADP-ribosyl)ation reaction, the accessibility of the DNA lesion could be increased by such a mechanism.

Future research will undoubtedly reveal whether this highly conserved enzyme has a positive influence on the maintenance of genome integrity or whether PARP has only an incidental and negative role in DNA excision repair as recently suggested by Satoh *et al* [35]. Whatever the scenario, we should soon evaluate the importance of poly (ADP-ribosyl)ation reactions *in vivo* in PARP-deficient mice and/ or in murine cell lines generated by targeted gene inactivation.

Acknowledgements

We are indebted to our colleagues Drs F. Simonin, G. Gradwohl, and S. Muller for their contribution to the work summarized in this review. We thank E. Le Cam for the electron microscopy experiments, Drs A. Bürkle, J. Hoeijmakers, M. O'Farrell, F. Althaus, M. Jacobson and H. Kim for their collaboration and Dr. E. Miranda for critical reading of the manuscript. This work was supported by Association pour la Recherche contre le Cancer (ARC) and by La Ligue Nationale Française Contre le Cancer.

References

1. Althaus F, Richter C: ADP-ribosylation of proteins. Enzymology and biological significance. Mol. Biol. Biochem, Biophys. 37: 1–126, 1987
2. Kameshita I, Matsuda Z, Tanigushi T, Shizuta Y: Poly(ADP-ribose) synthetase, Separation and identification of three proteolytic fragments as the substrate-binding domain, the DNA binding domain, and the automodification domain. J. Biol. Chem. 259: 4770–4776, 1984
3. Uchida K, Morita T, Sato T, Ogura T, Yamashita R, Mogushi S, Suzuki H, Nyunoya H, Miwa M, Sugimura T: Nucleotide sequence of a full-length cDNA for human fibroblast poly(ADP-ribose) polymerase. Biochem. Biophys. Res. Commun. 148: 617–622, 1987
4. Huppi K, Bhatia K, Siwarski D, Klinman D, Cherney B, Smulson M: Sequence and organization of the mouse poly(ADP-ribose) polymerase gene. Nucleic Acids Res. 17: 3387–3401, 1989
5. Saito I, Hatakeyama K, Kido T, Ohkubo H, Nakanishi S, Ueda K: Cloning of full-length cDNA encoding bovine thymus poly(ADP-ribose synthetase: Evolutionarily conserved segments and their potential functions. Gene 90: 249–254, 1990
6. Ittel M E, Garnier J M, Jeltsh J M, Niedergang C P: Chicken poly(ADP-ribose) polymerase: complete sequence deduced from cDNA, comparison with mammalian enzyme sequences. Gene 102: 157–164, 1991
7. Saulier B: Ph.D. Thesis: Isolement et caractérisation d'un ADN complementaire de la poly(ADP-ribose) polymerase chez l'amphibien *Xenopus Laevis.*, Université de Rennes, 1993
8. Uchida K, Hanai S, Ishikawa K, Ozawa Y, Uchida M, Sugimura T, Miwa M: Cloning of cDNA encoding Drosophila poly(ADP-ribose) polymerase: leucine zipper in the auto-modification domain. Proc. Natl. Acad, Sci. USA 90: 3481–34885, 1993
9. Ménissier-de Murcia J, Molinete M, Gradwohl G, Simonin F, de Murcia G: Zinc binding domain of poly(ADP-ribose) polymerase participates in the recognition of single strand breaks on DNA. J. Mol. Biol. 210: 229–233, 1989

10. Gradwohl G, Ménissier-de Murcia J, Molinete M, Simonin F, Koken M, Hoeijmakers J H J, de Murcia G: The second zinc-finger domain of poly(ADP-ribose) polymerase determines specificity for single-stranded breaks in DNA. Proc Natl. Acad. Sci. USA 87: 2990–2994, 1990
11. Schreiber V, Molinete M, Boeuf H, de Murcia G, Ménissier-de Murcia J: The human poly(ADP-ribose) polymerase nuclear localization signal is a bipartite element functionally separate from DNA binding and catalytic activity. EMBO J., 11: 3263–3269, 1992
12. Simonin F, Ménissier-de Murcia J, Poch O, Muller S, Gradwohl G, Molinete M, Penning C, Keith G, de Murcia G: Expression of site-directed mutagenesis of the catalytic domain of human poly(ADP-ribose) polymerase in Escherichia coli. J. Biol. Chem. 265: 19249–19256, 1990
13. Simonin F, Poch O, Delarue M, de Murcia G: identification of potential active-site residues in the poly(ADP-ribose) polymerase. J. Biol. Chem. 258: 8529–8535, 1993
14. Simonin F. Höfferer L, Panzeter P, Muller S, de Murcia G, Althaus F: The carboxy-terminal domain of human poly(ADP-ribose) polymerase: overproduction in *Escherichia coli*, large scale purification and characterization. J. Biol. Chem. 268: 13454–13461, 1993
15. Zahradka P, Ebisuzaki K: Poly(ADP-ribose) polymerase is a zinc metalloenzyme. Eur. J. Biochem. 142: 503–509, 1984
16. Mazen A, Ménissier-de Murcia J, Molinete M, Simonin F, Gradwohl G, Poirier G G, de Murcia G: Poly(ADP-ribose) polymerase: A novel Finger Protein. Nucleic Acids Res. 17: 4689–4698, 1989
17. Mazen A, Gradwohl G, de Murcia G: Zinc-binding protein detected by protein blotting. Anal. Biochem. 172: 39–42, 1988
18. Tanaka K, Miura N, Satokata I, Miyamoto I, Yoshida C, Satoh Y, Kond S, Yasui A, Okayama H,Okada Y: Analysis of a human DNA excision repair gene involved in group A xeroderma pigmentosum and containing a zinc-finger domain. Nature. 348: 73–76, 1990
19. O'Connor T, Graves R, de Murcia G, Castaing B, Laval J: Fpg protein of *Escherichia coli* is a zinc finger protein whose cystein residues have a structural and/or functional role. J. Biol. Chem. 268: 9063–9070, 1993
20. Benjamin RC, Gill DM: Poly(ADP-ribose) synthesis in vitro programmed by damaged DNA. J. Biol. Chem. 255: 10502–10508, 1980
21. Gradwohl G, Mazen A, de Murcia G: Poly(ADP-ribose) polymerase forms loops with DNA. Biochem, Biophys, Res. Commun. 148: 913–919, 1987
22. Sastry S, Kun E: The interaction of adenosine diphosphoribosyl transferase (ADPRT) with a cruciform DNA. Biochem. Biophys. Res. Commun. 167: 842–847, 1990
23. Molinete M, Vermeulen W, Bürkle A, Ménissier-de Murcia J, Küpper J, Hoeijmakers J, de Murcia G: Overproduction of the poly(ADP-ribose) polymerase DNA-binding domain blocks alkylation-induced DNA repair synthesis in mammalian cells. EMBO J., 5: 2109–2117, 1993
24. Ikejima M, Noguchi S, Yamashita R, Ogura T, Sugimura T, Gill M, Miwa M: The zinc fingers of human poly(ADP-ribose) polymerase are differentially required for the recognition of DNA breaks and nicks and the consequent enzyme activation. J. Biol. Chem. 265, 21907–21913, 1990
25. Schreiber V, de Murcia G, Ménissier-de Murcia J: Le transport des protéines dans le noyau: les signaux de localisation nucléaire et leurs récepteurs. Médicine/Science 8: 134–139, 1992
26. Buki K, Kun E: Polypeptide domains of ADP-ribosyltransferase obtained by digestion with plasmin. Biochemistry, 27: 5990–5995, 1988
27. Mazen A, Lamarre D, Poirier G, Gradwohl G, de Murcia, G: Localisation of the zinc-binding sites in the DNA-binding domain of the bovine poly(ADP-ribose) polymerase. In: ADP-ribose Transfer Reactions: Mechanisms and Biological significance (M.K. Jacobson and E.L. Jacobson, eds) Springer-Verlag, pp 89–93, 1989
28. Auer B, Nagl U, Herzog H, Schneider R, Schweiger M: Human Nuclear NAD+ ADP-ribosyltransferase (polymerizing): organization of the gene. DNA, 8: 575–580, 1989
29. Simonin F, Briand JP, Muller S, de Murcia G: Detection of poly(ADP-ribose) polymerase activity in crude extracts of higher and lower eukaryotes by activity-blot. Anal. Biochem. 195: 226–231, 1991
30. Naegeli H, Althaus FR: Regulation of Poly(ADP-ribose) polymerase. Histone-speechified adaptations of reaction products. J. Biol. Chem. 266: 10956–10601, 1991
31. Baker P J, Britton K L, Engel PC, Farrants G W, Lilley K S, Rice D W, Stillman TJ: Subunit assembly and active site location ion the structure of glutamate dehydrogenase. Proteins 12: 75–86, 1992
32. Rossmann M G, Liljas A, Brändén C-I, Banaszak L.J.: Evolutionary and structural relationships among dehydrogenases. In: "The Enzymes," 3rd ed. Vol. 11. (Boyer, P.D) (ed.). New York: Academic Press: 61–102, 1975
33. Marsischky G. T., Ikejima, M, Suzuki H, Sugimura T, Esumi H, Miwa M, and Collier R J: Directed mutagenesis of glutamic acid 988 of poly(ADP-ribose) polymerase. In: ADP-ribosylation reactions. Poirier, G. G. and Moreau, P., eds. Springer, New York p47–52, 1992
34. Küpper J-H, de Murcia G, Bürkle A: Inhibition of poly(ADP-ribosylation by overexpressing the Poly(ADP-ribose) polymerase DNA-binding domain. J. Biol. Chem., 265: 8721–18724, 1990
35. Satoh MS, Lindahl T: Role of poly(ADP-ribose) formation in DNA repair. Nature, 356: 356–358, 1992
36. de Murcia G, Jongstra-Bilen J, Ittel ME, Mandel P, Delain E: Poly(ADP-ribose) polymerase — automodification and interaction with DNA: electron microscopic visualization. EMBO J., 2: 543–548, 1983
37. Poirier G, de Murcia G, Jongstra-Bilen J, Niedergang C, Mandel, P: Poly ADP-ribosylation of polynucleosomes causes a relaxation of the chromatin structure. Proc. Natl. Acad. Sci. USA, 79: 3423–3426, 1982
38. de Murcia G, Huletsky A, Lamarre D, Gaudreau A, Pouyet J, Daune M, Poirier G: Modulation of chromatic superstructure induced by poly(ADP-ribose) synthesis and degradation. J. Biol. Chem. 261: 7011–7017, 1987
39. Le Cam E, Fack F, Ménissier-deMurcia J, Cognet J, Barbin A, Sarantoglou V, Révet B, Delain E, de Murcia G: Conformational analysis of a 139 base-pairs DNA fragment containing a single-stranded break, and its interaction with human poly(ADP-ribose) polymerase. J. Mol. Biol. 235: 1062–1071, 1994

Molecular and Cellular Biochemistry **138**: 25–32, 1994.

Poly(ADP-ribose) polymerase: Structural conservation among different classes of animals and its implications

Kazuhiko Uchida and Masanao Miwa
Department of Biochemistry, Institute of Basic Medical Sciences, University of Tsukuba, Tsukuba 305 Japan

Abstract

Poly(ADP-ribose) polymerase cDNAs have been isolated from different classes of animals. Cloning of genes from lower eukaryotes has allowed us to investigate directly the biological functions of poly(ADP-ribosyl)ation *in vivo*. The conservation of specific regions among mammals, chicken, *Xenopus laevis*, and *Drosophila melanogaster* reveals the essential structural elements required for recognition of breaks in DNA and for catalytic activity. Cys, His and basic residues in the zinc-finger consensus region are conserved. The carboxyl terminal region corresponding to an NAD-binding domain is strongly conserved. The dinucleotide-binding consensus sequence and β1-αA-β2, Rossmann fold structure, and β-sheet structures are completely conserved from mammals to insect. In *Drosophila*, a putative leucine-zipper motif has been identified, and other poly(ADP-ribose) polymerases also contain an α-helical, amphipathic structure in the auto-modification domain. In this article, we review the recent structural analyses of the functional domains of poly(ADP-ribose) polymerase in phylogenetically divergent species, and discuss the implications of structural conservation for its biological functions. (Mol Cell Biochem 138: 25–32, 1994)

Key words: NAD, evolution, polymerase chain reaction, zinc finger, leucine zipper

Abbreviations: aa – amino acid(s), *D. melanogaster* – *Drosophila melanogaster*, PARP – poly(ADP-ribose) polymerase [EC 2.4.2.30], PCR – polymerase chain reaction, *X. laevis* – *Xenopus laevis*

Introduction

The poly(ADP-ribosyl)ation reaction is thought to be ubiquitous in eukaryotic cells. The enzyme activity has been detected in lower eukaryotes like *Dictyostelium discoideum* [1] and *Crypthecodinium cohnii* [2]. The evidence for poly(ADP-ribose) polymerase (PARP) activity in yeast remains controversial. PARP has been purified extensively from human placenta [3,4], calf thymus [5], fish [6], and *Helix pomatia* [7]. Structural characterization of these PARP by Western blotting using a monoclonal antibody against human PARP [8] and by an activity gel assay [9] have been reported. Advances in a molecular biology have allowed characterization of PARP at a molecular level. The cDNA encoding human PARP was the first to be isolated [10–14]. Mouse [15], bovine [16], and chicken [17] cDNAs were cloned subsequently. A partial cDNA for rat PARP has also been cloned [18]. The polymerase chain reaction (PCR) has made it possible to efficiently clone PARP cDNA from other, phylogenetically distant species of eukaryotes like *Xenopus laevis* (*X. laevis*) [19,20], cherry salmon [20], and *Drosophila melanogaster* (*D. Melanogaster*) [21]. Gene cloning from these animals is an important step in elucidation of the biological roles of the poly(ADP-ribosyl)ation reaction, since 1) micro-injection of mRNA or the antibody to *Xenopus* and *Drosophila* oocytes is a powerful tool to study development and differentiation, and 2) thousands of mutants of *D. melanogaster* have been isolated and their phenotypes characterized, and it is easy to disrupt the target gene by P-element transfer or mutagenesis.

Comparison of the animo acid (aa) sequences deduced from the isolated cDNAs from divergent animals should

Address for offprints: M. Miwa, Department of Biochemistry, Institute of Basic Medical Sciences, University of Tsukuba, Tsukuba, Ibaraki 305, Japan

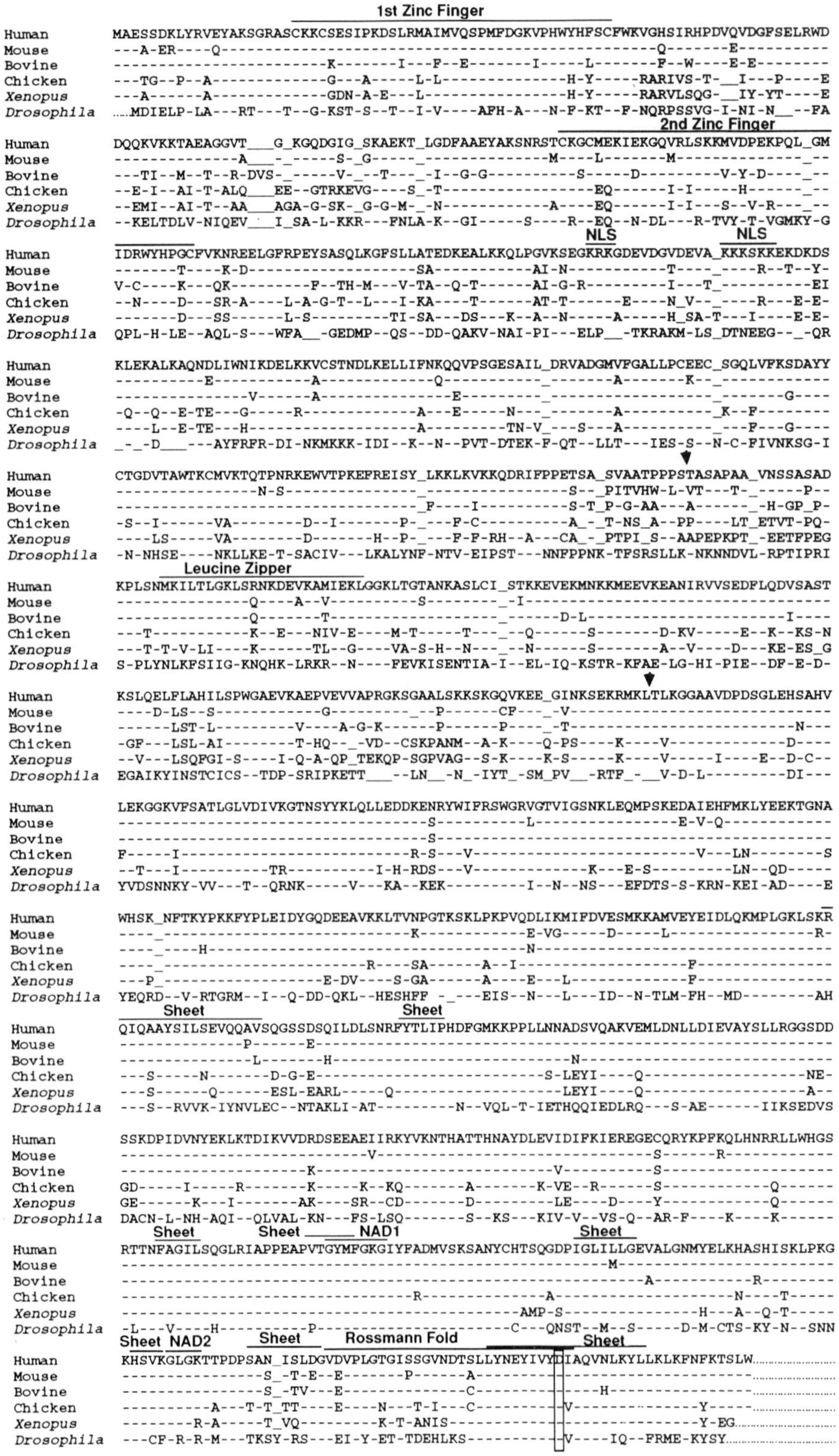

Fig. 1. Comparison of the deduced aa sequences of human [12], mouse [15], bovine [16], chicken [17], *Xenopus* [19] and *Drosophila* PARPs [21]. Only the differences from the aa sequence of the human PARP are shown. The boundaries of the proteolytic domains of human PARP are indicated by arrowheads. The motifs are indicated; NLS, nuclear location signal; NAD1 and NAD2, dinucleotide-binding consensus sequences; and Sheet, conserved β-sheet of the carboxyl half of the catalytic domain. Asp[993] at the end of β2 is boxed. The secondary structure was predicted by the Robson-Garnier method [22].

reveal the peptide sequences essential for enzymatic properties. The strictly conserved regions of primary and secondary structure should be those essential for DNA-recognition, association with the acceptor proteins, substrate binding and enzymatic activity. In this article, we compare the sequences of human, bovine, mouse, chicken, *X. laevis*, and *D. melanogaster* PARP cDNAs, with an emphasis on conservation of the functional elements, and expression of the PARP gene during development of *D. melanogaster*.

Cloning of cDNAs from different species using heterologous oligodeoxyribonucleotide consensus sequences

We have cloned human PARP cDNA by screening a cDNA library with an oligodeoxyribonucleotide probe corresponding to the aa sequence derived from purified PARP [11,12]. In lower eukaryotes, however, it was difficult to purify the enzyme in satisfactory quality and quantity because of endogenous protease activity or limitations in starting materials [7]. Homology probing and antibody probing are usual methods for isolation of cDNA. Although we tried such strategies, we failed to clone cDNA encoding PARP from *X. laevis*, *D. melanogaster*, and yeast. We surveyed the conserved aa sequences, which had lower complexity in codon usage, among mammals. Such consensus sequences were used to prepare mixed oligodeoxyribonucleotide primers, and DNA fragments were amplified by PCR. These PCR products had significant homology with mammalian PARP cDNAs [20]. The complete nucleotide sequences of *Xenopus* and *Drosophila* PARPs, and the partial cDNA sequence of cherry salmon PARP have been determined.

Table 1. Homology in nucleotide and amino acid sequences among PARPs of different species. H; human, M; mouse, B; bovine. C; chicken, X; *Xenopus*, D; *Drosophila*. aa; amino acid sequence homology. nt; nucleotide sequence homology. Amino acid sequence homology is indicated in similarity (and identity in parentheses) [23]

aa / nt	H	M	B	C	X	D
H		95 (92)	96 (90)	88 (79)	85 (75)	61 (43)
M	87		94 (88)	86 (77)	84 (73)	61 (43)
B	88	85		88 (77)	84 (72)	62 (43)
C	76	74	74		87 (78)	61 (43)
X	71	70	69	73		61 (43)
D	50	50	49	51	51	

Comparison of the deduced aa sequences of PARPs from mammals, chicken, *X. laevis* and *D. melanogaster*

Figure 1 shows the alignment of the aa sequences of human [12], mouse [15], bovine [16], chicken [17], *Xenopus* [19], and *Drosophila* [21] PARPs. Homology of the nucleotide and as sequences is summarized in Table 1 and Fig. 2. Homology in aa sequence is over 90% among mammals. In contrast, *Drosophila* PARP has 61% similarity (43% identity) in aa sequence to human PARP. *Drosophila* PARP is phylogenetically distant from vertebrates.

Zinc-finger motifs

Two zinc fingers repeated in the amino-terminus of PARP [12] are important for DNA binding and enzyme activity of PARP [24, 26]. PARP is activated by the presence of single

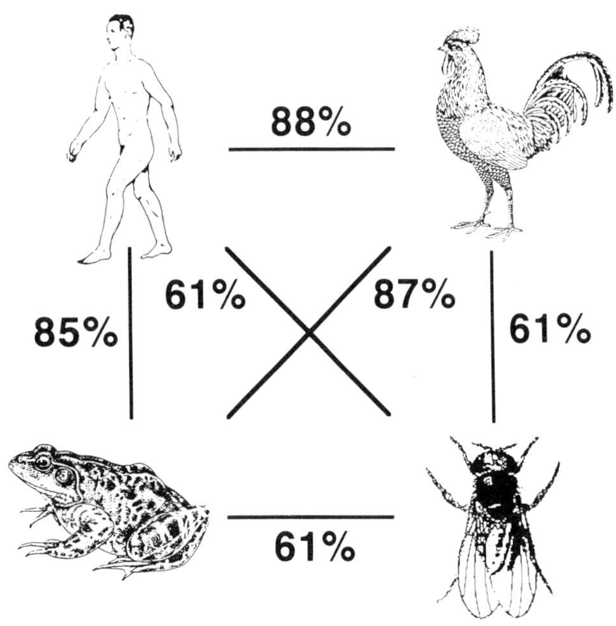

Fig. 2. Schematic representation of homology in aa sequence among different classes of animals. Similarity is determined by the maximum-match method [23].

```
                     +   *                +  *   *                           +     +
Human       1 MAESSDKLYRVEYAKSGRASCKKCSESIPKDSLRMAIMVQSPMFDGKVPHWYHFSCFWKVGHSIRH  66
Mouse       1 """A"ER"""""Q"""""""""""""""""""""""""""""""""""""""""""""Q""""  66
Bovine      1 """"""""""""""""""""""""K"""""""I"""F""E""""""""I"""""L""""""""F""W"  66
Chicken     1 """TG""P""A""""""""""""""G"""A"""""L"L"""""""""""""""H"Y""""""RARIVS"  66
Xenopus     1 """A"""""""A""""""""""""""GDN"A"E"""L"""""""""""""""""H"Y""""""RARVLSQ  66
Drosophila  1 .....MDIELP"LA"""RT"""T""G"KST"S""T""I"V""""AFH"A"""N"F"KT""F"NQRPSSV  64
                     +   *                  *   *                         +    +
Human     108 KT_LGDFAAEYAKSNRSTCKGCMEKIEKGQVRLSKKMVDPEKPQL_GMIDRWYHPGCFVKNREE 169
Mouse     108 ""_"""""""""""""""M"""L"""""""M"""""""""""""""""_"""""""""T"""""K"D" 169
Bovine    111 ""_"I""G"G""""""""""S""""D"""""""""V"Y"D"""""_""V"C""""K"""QK""" 172
Chicken   108 "S_"T""""""""""""""""""EQ""""""I"I""""""H"""""""_""""N"""D"""SR"A" 169
Xenopus   107 M"_"N"""""""""""""A""""EQ""""""I"I""""S""V"R""""_""""""""""D"""SS""" 168
Drosophila 105 NLA"K""GI"""""S"""R""EQ""N"DL"""R"TVY"T"VGMKY"GQPL"H"LE""AQL"S" 168
```

Fig. 3. Comparison of zinc-finger regions among human, mouse, bovine, chicken, *Xenopus*, and *Drosophila* PARPs. Boxes indicate consensus Cys and His in zinc fingers, and (″) indicates an identical aa to the human sequence and (_) indicates a gap. The conserved aa residues between the two zinc fingers and among species are marked with asterisks (basic residues) and (+).

```
                                  DNA-binding region          Connector        Leucine zipper
Transcription factors
  C/EBP            D K N S N E Y R V R R E R N N I A V R K S R D K A K Q R N V E T Q Q K V L E L T S D N D R L R K R V E Q L S R E L D T L
  CREB             E E A A R K R E V R L M K N R E A A R E C R R K K K E Y V K C L E N R V A V L E N Q N K T L I E E L K A L K D L Y C H K
  JUN              S Q E R I K A E R K R M R N R I A A S K C R K R K L E R I A R L E E K V K T L K A Q N S E L A S T A N M L T E Q V A Q L
  FOS              E E R R R I R R I R R E R N K M A A A K C R N R R R E L T D T L Q A E T D Q L E D K K S A L Q T E I A N L L K E K E K L
  GCN4             P E S S D P A A L K R A R N T E A A R R S R A R K L Q R M K Q L E D K V E E L L S K N Y H L E N E V A R L K K L V G E R
Human       Topo II T T V K F V V K M T E E K L A E A E R V G L H K V F K L Q T S L T C N S M V L F D H V G C L K K Y D T V L D I L R D L F
Drosophila  Topo II R R F P T A I D I L K E Y Y K L R R E Y Y A R R R D F L V G Q L T A Q A D R L S D Q A R F I L E K C E K K L V V E N K Q
Drosophila  PARP   S T F S R S L L K T N K N N D V L V R P T I P R I S P P L Y N L K F S I I G L K N Q H K E L R K I R E N L G G K F E V K
Xenopus     PARP   A A P T P I S P P A A P E P K P T V E E T F P E G K P L T N T K V L L I G K L S K N K D E V K T L I E G L G G K V A G S
Human       PARP   A S V A A T P P P S T A S A P A A V N S S A S A D K P L S N M K I L T L G K L S R N K D E V K A M I E K L G G K L T G T
                                               !     +     +         *       *       *       *
```

Fig. 4. Alignments of the leucine zipper region of the transcription factors, human and *Drosophila* DNA topoisomerase II, and PARPs from human, *Xenopus*, and *Drosophila*. Asterisks indicate the leucine residues in the heptad. Hydrophobic aa at the first position and the third or fourth position in the leucine repeats, the basic residues and the putative helix disrupting Pro and Asn in the neighboring region, are boxed. +, basic aa; !, helix disruptive Pro and Asn.

and double strand breaks of DNA [27,28]. The first (closer to the amino-terminus) finger is essential for PARP enzyme activity in response to nicked DNA or double strand DNA breaks. The second finger is essential for the activity sensitive to nicked DNA but not double strand DNA breaks [26]. Gradwohl et al. [25] expressed a 26 kDa DNA-binding domain of human PARP and demonstrated that disruption of the second finger results in loss of binding ability to nicks. Figure 3 shows that the duplicated zinc-finger regions at the amino-terminus is conserved in PARP from phylogenetically different species. In particular, the Cys and His residues of the fingers are completely conserved. Basic aa, Arg^{18}, Lys^{30}, Arg^{34}, Arg^{122}, Lys^{134} and Arg^{138}, and hydrophobic aa, Ala^{14}, Ile^{28}, Phe^{57}, Ala^{118}, Ile^{132}, Phe^{163}, are also conserved in both fingers among all species. These basic residues might be essential in the interaction with DNA. The conservation of specific aa also indicates that both zinc fingers participate coordinately in DNA binding, nick recognition, and enzyme activation.

Nuclear location signals

The bipartite nuclear location signals, KRK [29] and KKKIKK [13,14,29] are conserved from mammals to *Xenopus* PARP, but are not found at corresponding regions of *Drosophila* PARP [Fig. 1]. *Drosophila* PARP should be located in the nucleus, and $KRAK^{208-211}$, $KKQRLK^{223-228}$ or $KNKMKKK^{252-258}$ (numbers

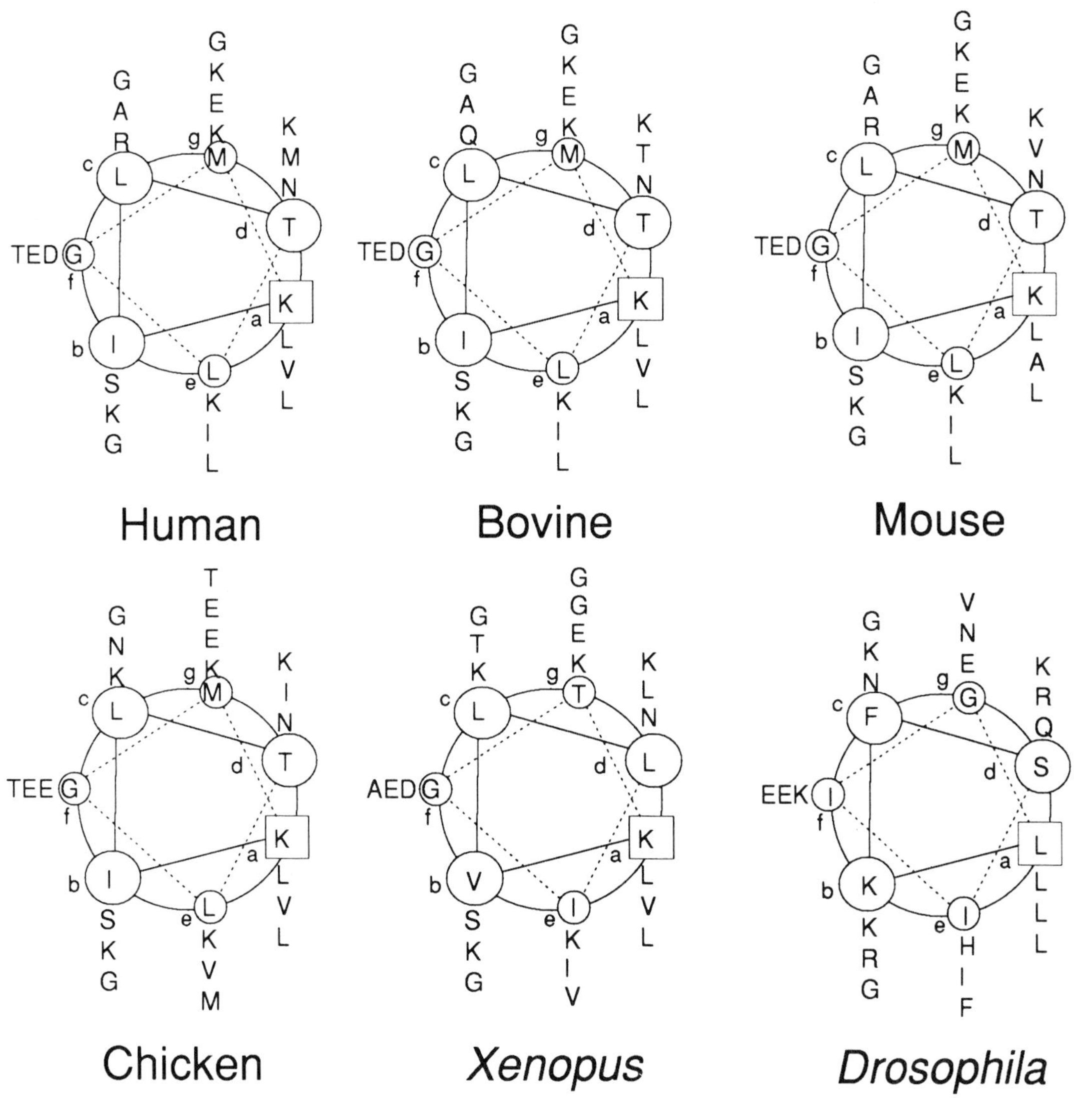

Fig. 5. Helical wheel representation of the leucine-zipper region of PARP, The analysis starts with the 1st Met in the heptad of mammalian and chicken PARP at position *g*, 1st Thr in the leucine zipper region of *Xenopus* PARP at position *g*. and 1st Leu in the heptad of *Drosophila* PARP at position *a* in the helical wheel.

correspond to *Drosophila* sequence) might serve as nuclear location signals in the absence of the usual bipartite signals.

Leucine-zipper motif

The putative leucine zipper of PARP was first found in *D. melanogaster* [21]. Interestingly, this motif is located in the auto-modification domain. The leucine zipper motif of PARP is conserved in all six species according to the following criteria. The region containing Leu repeats shows an amphipathic structure, the 3, 4 rule, and Pro is absent in this region [30,31] (Figs. 4, 5). However, Leu^{388} of *Drosophila* PARP is replaced by Lys in the other five species. The neighboring Val^{392} of *Xenopus* PARP and Met^{393} or Ile^{395} in mammalian and chicken PARP might be responsible for protein-interaction. Another possibility is that three heptad Leu residues may be sufficient for dimerization. The significance of these repeated Leu in PARP should be experimentally shown.

Although the auto-modification domain shows less conservation than the other two domains, Glu^{407} and Glu^{413} within the leucine-zipper motif are conserved in all six species. These Glu residues could be auto-modification sites for self-poly(ADP-ribosyl)ation [32]. Homo- or hetero-dimerization of PARP through the leucine zipper and regulation of dimerization by auto-modification should be demonstrated by West-western analysis using this leucine-zipper region as a protein probe.

NAD-binding structure

In the carboxyl terminal catalytic domain, PARP from all six species has the conserved dinucleotide binding motif, GXXXGKG (aa 888–894) [12–14] (aa numbers corresponding to the human PARP sequence) (Fig. 1). But another putative dinucleotide binding motif, GXGKT sequence (aa 950–954) [33], is not conserved in *Drosophila*. The regions from aa 695–702, 734–742, 764–779, 817–823, 859–908, and 984–1002 show great conservation. Several NAD-dependent dehydrogenases have a well-known particular structure for NAD binding, the Rossmann fold [34]. The Rossmann fold consists of a β1-αA-β2 arrangement. Simonin *et al.* first showed similarity between PARP and dehydrogenases by focusing on the Rossmann fold. Interestingly, *Drosophila* PARP retains the β1-αA-β2 structure (aa 967–994) like human [33] and chicken PARP [17] despite the absence of aa homology (Fig. 1). The hydrophobic aa and Asp^{993}, at the end of β2 is conserved. The GXG sequence lying in the β1-αA-β2 structure [33] is not conserved in *Drosophila*. A parallel six-stranded β-sheet structure with helix characteristic of dehydrogenases is important for the interaction with NAD (34). The conserved regions from aa 699–721, 734–742, 859–888, and 985–994 show β-sheet structure (Fig. 1). In the near future, X-ray or NMR analysis should provide a three-dimensional structure of PARP and identify the NAD-binding motif.

Expression of PARP during embryonic development

Figure 6 shows that in *Drosophila* embryos a high level of PARP mRNA is expressed 0–12 h after fertilization [21]. The PARP mRNA levels decrease 12–24 h after fertilization. Results of *in situ* hybridization are consistent with data of Northern blotting. Transcripts of the PARP gene were distributed homogeneously in *Drosophila* embryos (unpublished data). The PARP mRNA is transcribed and stored in eggs during oogenesis. After fertilization, the *Drosophila* PARP gene is constitutively transcribed.

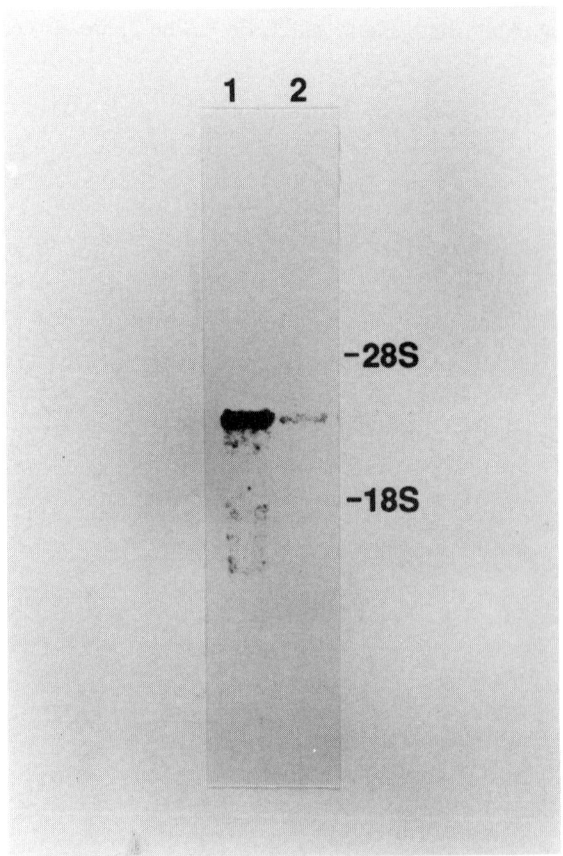

Fig. 6. Expression of PARP gene during embryonic development of *D. melanogaster* [21]. Total RNA (20 μg) from 0–12 h (lane 1) and 1224 h (lane 2) embryos were analyzed by Northern blotting. The arrow indicates *Drosophila* PARP mRNA of 3.2 kb. The positions of 28S and 18S human rRNA are shown as size markers.

Prospects

Isolation of a mutant with a deficient PARP gene would be a breakthrough for elucidating the function of PARP and poly(ADP-ribosyl)ation of proteins. Many *Drosophila* mutants have been isolated and well characterized. cDNA and genes for PARP isolated from lower eukaryotes should facilitate targeting of the PARP gene and analysis of its role(s) in development by introducing the anti-sense oligonucleotides into embryos.

Acknowledgements

We thank S. Hanai and K. Ishikawa for cloning and sequencing of *Drosophila* cDNA, and M. Uchida, Y. Ozawa and S. Hanai for cloning and sequencing of *Xenopus* cDNA. We also thank Drs. T. Sugimura, M. Okada, S. Kobayashi and Y. Taya for helpful comments and discussion, and Drs. A. Sarai and Y. Wang for computer analysis. This work was supported in part by Grants-in-aid for Scientific Research, Cancer Research and Tuskuba Project Research from the Ministry of Education, Science and Culture, Japan.

References

1. Rickwood D, Osman MS: Characterization of poly(ADP-ribose) polymerase activity in nuclei from the slime mould *Dictyostelium discoideum*. Mol. Cell Biochem 23: 79–84, 1979
2. Werner E, Sohst S, Gropp F, Simon D, Wagner H, Kröger H: Presence of poly(ADP-ribose) polymerase and poly(ADP-ribose) glycohydrolase in the dinoflagellate *Crypthecodinium cohnii*. Eur J Biochem 139: 81–86, 1984
3. Burtscher HJ, Auer B, Klocker H, Schweiger M, Hirsch-Kauffman, M: Isolation of ADP-ribosyltransferase by affinity chromatography. Anal Biochem 152: 285–290, 1986
4. Ushiro H, Yokoyama Y, Shizuta, Y: Purification and characterization of poly(ADP-ribose) polymerase from human placenta. J Biol Chem 262: 2352–2357, 1987
5. Mandel P, Okazaki H, Niedergang, C: Purification and properties of calf thymus polyadenosine diphosphate ribose polymerase, FEBS Lett 84: 331–336, 1977
6. Burtscher HJ, Schneider R, Klocker H, Auer B, Hirsch-Kauffman M, Schweiger M: ADP-ribosyltransferase is highly conserved: Purification and characterization of ADP-ribosyltransferase from a fish and its comparison with the human enzyme. J Comp Physiol 157: 567–572, 1987
7. Burtscher HJ, Klocker H, Schneider R, Auer B, Hirsch-Kauffman M, Schweiger M: ADP-ribosyltransferase from *Helix pomatia*: Purification and characterization. Biochem J 248: 859–864, 1987
8. Kameshita I, Yamamoto H, Fujimoto S, Shizuta Y: Antigenic determinant and interspecies cross-reactivity of a monoclonal antibody to poly(ADP-ribose) synthetase. FEBS Lett 182: 393–397, 1985
9. Scovassi AI, Izzo R, Franchi E, Bertazzoni U: Structural analysis of poly(ADP-ribose) polymerase in higher and lower eukaryotes. Eur J Biochem 159: 77–84, 1986
10. Alkhatib HM, Chen D, Cherney B, Bhatia K, Notario V, Giri C, Stein G, Slattery E, Roeder RG, Smulson ME: Cloning and expression of cDNA for human poly(ADP-ribose) polymerase. Proc Natl Acad Sci USA 84: 1224–1228, 1987
11. Suzuki H, Uchida K, Shima H, Sato T, Okamoto T, Kimura T, Miwa M: Molecular cloning of cDNA for human poly(ADP-ribose) polymerase and expression of its gene during HL-60 cell differentiation. Biochem Biophys Res Commun 146: 403–409, 1987
12. Uchida K, Morita T, Sato T, Ogura T, Yamashita R, Noguchi S, Suzuki H, Nyunoya H, Miwa M, Sugimura T: Nucleotide sequence of a full-length cDNA for human fibroblast poly(ADP-ribose) polymerase. Biochem Biophys Res Commun 148: 617–622, 1987
13. Kurosaki T, Ushiro H, Mitsuuchi Y, Suzuki S, Matsuda M, Matsuda Y, Katunuma N, Kangawa K, Matsuo H, Hirose T, Inayama S, Shizuta Y: Primary structure of human poly(ADP-ribose) synthetase as deduced from cDNA sequence. J Biol Chem 262: 15990–15997, 1987
14. Cherney B, McBride OW, Chen D, Alkhatib H, Bhatia K, Hensley P, Smulson ME: cDNA sequence, protein structure, and chromosomal location of the human gene for poly(ADP-ribose) polymerase, Proc Natl Acad Sci USA 84: 8370–8374, 1987
15. Huppi K, Bhatia K, Siwarski D, Klinman D, Cherney B, Smulson ME: Sequence and organization of the mouse poly(ADP-ribose) polymerase gene. Nucleic Acids Res 17: 3387–3401, 1989
16. Saito I, Hatakeyama K, Kido TG, Ohkubo H, Nakanishi S, Ueda K: Cloning of a full-length cDNA encoding bovine thymus poly(ADP-ribose) synthetase: evolutionarily conserved segments and their potential functions. Gene 90: 249–254, 1990
17. Ittel M-E, Garnier J-M, Jeltsh J-M, Niedergang, CP: Chicken poly(ADP-ribose) synthetase: complete deduced amino acid sequence and comparison with mammalian enzyme sequences. Gene 102: 157–164, 1991
18. Thibodeau J, Gradwohl G, Dumas C, Clairoux-Moreau S, Brunet G, Penning C, Poirier GG, Morea P: Cloning of rodent cDNA encoding the poly(ADP-ribose) polymerase catalytic domain and analysis of mRNA levels during the cell cycle. Biochem Cell Biol 67: 653–660, 1989
19. Uchida K, Uchida M, Hanai S, Ozawa Y, Ami Y, Kushida S, Miwa M: Isolation of poly(ADP-ribose) polymerase cDNA from *Xenopus laevis*: Phylogenetic conservation of the functional domain. Gene 137: 293–297, 1993
20. Ozawa Y, Uchida K, Uchida M, Ami Y, Kushida S, Okada N, Miwa M: Isolation of cDNAs encoding the catalytic domain of poly(ADP-ribose) polymerase from *Xenopus laevis* and cherry salmon using heterologous oligonucleotide consensus sequences. Biochem Biophys Res Commun 193: 119–125, 1993
21. Uchida K, Hanai S, Ishikawa K, Ozawa Y, Uchida M, Sugimura T, Miwa M: Cloning cDNA encoding *Drosophila* poly(ADP-ribose) polymerase: leucine zipper in the auto-modification domain. Proc Natl Acad Sci USA 90: 3481–3485, 1993
22. Garnier J, Osguthorpe DJ, Robson B: Analysis of the accuracy and implications of simple methods for predicting the secondary structure of globular proteins. J Mol Biol 120: 97–120, 1978
23. Needleman SB, Wunsch CD: A general method applicable to the search for similarities in the amino acid sequence of two proteins. J Mol Biol 48: 443–453, 1970
24. Ménissier-de Murcia J, Molinete M, Gradwohl G, Simonin F, and de Murcia G: Zinc-binding domain of poly(ADP-ribose) polymerase participates in the recognition of single strand breaks on DNA. J Mol Biol 210: 229–233, 1989
25. Gradwohl G, Ménissier-de Murcia J, Molinete M, Simonin F, Koken M, Hoeijmakers JH, and de Murcia G: The second zinc-finger domain of poly(ADP-ribose) polymerase determines specificity for single-stranded breaks in DNA. Proc Natl Acad Sci USA 87: 2990–2994, 1990

26. Ikejima M, Noguchi S, Yamashita R, Ogura T, Sugimura T, Gill DM, Miwa M: The zinc fingers of human poly(ADP-ribose) polymerase are differentially required for the recognition of DNA breaks and nicks and the consequent enzyme activation: Other structures recognize intact DNA. J Biol Chem 265: 21907–21913, 1990
27. Ohgushi H, Yoshihara K, Kamiya T: Bovine thymus poly(adenosine diphosphate ribose) polymerase: Physical properties and binding to DNA. J Biol Chem 255: 6205–6211, 1980
28. Benjamin RC, Gill DM: ADP-ribosylation in mammalian cell ghosts: Dependence of poly(ADP-ribose) synthesis on strand breakage in DNA. J Biol Chem 255: 10493–10501, 1980
29. Schreiber V, Molinete M, Boeuf H, de Murcia G, Ménissier-de Murcia J: The human poly(ADP-ribose) polymerase nuclear localization signal is a bipartite catalytic activity. EMBO J 11: 3263–3269, 1992
30. Landschulz WH, Johnson PF, McKnight SL: The leucine zipper: A hypothetical structure common to a new class of DNA binding proteins. Science 240: 1759–1764, 1988
31. O'Shea EK, Klemm JD, Kim PS, Alber T: X-ray structure of the GCN4 leucine zipper, a two-stranded, parallel coiled coil. Science 254: 539–544, 1991
32. Kawaichi M, Ueda K, Hayaishi O: Multiple autopoly(ADP-ribosyl)ation of rat liver poly(ADP-ribose) synthetase: Mode of modification and properties of automodified synthetase. J Biol Chem 256: 9483–9489, 1981
33. Simonin F, Ménissier-de Murcia J, Poch O, Muller S, Gradwohl G, Molinete M, Penning C, Keith G, de-Murcia G: Expression and site-directed mutagenesis of the catalytic domain of human poly(ADP-ribose) polymerase in *Escherichia coli*: Lysine 893 is critical for activity. J Biol Chem 265: 19249–19256, 1990
34. Rossmann MG, Moras K, Olsen KW: Chemical and biological evolution of a nucleotide-binding protein. Nature 250: 194–199, 1974

Molecular and Cellular Biochemistry **138:** 33–37, 1994.

Enzymology of ADP-ribose polymer synthesis

Rafael Alvarez-Gonzalez[1,2], Gustavo Pacheco-Rodriguez[2] and Hilda Mendoza-Alvarez[1,2]

The Departments of [1] Microbiology & Immunology and [2]Biochemistry & Molecular Biology, University of North Texas Health Science Center at Fort Worth, Fort Worth, Texas, 76107-2699, USA

Abstract

In this minireview, we summarize recent advances on the enzymology of ADP-ribose polymer synthesis. First, a short discussion of the primary structure and cloning of poly(ADP-ribose) polymerase (PARP) [EC 2.4.2.30], the enzyme that catalyzes the synthesis of poly(ADP-ribose), is presented. A catalytic distinction between the multiple enzymatic activities of PARP is established. The direction of ADP-ribose chain growth as well as the molecular mechanism of the automodification reaction catalyzed by PARP are described. Current approaches to dissect ADP-ribose polymer synthesis into individual reactions of initiation, elongation and branching, as well as a partial mechanistic characterization of the ADP-ribose elongation reaction at he chemical level are also presented. Finally, recent developments in the catalytic characterization of PARP by site-directed mutagenesis are also briefly summarized. (Mol Cell Biochem 138: 33–37, 1994)

Key words: poly(ADP-ribose) polymerase, structure, chemistry, kinetics, automodification, mechanism(s)

Introduction

Synthesis of chromatin-bound ADP-ribose polymers is catalyzed by poly(ADP-ribose) polymerase (PARP) [E.C. 2.4.2.30], an enzyme of 1014 amino acids that utilizes βNAD^+ as an ADP-ribose donor and is activated by DNA-strand breaks [1,2]. PARP is present in high amounts in the nucleus of actively dividing cells [1] and has been purified to homogeneity from a number of mammalian tissues [1].

Primary structure and cloning of PARP

Three distinct domains of PARP have been identified following partial proteolysis with papain and/or α-chymotrypsin [3–5]. These domains include (i) a 46 kDA DNA-binding fragment located at the amino-terminus; (ii) a 22 kDA central fragment containing the automodification sites; and (iii) a carboxy-terminal fragment of 54 kDA bearing the NAD-binding pocket.

The N-terminal DNA-binding domain contains a duplicated sequence (amino acid positions 2-97 and 106-207) in which 35 amino acid residues are highly conserved. More importantly, 'zinc-finger'-like motifs of the form C-Xaa_{28-30}-H-Xaa_2-C are contained in these sequences. Thus, this enzyme belongs to the 'zinc finger' DNA-binding protein family [6,7]. Zinc fingers I and II allow the enzyme to bind specifically double [8] and single strand-breaks [9] on DNA, respectively.

The automodification domain contains 15 highly conserved glutamate residues [10]. These glutamate residues may include some of the auto-poly(ADP-ribosyl)ation sites (*vide infra*). Recently, it has been reported that this domain in the *Drosophila melanogaster* gene [11] also contains a 'leucine zipper' DNA-binding motif. Therefore, it appears reasonable to speculate that this leucine zipper may allow for the homodimerization (self-macromolecular association) and heterodimerization of PARP with other chromatin proteins. In this dimerization process, protein-protein interactions may help stabilize $(PARP)_2$-DNA complexes and thus facilitate the characteristic intermolecular auto-poly(ADP-ribosyl)ation reaction [12].

Address for offprints: R. Alvarez-Gonzalez, Department of Microbiology and Immunology, Texas College of Osteopathic Medicine, 3500 Camp Bowie Boulevard, Forth Worth, TX76107-2699, USA

The carboxy-terminal region of PARP, the 54 kDA substrate-binding domain, contains the sequence GXXXGKG (aa residues 888-894) [11]. This amino acid sequence is highly homologous to the consensus sequence G/AXXXGKT/S/G typically associated with proteins that recognize and tightly bind to phosphoanhydride bonds of mono and dinucleotides (Rossmann fold), i.e., NAD.

Multiple enzyme activities of PARP

Whereas critical poly(ADP-ribose) covalent protein targets include histone proteins [13–15] and DNA-metabolizing enzymes [heterologous poly(ADP-ribosyl)ation] [16,17], the main poly(ADP-ribosyl)ated-polypeptide *in vitro* [18,19] and *in vivo* [20,21] is PARP itself [homologous poly(ADP-ribosyl)ation]. The efficient automodification reaction of PARP has been reported in bovine thymus [18], rat liver [19], calf thymus [22], and He-La Cells [23]. This reaction may involve anywhere between 15 [19] and 28 [24] amino acid acceptor sites on the enzyme molecule. Not surprisingly, this massive hyper-poly(ADP-ribosyl)ation reaction leads to the catalytic inactivation of PARP as a result of the electrostatic repulsion between the negatively charged enzyme-bound ADP-ribose polymers and DNA [22,25].

In addition to efficiently catalyzing both an automodification reaction and a heterologous poly(ADP-ribosyl)ation reaction, PARP appears to be responsible for the catalysis of three chemically distinct enzymatic steps required for poly(ADP-ribose) synthesis:

(i) the glutamate-specific attachment of one ADP-ribose unit to an acceptor protein (initiation reaction) (26);

(ii) the characteristic (2'-1") ribose-ribose glycosidic bond formation (ADP-ribose chain elongation reaction) [1]; and

(iii) the (2"-1"') ribose-ribose bonding between ADP-ribose units (branching reaction) [27].

While the initiation reaction occurs via a quasi-distributive mechanism [12,19], ADP-ribose polymerization (elongation and branching reactions) appears to take place via a highly processive mechanism [28].

Direction of ADP-ribose chain growth and molecular mechanism of enzyme automodification

The elongation of ADP-ribose chains has been shown to occur via a protein-distal elongation mechanism [29–31] where the 2'-hydroxyl group of the non-reducing end of the growing ADP-ribose chain (polymer tail) is utilized as the acceptor for the covalent binding of the newly added ADP-ribose unit. This mechanism of ADP-ribose polymerization predicts that the automodification reaction of PARP is bimolecular (intermolecular) in nature and that initial rates of auto-poly(ADP-ribosyl)ation should increase with the square of the enzyme concentration. In agreement with this hypothesis, we have recently observed that initial rates of automodification increase with second order kinetics as a function of the concentration of PARP at saturating concentrations of DNA and $MgCl_2$ [12]. In fact, second order kinetics was observed during both the auto-mono(ADP-ribosyl)ation (at nM [NAD]) and the auto-poly(ADP-ribosyl)ation (at µM [NAD]) reactions of PARP [12]. Thus, our results are consistent with the conclusion that PARP performs the automodification reaction as a catalytic dimer [12]. Furthermore initial rates of automodification were also observed to increase with second order kinetics as a function of the concentration of NAD at nM levels [12] suggesting that at least one molecule of βNAD^+ must be bound to each monomeric unit of PARP for efficient automodification to occur. Further experiments are in progress to determine whether this kinetic behavior may also be indicative of cooperativity between monomeric units of the enzyme and/or allosteric regulation of this enzyme.

Experimental evidence supporting the self-macromolecular association of PARP was also recently obtained by molecular sieve chromatography, non-denaturing polyacrylamide gel electrophoresis, and chemical cross-linking experiments [32]. Furthermore, as indicated above, Uchida, *et. al.*, [10] reported that a cDNA clone encoding the *Drosophila* poly(ADP-ribose) polymerase gene contains a leucine zipper in the automodification domain. The periodic repetition of leucine residues at every seventh amino acid residue (leucine zipper) in the automodification domain, and the basic animo acid-rich region adjacent to the leucine residues of this sequence, may mediate the homo- and hetero-dimerization of PARP with itself and other DNA-binding proteins, respectively.

We have also observed that the catalytic dimerization of PARP is dependent on the time and temperature of incubation as well as enzyme concentration (unpublished results). Furthermore, successive dilutions of this enzyme down to low nM concentrations in the absence of DNA results in the preferential monomerization of PARP which possesses a limited ability to generate free ADP-ribose, due to the absence of protein acceptor molecules. Instead, it behaves as an 'abortive' NAD glycohydrolase.

Dissection of ADP-ribose polymer synthesis into individual reactions of initiation, elongation, and branching

In 1980, Smulson and co-workers [33] demonstrated that the average chain length of protein-bound ADP-ribose polymers synthesized in mammalian oligonucleosomal fractions [heterologous and homologous poly(ADP-ribosyl)ation combined] increased with the concentration of βNAD$^+$. However, under their *in vitro* conditions, the contribution of endogenous mono(ADP-ribosyl)transferase activity [34], was not taken into consideration.

Based on these observations, we decided to evaluate the effect of decreasing the concentration of NAD$^+$, from the typically used high μM levels [1] to low nM levels, on the chain length of polymers of ADP-ribose synthesized with highly purified enzyme. In these experiments, we were able to dissect ADP-ribose polymerization into initiation, elongation, and branching reactions. In these experiments, while nM [NAD] (50–400 nM) favored the auto-mono(ADP-ribosyl)ation of PARP, 1–2 μM [NAD] favored ADP-ribose elongation (and some initiation), and 10–200 μM [NAD] resulted in the exclusive formation of highly branched polymers of ADP-ribose [12]. It is of interest to note that the variation in ADP-ribose chain lengths as function of the βNAD$^+$ concentration may have physiological relevance during the repair of DNA-damage, since under these conditions, the NAD pools are quickly lower to 10% or less of the controls as a result of the dynamic turnover of chromatin-bound poly(ADP-ribose) [35–37]. Indeed, a direct relationship between the intracellular concentration of βNAD$^+$ and modulation of chromatin structure and function has also been postulated [38].

ADP-ribose chain elongation

As indicated above, PARP catalyzes the initiation, elongation, and branching reactions of ADP-ribose polymer synthesis (26–31). It has been estimated that for every initiation step [protein-mono(ADP-ribosyl)ation], this enzyme is capable of catalyzing 5–7 branching reactions and over 200 elongation events [39,40]. Thus, the ADP-ribose chain elongation reaction appears to be the main catalytic function of this protein.

We have recently focused on the molecular, chemical, and enzymatic characterization of this reaction. To better determine the mechanisms of ADP-ribose chain elongation, we have synthesized model compounds that function as the ADP-ribose donor(s), i.e., β3′-deoxyNAD$^+$ [31,41,42], and the ADP-ribose acceptor, i.e., arginine methyl ester-(ADP-ribose) [43], of the transfer reaction. Whereas β3′-deoxyNAD$^+$ has proven to be instrumental in studying the direction of ADP-ribose chain growth [31], it can also be used to avoid the self-inactivation of PARP as a result of auto-poly(ADP-ribosyl)ation [31]. On the other hand, the mono(ADP-ribosyl)ated-amino acid has been useful to bypass the initiation step of ADP-ribose polymer synthesis [43] and provides PARP with a "primed" amino acid-mono(ADP-ribose) conjugate ready for ADP-ribose chain elongation. We have recently reported the enzymatic synthesis of arginine methyl ester-[^{32}P](ADP-ribose) with an arginine specific mono(ADP-ribosyl)transferase utilizing βNAD$^+$ as the ADP-ribose donor [43] and arginine methyl ester as the acceptor molecule. Following purification of arginine methyl ester-mono(ADP-ribose) by affinity chromatography and HPLC, we observed that it was utilized as an ADP-ribose elongation acceptor [43]. Therefore, it appears that PARP, as any enzyme with transferase properties, possesses at least two substrate binding sites: one for the ADP-ribose donor (βNAD$^+$ binding site) and one for the ADP-ribose elongation acceptor (previously undescribed).

Should this hypothesis be correct, one should also be able to block ADP-ribose polymer synthesis with competitive inhibitors of βNAD$^+$. i.e., benzamide and its chemical derivatives [1]. In addition, one should also be able to develop model compounds that inhibit PARP by competing with the ADP-ribose elongation acceptor (ADP-ribose growing chain) for the second substrate-binding site. Preliminary experimental evidence demonstrating that this is the case was observed with β2′-deoxyNAD$^+$ [43]. This βNAD$^+$ analog lacks the 2′-OH group utilized as the target for elongation and effectively competes with the ADP-ribose elongation acceptor substrate, i.e., the protein-distal ADP-ribose unit of the growing polymer. In these experiments [44], we observed that β2′-deoxyNAD$^+$ was a potent non-competitive inhibitors of βNAD$^+$. More recently, we have synthesized additional model compounds with chemical structures that may interfere with the putative second substrate-binding site of PARP. In support of the idea that this enzyme actually recognizes an ADP-ribose elongation acceptor substrate, we have observed that agmatine-(ADP-ribose), a mono(ADP-ribosyl)ated molecule enzymatically synthesized with cholera toxin, is a potent and specific inhibitor of the ADP-ribose polymerization reaction (Pacheco-Rodriguez, G. and Alvarez-Gonzalez, R., manuscript in preparation).

Utilization of site-directed mutagenesis to study the catalytic mechanism of PARP

Utilization of site-directed mutagenesis has recently proven to be a powerful technique to determine the catalytic role of specific amino acid residues on the structure and function of proteins. The availability of bacterial expression systems [45,46] to quickly generate specific point mutations on the PARP gene has allowed for the identification of potential active site amino acid residues in the human PARP gene [46]. Simonin, *et al.*, [46] have recently reported that non-conservative mutations of Lys-893 (K893I) and Asp-993 (D993A), two amino acid residues of the substrate-binding domain, presumed to be involved in the recognition of βNAD$^+$, completely inactivate human PARP. By contrast, conservative and non-conservative mutations of Asp-914 (D914E and D914A, respectively) and Lys 953 (K953R and K953I, respectively) only partially affect the activity of PARP [46]. Thus, it appears that amino acid residues Lys-893 and Asp-993 might be involved in the covalent attachment of the first ADP-ribosyl residue from βNAD$^+$ to the acceptor molecule (catalysis of the initiation reaction) [46]. Alternatively, these residues might just be involved in the correct stereochemical positioning of βNAD$^+$ prior to catalysis. Further experimentation utilizing site-directed mutagenesis as well as the determination of the kinetic and chemical mechanisms of ADP-ribose polymer synthesis utilizing pre-steady state, steady state kinetics, as well as pH studies, is required to accurately determine the catalytic role of specific amino acid residues on this multi-domain and multi-catalytic enzyme of 1014 amino acid residues.

Acknowledgements

The support of this research project by the Petroleum Research Fund, administered by the American Chemical Society, is gratefully acknowledged.

References

1. Althaus F R, Richter C R: ADP-ribosylation of Proteins. Mol Biol Biochem Biophys 37: 1-126, 1987
2. Benjamin R C, Gill D M: Poly(ADP-ribose) Synthesis *in vitro* Programmed by Damaged DNA. A Comparison of DNA Molecules Containing Different Types of Strand Beaks. J Biol Chem 255: 10501-10508, 1980
3. Nishikimi N, Ogasawara K, Kameshita I, Taniguchi T, Shizuta Y: Poly(ADP-ribose) Synthetase. The DNA Binding Domain and the Automodification Domain. J Biol Chem 257: 5102-6105, 1982
4. Kameshita I, Matsuda Z, Taniguchi T, Shizuta Y: Poly(ADP-ribose) Synthetase. Separation and Identification of the Three Proteolytic Fragments as the Substrate Binding Domain, the DNA-Binding Domain, and the Automodification Domain. J Biol Chem 259: 4770-4776, 1984
5. Kameshita I, Matzuda M, Nishikimi N, Ushiro H, Shizuta Y: Reconstitution and Poly(ADP-ribosyl)ation of Proteolytically Fragmented Poly(ADP-ribose) Synthetase. J Biol Chem 262: 3863-3868, 1986
6. Mazen A, Menessier-DeMurcia J, Molinete M, Simonin F, Gradwohl G, Poirier G G, de Murcia G: Poly(ADP-ribose) Polymerase: A Novel Finger Protein. Nucl. Acids Res 17: 4689-4698, 1989
7. Menessier-DeMurcia J, Molinete M, Gradwohl G, Simonin F, DeMurcia G: Zinc Binding Domain of Poly(ADP-ribose) Polymerase Participates in the Recognition of Single Strand Breaks on DNA. J Mol Biol 210: 229-233, 1989
8. Ikejima M, Noiguchi S, Yamashita R, Ogura T, Sugimura T, Gill D M, Miwa M: The Zinc Fingers of Human Poly(ADP-ribose) Polymerase Are Differentially Required for the Recognition of DNA Breaks and Nicks and the Consequent Enzyme Activation. J Biol Chem 266: 21907-21913, 1991
9. Gradwohl G, Menessier de Murcia G, Molinete M, Simonin F, Koken M, Hoeijmakers J H J, deMurcia G: The Second Zinc Finger Domain of Poly(ADP-ribose) Polymerase Determines Specificity for Single-Stranded Breaks in DNA. Proc Natl Acad Sci USA 87: 2990-2994, 1990
10. Uchida K, Hanai S, Ishikawa K-I, Ozawa Y-I, Uchida M, Sugimura T, Miwa M: Cloning of cDNA encoding *Drosophila* Poly(ADP-ribose) Polymerase: Leucine Zipper in the Automodification Domain. Proc Natl Acad Sci USA 90: 3481-3484, 1993
11. Kurasaki T, Ushiro H, Mitsuchi Y, Suzuki S, Matsuda M, Matsuda Y, Katunuma N, Kangawa K, Matsuo H, Hirose T, Inayama S, Shizuta Y: Primary Structure of Human Poly(ADP-ribose) Synthetase as Deduced from cDNA Sequence. J Biol Chem 262: 15990-15997, 1987
12. Mendoza-Alvarez H, Alvarez-Gonzalez R: Poly(ADP-ribose) Polymerase is a Catalytic Dimer and the Automodification Reaction is intermolecular. J Biol Chem 268: 22575–22580, 1993
13. Huletsky A, Niedergang C, Frechette A, Aubin R, Gaudreau A, Poirier G G: Sequential ADP-ribosylation Pattern of Nucleosomal Histones. Eur J Biochem 146: 277-285, 1985
14. Boulikas T: At Least 60 ADP-ribosylated Variant Histones are Present in Nuclei from Dimethylsulphate-treated and Untreated Cells. EMBO J 7: 57-67, 1988
15. Boulikas T: Poly(ADP-ribosylated) Histones in Chromatin Replication. J Biol Chem 265: 14638-14647, 1990
16. Yoshihara K, Itaya A, Tanaka Y, Ohashi Y, Ito K, Teraoka H, Tsukuda K, Matsukage A, Kamiya T: Inhibition of DNA Polymerase α, DNA Polymerase ß, Terminal Deoxynucleotidyltransferase and DNA Ligase II by Poly(ADP-ribosyl)ation Reaction *in vitro*. Biochem Biophys Res Commun 128: 61-67, 1985
17. Ohahi Y: Effect of Ionic Strength on Chain Elongation in ADP-ribosylation of Various Nucleases. J Biochem 99: 971-979, 1986
18. Yoshihara K, Hashida T, Tanaka Y, Oghushi H: Enzyme-bound Early Product of Purified poly(ADP-ribose) Polymerase. Biochem Biophys Res Commun 78:1281-1288, 1977
19. Kawaichi M, Ueda K, Hayaishi O: Multiple Auto-poly(ADP-ribosyl)ation of Rat Liver Poly(ADP-ribose) Synthetase: Mode of Modification and Properties of Automodified Synthetase. J Biol Chem 256: 9483–9489, 1981
20. Adamietz P: Poly(ADP-ribose) Synthetase is the Major Endogenous Non-histone Acceptor for Poly (ADP-ribose) in Alkylated Rat Hepatoma Cells. Eur J Biochem 169: 365-372, 1987
21. Kreimeyer A, Wielckens K, Adamietz P, Hilz H: DNA-repair Associated ADP-ribosylation *in vivo*. Modification of Histone H1 Differs from

that of the Principal Acceptor Proteins. J Biol Chem 259: 890-896, 1984

22. Zahradka P, Ebisuzaki K: A Shuttle Mechanism for DNA-Protein Interactions. The Regulation of Poly(ADP-ribose) Polymerase Eur J Biochem 127:579-585, 1982
23. Jump D B, Smulson M: Purification and Characterization of the Major Nonhistone Protein Acceptor for Poly(adenoisine diphosphate ribose) in He-La cell nuclei. Biochem 19: 1024-1030, 1980
24. Desmarais Y, Menard L, Lagueux J, Poirier G G: Enzymological Properties of Poly(ADP-ribose)polymerase. Characterization of Automodification Sites and NADase Activity. Biochem Biophys Acta 1078: 179-186, 1991
25. Ferro A M, Olivera B M: Poly(ADP-ribosylation) *in vitro*: Reaction parameters and Enzyme Mechanism. J Biol Chem 257: 7808-7813, 1982
26. Kawaichi M, Ueda K, and Hayaishi O: Initiation of Poly(ADP-ribosyl)-Histone Synthesis of Poly(ADP-ribose) Synthetase. J Biol Chem 255: 816-819, 1980
27. Miwa M, Saikawa N, Yamaizumi Z, Nishimura S, Sugimura T: Structure of Poly(adenosine diphosphate ribose): Identification of 2'-(1''-ribosyl-2''- (or 3''-) (1'''-ribosyl)adenosine-5',5'',5'''-tris(phosphate) as a Branch Linkage. Proc Natl Acad Sci USA 76: 595-599, 1979
28. Naegeli H, Loetscher P, Althaus F R: Poly ADP-ribosylation of Proteins: Processivity of Post-translational Modification. J Biol Chem 264: 14382-14384, 1989
29. Ueda K Kawaichi M, Okayama H, Hayaishi O: Poly(ADP-ribosyl)ation of Nuclear Proteins: Enzymatic Elongation of Chemically Synthesized ADP-ribose Histone Adducts. J Biol Chem 254: 679-687, 1979
30. Taniguchi T: Reaction Mechanism for Automodification of Poly(ADP-ribose) Synthetase: Biochem Biophys Res Commun 147: 1008-1012, 1987
31. Alvarez-Gonzalez R:L 3'-deoxy-NAD$^+$ as a Substrate for Poly(ADP-ribose) Polymerase and the Reaction Mechanism of Poly(ADP-ribose) Elongation. J Biol Chem 263: 17690-17696, 1988
32. Bauer P I, Buki K G, Hakam A, Kun E: Macromolecular Association of ADP-ribosyltransferase and its Correlation with Enzymic Activity. Biochem J 270: 17-26, 1990
33. Butt T R, Smulson M: Relationship Between Nicotinamide Adenine Dinucleotide and in Vitro Synthesis of Poly(adenosine Diphosphate ribose) on Purified Nucleosomes. Biochem 19: 5235-5242, 1980
34. Tanigawa Y, Tsuchiya M, Imai Y, Shimoyama M: ADP-ribo-syltransferase from Hen Liver Nuclei. J Biol Chem 259: 2022-2029, 1984
35. Wielckens K, Schmidt A, George E, Bredehorst R, Hilz H: DNA Fragmentation and NAD Depletion: Their Relation to the Turnover of Endogenous Mono(ADP-ribosyl) and Poly(ADP-ribosyl) Proteins. J Biol Chem 257: 12872–12877, 1982
36. Jacobson E L, Antoe K M, Juarez-Salinas H, Jacobson M K: Poly(ADP-ribose) Metabolism in Ultraviolet Irradiated Human Fibroblasts. J Biol Chem 258: 103-107, 1983
37. Alvarez-Gonzalez R, Althaus, F R: Poly (ADP-ribose) Catabolism in mammalian Cells Exposed to DNA-damaging Agents. Mutation Res 218:67-74, 1989
38. Loetscher P. Alvarez-Gonzalez R, Althaus F R: Poly(ADP-ribose) May Signal Changing Metabolic Conditions to the Chromatin of Mammalian Cells. Proc Natl Acad Sci USA 84: 1286–1289, 1987
39. Alvarez-Gonzalez R, Jacobson M K: Characterization of Polymers of Adenosine Diphosphate Ribose Generated *in vitro* and *in vivo*. Biochemistry 26: 3218-3224, 1987
40. Kiehlbauch C C, Aboul-Ela N, Jacobson E L, Ringer S P, Jacobson M: High Resolution Fractionation and Characterization of ADP-ribose Polymers. Anal. Biochem. 208: 26-34, 1993
41. Alvarez-Gonzalez R: Synthesis and Purification of Deoxyribose Analogues of NAD$^+$ by Affinity Chromatography and SAX-HPLC. J Chromatography 444: 89-85, 1988
42. Alvarez-Gonzalez R, Panzeter P, Ringer D P, Mendoza-Alvarez H O: Poly(3'-dADP-ribosyl)ation of Proteins in Liver Chromatic Isolated from Rats Fed with Hepatocarcinogens. In: Poirier, G.G. and Moreau, P., eds, ADP-ribosylation Reactions, Springer-Verlag, New York, Berlin-Heidelberg, pp 149-252, 1992
43. Pacheco-Rodriguez G, Moss J, Alvarez-Gonzalez R: ADP-ribose Elongation of Mono-(ADP-ribosyl)-arginine Methyl Ester by Poly(ADP-ribose) Polymerase. FASEB Journal 7: A1196, 1993
44. Alvarez-Gonzalez R, Moss J, Niedergang C, Althaus F R: Selective probing of ADP-ribosylation reactions with Oxidized 2'-Deoxy-nicotinamide Adenine Dinucleotide Biochemistry 27: 5378-5383, 1988
45. Simonin F, Menesseier-de Murcia J, Poch O, Muller S, Gradwohl G, Molinete M, Penning C, Keith G, de Murcia G: Expression and Site Directed Mutagenesis of The Catalytic Domain of Human Poly(ADP-ribose) Polymerase in *Echerichia coli*. J Biol Chem 265:19249-19256, 1990
46. Simonin F, Poch O, Delarue M, de Murcia G: Identification of Potential Active-site Residues in the Human Poly(ADP-ribose) Polymerase. J Biol Chem 268: 852908535, 1993

Molecular and Cellular Biochemistry **138**: 39–44, 1994.

Interaction of poly(ADP-ribose)polymerase with DNA polymerase α

Shonen Yoshida and Cynthia Marie G. Simbulan

Laboratory of Cancer Cell Biology, Research Institute for Disease Mechanism and Control, Nagoya University School of Medicine, Showa-ku, Nagoya 466, Japan

Abstract

Homogeneously purified poly(ADP-ribose) polymerase (PARP) specifically stimulated the activity of immunoaffinity-purified calf or human DNA polymerase α by about 6 to 60-fold. Apparently, poly(ADP-ribosyl)ation of DNA polymerase α was not necessary for the stimulation. The effects of PARP on DNA polymerase α were biphasic: at very low concentrations of DNA, it rather inhibited its activity, whereas, at higher DNA concentrations, PARP greatly stimulated it. The autopoly(ADP-ribosyl)ation of PARP suppressed both its stimulatory and inhibitory effects. By immunoprecipitation with an anti-DNA polymerase α antibody, it was clearly shown that PARP may be physically associated with DNA polymerase α. Stimulation of DNA polymerase α may be attributed to the physical association between the two, rather than to the DNA-binding capacity of PARP, since the PARP fragment containing only the DNA binding domain showed little stimulatory activity. The existence of PARP-DNA polymerase α complexes were also detected in crude extracts of calf thymus. (Mol Cell Biochem **138**: 39–44, 1994)

Key words: DNA polymerase α, poly(ADP-ribose)polymerase, immunoprecipitation, autopoly(ADP-ribosyl)ation

Introduction

Accumulating evidence suggests that DNA synthesis may closely be correlated with an increased level of PARP in several tissues [1–6]. Newly replicated chromatin [7] as well as DNA fragments rich in replication forks [4] were co-purified with PARP. On the other hand, addition of a PARP inhibitor arrested Hela cell proliferation [8]. These effects have been correlated with poly(ADP-ribosyl)ation of nuclear proteins, such as histones [9], endonucleases [10], reverse transcriptase [11], DNA polymerase β [12], topoisomerase 1 [13, 14], DNA ligase II [15], and SV40 T-antigen [16, 17]. In contrast, Eki and Hurwitz recently demonstrated that PARP strongly suppressed the *in vitro* replication of SV40 DNA, without poly(ADP-ribosyl)ation of proteins [18]. The role(s) of PARP in DNA replication, therefore, has yet to be elucidated.

Increasing evidence also suggests that PARP participates in DNA repair by binding to DNA strand breaks, constituting a signal that switches off DNA synthesis temporarily to ensure that lesions are not replicated before repair [19]. As a signal for repair, PARP might also interact with repair enzymes, such as DNA polymerases, but this possibility has not yet been examined. In the present study, we demonstrate for the first time that PARP can bind physically to DNA polymerase α and directly affect its activity [20].

Results and discussion

PARP Specifically Stimulates DNA Polymerase α

PARP was purified from calf thymus by affinity chromatography on a 3-aminobenzamide Affi-gel column, followed by hydroxylapatite and Sephadex G-200 column chromatographies [21], with the final preparation showing a single polypeptide of 116 kDa after SDS-PAGE. The DNA polymer-

Address for correspondence: S. Yoshida, Laboratory of Cancer Cell Biology, Research Institute for Disease Mechanism and Control, Nagoya Uuiversity School of Medecine, Showa-ku, Nagoya 466, Japan

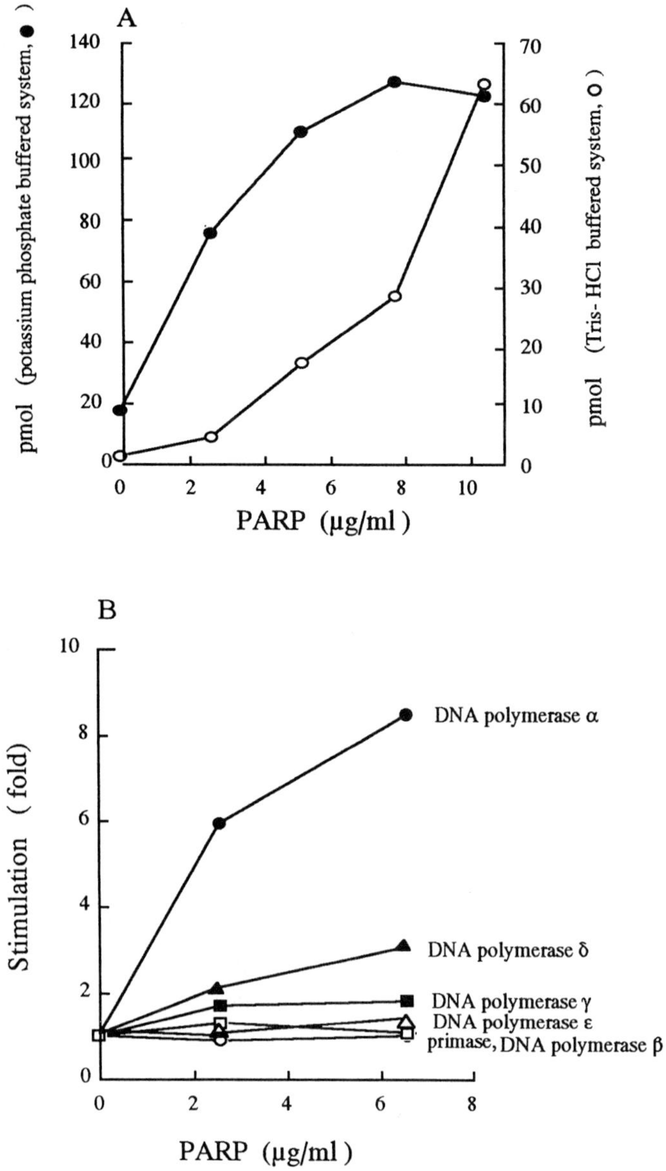

Fig. 1. DNA polymerase α is stimulated specifically and dose-dependently by PARP. A. Dose-dependent stimulation of DNA polymerase α activity by PARP, in potassium phosphate-buffered system (●) and in Tris-HCl-buffered system (o). Incubation was carried out at 37°C for 60 minutes. The reaction mixture (50 μl) contained 40 mM potassium phosphate (pH 7.2) or Tris-HCl (pH 7.5), 4 mM 2-mercaptoethanol, 40 μM each of dATP, dGTP, dCTP, 20 μM [^{3}H]-dTTP (500 cpm/nmol), 4 mM $MgCl_2$, 5 ´μg BSA, 5 μg activated calf thymus DNA, and 0.05 unit of calf thymus DNA polymerase α-primase complex. (from [20]) B. Enzyme specificity of inhibition. DNA polymerase β (o), DNA polymerase γ (■) [26] and DNA primase (□) [27], as well as calf thymus DNA polymerase α (●), δ(▲), and ε (Δ) were assayed, in the presence of indicated amounts of PARP (from [20] and unpublished results).

ase α-primase complex was also purified from calf thymus on an immunoaffinity Sepharose 4B column conjugated with monoclonal antibody (MT17) directed against the DNA polymerase α catalytic subunit [22]. The final preparation showed 4 subunits of 150, 70, 53, and 45 kDa [23]. Addition of homogeneously purified PARP (5–10 μg/ml) to the reaction with immunoaffinity-purified DNA polymerase α dose-dependently stimulated DNA polymerase α activity in either potassium phosphate-buffered assays or Tris-HCl-buffered assays. Stimulation was 6 to 8-fold in potassium phosphate and up to 60 fold in Tris-HCl (Fig. 1A), since the immunoaffinity-purified DNA polymerase α enzyme showed little activity in Tris-HCl [23]. This stimulation was evidently specific for DNA polymerase α because DNA polymerases β and γ, DNA primase, as well as DNA polymerase ε were not stimulated at all, while DNA polymerase δ was also dose-dependently stimulated, but to a much lesser extent (Fig. 1B).

An anti-PARP antibody, which recognizes the catalytic domain [24], conjugated to CNBr Sepharose, immunoprecipitated PARP, resulting in a concomitant decrease in stimulatory activity. Therefore, it was concluded that the DNA polymerase α stimulation is indeed due to PARP itself.

Mode of Action

As shown in Fig. 2, the curve of the activity versus DNA concentration became sigmoidal in the presence of PARP. Maximum stimulation was observed at concentrations higher than the saturation level (100 μg/ml, the reaction was rather inhibited by the addition of PARP. At DNA doses less than 10 μg/ml, PARP apparently inhibited the reaction in competition with the DNA template-primer (Fig. 2B). This is consistent with results of Eki and Hurwitz [18], wherein the binding of PARP to DNA termini also inhibited the elongation of Okazaki fragments by DNA polymerase α in the SV40 DNA replication system. In this system, they used 0.3 μg/ml of supercoiled [18], which is approximately 16-fold less than the concentrations of activated DNA used in our usual assay system. At low DNA concentrations, the inhibitory effect of PARP caused by DNA binding might become predominant. On the other hand, at high DNA concentrations, a 4-fold increase in V_{max} was observed, suggesting an entirely different mechanism for stimulation. Interestingly, although essentially the same results were observed with another DNA primed-template, primed M13 DNA, when an RNA-primed DNA template, poly(dT)-oligo(rA), was used as template, there was no inhibition by PARP at low DNA concentrations, and stimulation was observed at all DNA concentrations. This suggests that PARP does not compete for binding to RNA primer ends, thus the stimulatory effect of PARP was predominant at all DNA concentrations.

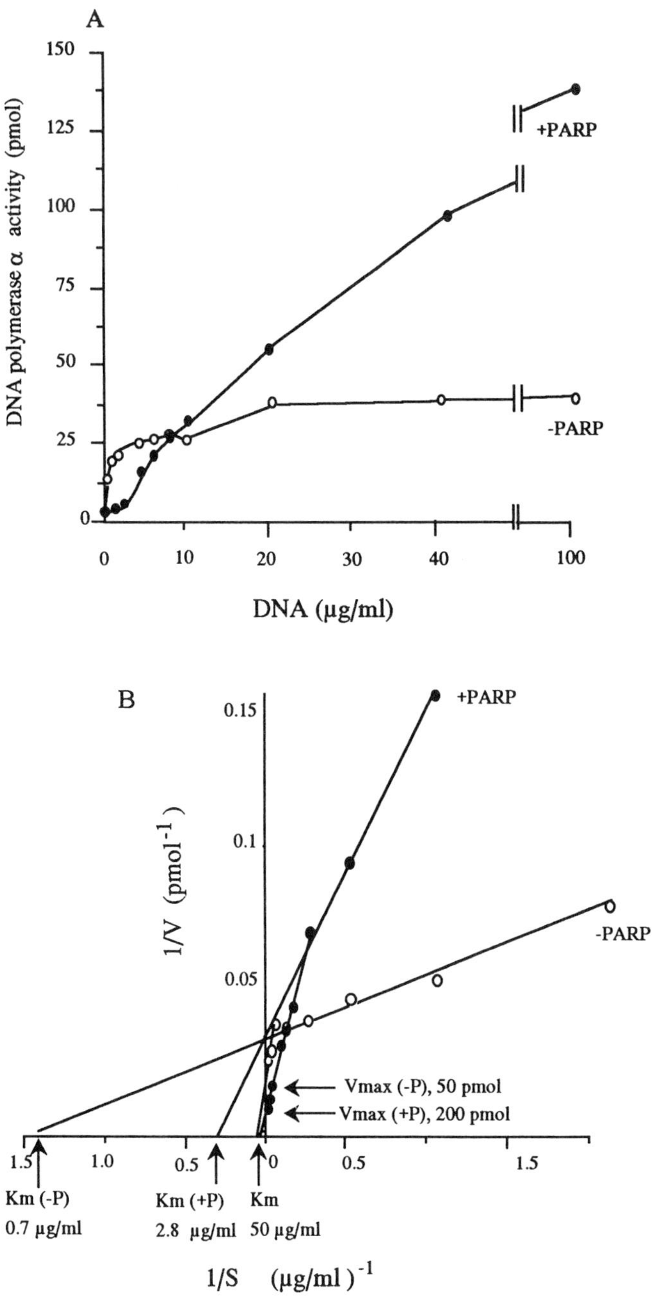

Fig. 2. Mode of inhibition: effects on Km and Vmax for the template-primer. A. The activity of DNA polymerase α was assayed using indicated concentrations of activated DNA as template-primer, in the presence of 5.2 μg/ml PARP (●) or without PARP (o). All experimental points are means of 3 independent experiments, with the standard deviation routinely less than 10% of the mean. B. Lineaweaver-Burk plots were made based on the data from A. Symbols used in A were the same as those in B (from [20]).

Effects of Autopoly(ADP-Ribosyl)ation and PARP Fragments

As shown in Fig. 3A, the stimulatory effect of PARP on DNA polymerase α activity was cancelled by its autopoly(ADP-ribosyl)ation. Kinetic analysis further showed that automodified PARP did not affect the kinetic parameters at either high or low DNA concentrations (see [20]).

Purified PARP (116 kDa) was cleaved with papain into 46 kDa DNA-binding and 74 kDa catalytic domain fragments ([25]; Fig. 3C) and then tested for any stimulatory activity on DNA polymerase α (Fig. 3B). Both fragments showed only slight stimulation, suggesting that an intact molecule of PARP is required for full stimulation.

PARP is Physically Associated with DNA Polymerase α

After preincubation of PARP with DNA polymerase α for 60 min at 0 °C, immunoprecipitation was performed with a monoclonal antibody, MT17 [22], which is directed against the 150 kDa polypeptide of DNA polymerase α, conjugated to CNBr-Sepharose 4B. The immunocomplex was then subjected to SDS-polyacrylamide gel electrophoresis. It was clearly shown that PAPR was coimmunoprecipitated with DNA polymerase α (Fig. 4). Consistently, the stimulatory activity by PARP in the supernatant was also depleted by anti-DNA polymerase α antibody but not with control IgG. These results demonstrate that PARP can bind directly to DNA polymerase a, thereby stimulating its activity. To determine whether any DNA polymerase α-PARP complexes exist in cell extracts, gel filtration chromatography was performed using crude fractions. Repeated gel filtrations revealed large complexes of 400 kDa and 700 kDa containing both PARP and DNA polymerase α (Fig. 5), indicating the existence of DNA polymerase α-PARP complexes *in vivo*.

Conclusion

The remarkable stimulation of DNA polymerase α by PARP, especially in the Tris-HCl-buffered system (about 60-fold) may suggest that PARP positively regulates DNA replication, in agreement with observations that PARP is induced by proliferation stimuli. On the other hand, the physical association of PARP with DNA polymerase α may be implicated in DNA repair. Accumulating evidence suggest that PARP participates in DNA repair, as a sensor for DNA damage, by binding to DNA strand breaks and locating the DNA repair enzyme complex to the damaged sites [19]. The physical binding between PARP and DNA polymerase α described here might play a role in this process. A model illustrating

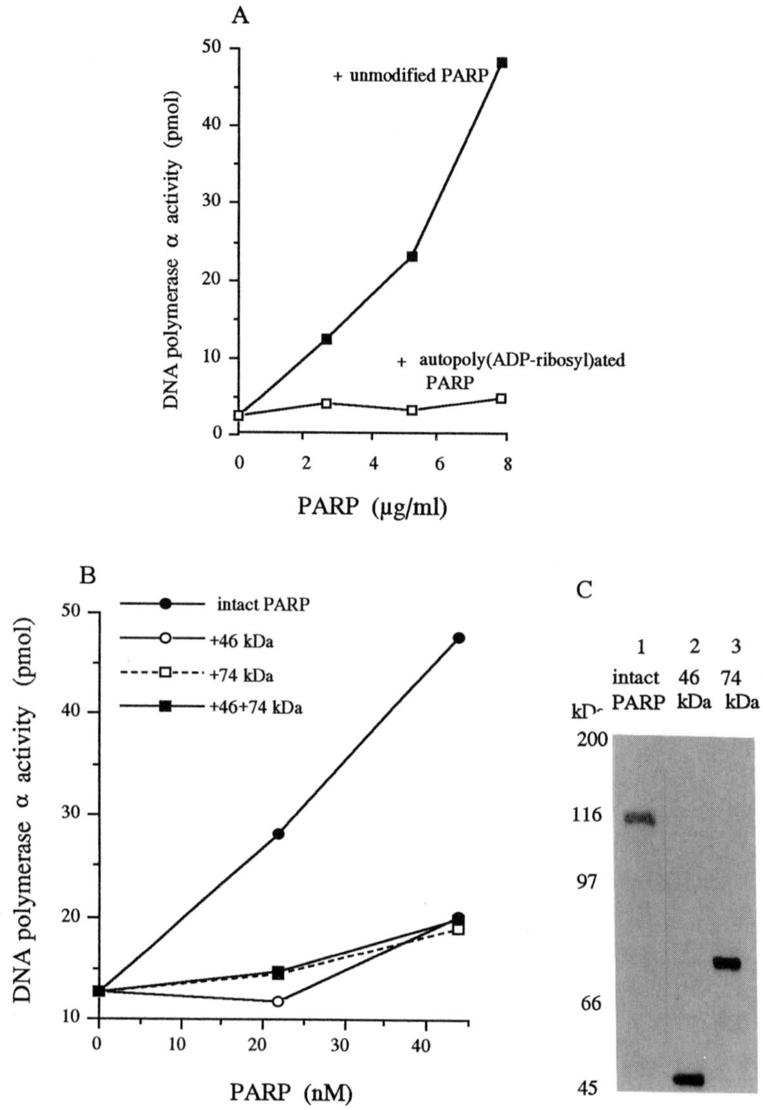

Fig. 3. Effects of autopoly(ADP-ribosyl)ation and PARP fragments. A. Autopoly(ADP-ribosyl)ation of PARP was carried out by incubating purified PARP (1.32 μg) at 25 °C for 20 min in 10 μl of 100 mM Tris HC1 (pH 8.0), containing 2 mM dithiothreitol, 10 mM $MgCl_2$, 100 μM NAD, 10% glycerol, and 100 μg/ml activated DNA (25), resulting in an average chain length of more than 10 ADP-ribose units. Indicated amounts of autopoly(ADP-ribosyl)ated PARP (○,□), as well as untreated PARP (●,■), were then assayed for stimulation of DNA polymerase α activity in Tris-HC1 buffered system (from Ref. 20). B. Effect of PARP fragments on DNA polymerase a activity, as compared to intact enzyme. Equimolar amounts of purified 46 kDa DNA binding fragments (○), the 74 kDa catalytic domain fragments (□), as well as the intact 116 kDa PARP enzyme (●), were assayed for stimulation of DNA polymerase α activity. Reconstitution was also tried with equimolar amounts of both fragments added to the reaction mixture (■) and then assayed for stimulatory activity. C. SDS Polyacrylamide gel electrophoresis of intact 116 kDa PARP, the 46 kDa DNA binding and the 74 kDa catalytic domain fragments. C. (from [20]).

the interaction of PARP with DNA polymerase α-primase (Fig. 6) proposes that both enzymes, being localized near DNA rich in replication forks or damaged DNA, could physically associate and bind to each other, resulting in a conformational change at the catalytic site of DNA polymerase α, which stimulates its activity after binding to DNA ends (if not in limiting amounts) or to RNA primer ends. In the same way that automodification of PARP reduces its affinity for DNA, it may also lose its ability to bind to DNA polymerase α, thereby releasing it from the complex, to be later regenerated into unmodified PARP by the action of specific glycohydrolases. When DNA termini are limiting, however, PARP and DNA polymerase α compete for binding to the DNA, and PARP temporarily blocks access of the α enzyme to these DNA ends; at this time, association between the two enzymes may localize the α enzyme to this site, until automodification of PARP releases it from the DNA, allowing the α enzyme to bind and later elongate the DNA strand.

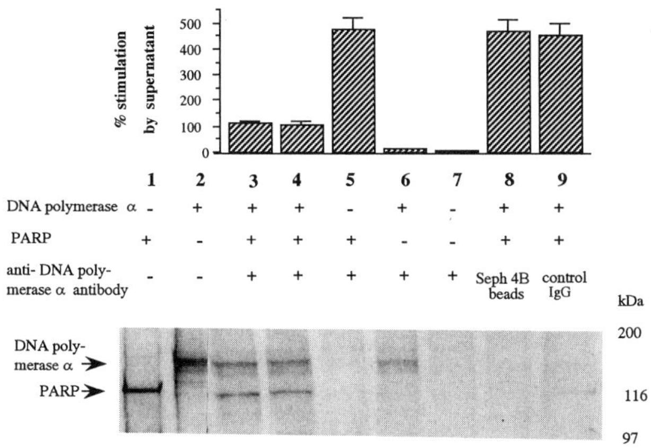

Fig. 4. Immunoprecipitation with anti-DNA polymerase α monoclonal antibody. Equimolar amounts of purified PARP and DNA polymerase α were pre-incubated for 60 minutes on ice, and immunoprecipitated with anti-DNA polymerase α antibody (MT17) conjugated to CNBr-activated Sepharose 4B (lanes 3 to 7), with control IgG (anti- Rb protein antibody, lane 9), or with Sepharose 4B beads alone (lane 8). The pellets were washed extensively, resuspended in electrophoresis buffer, and loaded on a 6.5% polyacrylamide gel. The immunocomplex bound to the beads was then subjected to SDS polyacrylamide gel electrophoresis and visualized by silver-staining (lower panel). Lane 1, input PARP; lane 2, input DNA polymerase α. Ten µl aliquots of the supernatants were also assayed for DNA polymerase α stimulation (upper panel).

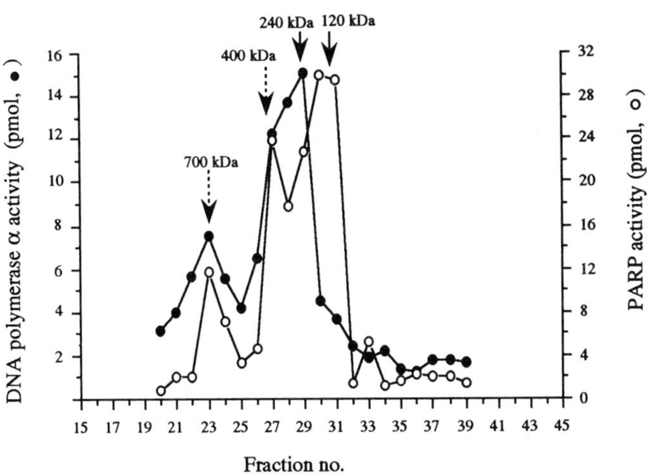

Fig. 5. Partial purification of DNA polymerase α-PARP complexes from calf thymus crude extracts by gel filtration chromatography. Calf thymus crude extract (10 ml, 22) was concentrated on a small phosphocellulose column and by overnight dialysis in solid sucrose. This concentrated fraction (0.2 ml) was applied to a Superose 6 gel filtration column (24 ml, FPLC), and fractions containing both PARP and DNA polymerase α activity, which eluted at 300–500 kDa, were rechromatographed on the same Superose 6 column. Both activities were measured in the same way. Recoveries of enzyme activities from the column were over 90 % (from [20]).

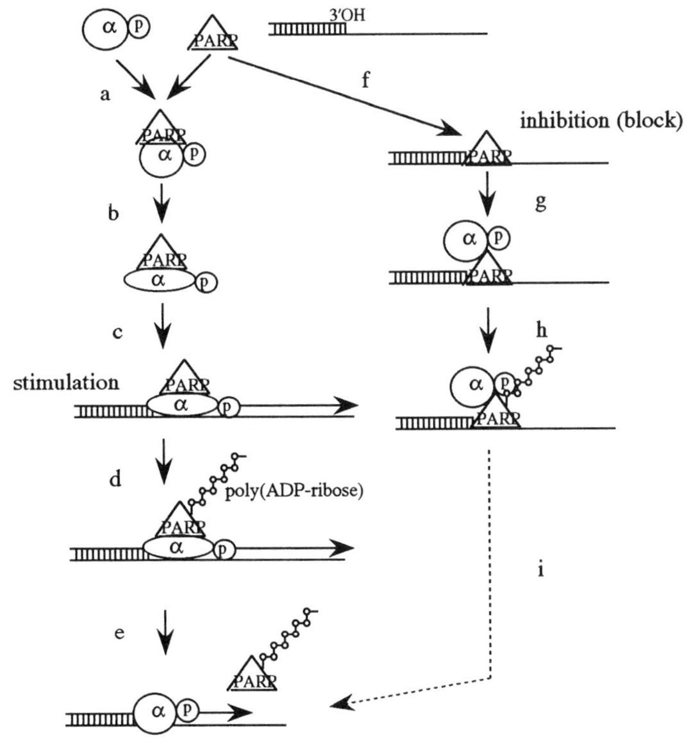

Fig. 6. Model for the interaction of PARP with DNA polymerase α-primase. Free DNA polymerase α-primase complex(α-p) and free PARP, both localized near DNA rich in replication forks, physically associate (a); PARP binds at an allosteric site, resulting in conformational changes at the catalytic site of DNA polymerase α (b), thereby stimulating its activity upon binding to the DNA template-primer (c); autopoly(ADP-ribosyl)ation of PARP (d) releases the modified PARP from the complex (e), and unmodified PARP is later regenerated. At low DNA concentrations, PARP and DNA polymerase α compete for binding to the limited 3′ OH ends of DNA; although binding of PARP blocks the binding of the α enzyme to the DNA (f), it specifically localizes DNA polymerase α to this site (g); automodification of PARP reduces its affinity to DNA (h), releasing the modified PARP (i), and allowing access to the 3′ OH end for DNA polymerase α-primase, which then proceeds to elongate the DNA strand.

The fact that poly(ADP-ribosyl)ation function of PARP accounts for only a few percent of its molecular activity in intact cells [28] indicates that its DNA binding function, as well as association with other proteins, such as DNA polyase α, may also represent other potentially significant biological roles of PARP.

Acknowledgements

The large part of experiments described here was recently published [20] in collaboration with Drs. Yutaka Shizuta and Kaoru Miyahara of Kochi Medical School, and Drs. Motoshi Suzuki, Shunji Izuta, Takeshi Sakurai, Ericka Savoysky, and Kiyohide Kojima of Nagoya University School of Medicine.

This work was supported by the Grant-in-Aid for Cancer

Research and for Scientific Research from the Ministry of Education, Science and Culture of Japan.

References

1. Gill DM: Poly(adenosine diphosphate ribose) synthesis in soluble extracts of animal organs. J Biol chem 247: 5964–5971, 1972
2. Burzio L, Koide SS: In vitro effect of NAD on DNA synthesis in isolated nuclei from regenerating rat liver and Novikoff hepatoma. FEBS Lett 20: 29–32, 1972
3. Lehman AR, Kirk-Bell S, Shall S, Whish WJ: The relationship between cell growth, macromolecular synthesis and poly(ADP ribose)polymerase in lymphoid cells. Exp Cell Res 83: 63–72, 1974
4. de Murcia G, Jongstra-Bilen J, Ittel M.E, Mandel P, Delain E: Poly (ADP-ribose) polymerase automodification and interaction with DNA: electron microscopic visualization. EMBO J 2: 543–548, 1983
5. Cesarone CF, Scarabelli L, Scovassi I, Izzo R, Menegazzi M, DeProti AC, Orunesu M, Bertazzoni U: Changes in activity and mRNA levels of poly(ADP-ribose) polymerase during rat liver regeneration. Biochim Biophys Acta 1087: 241–246, 1990
6. Cesarone CF, Scarabelli L, Giannoni P, Gallo G, Orunesu M: Relationship between poly(ADP-ribose) polymerase activity and DNA synthesis in cultured hepatocytes. Biochim Biophys Res Commun 171: 1037–1043, 1990
7. Anachkova B, Russev G, Poirier GG: DNA replication and poly(ADP-ribosyl)ation of chromatin. Cytobios 59: 19–28, 1989
8. Stone PR, Shall S: Poly(ADP-ribose) polymerase activity during growth cycle of mouse fibroblast (LS cells). Exp Cell Res 91: 95–100, 1975
9 Boulikas T: Poly(ADP-ribosylated) histones in chromatin replication. J Biol Chem 265: 14638–14647, 1990
10. Tanaka Y, Yoshihara K, Itaya A, Kamiya T, Koide S: Mechanism of inhibition of Ca^{2+}, Mg^{2+}-dependent endonuclease of bull seminal plasma by ADP-ribosylation. J Biol Chem 259: 6579–6585, 1990
11. Buki KG, Bauer PI, Kun E: Inhibitory binding of adenosine diphospho-ribosyl transferase to DNA primer site of reverse transcriptase templates. Biochem Biophys Res Commun 180: 496–503, 1991
12. Ohashi Y, Itaya A, Tanaka Y, Yoshihara K, Kamiya T, Matsukage A: Poly(ADP-ribosyl)ation of DNA polymerase ß in vitro. Biochem Biophys Res Commun 140: 666–673, 1986
13. Ferro AM, Olivera B.M: Poly(ADP-ribosylation) of DNA topoisomerase I from calf thymus. J Biol Chem 259: 547–554, 1984
14. Kasid U, Halligan B, Liu L, Dritschilo A, Smulson M: Poly(ADP-ribose)-mediated post-translational modification of chromatin-associated human topoisomerase I. J Biol Chem 264: 18687–18692, 1989
15. Creissen D, Shall S: Regulation of DNA ligase activity by poly(ADP-ribose). Nature 296: 271–272, 1982
16. Goldman N, Brown M, Khoury G: Modification of SV40 T-antigen by poly(ADP-ribosylation). Cell 24: 567–572, 1981
17. Baksi K, Alkhatib H, Smulson M: In vivo characterization of the poly(ADP-ribosylation) of SV40 chromatin and large T-antigen by immunofractionation. Exp Cell Res 172: 110–123, 1987
18. Eki T, Hurwitz J: Influence of poly(ADP-ribose) polymerase on the enzymatic synthesis of SV40 DNA. J Biol Chem 266: 3087–3100, 1991
19. Satoh M, and Lindahl T: Role of poly(ADP-ribose) formation in DNA repair. Nature 356: 356–358, 1992
20. Simbulan C.M.G, Suzuki M, Izuta S, Sakurai T, Savoysky E, Kojima K, Miyahara K, Shizuta Y, Yoshida S: Poly(ADP-ribose) polymerase stimulates DNA polymerase α by physical association. J Biol Chem 268: 93–99, 1993
21. Itoh S, Shizuta Y, Hayishi O: Purification and characterization of poly(ADP-ribose) synthetase from calf thymus. J Biol Chem 254: 3647–3651, 1979
22. Tamai K, Kojima K, Hanaichi T, Masaki S, Suzuki M, Umekawa H, Yoshida S: Structural study of immunoaffinity-purified DNA polymerase α-DNA primase complex from calf thymus. Biochim Biophys Acta 950: 263–273, 1988
23. Yoshida S, Tamai K, Umekawa H, Suzuki M, Kojima K: A novel stimulating protein of mammalian DNA polymerase α. J Biochem 106: 389–395, 1989
24. Ushiro H, Yokoyama Y, Shizuta Y: Purification and characterization of poly(ADP-ribose) synthetase from human placenta. J Biol Chem 262: 2352–2357, 1987
25. Kameshita I, Matsuda Z, Taniguchi T, Shizuta Y: Poly(ADP-ribose) synthetase. J Biol Chem 259: 4770–4776, 1984
26. Yoshida S, Yamada M, Masaki S: Novel properties of DNA polymerase ß with poly(rA)-oligo(dT) template primer. J Biochem 85: 1387–1395, 1979
27. Yoshida S, Suzuki R, Masaki S, Koiwai O: DNA primase associated with 10S DNA polymerase α from calf thymus. Biochim Biophys Acta 741: 348–357, 1983
28. Yamanaka H, Penning CA, Willis EH, Wasson DB, Carson DA: Characterization of human poly (ADP-ribose) polymerase with auto-antibodies. J Biol Chem 263: 3879–3883, 1983

Molecular and Cellular Biochemistry **138:** 45–52, 1994.

Poly(ADP-ribose) catabolism in mammalian cells

Jean Lagueux, Girish M. Shah, Luc Ménard[1], Hélène Thomassin[2], Caroline Duchaine, Christoph Hengartner[3] and Guy G. Poirier

Poly(ADP-ribose) metabolism group, Molecular Endocrinology Research Center, CHUL Research Center, 2705 blvd, Laurier, Ste-Foy, Québec, Canada, G1V 4G2.

Present addresses: [1]*254, Jones Building Research Drive, Box 3680, Duke University Medical Center, Durham, NC, 27710, USA;* [2]*CNRS, Centre de recherche sur l'endocrinologie moléculaire et le développement, 9 rue Jules Hetzel, 92190 Meudon-Bellevue, France;* [3]*Whitehead Institute for Biomedical Research, Nine Cambridge Center, MA 02142, USA*

Abstract

Poly(ADP-ribose) catabolism is a complex situation involving many proteins and DNA. We have developed an *in vitro* turnover system where poly(ADP-ribose) metabolism is monitored in presence of different relative amounts of two principal enzymes poly(ADP-ribose) transferase and poly(ADP-ribose) glycohydrolase along with other proteins and DNA. Our current results reviewed here show that the quality of polymer, i.e. chain length and complexity, as well as preference for the nuclear substrate varies depending upon the availability of poly(ADP-ribose) glycohydrolase. These results are interpreted in the light of the recent data implicating poly(ADP-ribose) metabolism in DNA-repair. (Mol Cell Biochem 138: 45–52, 1994)

Key words: poly (ADP-ribose) metabolism, turnover, chromatin, DNA-repair, histones

Introduction

The metabolic processes associated with DNA have always been a challenge to analyze because of the complex structural organization of chromatin and the dynamic interactions among its component enzymes. Each of these enzymes, that is DNA polymerase, DNA topoisomerase, repair enzymes, poly(ADP-ribose)transferase, etc., have a key role to play within a precise time frame during these metabolic processes.To understand more fully such complex processes, one must study the individual component in an *in vitro* system, decipher the fine mechanisms related to the role of each component, and then try to reevaluate the contribution of the component at the cellular level in conjunction with other processes.

Poly(ADP-ribosyl)ation, a post-translational modification, is involved in various DNA-related processes, such as chromatin decondensation [1,2], DNA replication, DNA repair [3–5, for review 6,7 and S. Shall, in this issue], gene expression [8], and cellular differentiation [9,12]. In these processes, the role of poly(ADP-ribosyl)ation appears to be correlated with base excision repair [4,5]. We have been studying the contribution of polymer metabolism in these processes using an *in vitro* model that analyzes the interactions between poly(ADP-ribose) transferase (pADPRT), (E.C. 2.4.2.30) and poly(ADP-ribose)glycohydrolase (pADPRG).

Role of poly(ADP-ribose) in DNA related metabolism

Chromatin organization

Access to a DNA strand is important in DNA repair, replication, and gene expression. Therefore, alteration of chromatin structure, as observed with poly(ADP-ribosyl)ation-mediated decondensation of chromatin, could control the accessibility to DNA by other enzymes. Studies by our group, in collaboration with de Murcia, have shown a direct relationship between the relaxation of polynucleosome structure and poly(ADP-ribosyl)ation [1]. The recondensation of

Address for offprints: G.G. Poirier, Poly(ADP-ribose) metabolism group, Molecular Endocrinology Research Center, CHUL Research Center, 2705 blvd, Laurier, Ste-Foy, Québec, Canada G1V 4G2

polynucleosomes by the addition of the catabolic enzyme pADPRG [1] demonstrated the involvement of polymer metabolism in chromatin structure and the need to study in detail the dynamic interaction between pADPRT and pADPRG to control the organization of chromatin. These studies and others [13,14] suggested that a covalent linkage of polymers to histones H1 and H2B could explain the decondensation of DNA. More recently, however, Althaus's group suggested a non-covalent modification of histone to explain their shuttling from DNA [15]. Thus, more work is necessary to elucidate the exact mechanisms of decondensation. Nevertheless, poly(ADP-ribosyl)ation has an important role in modulating chromatin structure.

DNA integrity

The amount of data that relates poly(ADP-ribosyl)ation with DNA repair processes continues to increase (The role of poly(ADP-ribose) metabolism in DNA repair is covered elsewhere in this issue). One of the most interesting studies on the subject was the correlation between pADPRT activity in mononuclear leukocytes of mammalian species and the species-specific life span as reported by Grube and Bürkle (16, see elsewhere in this issue). This work extends the findings of Pero *et al.* [17] describing a positive correlation between pADPRT activity after γ-irradiation and the life span of different mammalian species. Thus, pADPRT activity is implicated in DNA integrity through its involvement in DNA repair functions.

Satoh and Lindahl [3] have shown that automodification of pADPRT leads to an increase in DNA repair of plasmids containing single-stranded breaks. In a human cell extract DNA repair system, they were able to modulate pADPRT activity by adding or removing NAD^+, causing pADPRT to block the DNA strand interruption. They were thus able to initiate or shut off the repair of the damaged plasmid by manipulating polymer metabolism. They refined their work and found that pADPRT interferes with base excision repair [4]. Recently, Molinete *et al.* [5] showed that the DNA-binding domain is a non-competitive inhibitor of pADPRT in DNA repair and it interferes with base-excision repair. Thus pADPRT has a direct role in the repair of DNA strand breaks by a mechanism that remains to be elucidated.

Poly(ADP-ribose)transferase

Poly(ADP-ribose) transferase (pADPRT) is the eukaryotic enzyme that catalyzes the formation of ADP-ribose polymers linked to nuclear proteins by using NAD^+ as a substrate and releasing nicotinamide as a byproduct [1,18,19]. This enzyme has three functional domains: a DNA-binding domain, which comprises two zinc fingers; an automodification domain; and a NAD^+-binding domain. Deletion or site-directed mutagenesis experiments demonstrated the requirement for both zinc fingers for single-stranded break recognition, but only the finger nearer to the N-terminal of the protein was required for double-stranded break recognition [20,21]. Interestingly, pADPRT in the cell seems to be the main acceptor for the polymer [22], and this automodification process modulates its own activity. It has been demonstrated that pADPRT has an absolute dependence on DNA strand breaks to activate its polymerizing activity [23,24], and DNA recognition by the enzyme is not sequence specific [21,25].

Poly(ADP-ribose)glycohydrolase

Poly(ADP-ribose) is degraded exoglycosidically by poly(ADP-ribose)glycohydrolase [26,41] to the last ADP-ribose on the acceptor protein; the final produce of this reaction is ADP-ribose. pADPRG has been known for years, and a few attempts to characterize it has been made. However, inherent difficulties have prevented separation of its functional domains, and neither its gene nor the cDNA has been cloned. Even its mode of action is under extensive study and some controversy remains as to whether it acts endoglycosidically or exoglycosidically, since the Km of pADPRG for the polymer varies as a function of its length [26]. Therefore, the rate of shortening of polymer chain by pADPRG will vary with length.

Distribution of poly(ADP-ribose) metabolizing enzymes in the cell

Several groups have estimated the number of pADPRT molecules in the cell. Yamanaka *et al.* [27] measured pADPRT in several cell lines and their estimates varied from 50,000 molecules in human monocytes to 2,000,000 molecules in human CEM. More than 95% of pADPRT is localized in the nucleus, and a preferential nucleolar distribution has been observed [27–29]. In contrast, poly(ADP-ribose)glycohydrolase is about 13- to 50-fold less abundant [26,30], but its specific catalytic activity is about 50- to 70-fold higher, presuming that 1% of pADPRT is activated [27]. Thus, a balance exists between the anabolism and catabolism of poly(ADP-ribose) in the cell. The number of pADPRT molecules in the cell does not seem to be rate limiting in the process of DNA repair. Smulson's group [31] has investigated the expression of pADPRT in response to DNA strand breakage and has found no increase in the transcript level of pADPRT mRNA. They also showed that overexpression of pADPRT cDNA enhanced the rate of DNA repair without altering the extent of DNA repair.

Turnover of poly(ADP-ribose) in the cell

Since the poly(ADP-ribose) metabolism is implicated in several essential cellular processes, it is expected that polymer level is kept within very narrow limits in the cell. This specific equilibrium could change in response to a stress situation and then, after the transient response, the equilibrium would be reestablished. The half-life of poly(ADP-ribose) (pADPR) is close to 1 min in a stress situation [32]. *In vitro*, the half-life of pADPR varies as a function of the chain length and the complexity and nature of acceptor proteins. Since the half-life of pADPR represents a balance between its synthesis by pADPRT and its degradation by pADPRG, a transient increase of pADPR levels could be achieved by increasing the activation of pADPRT in response to strand breaks or by inhibition of pADPRG.

Heat shock
During heat shock, the cellular concentration of the polymer increases [33], probably by inhibition of pADPRG [34]. Effectively, the pADPRG activity transiently drops 10-fold, resulting in a fivefold increase in the cellular polymer level. After recovery, pADPRG preferentially degrades the long polymer as previously demonstrated by Hatakeyama [26].

Alkylating agents
Following cellular exposure to alkylating agents (MNNG), there is an increase in the pADPR level: a major polymer population has a half-life of less than 1 min [32], and a minor population having a half-life of several hours. In this case, the generation of strand breaks leads to the stimulation of pADPRT.

Oxidative stress
Preliminary results from our laboratory show a rapid increase in pADPR levels in C3H 10T1/2 cells exposed to oxidative stress produced by 300 μM H_2O_2 for 30 min. Again, pADPRT is stimulated by the creation of strand breaks, but this increase is partly due to a 20 to 30% transient decrease in pADPRG activity, which is then followed by a rapid recovery to basal level 15 min after the treatment ends.

Complex situation

The multiple roles of poly(ADP-ribosyl)ation in various DNA processes make it difficult to characterize pADPRT-pADPRG activities in the cell. Numerous cellular inhibitors of pADPRT are available [35], but inhibitors for pADPRG are lacking. An *in vitro* minimal system is needed that would isolate this interaction using only these two enzymes in the presence of DNA with strand breaks. This system could be made more complex by adding other nuclear components, like histones or DNA-repair enzymes. The *in vitro* study of polymer turnover will thus permit evaluation of the roles of pADPRT and pADPRG in chromatin reconstitution, replication, transcription, and repair and the influence of different DNA conformations or other enzymes on the action of pADPRT and pADPRG.

Experimental model

Equation

There is a close coordination between pADPRT and pADPRG in the nucleus. *In vitro* and *in vivo* experiments [33,34,36-40] have shown that the steady state of pADPR is controlled within very narrow limits by pADPRG. Poly(ADP-ribose) turnover can be viewed as a result of the sequential action of pADPRT and pADPRG. Such a mechanism is depicted schematically in Equations 1 and 2.

$$\text{pADPRT} + \text{NAD}^+ \rightarrow \text{pADPRT-NAD}^+ \rightleftharpoons \text{pADPRT} + \text{pADPR} + \text{nicotinamide} + \text{ADPR}_{ns} \quad (1)$$

$$\text{pADPRG} + \text{pADPR} \rightarrow \text{pADPRG-pADPR} \rightleftharpoons \text{pADPRG} + \text{ADPR} \quad (2)$$

These equations represent the simplest *in vitro* situation, where pADPRT is the only acceptor of its product. In equation (1), $ADPR_{ns}$ represents the non-specific ADP-ribose produced by the abortive NADase reaction [41,42]. This reaction results from the hydrolysis of NAD^+ by pADPRT by enzyme during elongation of the polymer chain. This phenomenon is modulated the same way as the main polymerizing reaction, that is, by the automodification process [41 and Poirier *et al.* unpublished observations].

Definition of the turnover system

The study of the relationship between pADPRT and the pADPRG led our laboratory to design an *in vitro* turnover system in which both enzymes act simultaneously. The conditions for the optimum activity of these enzymes are quite different, and the final assay system represents a balance between the activities of these two enzymes [22,43]. To reflect the cellular concentration, 200 μM NAD^+ was used. The source of activating DNA could be standard activated DNA, nucleosomal DNA, or nicked plasmid DNA [22,43,44]. Other acceptor proteins could also be added. Using this system, it was easy to study the effects of other components added during the reaction, for example, specific inhibitors of the enzymes.

Turnover modulation

Influence of pADPRG on poly(ADP-Ribosyl)ation

The design of such an *in vitro* turnover system allowed exploration of the equilibrium between pADPRT and pADPRG and the variables controlling this equilibrium. We established conditions to obtain half-life values for pADPR close to that in the cell. Ménard *et al.* [45] showed that pADPRG can decrease free pADPR levels in a linear fashion proportional to its concentration; in contrast, in the turnover system where pADPR is attached to protein, the rate of degradation is proportional to twice the concentration of pADPRG [22]. By varying the amount of pADPRG with respect to pADPRT, pADPR half-lives increased from 26s to 55s; these values are close to the cellular half-life of about 1 min [32]. Interestingly, pADPRG modulated not just the total amount of pADPR in the system but also the pattern of distribution of pADPR on proteins. In the absence of pADPRG, pADPRT is preferentially automodified even in presence of histone H1; in contrast, in the turnover situation, where pADPRG is present, up to 40% of the total pADPR is on the histone H1 fraction [22].

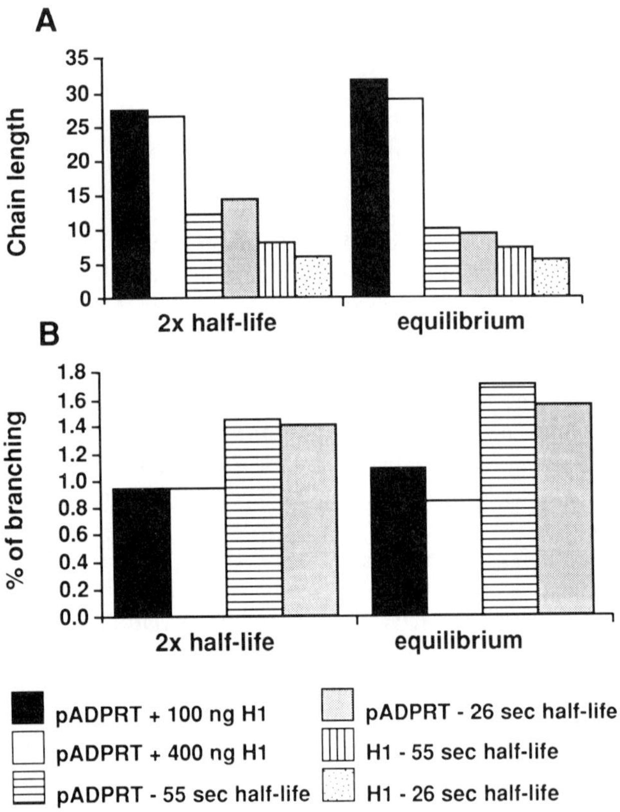

Fig 1. Effect of turnover on the complexity of pADPR. Panel A: Analysis of chain length at the time corresponding to 2 X half-life and to the equilibrium phase. Histones did not modulate the average chain length. The pADPRT in both turnover situations showed a reduced chain length at 2X half-life and an even greater reduction at equilibrium. In the same turnover situation, histone H1 was not affected. Panel B: Percentage of branching. A more branched pADPR structure occurs in both turnover situations even though the polymer level is dramatically reduced (40%) in the faster turnover situation.

The first effect we wanted to verify was modulation of the chain lengths and branching density of pADPR by pADPRG. The turnover system was modified to contain histone H1, which is involved in the decondensation of chromatin by poly(ADP-ribosyl)ation [1,13,14]. Using pADPRT alone with increasing amounts of histone H1, neither the chain length nor the percentage of branching was affected (Fig 1). Under the same conditions, pADPRG was added to obtain half-lives of 55 or 26 s; the average polymer length was decreased both at 2 times half-life and at equilibrium (Fig. 1a). The average polymer size is decreased at equilibrium, reflecting the cellular situation where an initial burst of pADPR is then rapidly degraded yielding an effective half-life of approximately 1 min. The pADPRG hydrolyzes large polymers to short pADPR and permits an increase in the proportion of branched pADPR (Fig 1b). In these experiments, the branching of pADPR on histones was not determined because of the technical limitation of the HPLC method. Such a short polymer would have an effect on the local environment of the DNA strand breaks. Adding negative charges would alter the pADPRT affinity for DNA, allowing its release from the strand break as proposed by Satoh and Lindahl [3].

Effects of other acceptor proteins on the equilibrium pADPRT-pADPRG

Poly(ADP-ribose) turnover experiments were also conducted using chromatin (polynucleosomes), a situation which is more analogous to the nuclear environment of these enzymes. In this system, the acceptor proteins are the pADPRT itself, histone H1, and the core histones. Two phases of the turnover were examined. The synthesis phase, in which both enzymes are active simultaneously and the degradation phase, in which pADPRT was inhibited at the plateau phase of the synthesis by adding 3-aminobenzamide. Catabolic activity was varied from non-turnover situation of 1:0 (*i.e.*, no pADPRG) to 1:10 (1 unit of pADPRT with 10 units of pADPRG) to mimic the conditions of pADPR turnover in the intact cell [32].

Synthesis phase. The amount of the total polymer in the system is inversely proportional to the amount of pADPRG. Variations in the ratio of pADPRT to pADPRG also affects the distribution of the labeling on each type of protein. In the non-turnover situation, pADPRT itself is the main acceptor protein, with 50% of the polymer, followed by histone H1 at 27% and the core histones at 24% (Fig. 2) [43]. At the 1:0.1 ratio, an increase in the labeling of pADPRT (up to 56%) occurs, and the core histones drop to 14% of the

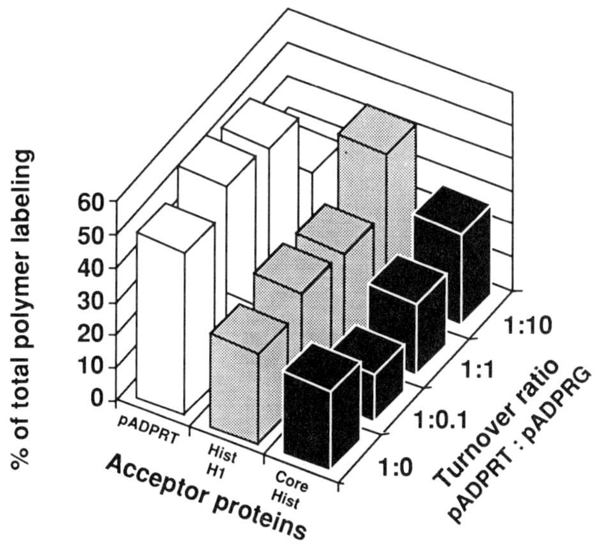

Fig. 2. Relative labeling of nucleosomal proteins at different enzyme ratios in the synthesis phase of turnover. The total pADPR will vary dramatically on pADPRT from 5.5 pmoles in the 1:0 ratio, to 5 pmoles at 1:0.1, 1.24 pmole at 1:1 ratio and 0.11 pmole at 1:10 ratio, reflecting the activity of pADPRG. In this figure it can be noticed that not only is the total pADPR affected but also the relative distribution of pADPR. The labeling of pADPRT is 50-56% at the lower ratio and drops to 30% at the 1:10 ratio with a shift to histone H1.

total polymer. At the 1:10 ratio, histone H1 becomes the main acceptor protein, in accordance with the previous work of Ménard *et al.* [22] using a simpler system.

In the synthesis phase, the relative labeling of acceptor proteins does not vary with time for the ratios 1:0, 1:0.1, and 1:1. Figure 3 gives an example of the 1:0.1 ratio that is also representative of the other conditions. The ratio of pADPRG affects the distribution pattern of pADPR at the beginning of the reaction; the proportion remains constant throughout the time course.

On the other hand, in the higher catabolism ratio 1:10, a shift clearly occurs in the relative labeling of each class of proteins during the time course (Fig 4). With time the relative labeling shifts from pADPRT to histone H1, whereas polymers on the core histones remain stable. The polymer is preferentially removed from the pADPRT by the pADPRG during the synthesis phase.

Degradation phase. In the turnover situation, when pADPRT is inhibited in the 1:0.1 system, the relative labeling on each protein remains constant; that is, the degradation is uniform for each protein category (Fig 5a). Under these conditions, about 40% of the initial polymer is removed. In a more cellular-like condition, when the ratio is 1:1, a preferential degradation is first performed on the polymer bound to pADPRT for the first 10 min and then to the polymer bound to histone H1 class protein (Fig 5b).

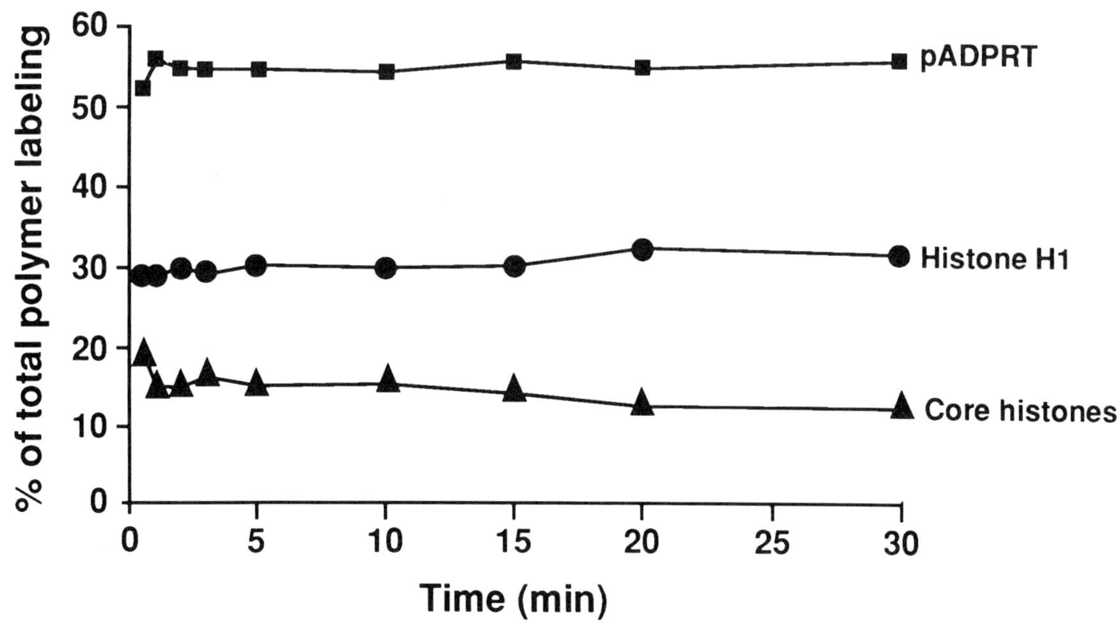

Fig. 3. Relative labeling of nucleosomal proteins during the synthesis phase of turnover. The data shown here at the ratio 1:0.1 is representative of the 1:0 and 1:1 ratios. The distribution of pADPR remains the same during the synthesis phase showing no preference for a protein in the synthesis phase. In the 1:0.1 and 1:1 ratios, the initial percentage will be different but the shape of the time course are the same.

Conclusion

In a putative DNA repair situation such as one described by Satoh and Lindahl [3], the role of pADPR turnover could be globally evaluated.

pADPRT recognises and binds to DNA strand break resulting in the activation of the enzyme and inaccessibility of the nick for DNA binding enzymes; for example, DNA polymerase β [19]. Automodification of the pADPRT and poly(ADP-ribosyl)ation of the nucleosomal proteins by pADPRT then allows the decondensation of the DNA, as reported by de Murcia *et al.* [1], for access to the damaged region. In the light of date obtained during heat shock [33,34] or oxidative stress, the activity of pADPRG is then partially inhibited, corresponding to a ratio of 1:0 in our turnover system. The reactivated pADPRG then modulates the distribution pattern of pADPR on the nucleosomal proteins, as we have shown [22,43]. By preferentially removing pADPR from pADPRT and keeping the chain length short, pADPRG will probably prevent the interaction of pADPR with other DNA binding proteins near the region of decondensation. The second phase, leading to the recondensation of DNA, begins with inhibition of pADPRT due to automodification, which in our system is also accomplished by a specific inhibitor. In the cell, this could occur through competition for the nick by DNA repair enzymes or by repulsion of automodified pADPRT from the DNA by the addition of the negatively charged polymer. Satoh *et al.* [3,4] demonstrated in a cell-free system that pADPRT binds to the nick in the absence of NAD^+ and prevents the access of repair enzymes. Adding NAD^+ allows automodification of pADPRT, and subsequent resumption of repair; these data suggest that the repair process, truly more complex in the cell, could use such a process. The idea of competition between pADPRT and repair enzymes for the nick is also supported by the data of Sugimura *et al.* [19], who suggested that pADPRT inhibits DNA polymerase β either by direct competition for the nick or indirectly by poly(ADP-ribosyl)ation of DNA polymerase β. The data obtained with an inactivated recombinant pADPRT suggest competition for the nick is the most likely mechanism [19]. Our turnover data indicates the pADPR is preferentially removed from pADPRT at that step by pADPRG. Another possibility is that the preferential removal of pADPR is related to the higher affinity of pADPRG for long polymer as shown by Hatakeyama *et al.* [26]; histones having shorter pADPR, the pADPRG will preferentially act on longer polymer attached to pADPRT. In the final step, the polymer from the histones is removed, and recondensation of the repaired DNA with histones occurs to form intact chromatin.

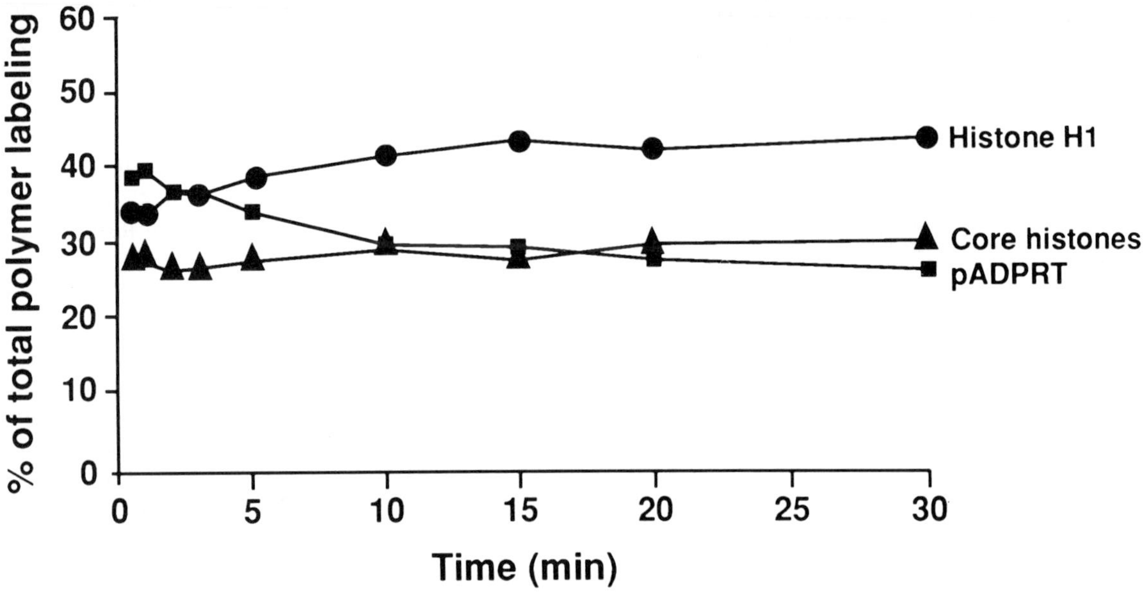

Fig 4. Relative labeling on nucleosomal proteins during the time course of the synthesis phase of turnover for the ratio 1:10. In this turnover ratio there is a clear shift of relative labeling from pADPRT to histone H1 showing a preferential removing of pADPR from pADPRT by pADPRG.

The *in vitro* turnover system is a powerful tool for dissecting the involvement of pADPR in different situations. We will soon be presenting a mathematical model describing the equilibrium of pADPRT and pADPRG in the turnover. More work needs to be done on the specific affinity of pADPRT for DNA and its ability to compete with other enzymes active in DNA metabolism.

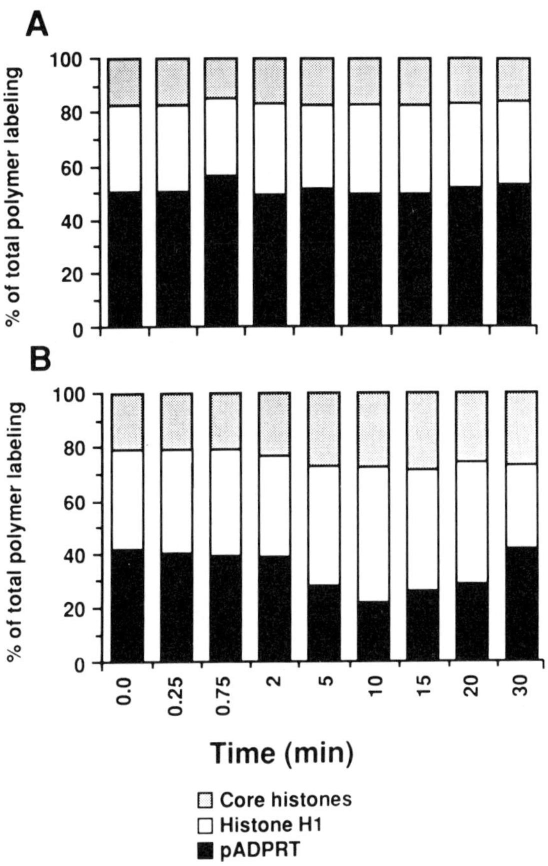

Fig. 5. Relative labeling of nucleosomal proteins in the degradation phase of turnover. In the final step of the metabolism of pADPR, the pADPRT was inhibited by 3-AB and the degradation of pADPR on each protein was monitored. In the 1:0.1 ratio (panel A) the degradation occurs equally on each of the proteins when in the 1:1 ratio (Panel B), the degradation seems to be made preferentially on pADPRT after a lag time. Histone H1 will then have more labeling than pADPRT. In a last step, pADPR will be removed from histone H1.

References

1. de Murcia G, Huletsky A, Lamarre D, Gaudreau A, Pouyet J, Daune M, Poirier GG: Modulation of chromatin superstructure induced by poly(ADP-ribose) synthesis and degradation. J Biol Chem 261: 7011-7017, 1986
2. Thibault, L, Hengartner, M, Lagueux, J, Poirier, GG, Muller, S: Rearrangements of the nucleosome structure in chromatin by poly(ADP-ribose). Biochim Biophys Acta 1121:317-324, 1992
3. Satoh MS, Lindahl T: Role of Poly(ADP-ribose) Formation in DNA Repair. Nature 356: 356-358, 1992
4. Satoh MS, Poirier GG, Lindahl T: NAD+-Dependent Repair of Damaged DNA by Human Cell Extracts. J Biol Chem 268: 5480-5487, 1993
5. Molinete M, Vermeulen W, Bürkle A, Ménissier-de Murcia J, Küpper JH, Hoeijmakers, JHJ, de Murcia G: Overproduction of poly(ADP-ribose) polymerase DNA-binding domain blocks alkylation-induced DNA repair synthesis in mammalian cells. Embo J 12: 2109-2117, 1993
6. Althaus FR, Richter C: ADP-ribosylation of proteins. Enzymology and biological significance. Mol Biol Biochem Biophys 37: 1-237, 1987
7. Cleaver JE, Morgan WF: poly(ADP-ribose)polymerase: a perplexing participant in cellular responses to DNA breakage. Mutat Res 257: 1-18, 1991
8. Ghani QP, Hussain MZ, Zhang J, Hunt KH: Control of procollagen gene transcription and prolyl hydroxylase activity by poly(ADP-ribose). In: GG Poirier and P Moreau (eds) ADP-Ribosylation Reactions, Springer-Verlag, New-York, 1992, pp211-217
9. Althaus FR, Lawrence SD, Sattler GL, Longfellow DG, Pitot HC: Chemical quantification of unscheduled DNA synthesis in cultured hepatocytes as an assay for the rapid screening of potential chemical carcinogens. Cancer Research 42:3010-3015, 1982
10. Caplan AI, Rosenberg MJ: Interrelationship between poly(ADP-ribose) synthesis, intracellular NAD levels, and muscle or cartilage differentiation from mesodermal cells of embryonic chick limb. Proc Natl Acad Sci USA 72: 1852-1857, 1975
11. Farzaneh F, Zalin R, Brill D, Shall SD: DNA strand breaks and ADP-ribosyl transferase activation during cell differentiation. Nature 300:362-366, 1982
12. McNerney R, Tavasolli M, Shall S, Brazinski A, Johnstone A: Changes in mRNA levels of poly(ADP-ribose) polymerase during activation of human lymphocytes. Biochem Biophys Acta 1009: 185-187, 1989
13. Niedergang C, De Murcia G, Ittel ME, Pouyet J, Mandel P: Time course of polynucleosome relaxation and ADP-ribosylation. Correlation between relaxation and histone H1 hyper-ADP-ribosylation. Eur J Biochem 146: 185-191, 1985
14. Boulikas T: At least 60 ADP-ribosylated variant histones are present in nuclei from dimethylsulfate-treated and untreated cells. EMBO J 7: 57-67, 1988
15. Panzeter PL, Realini CA, Althaus FR: Noncovalent Interactions of Poly(adenosine Diphosphate Ribose) with Histones. Biochemistry 31: 1379-1385, 1992
16. Grübe K, Bürkle A: Poly(ADP-ribose) Polymerase Activity in Mononuclear Leukocytes of 13 Mammalian Species Correlates with Species-Specific Life Span. Proc Natl Acad Sci USA 89: 11759-11763, 1992
17. Pero RW, Holmgren K, Persson L: γ-radiation induced ADP-ribosyl transferase activity and mammalian longevity. Mutation Res 142: 69-73, 1985
18. Kreimeyer A, Wielckens K, Adamietz P, Hilz H: DNA-repair-associated ADP-ribosylation *in vivo*. J Biol Chem 259: 890-896, 1984
19. Sugimura T, Masutani M, Ogura T, Takenouchi N, Ikejima M, Esumi H: Regulation of DNA polymerase β by Poly(ADP-ribose) polymerase. In: GG Poirier and P Moreau (eds). ADP-Ribosylation Reactions. Springer-Verlag, New-York, 1992, pp 276-281
20. Ménissier-de Murcia J, Molinete M, Gradwohl G, Simonin F, de Murcia G: Zinc-binding domain of poly(ADP-ribose)polymerase participates in the recognition of single strand breaks on DNA. J Mo. Biol 210: 229-233, 1989
21. Ikejima M, Noguchi S, Yamashita R, Ogura T, Sugimura T, Gill DM, Miwa M: The zinc fingers of human poly(ADP-ribose) polymerase are differentially required for the recognition of DNA breaks and nicks and the consequent enzyme activation. Other structures recognize intact DNA. J Biol Chem 265:21907-21913, 1990
22. Ménard L, Thibault L, Poirier GG: Reconstitution of an in vitro poly(ADP-ribose) turnover system. Biochim Biophys Acta 1049: 45-58, 1990
23. Benjamin RC, Gill DM: Poly(ADP-ribose) synthesis in vitro programmed by damaged DNA. A comparison of DNA molecules containing different types of strand breaks. J Biol Chem 255: 10502-10508, 1980
24. Ohgushi H, Yoshihara K, Kamiya T: Bovine thymus poly(adenosine diphosphate ribose) polymerase physical properties and binding to DNA. J Biol Chem 255: 6205-6211, 1980
25. Herzog H, Zabel BU, Schneider R, Auer B, Hirsch-Kauffmann M,

Schweiger M: Human nuclear NAD+ ADP-ribosyltransferase: localization ofthe gene on chromosome 1q41-q42 and expression of an active human enzyme in Escherichia coli. Proc Natl Acad Sci USA 86: 3514-3518, 1989

26. Hatakeyama K, Nemoto Y Ueda K, Hayaishi O: Purification and characterization of poly(ADP-ribose) glycohydrolase. Different modes of action on large and small poly(ADP-ribose) J Biol Chem 261: 14902-14911, 1986
27. Yamanaka H, Penning CA, Willis EH, Wasson DB, Carson DA: Characterization of human poly(ADP-ribose) polymerase with autoantibodies. J Biol Chem 263: 3879-3883, 1988
28. Lamarre D, Talbot B, de Murcia G, Laplante C, Leduc Y, Mazen A, Poirier GG: Structural and functional analysis of poly(ADP-ribose) polymerase: an immunological study. Biochim Biophys Acta 950: 147-160, 1988
29. Kaufmann SH, Brunet G, Talbot B, Lamarre D, Dumas C, Shaper JH, Poirier GG: Association of poly(ADP-ribose) polymerase with the nuclear matrix: the role of intermolecular disulfide bond formation, RNA retention, and cell type. Exp Cell Res 192: 524-535, 1991
30. Tanuma S, Kawashima K, Endo H: Acceptor proteins for (ADP-ribose)n in the HeLa S3 cell cycle. J Biochem (Tokyo) 99: 915-922, 1986
31. Bhatia K, Pommier Y, Giri C, Fornace AJ, Imaizumi M, Breitman TR, Cherney BW, Smulson ME: Expression of the poly(ADP-ribose) polymerase gene following natural and induced DNA strand breakage and effect of hyperexpression on DNA repair. Carcinogenesis 11: 123-128, 1990
32. Alvarez-Gonzalez R, Althaus FR: Poly(ADP-ribose)catabolism in mammalian cells exposed to DNA-damaging agents. Mutation Res 218: 67-74, 1989
33. Jonsson GG, Jacobson EL, Jacobson MK: Mechanism of alteration of poly(adenosine diphosphate-ribose) metabolism by hyperthermia. Cancer Res 48: 4233-4239, 1988
34. Jonsson GG, Ménard L, Jacobson EL, Poirier GG, Jacobson MK: Effect of hyperthermia on poly(adenosine diphosphate-ribose) glycohydrolase. Cancer Res 48: 4240-4243, 1988
35. Banasik M, Komura H, Shimoyama M, Ueda K: Specific Inhibitors of Poly(ADP-Ribose) Synthetase and Mono(ADP-ribosyl transferase. J Biol Chem 267: 1569-1575, 1992
36. Ueda K, Oka J, Narumiya S, Miyakawa N, Hayaishi O: Poly ADP-ribose glycohydrolase from rat liver nuclei, a novel enzyme degrading the polymer. Biochem Biophys Res Commun 46: 516-523, 1972
37. Zahradka P, Ebisuzaki K: A shuttle mechanism for DNA-protein interactions. The regulation of poly ADPr polymerase. Eur J Biochem 127: 579-585, 1982
38. Shall S: ADPR in DNA repair: a new component of DNA excision repair. Adv Rad Biol 11: 1-69, 1984
39. Wielckens K, George E, Pless T, Hilz H: Stimulation of poly (ADPribosyl)ation during Ehrlich ascites tumor cell 'starvation' and suppression of concomitant DNA fragmentation by benzamide. J Biol Chem 258: 4098-4104, 1983
40. Juarez-Salinas H, Sims JL, Jacobson MK: Poly (ADP-ribose) levels in carcinogen-treated cells. Nature 282: 740-741, 1979
41. Desmarais Y, Ménard L, Lagueux J, Poirier GG: Enzymological properties of poly(ADP-ribose) polymerase: characterization of automodification sites and NADase activity. Biochim Biophys Acta 1078: 179-186, 1991
42. Bauer PI, Buki KG, Hakam A, Kun E: Macromolecular association of ADP-ribosyltransferase and its correlation with enzymic activity. Biochem J 270:17-26, 1990
43. Thomassin H, Ménard L, Hengartner C, Kirkland JB, Poirier GG: Poly(ADP-ribosyl)ation of chromatin in an in-vitro poly(ADP-ribose)-turnover system. Biochim Biophys Acta 1137: 171-181, 1992
44. Hengartner C, Lagueux J, Poirier GG: Analysis of the Activation of Poly(ADP-Ribose) Polymerase by Various Types of DNA. Biochemistry and Cell Biology 69: 577-580, 1991
45. Ménard L, Poirier GG: Rapid assay of poly(ADP-ribose) glycohydrolase. Biochem Cell Biol 65: 668-673, 1987

Molecular and Cellular Biochemistry **138**:53–59, 1994.

Histone shuttling by poly ADP-ribosylation

Felix R. Althaus[1], Liane Höfferer, Hanna E. Kleczkowska, Maria Malanga, Hanspeter Naegeli, Phyllis L. Panzeter and Claudio A. Realini

University of Zürich-Tierspital, Institute of Pharmacology and Toxicology, Winterthurerstrasse 260, CH-8057 Zürich, Switzerland

Abstract

The enzymes poly(ADP-ribose)polymerase and poly(ADP-ribose) glycohydrolase may cooperate to drive a histone shuttle mechanism in chromatin. The mechanism is triggered by binding of the N-terminal zinc-finger domain of the polymerase to DNA strand breaks, which activates the catalytic activities residing in the C-terminal domain. The polymerase converts into a protein carrying multiple ADP-ribose polymers which displace histones from DNA by specifically targeting the histone tails responsible for DNA condensation. As a result, the domains surrounding DNA strand breaks become accessible to other proteins. Poly(ADP-ribose)glycohydrolase attacks ADP-ribose polymers in a specific order and thereby releases histones for reassociation with DNA. Increasing evidence from different model systems suggests that histone shuttling participates in DNA repair *in vivo* as a catalyst for nucleosomal unfolding. (Mol Cell Biochem 138: 53–59, 1994)

*Key words:*poly(ADP-ribose)polymerase, poly(ADP-ribose)glycohydrolase, DNA repair, chromatin, nucleosomal unfolding, NAD^+

Introduction

Poly(ADP-ribose) is a homopolymer of repeating adenosine diphosphate ribose units. Polymers containing up to 200 ADP-ribose residues and multiple branches may be found in possibly all eukaryotic cells except yeast, where the existence of poly(ADP-ribose) remains a controversial issue. The reducing end of ADP-ribose polymers is covalently bound to nuclear proteins and a single polymer may add a relative molecular mass of more than 100 000 to its acceptor protein. The predominant acceptor is the nuclear enzyme poly(ADP-ribose)polymerase which may carry up to 28 polymers of various sizes and complexities [1-5]. While the covalent protein link seems to be pre-requisite for its biological activity, poly(ADP-ribose) may function primarily via *non-covalent* interactions with other macromolecules [1]. Focusing on this possibility we have found that the enzymes of the poly ADP-ribosylation system may cooperate to shuttle histones off and back onto DNA [6-8].

Poly(ADP-ribose)polymerase

Purified poly(ADP-ribose)polymerase binds specifically to single and double strand DNA breaks [9-11]. Two zinc finger motifs in the N-terminal domain of the polymerase determine these binding specificities [13-14]. In DNA excision repair of living cells, such sites appear either directly as a consequence of DNA damage (e.g. ionizing radiation), or indirectly following enzymatic incision [for review see 15,16].

The four different catalytic activities required for poly (ADP-ribose) synthesis reside in the 40-kDa C-terminal domain of the polymerase. Recent studies show that this domain functions autonomously in the absence of the

Address for offprints: F.R. Althaus, University of Zürich-Tierspital, Institute of Pharmacology and Toxicology, Winterthurerstrasse 260, CH-8057 Zürich, Switzerland

DNA-binding and automodification domain. The activities are stimulated 500-fold upon binding of the N-terminal zinc finger domain to DNA strand breaks [17]. *In vivo*, poly(ADP-ribose)polymerase is activated concomitant with the appearance of DNA strand breaks [e.g.18-21], and is the predominant acceptor for ADP-ribose polymers [22,23]. Thus, upon activation, the enzyme poly(ADP-ribose)polymerase converts into a protein carrying up to 28 ADP-ribose polymers [24] of various sizes and branching frequencies ('automodification reaction'). The functional consequences of automodification are loss of DNA binding affinity and inactivation of the polymerase [25-28].

The molecular mode of automodification is as follows: after activation by DNA strand breaks, purified poly(ADP-ribose)polymerase produces a distinct pattern of polymers which are bound to a small number of polymerase molecules. As the reaction progresses, more enzyme molecules become automodified with a polymer pattern identical to the one produced in the first round of synthesis. An exception to this is the fraction of branched polymers which increases during the first 10 min of automodification. Finally, when the reaction comes to a halt, all available polymerase molecules are automodified. Thus, *in vitro*, automodification of the polymerase follows a processive reaction mode [29]. Processivity has also been observed in more complex model systems, such as in nucleosomal core particles, in isolated nuclei [30], and in intact cells [31,32]. However, the polymer patterns differ from those produced by the polymerase *in vitro*. We have identified histones H1, H2A, H2B, H3 and H4 as the key factors regulating the numbers, size and branching frequencies of polymers. By binding to nascent polymers, these histones affect polymer termination in a histone-type specific way [30]. Thus, poly(ADP-ribose)polymerase operates processively, and when it encounters DNA-bound histones, it adapts by producing histone-type specific patterns of ADP-ribose polymers. No adaptation occurs in the presence of other DNA-binding or non-binding proteins [30, 33].

Poly(ADP-ribose)Glycohydrolase

DNA repair *in vivo* is coupled with a stimulation of polymer catabolism. At high levels of DNA damage, the catabolic half life of individual polymer molecules may be less than 30 seconds [31,32]. The rapid turnover contrasts with the slower catabolism of a constitutive polymer fraction exhibiting a half life of 7.7 hours in undamaged cells [34]. An explanation for this differential catabolism has emerged from *in vitro* studies of Hatakeyama et al. [35]: purified poly(ADP-ribose)glycohydrolase operates in a biphasic, bimodal reaction mode. Briefly, large polymers are degraded to smaller polymers in a fast and processive reaction. Further degradation then proceeds more slowly and in a distributive reaction mode. Rapid initial degradation of large polymers may be further accelerated by initial endoglycosidic incision [36]. Based on these studies, polymer catabolism emerges as a highly organized process, in which the order and kinetics of degradation is largely determined by the size (and structural complexity?) of the substrate. In view of the shuttle mechanism to be discussed below, it is important to note that poly(ADP-ribose)glycohydrolase plays an important role in reversing the automodified state of poly(ADP-ribose)polymerase. This reactivates the polymerase and restores DNA binding [26,28].

Cooperation of the Two Enzymes in the Histone Shuttle Mechanism

In nucleosomal core particles retaining endogenous poly(ADP-ribose)polymerase in active form, the automodification reaction causes dissociation of the 146 bp core DNA fragment from the core histones [37]. DNA dissociation and polymer synthesis follow parallel time courses and inhibition of polymer synthesis with competitive inhibitors or with α-anomeric NAD (which is not a substrate for the enzyme), prevents the release of DNA. This demonstrates the potential of the automodification reaction of poly(ADP-ribose)polymerase to interfere with histone-DNA interactions at the nucleosomal level of chromatin.

Purified poly(ADP-ribose)polymerase dissociates from DNA upon automodification [25,26,28]. The polymers attached to the polymerase are more acidic than DNA (2 negative charges per monomer unit as compared to only one per nucleotide of DNA) and therefore should compete with DNA for histone binding. In fact, filter binding studies suggest competition of free ADP-ribose polymers with DNA for binding of histones H1, H3 and H4 [38]. Thus, electrostatic interactions of polymerase-bound polymers with basic proteins such as histones could account for the release of DNA from nucleosomal core particles. However, the experimental evidence shows that the interaction of histones with ADP-ribose polymers is far stronger and more specific than would be expected on the basis of electrostatic interactions [33]. Exceptionally strong interactions are indicated by the fact that histone-polymer complexes resist phenol partitioning, strong acid, detergents, and high salt concentrations. The following rules define these interactions: branched polymers exhibit the highest binding affinities, followed by long linear polymers and short linear polymers, and among histones the hierarchy of binding is H1>H2A>H2B =H3>H4 [33]. This surprising binding specificity of histones for polymers is further supported by three additional lines of evidence: first, among 40 basic as well as acidic DNA-binding and non-binding proteins tested, only histones bound to ADP-ribose polymers [33, and unpublished observations]; second, when nuclear lysates were run over agarose-poly(ADP-ribose) affinity columns, only histones bound to the resin [39]; finally, using a novel blot technique (i.e. a Western blot of nuclear proteins probed with radiolabeled polymers instead of antibodies), the histone species gave the predominant signals, while a few other polypeptides of yet unknown identity scored also positive in this analysis [40]. Binding was confined to specific regions, and most surprisingly, was targeted precisely to those histone tails which are responsible for DNA condensation. Specifically, targeting of polymers was to the C-terminal region of H1 and the N-terminal regions of H3 and H4. These regions are most easily accessible for macromolecular interactions in a chromatosomal configuration [41,42], and bend linker DNA to bring nucleosomes together. Their absence results in internucleosomal unwinding [43,44].

Thus, the polymers of automodified poly(ADP-ribose)polymerase may target histones for dissociation from DNA. In fact, ADP-ribose polymers covalently bound to the polymerase are about 10 times more potent in dissociating histones from DNA as compared to free polymers [7,8]. Conversely, histones bind with strong preference to ADP-ribose polymers when automodified poly(ADP-ribose)polymerase and DNA are presented as binding competitors. Under these conditions, histone-DNA complexes are formed only after all polymer binding sites are saturated. The branching points of ADP-ribose polymers turned out to be the sites with highest binding affinities for histones [7,33].

Like the automodification reaction of poly(ADP-ribose)polymerase, the dissociation of histone-DNA complexes proceeds in a processive manner, i.e. in the first round of automodification, a small number of histone-DNA complexes is completely stripped of histones, and as more automodified polymerase molecules are formed, another subset of complexes is stripped until finally all complexes are dissociated. Calculations of the reaction stoichiometry revealed that 40 ADP-ribose residues bound to the polymerase can dissociate the entire histone complement of a chromatosome [i.e. a chromatin particle containing 165 bp DNA, eight core histones, and one molecule of histone H1 [45]. Compared to other naturally occurring polyanions, such as poly(A), tRNA, and heparin, poly(ADP-ribose) is 100 to 1000 times more potent in dissociating histone-DNA complexes when calculated for equivalent numbers of negative charges [7,33].

The automodification reaction of poly(ADP-ribose)polymerase affects histone-DNA complexes in other ways. The DNA becomes accessible to a number of proteins and this phenomenon parallels the increase in automodified polymerase molecules. Like histone-DNA complex dissociation, development of protein accessibility occurs processively and is prevented by inhibitors of poly(ADP-ribose)polymerase. Accessibility is established for DNase I, micrococcal nuclease, DNA helicase A, and DNA topoisomerase I [7,8]. Because of their roles in mammalian DNA replication and DNA repair, DNA helicase A [46] and DNA topoisomerase I [15] are of particular interest. The strand separation reaction catalyzed by DNA helicase A is completely inhibited when histone H1 or more histones are present on the DNA template. Sequestration of histones by ADP-ribose polymers establishes template accessibility and reactivation of DNA helicase [47]. Similarly DNA topoisomerase I is unable to access DNA in a DNA-histone complex [48,49] unless the histones are removed by automodifying poly(ADP-ribose)polymerase [8,47]. Thus, the automodification of poly(ADP-ribose)polymerase may directly affect histone-DNA interactions such that DNA becomes locally accessible to other proteins [7].

The reverse reaction, i.e. the reassembly of dissociated histone-DNA complexes is effected by poly(ADP-ribose)glycohydrolase [7]. Several aspects of the poly(ADP-ribose)glycohydrolase reaction are noteworthy. Initially, polymers are degraded in an endoglycosidic and exoglycosidic manner, and linear polymers are degraded faster than branched polymers [50; Braun and Althaus, unpublished observations]. The slower degradation of branched polymers is attributable to the strong binding of histones to the sites of branching. Thus, the degradation of polymerase-bound polymers and the concomitant reassembly of histone-DNA complexes emerges as a highly organized process which determines the order of reassociation of different histone species with DNA. Various factors, such as the numbers,sizes, and complexities of polymers, the differential polymer binding affinities of different histone species, and the differential processing of polymers by poly(ADP-ribose)glycohydrolase seem important in this process [8].

In summary poly(ADP-ribose)polymerase and poly(ADP-ribose) glycohydrolase may act in concert to temporarily remove histones from sites of DNA strand breaks and put them back on. This allows other proteins to temporarily access DNA at the site of shuttling. The shuttle mechanism operates in a processive, adaptive, target specific, and reversible manner. This indicates that *in vivo*, poly(ADP-ribose)polymerase is indeed specifically targeted to sites of

DNA strand breaks.

From In Vitro *to* In Vivo

Nucleotide excision repair of DNA by human cell extracts is suppressed in reconstituted nucleosomes [51]. In the same model system, rejoining of DNA single strand breaks depends on the availability of NAD for poly(ADP-ribose) synthesis [52, 53]. Expression of the subcloned DNA-binding domain of poly(ADP-ribose)polymerase in transfected monkey cells blocks postincisional activation of the resident enzyme following carcinogen treatment [54]. Furthermore, the microinjected DNA-binding domain blocks base excision repair in human fibroblasts [55]. This indicates that *in vivo*, poly(ADP-ribose)polymerase is indeed specifically targeted to sites of DNA strand breaks and is required for further repair processing. The molecular steps leading to activation of poly(ADP-ribose)polymerase after DNA damage have also been demonstrated *in vivo* [cf. 56-59], and the formation of branched polymers (with high affinity for histones) has been confirmed in living cells [60]. Finally, there is evidence for sequential automodification of the polymerase *in vivo* [61] and the formation of a strictly conserved polymer size pattern during base excision repair in living cells [31,32]. These findings were predicted from *in vitro* analyses demonstrating a processive mode of polymer formation. The polymer pattern synthesized in mammalian cells in response to carcinogen treatment is identical to that formed in nuclei and isolated core particles [31,32], suggesting that histones are also the predominant regulators of polymer sizes and branching *in vivo* ('adaptiveness'). Degradation of ADP-ribose polymers in repairing cells [34] exhibits the biphasic pattern characteristic of poly(ADP-ribose)glycohydrolase action *in vitro* [35], and the extent of stimulation of polymer turnover depends on the level of DNA damage [34]. Dissociation of histones from DNA has also been observed in mammalian cells engaged in DNA excision repair [for review see 62]. DNA domains engaged in repair become accessible to chemical and enzymatic probes as occurs following automodification of poly(ADP-ribose)polymerase *in vitro* [7]. Finally, unfolding of nucleosomes is blocked in poly(ADP-ribose)-depleted cells, and this leaves bulky DNA adducts unexcised [63]. These results strongly suggest that histone shuttling is involved in nucleosomal unfolding in DNA repair *in vivo* [31,32].

In summary, two types of conclusions emerge from the evidence discussed above. First, the *in vitro* reconstitution approach has clearly demonstrated the *potential* of the poly ADP-ribosylation system to function as a histone shuttle mechanism on DNA. In addition, this approach has brought to light several key features of the mechanisms involved, such as targetability, processivity, adaptiveness, specificity, and reversibility. Several of these features are also expressed when the poly ADP-ribosylation system participates in DNA repair *in vivo*. Apart from this evidence, a number of striking parallels have been noted, such as the coincidence of postincisional activation of the poly ADP-ribosylation system and the onset of nucleosomal unfolding, and the coincidence of nucleosomal refolding with deactivation of the poly ADP-ribosylation system. Moreover, the stimulation of poly(ADP-ribose) turnover and nucleosomal unfolding are both dependent on the level of DNA damage. Finally, inhibition of poly(ADP-ribose)polymerase in mammalian cells blocks nucleosomal unfolding coincident with the repair of bulky DNA adducts, suggesting that the poly ADP-ribosylation system is indeed involved in dissociating histones from nucleosomal DNA.

Outlook

Histone shuttling by the poly ADP-ribosylation system may be involved in other chromatin functions involving processing of DNA breaks, such as DNA recombination, sister-chromatid exchange, plasmid insertion into the genome, and gene rearrangements [65]. Future studies will have to consider that other proteins may compete with the polymerase for binding to DNA strand breaks [e.g.66]. Apart from DNA strand breaks, poly(ADP-ribose)polymerase may directly target proteins by means of its leucine-zipper motif [67]. Transcription factors and oncogenic proteins containing matching zipper domains are likely candidates for such interactions. Another aspect to be considered is the possible association of the poly ADP-ribosylation system with nuclear matrix structures [e.g. 68-70]. Furthermore, several aspects of poly ADP-ribosylation are not explained by the shuttle mechanism. For example, a minor fraction of protein-bound polymers arises from the heteromodification of histones and various nuclear enzymes [for review see 1,71]. The molecular conditions for heteromodification are poorly understood and the biological role of this reaction remains to be elucidated. Molecular genetic approaches targeting the roles of poly(ADP-ribose)glycohydrolase and ADP-ribosyl protein lyase will bring significant advances, since both these enzymes are not easily accessible for biochemical characterization in tissues and cells. Finally, specialized techniques, such as the incorporation of photoreactive substituents in ADP-ribose polymers, could prove useful for the analysis of macromolecular interactions of ADP-ribose polymers with chromatin proteins *in vivo*.

Acknowledgements

Research performed in our laboratory was supported by the Swiss National Foundation for Scientific Research, the Krebsliga, and the Kanton Zürich.

References

1. Althaus FR and Richter C: ADP-Ribosylation of Proteins: Enzymology and Biological Significance. Springer-Verlag, Berlin Heidelberg New York Tokyo, 1987, 237 pp
2. De Murcia G, Huletsky A and Poirier GG: Review: Modulation of chromatin structure by poly(ADP-ribosyl)ation. Biochem Cell Biol 66: 6260635, 1988
3. Jacobson MK and Jacobson EL (eds): ADP-ribose Transfer Reactions: mechanisms and Biological Significance. Springer, New York, 1989, 527 pp
4. De Murcia G. Ménissier-De Murcia J and Schreiber V: Poly(ADP-ribose)polymerase: molecular biological aspects. BioEssays 13: 455-462, 1991
5. Poirier GG and Moreau P (eds): ADP-Ribosylation Reactions. Springer, New York, 1992, 410 pp
6. Althaus FR, Collinge M, Loetscher P. Mathis G, Neigeli H, Panzeter P and Realini C: The poly ADP-ribosylation system of higher eukaryotes: How can it do what: In: G. Obe and A.T. Natarajan (eds). Chromosomal Aberrations — Basic and Applied Aspects, Springer, Berlin Heidelberg, 1989, pp 22-30
7. Realini C and Althaus FR: Histone shuttling by poly ADP-ribosylation, J. Bol Chem 267: 18858-18865, 1992
8. Althaus FR: Poly ADP-ribosylation: A histone shuttle mechanism in DNA excision repair. J Cell Science 102: 663-670, 1992
9. Benjamin RC and Gill DM: ADP-ribosylation in mammalian cell ghosts. Dependence of poly(ADP-ribose)synthesis on strand breakage in DNA. J Biol chem 255: 109493-10501, 1980
10. Benjamin RC and Gill DM: Poly(ADP-ribose)synthesis *in vitro* programmed by damaged DNA. A comparison of DNA molecules containing different types of strand breaks. J Biol Chem 255: 10502-10508, 1980
11. Ménissier-De Murcia J, Molinete M. Gradwohl G. Simonin F and De Murcia G: Zinc-binding domain of poly(ADP-ribose)polymerase participates in the recognition of single strand breaks on DNA. J Mol Biol 210: 229-233, 1989
12. Mazen A, Ménissier-De Murcia J Molinete M, Simonin F, Gradwohl G, Poirier GG and De Murcia G: Poly(ADP-ribose)polymerase: a novel finger protein. Nucl Acids Res 17: 4689-4598, 1989
13. Gradwohl G, Ménissier-De Murcia J, Molinete M, Simonin F, Koken M., Hoeijmakers JHJ and De Murcia G: The second zinc-finger domain of poly(ADP-ribose)polymerase determines specificity for single-stranded breaks in DNA. Proc Natl Acad Sci USA 87: 2990-2994, 1990
14. Ikejima M, Noguchi S, Yamashita R, Ogura T, Sugimura T, Gill DM and Miwa M: The zinc fingers of human poly(ADP-ribose)polymerase are differentially required for the recognition of DNA breaks and nicks and the consequent enzyme activation. J Biol Chem 265: 21907-21913, 1990
15. Friedberg EC: DNA Repair. W. H. Freeman, New York, 1985, 614 pp
16. Lambert MW and Laval J (eds): DNA repair mechanisms and their biological implications in mammalian cells. Plenum Press, New York, 1989, 667 pp
17. Simonin F, Höfferer L, Panzeter PL, Muller S, de Murcia G and Althaus FR: The carboxyl-terminal domain of human poly(ADP-ribose)polymerase. Overproduction in *Escherichia coli*, large scale purification, and characterization. J Biol Chem 268: 13454-13461, 1993
18. Cohen JJ and Berger NA: Activation of poly(adenosine diphosphate ribose)polymerase with UV irradiated and UV endonuclease treated SV40 minichromosomes. Biochem Biophys Res Commun 98: 268-274, 1981
19. Berger NA and Sikorski GW, Poly(adenosine diphosphoribose)synthesis in ultraviolet-irradiated xeroderma pigmentosum cells reconstituted with micrococcus luteus UV endonuclease. Biochemistry 20: 3610-3614, 1981
20. McCurry LS and Jacobson MK: Poly(ADP-ribose) synthesis following DNA damage in cells heterozygous or homozygous for the Xeroderma pigmentosum genotype. J Biol Chem 256: 551-553, 1981
21. Cohen JJ, Catino DM, Petzold SJ and Berger NA: Activation of poly(adenosine diphosphate ribose)polymerase by SV40 minichromosomes: Effects of deoxyribonucleic acid damage and histone H1. Biochemistry 21: 493-4940, 1982
22. Ogata N, Ueda K, Kawaichi M and Kayaishi O: Poly(ADP-ribose) synthetase, a main acceptor of poly(ADP-ribose) in isolated nuclei. J Biol Chem 256: 4135-4137, 1981
23. Adamietz P: Poly(ADP-ribose)synthase is the major endogenous nonhistone acceptor for poly(ADP-ribose) in alkylated rat hepatoma cells. Eur J Biochem 169: 365-672, 1987
24. Desmarais Y, Ménard L, Lagueux J and Poirier GG: Enzymological properties of poly(ADP-ribose)polymerase: characterization of automodification sites and NADase activity. Biochem Biophys Acta 1078: 179-186, 1991
25. Ferro AM and Olivera BM: Poly(ADP-ribosylation) *in vitro*. Reaction parameters and enzyme mechanism. J Biol Chem 257: 7808-7813, 1982
26. Zahradka P and Ebisuzaki K: A shuttle mechanism for DNA-protein interactions. The regulation of poly(ADP-ribose)polymerase. Eur J Biochem 127: 579-585, 1982
27. De Murcia G, Jongstra-Bilen J, Ittel ME, Mandel P and Delain E: Poly(ADP-ribose)polymerase automodification and interaction with DNA: electron microscopic visualization. EMBO J 2 4543-548, 1983
28. Gaudreau A, Ménard L., De Murcia G and Poirier GG: Poly(ADP-ribose) accessibility to poly(ADP-ribose)glycohydrolase activity on poly (ADP-ribosyl)ated nucleosomal proteins. Biochem Cell Biol 64: 146-153, 1986
29. Naegeli H Loetscher P and Althaus FR: Poly ADP-ribosylation of proteins. Processivity of posttranslational modification. J Biol Chem 264: 14382-14385, 1989
30. Naegeli H and Althaus FR: Regulation of poly(ADP-ribose)polymerase. Histone-specific adaptations of reaction products. J Biol Chem 266: 10596-10601, 1991
31. Malanga M and Althaus FR: Poly(ADP-ribose) molecules formed during DNA repair *in vivo*. J Biol Chem, in press, 1994

32. Malanga M and Althaus FR: Effect of DNA damage on the numbers and size distributions of ADP-ribose polymers *in vivo*. Experientia 49: 2, 1993
33. Panzeter PL, Realini CA and Althaus FR: Poly(ADP-ribose) molecules formed during DNA repair *in vivo*. J Biol Chem, in press, 1994
34. Alvarez-Gonzalez R and Althaus FR: Poly(ADP-ribose) catabolism in mammalian cells exposed to DNA-damaging agents. Mutat Res 218: 67-74, 1989
35. Hatakeyama K, Menoto YU, Ueda K, Hayaishi I: Purification and characterization of poly(ADP-ribose)glycohydrolase. Different modes of action on large and small poly(ADP-ribose). J Biol Chem 261: 14902-14911, 1986.
36. Ikejima M and Gill DM: Poly(ADP-ribose) degradation by glycohydrolase starts with an endonucleolytic incision. J Biol Chem 263: 11037-11040, 1988
37. Mathis G and Althaus FR: Release of core DNA from nucleosomal core particles following (ADP-ribose)n-modification *in vitro*. Biochem Biophys Res Commun 143: 1049-1057, 1987
38. Wesierska-Gadek J and Sauermann G: The effect of poly(ADP-ribose) on interactions of DNA with histones H1, H3, and H4. Eur J Biochem 173: 675-679, 1988
39. Panzeter PL, Zweifel B and Althaus FR: Synthesis of poly(ADP-ribose)-agarose beads: An affinity resin for studying (ADP-ribose)n-protein interactions. Anal Biochem 207:157-162, 1992
40. Panzeter PL, Zweifel B,Malanga M, Waser SH, Richard MC and Althaus FR: Targeting of histone tails of poly(ADP-ribose). J Biol Chem 17662-17664 *in press*, 1993
41. Böhn L and Crane-Robinson C: Proteases and structural probes for chromatic: the domain structure of histones. Bioscience Rep 4: 365-386, 1984
42. Wood MJ, Yau P, Imai BS, Bolderg MW, Lambert SJ, Fowler AG, Baldwin JP, Godfrey JE, Moudrianakis EN, Koch MHJ, Ibel K, May RP and Bradbury EM: Neutron and X-ray scatter studies of the histone octamer and animo and carboxyl domain trimmed octamers. J Biol Chem 266: 5969-5702, 1991
43. Rodriguez AT, Pérez L, Moràn F, Montero F and Suau P: Cooperative interaction of the C-terminal domain of histone H1 with DNA. Biophys Chem 39: 145-152, 1991
44. Garcia Ramirez M, Dong F and Ausio J: Role of the histone 'tails' in the folding of oligonucleosomes depleted of histone H1, J Biol Chem 267: 19587-19595, 1992
45. Simpson RT: Structure of the chromatosome, a chromatin particle containing 160 base pairs of DNA and all the histones. Biochemistry 17: 5524-5531, 1978
46. Thömmes P and Hübscher U: DNA helicase from calf thymus. Purification to apparent homogeneity and biochemical characterization of the enzyme. J Biol Chem 265: 14347-14354, 1990
47. Thömmes P, Strack B, Ferrari E, Realini C, Althaus FR and Hübscher U: Eukaryotic DNA helicases. Experientia 48: A82, 1992
48. Richter A and Kapitza M: Histone H1 inhibits eukaryotic DNA topoisomerase I. FEBS Letts 294: 125-128, 1991
49. Richter A and Ruff J: DNA topoisomerase I cleavage sites in DNA and nucleoprotein complexes. Biochem 30: 9741-9748, 1991
50. Braun S, Panzeter PL, Collinge M, and Althaus FR: Endoglycosidic cleavage of branched polymers by poly(ADP-ribose) glycohydrolase. Eur J Biochem 220: 369–375, 1994
51. Wang Z, Wu X and Friedberg EC: Nucleotide excision repair in DNA by human cell extracts is suppressed in reconstituted nucleosomes. J Biol Chem 266: 22472-22478, 1991
52. Satoh MS and Lindahl T: Role of poly(ADP-ribose) formation in DNA repair. Nature 356:356-358, 1992
53. Satoh MS, Poirier GG and Lindahl T, NAD-dependent repair of damaged DNA by human cell extracts. J Biol Chem 268: 5480-5487, 1993
54. Küpper JH, De Murcia G and Bürkle A: Inhibition of poly(ADP-ribosyl)ation by overexpressing the poly(ADP-ribose)polymerase DNA-binding domain in mammalian cells. J Biol Chem 265: 1872-1874, 1990
55. Molinete M, Vermeulen W, Bürkle A, Ménissier-de Murcia J, Küpper JH, Hoeijmakers JHJ and de Murcia G: Overproduction of the poly(ADP-ribose)polymerase DNA-binding domain blocks alkylation-induced DNA repair synthesis in mammalian cells. EMBO J 12: 2109-2117, 1993
56. Skidmore CJ, Davies MI, Goodwin POM, Halldorson H, Lewis PJ, Shall S, Zia'ee A: The involvement of poly(ADP-ribose)polymerase in the degradation of NAD caused by gamma-radiation and N-methyl-N-nitrosourea. Eur J Biochem 101: 135-142, 1979
57. Skidmore CJ, Davies MI, Omidiji O, Zia'ee AA and Shall S: Poly(ADP-ribose)polymerase —An enzyme responsive to DNA damage. In: ME Smulson and T. Sugimura (eds). Novel ADP-ribosylations of regulatory enzymes and proteins. Elsevier, Amsterdam New York, 1980, pp. 197-205
58. Juarez-Salinas H, Sims JL and Jacobson MK, Poly(ADP-ribose) levels n carcinogen-treated cells. Nature 282, 740-741, 1979
59. Durkacz B, Omidiji O, Gray DA and Shall S, [ADP-ribose]n participates in DNA excision repair. Nature 283: 593-596, 1980
60. Juarez-Salinas H, Levi V, Jacobson EL and Jacobson MK: Poly(ADP-ribose) has a branched structure *in vivo*, J Biol Chem 257: 607-609, 1982
61. Cole GA, Bauer G, Kirsten E, Mendeleyev J, Bauer PI, Buki KG, Hakam A and Kun E: Inhibition of HIV-1 IIIb replication in AA-2 and MT-2 cells in culture by two ligands of poly(ADP-ribose)polymerase: 6-amino-1,2-benzopyrone and 5-iodo-6-amino-1,2-benzyopyrone. Biochem Biophys Res Commun 180: 504-514, 1991
62. Smerdon MJ, DNA excision repair of the nucleosome level of chromatin. In: M.W. Lambert and J. laval (eds). DNA Rep[air Mechanisms and Their Biological Implications in Mammalian Cells. Plenum Press, New York London, 1989, pp 271-294
63. Mathis G and Althaus FR: Uncoupling of DNA excision repair and nucleosomal unfolding in poly(ADP-ribose)-depleted mammalian cells. Carcinogenesis 11: 1237-1239, 1990
64. Sidik K and Smerdon MJ: Correlation between repair patch ligation and nucleosome rearrangement in human cells treated with bleomycin, UV radiation or methyl methanesulfonate. Carcinogenesis 13: 135-138, 1992
65. Bertazzoni U, Scovassi AI and Shall S: Fourth European Meeting on ADP-ribosylation of proteins, Pavia/Italy, 20-23 April 1989. Mutat Res 219: 303-307, 1989
66. Blier PR, Griffith AJ, Craft J and Hardin JA: Binding of Ku protein to DNA: measurement of affinity for ends and demonstration of binding to nicks. J Bio Chem 268: 7594-7601, 1993
67. Uchida K, Hanai S, Ishikawa K, Ozawa Y, Uchida M, Sugimura T and Miwa M: Cloning of cDNA encoding *Drosophila* poly(ADP-ribose)polymerase: leucine zipper in the automodification domain. Proc Natl Acad Sci, USA, 90: 3481-3485, 1993
68. Wesierska-Gadek J and Sauermann G: Modification of nuclear matrix proteins by ADP-ribosylation: association of nuclear ADP-ribosyltransferase with the nuclear matrix. Eur J Biochem 153: 421-428, 1985
69. Cardenas-Corona ME, Jacobson EL and Jacobson MK: Endogenous polymers of ADP-ribose are associated with the nuclear matrix. J Biol Chem 262: 14863-14866, 1987
70. Kaufmann SH, Burnet G, Talbot B, Lamarre D, Dumas C, Shaper JH

and Poirier GG: Association of poly(ADP-ribose)polymerase with the nuclear matrix. Exp Cell Res 192: 524-535, 1991

71. Ueda K: Nonredox reactions of pyridine nucleotides. In: D. Dophin R. Poulson O. Avranovic (eds. Pyridine Nucleotides, Wiley, New York 1987,pp 549-598

Molecular and Cellular Biochemistry **138:** 61–69, 1994.

Growth-phase-dependent response to DNA damage in poly(ADP-ribose) polymerase deficient cell lines: basis for a new hypothesis describing the role of poly(ADP-ribose) polymerase in DNA replication and repair

Satadal Chatterjee and Nathan A. Berger

Department of Medicine and Ireland Cancer Center, University Hospitals of Cleveland, Case Western Reserve University, 2074 Abington Road, Cleveland, OH 44106, USA

Abstract

We have studied the role of poly(ADP-ribose) polymerase in the repair of DNA damage induced by x-ray and N-methyl-N′-nitro-N-nitrosoguanidine (MNNG) by using V79 chinese hamster cells, and two derivative mutant cell lines, ADPRT54 and ADPRT351, that are deficient in poly(ADP-ribose) polymerase activity. Under exponentially growing conditions these mutant cell lines are hypersensitive to x-irradiation and MNNG compared to their parental V79 cells which could be interpreted to suggest that poly(ADP-ribose) polymerase is involved in the repair of DNA damage. However, the level of DNA strand breaks induced by x-irradiation and MNNG and their rates of repair are similar in all the cell lines, thus suggesting that it may not be the difference in strand break formation or in its rate of repair that is contributing to the enhanced cell killing in exponentially growing poly(ADP-ribose) polymerase deficient cell lines. In contrast, under growth-arrested conditions, all three cell lines become similarly sensitive to both x-irradiation and MNNG, thus suggesting that poly(ADP-ribose) polymerase may not be involved in the repair of DNA damage in growth-arrested cells. These paradoxical results could be interpreted to suggest that poly(ADP-ribose) polymerase is involved in DNA repair in a cell-cycle-dependent fashion, however, it is functionally active throughout the cell cycle. To resolve this dilemma and explain these results and those obtained by many others, we propose that the normal function of poly(ADP-ribose) polymerase is to prevent DNA recombination processes and facilitate DNA ligation. (Mol Cell Biochem **138:** 61–69, 1994)

Key words: poly(ADP-ribose) polymerase, poly(ADP-ribose) polymerase deficiency, X-ray damage, MNNG damage, DNA replication and growth arrest, DNA recombination and repair

Introduction

Poly(ADP-ribose) polymerase is a nuclear enzyme of molecular weight 116,000. The enzyme binds to and is activated by DNA strand breaks; it then utilizes nicotinamide adenine dinucleotide (NAD) for the synthesis of ADP-ribose polymers which are covalently attached to several nuclear proteins including the enzyme itself [1,2]. Both single- and double-strand DNA breaks activate poly(ADP-ribose) polymerase [3,4]. These breaks can be produced directly by agents such as x-rays or indirectly following enzymatic excision of damaged DNA [2,5,6]. Specific types of DNA damage that cannot be acted on by excision repair processes, such as ultraviolet (UV) radiation damage in xeroderma pigmentosum cells, do not activate poly(ADP-ribose) polymerase, whereas the same damage in excision

Address for offprints: N. Berger, Department of Medicine and Ireland Cancer Center, University Hospitals of Cleveland, Case Western Reserve University, 2074 Abington Road, Cleveland, OH 44106, USA

repair competent cells successfully activates the enzyme [7,9]. Using purified in vitro systems, we have shown that closed circular SV40 minichromosomes do not activate poly(ADP-ribose) polymerase whereas SV40 microchromosomes, pretreated with restriction endonucleases to produce DNA strand breaks, do activate the enzyme [10]. We have also shown that UV-irradiated SV40 minichromosomes do not activate purified poly (ADP-ribose) polymerase, whereas UV-irradiated minichromosomes pretreated with M. luteus UV endonuclease successfully activate the enzyme [11].

While poly-(ADP-ribose) polymerase is a damage-responsive enzyme, activated specifically by DNA strand breaks, it is, however, different from the class of damage-inducible gene products. Thus, poly(ADP-ribose) polymerase is a constitutive enzyme whose gene transcription and mRNA levels remain constant after DNA damage, whereas transcription of damage-inducible genes is increased dramatically following DNA damage [12].

Using enzyme inhibitors to interfere with poly(ADP-ribose) synthesis results in a dose-dependent increase in sister chromatid exchanges (SCE) [13-19]. Furthermore, inhibition of enzyme activity by poly(ADP-ribose) polymerase inhibitors in cells exposed to DNA damage results in an even greater frequency of sister chromatid exchanges and produces a variety of additional cellular effects including increased cell killing, increased frequency of single-strand breaks, and increased repair replication [2,20-28]. These alterations have been interpreted as demonstrating that interference with poly(ADP-ribose) polymerase interferes with DNA repair and thus provides the basis to suggest that poly(ADP-ribose) polymerase has a direct role in the DNA repair process [23,29]. Much of the evidence for the role of poly(ADP-ribose) polymerase in the DNA repair process is based on the effects of poly(ADP-ribose) polymerase inhibitors. However, since these agents lack specificity, the concept of the involvement of this enzyme in DNA repair has been questioned [26,30-33]. Thus, despite the availability of a great deal of experimental results in this field, there is still no generally agreed biological role for poly(ADP-ribose) polymerase in DNA repair.

To better understand the role of poly(ADP-ribose) polymerase in cellular processes and to evaluate its function in the absence of inhibitors, we isolated two mutant cell lines, deficient in enzyme activity. The ADPRT54 and ADPRT351 cell lines are respectively 89% and 95% deficient in poly(ADP-ribose) polymerase activity relative to their parental V79 Chinese hamster cell lines [34,35]. In this report we studied the response of these cell lines to two different types of DNA damaging agents, x-irradiation and MNNG, under conditions of logarithmic and density inhibited growth. X-irradiation was used to produce direct DNA strand breaks, whereas the monofunctional alkylating agent, MNNG, was used to indirectly produce DNA strand breaks through the excision repair process. We found that under logarithmic growth conditions ADPRT54 and ADPRT351 cells were hypersensitive to x-ray and MNNG compared to the parental V79 cells. However, under growth-arrested conditions, the mutant cell lines become more resistant to x-ray and MNNG and show essentially the same sensitivity as do the parental V79 cells. Furthermore, as opposed to the results found with enzyme inhibitors, our studies show that the level of DNA strand breaks induced by x-ray or MNNG and their rate of repair in poly(ADP-ribose) polymerase deficient cell lines is similar to that in the parental V79 cells. These and previous studies can be explained by the hypothesis that the normal function of poly(ADP-ribose) polymerase is to protect DNA ends from recombinational events and that when the system is stressed by too many DNA strand ends or by too little enzyme activity, the result is an increase in DNA recombination with increased SCE, increased homologous and nonhomologous recombination, with the latter leading to toxicity and cell death.

Results

The clonogenic survival curves of the exponentially growing V79, ADPRT54, and ADPRT351 cell lines following exposure to variable doses of x-ray are illustrated in Figure 1. The V79 survival curve exhibits a shoulder at low doses of x-ray followed by an exponential region at increasing doses. This shoulder is reduced in ADPRT54 and ADPRT351 cells. More importantly, the poly(ADP-ribose) polymerase deficient cell lines exhibit considerable hypersensitivity in the exponential region compared to the parental V79 cells. Thus, exponentially growing cells with impaired poly(ADP-ribose) polymerase activity are hypersensitive to x-irradiation.

In subsequent experiments shown in Figure 2 we evaluated the cytotoxic effects of x-irradiation on V79, ADPRT54 and ADPRT351 cell lines under conditions of density dependent growth arrest. V79 cells and the poly(ADP-ribose) polymerase deficient cell lines show a similar shoulder at low doses of x-ray. In addition, all the cell lines show similar exponential regions at high doses indicating that under growth-arrested conditions, all the cell lines are similarly sensitive to x-irradiation. Thus, under growth-arrested conditions, cellular sensitivity to x-irradiation becomes independent of poly(ADP-ribose) polymerase activity.

Next we evaluated the repair of x-ray-induced DNA single strand breaks in exponentially growing V79 and poly(ADP-ribose) deficient ADPRT351 cell lines by the

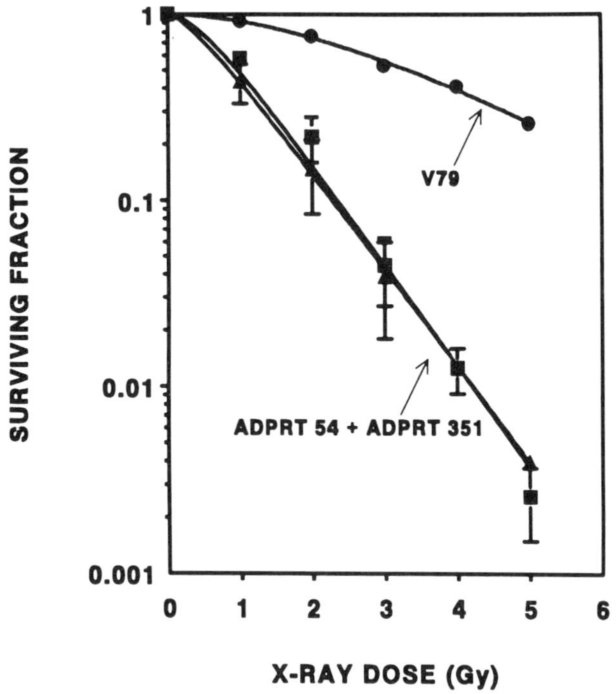

Fig 1. X-ray survival response of exponentially growing Chinese hamster V79 (●), ADPRT54 (■), and ADPRT351 (▲) cell lines. Description and maintenance of the cell lines was described previously [35-38]. Briefly, the mutant cell lines ADPRT54 and ADPRT351 have 5% to 11% poly(ADP-ribose) polymerase activity compared to their parental V79 chinese hamster cell line [35]. All cell lines were maintained in α-modified Eagle's medium buffered to pH 7.2 with 25mM N-2-hydroxyethyl-piperazine-N'-2-ethanesulfonic acid (HEPES) and supplemented with 100 units/ml penicillin, 100μg/ml streptomycin and 5% heat inactivated fetal calf serum. For x-irradiation, cells were irradiated in a Gammacell 1000 (Nordion International, Inc., Kanata, Ontario, Canada) at a dose rate of 8.07 Gy/min. The cells were exposed to indicated doses of x-rays at room temperature and the fraction of surviving cells was determined by clonogenic survival assays as described previously [42-45]. Points represent mean survival from at least two independent experiments. Error bars represent standard deviations and are shown when larger than the size of the points as drawn.

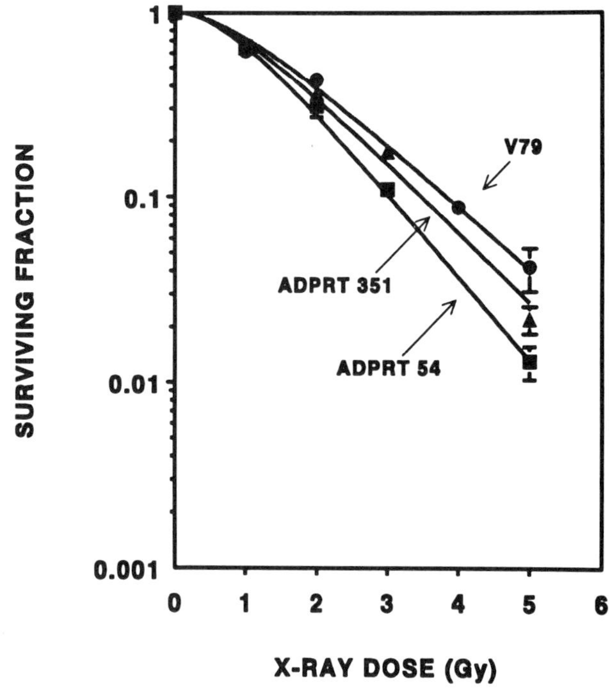

Fig 2. X-ray survival response of growth-arrested Chinese hamster V79 (●), ADPRT54 (■), and ADPRT351 (▲) cell lines. Cell cultures and irradiation were performed as in Figure 1. The cells were exposed to indicated doses of x-rays at room temperature and the fraction of surviving cells was determined by clonogenic survival assays as described previously [42-45]. Plateau phase cultures were obtained by growing the cells to confluence with daily medium changes. DNA synthesis rate of plateau phase cells was measured by labeling cells with ^{3}H-thymidine and counting radioactivity in TCA insoluble fractions and was 1% to 4% compared to that of exponentially growing cells. Points represent mean survival from at least two independent experiments.Error bars represent standard deviations and are shown when larger than the size of the points as drawn.

alkaline unwinding technique. The cells were exposed to 10Gy of x-irradiation and subsequently incubated at 37°C for various time periods to recover from x-ray-induced damage. Figure 3 shows that immediately following x-ray exposure, both the cell lines show similar levels of induction of DNA strand breaks. During the first 2 minutes incubation, following x-ray exposure, ADPRT351 cells consistently show a small increase in breaks. In contrast, during this same period, the strand break level dropped in the V79 cells, indicating progress in repair. However, after the first 2 minutes, the rate of strand break repair in both cells lines appears to be similar. Figure 3 suggests that repair of x-ray-induced DNA strand breaks occurs in two components. A fast component occurs for approximately the first 15 minutes following x-ray exposure, whereas, a slow component subsequently becomes prominent and remains operational until repair is complete. As shown in Figure 3, no noticeable differences in repair of x-ray-induced DNA strand breaks was observed between V79 and ADPRT351 cell lines except for the first two minutes following x-irradiation.

We also examined the effects of MNNG, a monofunctional alkylating agent, on the different cell lines. Figure 4 shows the clonogenic survival of exponentially growing V79, ADPRT54, and ADPRT351 cell lines following treatment with MNNG. These studies clearly show that poly(ADP-ribose) deficient cell lines ADPRT54 and ADPRT351 are hypersensitive to MNNG relative to their parental V79 cells, thus indicating a possible role of poly(ADP-ribose) synthesis in proliferating cells in counteracting MNNG-induced cytotoxicity.

Subsequently, we evaluated the cytotoxic effects of MNNG in density dependent, growth-arrested V79, ADPRT54, and ADPRT351 cell lines. Figure 5 shows that

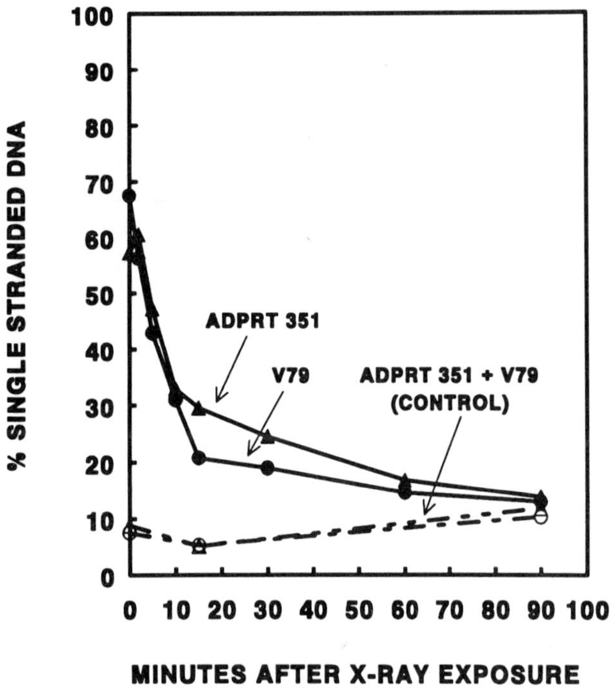

Fig 3. Repair of x-ray-induced strand breaks in exponentially growing Chinese hamster V79 (●), and ADPRT351 (▲) cell lines. Cell cultures and irradiation were performed as in Fig 1. The cells were exposed to 10 Gy x-rays at 0°C and subsequently incubated at 37°C for various time periods to recover from x-ray-induced damage. Formation and repair of DNA strand breaks were measured by alkaline unwinding technique as described by Ahnström and Erixon [46]. V79 (○) and ADPRT351 (Δ) represent unirradiated controls.

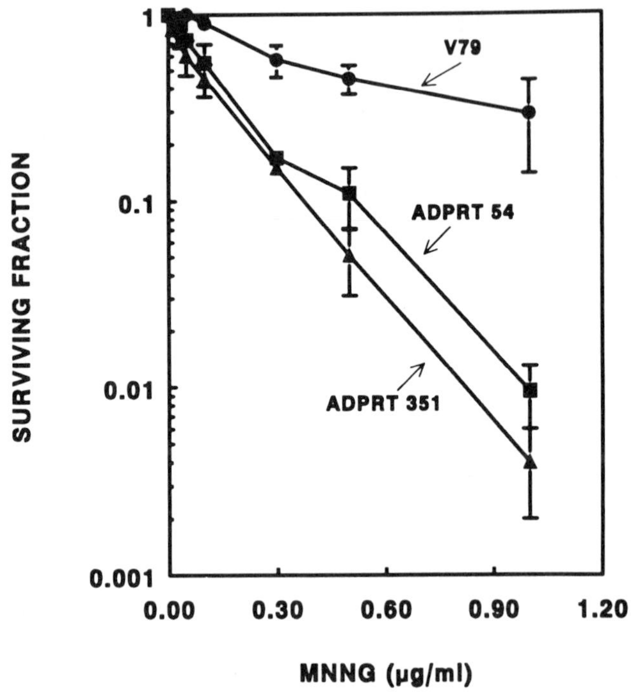

Fig. 4. Cytotoxicity of MNNG in exponentially growing V79 (●), ADPRT54 (■), and ADPRT351 (▲) cell lines. Cell culture were maintained as in Figure 1. Cells were treated with indicated drug concentrations for 1 hour at 37°C and the fraction of surviving cells was determined by the clonogenic survival assays as described previously [42-45]. MNNG was purchased from the Sigma Chemical Co. (St. Louis, MO), MNNG stock solution, 10mg/mL, in 100% DMSO was prepared fresh before each experiment and subsequently diluted with 0.9% NaCl. All drug treatments were carried out in growth medium at 37°C. Techniques for determination of clonogenic survival were previously described [42,45]. Points represent mean survival from at least two independent experiments. Error bars represent standard deviations and are shown when larger than the size of the points as drawn.

in contrast to logarithmically growing cells, growth-arrested poly(ADP-ribose) polymerase deficient cell lines are significantly more resistant to MNNG. Furthermore, under growth-arrested conditions, all the cell lines are similarly sensitive to MNNG. Thus, under growth-arrested conditions, all the cell lines exhibit similar levels of sensitivity to MNNG, indicating that under growth-arrested conditions, MNNG-induced cytotoxicity is independent of poly(ADP-ribose) synthesis.

Next we examined MNNG-induced DNA strand break repair in the exponentially growing V79 and ADPRT351 cells. Following exposure to MNNG, cells were incubated at 37°C and at sequential time intervals, DNA strand breaks were measured by the alkaline unwinding technique. Immediately after MNNG exposure, the level of single strand breaks in V79 and ADPRT351 cells were 56% and 41% respectively. As shown in Figure 5, the subsequent rates of repair of breaks appeared to be similar in both V79 and ADPRT351 cell lines, thus suggesting that differences in the repair rates did not contribute to differences in MNNG susceptibility.

Discussion

Previous studies with inhibitors of poly(ADP-ribose) polymerase have shown that when these agents are given in association with DNA damaging agents they result in increased cell killing, increased frequency of DNA strand breaks and increased DNA repair replication [2,21,23-25,27-29]. These observations have been used to provide important support for the notion that poly(ADP-ribose) polymerase is involved in DNA repair.

In our studies using cells deficient in poly(ADP-ribose) polymerase activity, we have shown that (1) deficiency in enzyme activity potentates the cytotoxicity of x-irradiation and MNNG in proliferating cells but not in growth-arrested cells; and (2) the levels of DNA strand breaks induced by x-ray and MNNG and their rates of repair are similar in enzyme-deficient cells and parental V79 cells. We have also

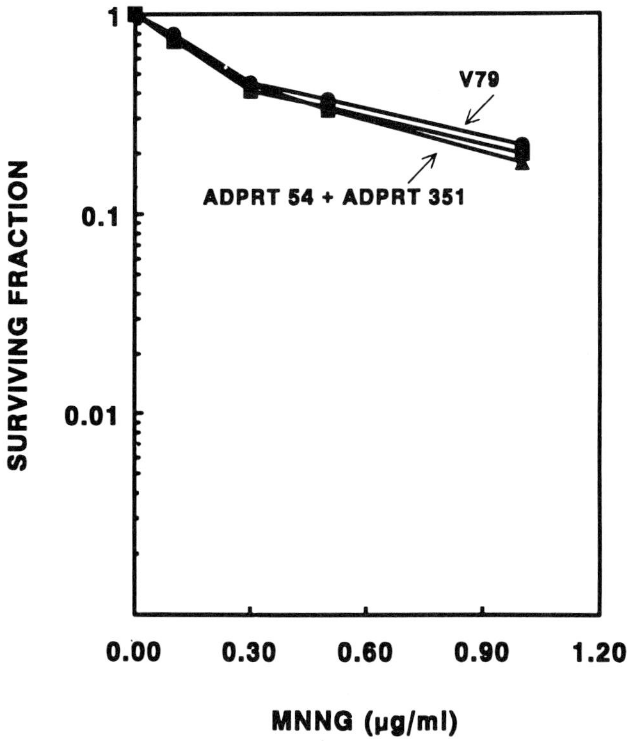

Fig. 5. Cytotoxicity of MNNG in growth-arrested V79 (●), ADPRT54 (■) and ADPRT351 (▲) cell lines. Cell cultures were maintained as in Figure 1, plateau phase cultures were obtained as in Figure 2 and MNNG treatment was done as in Figure 4. Cells were treated with indicated drug concentrations for 1 hour at 37°C and the fraction of surviving cells was determined by the clonogenic survival assays as described previously [42-45]. Points represent mean survival from at least two independent experiments. Error bars represent standard deviations and are shown when larger than the size of the points as drawn.

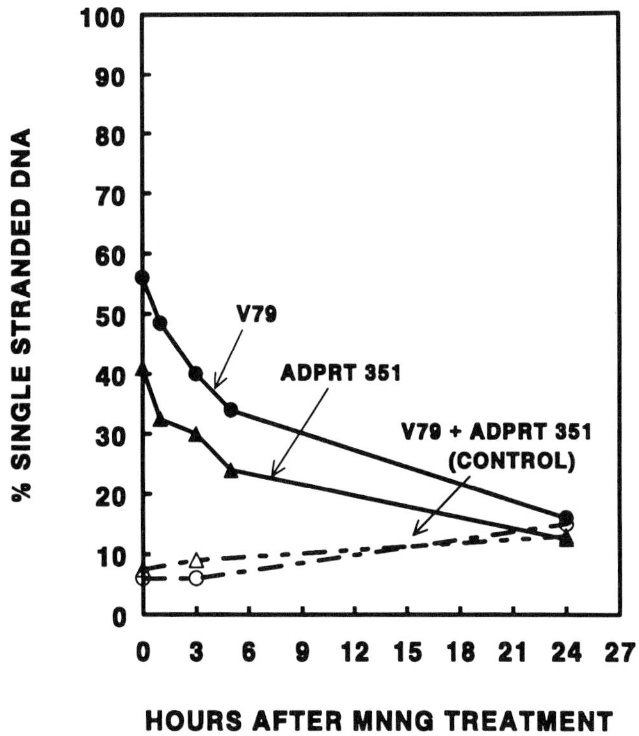

Fig. 6. Repair of MNNG-induced strand breaks in exponentially growing Chinese hamster V79 (●) and ADPRT351 (▲) cell lines. Cell cultures were maintained as in Fig. 1 and MNNG treatment was done as in Fig. 4. Formation and repair of DNA strand breaks were measured by alkaline unwinding technique as described previously [46]. The cells were treated with 2μ/ml MNNG for 1 hour at 37°C to recover from MNNG-induced damage. V79 (○) and ADPRT351 (Δ) represent untreated controls. Percent single-stranded DNA was measured as in Fig. 3.

shown that the ADPRT54 and ADPRT351 cell lines are deficient in poly(ADP-ribose) polymerase activity during both growth-arrested and proliferating states. Thus, under growth-arrested conditions poly(ADP-ribose) synthesis deficient cell lines and the parental cells with normal enzyme activity show the same sensitivity to x-irradiation and MNNG. Thus, our findings suggest that normal activity of poly(ADP-ribose) polymerase may facilitate cell survival following DNA damage in proliferating cells but not in growth-arrested cells. These results create the apparent paradox that poly(ADP-ribose) polymerase is involved in DNA repair in proliferating cells but not in growth-arrested cells. This raises the dilemma of explaining how an enzyme can function in the DNA repair process in only certain growth phases, when it is functionally active throughout the different growth phases.

As illustrated in Figure 7, we have generated a hypothesis to explain the results reported above and also the fact that the one consistently agreed-upon cellular consequence of poly(ADP-ribose) polymerase inhibition is an increase in SCE frequency. This increased SCE frequency is observed in: (1) cells deficient in poly(ADP-ribose) polymerase [35]; (2) cells with impaired NAD synthesis resulting in low substrate levels for the enzyme [35]; and (3) cells grown in the presence of poly(ADP-ribose) polymerase inhibitors [13-19].

Under normal growth conditions, semiconservative DNA replication consists of three steps that include: (1) DNA strand separation; (2) discontinuous DNA synthesis; and (3) ligation of newly synthesized DNA fragments. In addition, cells always undergo some recombination during the normal DNA replication process, as demonstrated by the low but constant level of background SCE in normal untreated cells [13-19,34,35]. These recombination events may function as a bypass mechanism for spontaneously damaged DNA and/or they may function to close gaps between fragments that are not acted upon by the normal ligation process.

Under normal circumstances, the recombination process occurs infrequently despite the multitude of strand ends created by the discontinuous nature of DNA synthesis. We

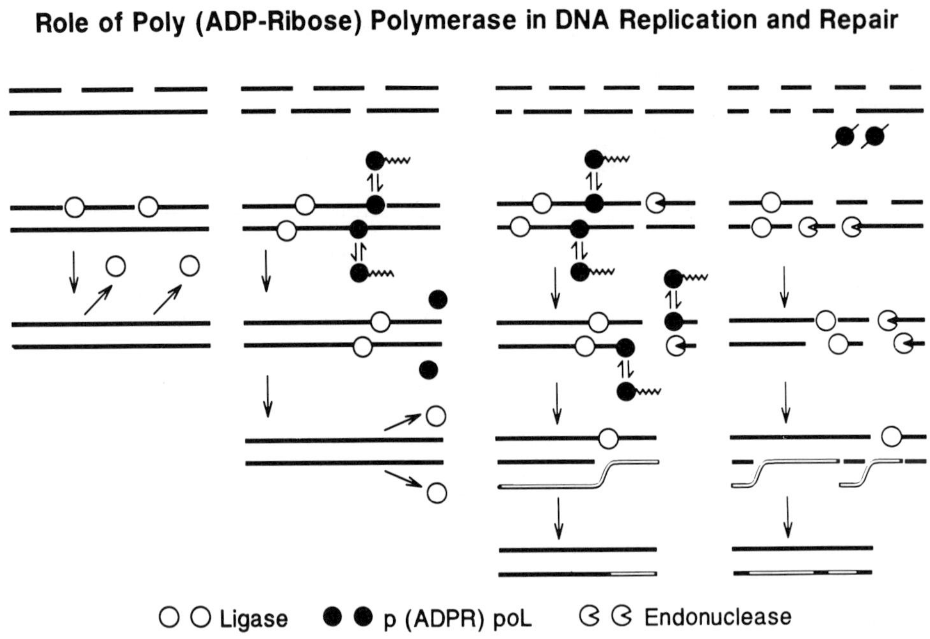

Fig. 7. Model of poly(ADP-ribose) polymerase role in DNA replication and repair. Open circles indicate ligase, closed circles represent poly(ADP-ribose) polymerase, jagged lines represent poly(ADP-ribose), and notched circles represent endonuclease. This model assumes a limited number of molecules of DNA ligase and poly(ADP-ribose) polymerase. Although we have indicated two molecules of each enzyme, we do not mean to imply that there are equal numbers of each enzyme present, rather we suggest that each enzyme is present in limited quantities. The first panel on the left shows DNA with a limited number of strand ends. These could be the result of direct breaks, the excision repair process or replication intermediates resulting from replicative DNA synthesis. In this situation, there are sufficient ligase molecules to directly seal all DNA strand disruptions and restore DNA integrity.

In the second column, there are an increased number of DNA strand ends that exceed the available number of ligase molecules. In this situation, all DNA ligase molecules are engaged. The DNA strand ends that exceed the ligase capacity are protected by poly(ADP-ribose) polymerase which is activated by binding DNA strand ends to synthesize poly(ADP-ribose). Polymer synthesis causes the enzyme to dissociate from the DNA. Subsequent degradation of poly(ADP-ribose) by poly(ADP-ribose) glycohyrolase allows the enzyme to return, bind another DNA strand end and reinitiate polymer synthesis. Poly(ADP-ribose) polymerase cycles on and off of DNA strand ends until ligase is available to join the disruption, and restore DNA integrity.

The third column illustrates the situation that occurs when the number of DNA strand ends increase to the point that they are in excess of the number of ligase and poly(ADP-ribose) molecules. This could occur following DNA damage and excision repair in replicating cells where strand ends are already increased due to the presence of replication intermediates. Under these conditions, endonucleases begin to act on some of the free ends and the recombination process is used to repair other sections. This process results in DNA deletions and recombination. The latter process appears as sister chromatid exchanges which usually represent homologous recombination. However, as the frequency of SCE and DNA deletions increase, there is an increased probability for the occurrence of chromosomal aberrations, non-homologous recombination events, gene deletions, and cytotoxicity.

The fourth column illustrates the situation with increased DNA strand breaks and inactive poly(ADP-ribose) polymerase. The latter condition could be due to mutations which inactivate the enzyme, the presence of enzyme inhibitors or the absence of sufficient NAD to provide substrate for enzyme activity. In this situation, the number of free strand ends, far exceeds the capacity to be joined by the existing number of ligase molecules but poly(ADP-ribose) polymerase is not available to protect the excess free ends. As a result of the lack of protection, DNA strand ends are subject to more extensive endonuclease action resulting in DNA deletions. DNA strand ends and gaps are ultimately rejoined by ligation and recombination, gene deletion, and increased cytotoxicity. The latter condition explains the observations in poly(ADP-ribose) polymerase deficient cell lines treated with DNA damaging agents during logarithmic growth. When the increased strand ends occur during the replication process and represent replication intermediates, the absence of poly(ADP-ribose) polymerase could result in more strand ends available for recombination and SCE accounting for the increased frequency of SCE in poly(ADP-ribose) polymerase deficient cell lines. In contrast, the diagram in the left column explains the situation in poly(ADP-ribose) polymerase deficient cell lines which are treated with DNA damaging agents during growth arrest. These cells have a limited number of DNA strand breaks relative to ligase, since during growth arrest, there are no ends due to replication intermediates. In this case poly(ADP-ribose) polymerase is not required to protect DNA strand ends and the DNA integrity is restored even though the cells are poly(ADP-ribose) polymerase deficient.

propose that the normal function of poly(ADP-ribose) polymerase is to transiently bind to these DNA strand ends and protect them from the action of endonucleases or entering into recombination processes. Binding of the enzyme to DNA strand ends activates synthesis of poly(ADP-ribose) and subsequent dissociation of the enzyme from DNA allows ligation to take place. Thus, we propose that poly(ADP-ribose) polymerase protects DNA strand ends from endonuclease activity and/or recombination processes. However, interruption or interference with any of the steps involved in the normal function of poly(ADP-ribose) polymerase will facilitate the recombination process and endonuclease action leading to the appearance of SCE and genomic deletions.

Assuming that replicating cells contain an excess of strand ends relative to ligase, then many ends will undergo recombination unless they are shielded from their process until free ligase becomes available. This explains why cells from Bloom's syndrome patients, deficient in ligase I activity, exhibit an increase in spontaneous SCE [36]. We suggest that poly(ADP-ribose) polymerase protects the DNA strand ends, cycles on and off of these ends in a fashion that controls their rate of availability and favors their utilization in the ligase reaction relative to the recombination process. However, in the absence of normal poly(ADP-ribose) polymerase activity, the DNA strand ends will not be protected and this will predispose to recombination processes and an increase in SCE. This increase in SCE is exactly what we observe in cells with impaired poly(ADP-ribose) synthesis based on deficiency of the enzyme poly(ADP-ribose) polymerase or in those cells treated with poly(ADP-ribose) polymerase inhibitors during the normal replication process [13–19,35]. Under these conditions, it is probable that most SCE represent homologous recombination and therefore the cells experience little or no toxicity associated with the inhibition of poly(ADP-ribose) polymerase and the increased SCE.

Exposure of cells to DNA damaging agents initiates the DNA repair process which includes: (1) creation of DNA strand breaks; (2) excision of damaged DNA; (3) synthesis of repair patches; and (4) ligation of strand breaks and/or repair patches to parental DNA. Under these conditions, persistent DNA damage or errors introduced into the DNA sequence during this repair process can led to mutations and/or cell death. However, when replicating cells are exposed to DNA damaging agents, the situation becomes more complicated. In addition to the normal replication/repair process described above, there is an increase in recombination processes, as demonstrated by the dose-dependent increase in SCE formation that occurs in response to treatment with DNA damaging agents [21,22]. According to our hypothesis, this increase in SCE occurs because the number of DNA strand ends generated by the combination of semiconservative replication and the excision repair process exceeds the capacity of poly(ADP-ribose) polymerase to completely protect against recombination events.

SCE, in general, represent homologous recombination events and thus are not cytotoxic. However, as the endonuclease susceptibility and frequency of recombination increase, it is likely that there is a proportional increase in imperfect (nonhomologous) recombination. The latter produce DNA rearrangements and deletions with the potential for causing mutations and/or cell death. Thus, when replicating cells are subjected to DNA damage, two different processes including: (1) errors during replication repair, and (2) nonhomologus DNA recombination contribute to cell death, whereas, when growth-arrested cells are subjected to DNA damage, only one process of faulty repair contributes to cell death.

Replicating cells with impaired poly(ADP-ribose) polymerase activity show a ten-to twentyfold increase in sister chromatid exchanges compared to normal, untreated cells [13-19,35]. This can be explained by an insufficient amount of poly(ADP-ribose) polymerase relative to the number of DNA strand ends produced by discontinuous DNA synthesis. When these cells are further stressed by increasing doses of DNA damaging agents, the number of recombinational events is further increased and the chance of nonhomologus recombination also increases. Thus, cells with impaired poly(ADP-ribose) polymerase activity will be more susceptible to the induction of recombination and will undergo both homologous and nonhomologous recombination at much lower doses of DNA damaging agents compared to cells with normal poly(ADP-ribose) polymerase activity. Consequently, treatment of cells with a given dose of DNA damaging agent will result in enhanced cell killing in poly(ADP-ribose) deficient cells relative to the cells with normal poly(ADP-ribose) synthetic activity. Since only replicating cells can undergo recombination, we hypothesize that impaired poly(ADP-ribose) polymerase activity will result in increased cytotoxicity in replicating cells treated with DNA damaging agents, but not in growth-arrested cells with impaired poly(ADP-ribose) synthesis.

The proposal put forth in this manuscript explains the observations in our present experiments and fits well with a wide variety of studies examining the interaction between poly(ADP-ribose) polymerase inhibitors and DNA damaging agents. Our studies are in agreement with the suggestion by Creissen and Shall that poly(ADP-ribose) synthesis facilitates the DNA ligation process, however, our results do not support direct ADP-ribosylation and subsequent activation of ligase [37]. Our proposal provides an explanation for the observations by Jacobson et al, that poly(ADP-ribose) polymerase inhibitors potentiate the effect of MNNG in proliferating cells but not in quiescent cells [38,39]. It also provides an explanation for the demonstration by Boorstein and Pardee that methylmethane sulfonate-induced cytotoxicity can be maximally potentiated by 3-aminobenzamide during the S phase of the cell cycle [40]. In addition, our hypothesis explains the observation by Wiencke and Morgan that inhibitors of poly(ADP-ribose) polymerase in combination with DNA damaging agents produce chromatid aberrations in a cell-cycle-dependent fashion with maximal aberrations occurring in cells treated during S phase [41].

Acknowledgements

This work was supported by NIH Grants PO1-CA51183 and P30-CA43703.

References

1. Ueda K, Hayaishi O. ADP-ribosylation. Annu Rev Biochem 54: 73-100, 1985
2. Berger NA: Poly(ADP-ribose) in the cellular response to DNA damage. Radiat Res 101: 4-15, 1985
3. Benjamin RC, Gill DM: ADP-ribosylation in mammalian cell ghosts. Dependence of poly(ADP-ribose) synthesis on strand breakage in DNA. J Biol Chem 255: 10493-10501, 1980
4. Benjamin RC, Gill DM: Poly(ADP-ribose) synthesis in vitro programmed by damaged DNA. A comparison of DNA molecules containing different types of strand breaks. J Biol Chem 255 10502-10508, 1980
5. Cleaver JE, Bodell WJ, Morgan WF, Zelle B: Differences in the regulation by poly(ADP-ribose) of repair of DNA damage from alkylating agents and ultraviolet light according to cell type. J Bio Chem 258: 9059-9068, 1983
6. Nduka N, Skidmore CF, Shall S: The enhancement of cytotoxicity of *N*-methyl-*N*-nitrosourea and of γ-radiation by inhibitors of poly(ADP-ribose) polymerase. Eur J Biochem 105: 525-530, 1980
7. McCurry LS, Jacobson MK: poly(ADP-ribose) synthesis following DNA damage in cells heterozygous or homozygous for the xeroderma pigmentosum genotype. J Biol Chem 256: 551-553, 1981
8. Berger NA, Sikorski GW, Petzold SJ, Kurohara KK: Defective poly(adenosine diphosphoribose) synthesis in xeroderma pigmentosum. Biochem 19: 289-293, 1980
9. Berger NA, Sikorski GW: Poly(adenosine diphosphoribose) synthesis in ultraviolet-irradiated xeroderma pigmentosum cells reconstituted with *micrococcus luteus* UV endonuclease. Biochem 20: 3610-3614, 1981
10. Cohen JJ, Catino DM, Petzold SJ, Berger NA: Activation of poly(adenosine diphosphate ribose) polymerase by SV40 minichromosomes: Effects of deoxyribonucleic acid damage and histone H1. Biochem 21: 4931-4940, 1982
11. Cohen JJ, Berger NA: Activation of poly(adenosine diphosphate ribose) polymerase with UV irradiated and UV endonuclease treated SV40 minichromosome. Biochem Biophys Res Commun 98: 268-274, 1981
12. Bhatia K, Pommier Y, Giri C, Fornace AJ, Imaizumi M, Breitman TR, Cherney BW, Smulson ME: Expression of poly(ADP-ribose) polymerase gene following natural and induced DNA strand breakage and effect of hyperexpression on DNA repair. Carcinogenesis 11: 123-128, 1990
13. Schwartz JL, Morgan WF, Kapp LN, Wolff S: Effects of 3-aminobenzamide on DNA synthesis and cell cycle progression in Chinese hamster ovary cells. Exp Cell Res 143: 377-382, 1983
14. Utakoji T, Kosoda K, Umezawa K, Sawamura M, Matshushina T, Miwa M, Sugimura T: Induction of sister chromatid exchanges by nicotinamide in Chinese hamster lung fibroblasts and human lyphoblastoid cells. Biochem Biophys Res Commun 90: 1147-1152, 1979
15. Oikawa A, Tohda H, Kanai M, Miwa M, Sugimura T: Inhibitors of poly(ADP-ribose) polymerase induce sister chromatid exchanges. Biochem Biophys Res Commun 97: 1311-1316, 1980
16. Hori T-A: High incidence of sister chromatid exchanges and chromatid interchanges in the conditions of lowered activity of poly(ADP-ribose) polymerase. Biochem Biophys Res Commun 102: 38-45, 1981
17. Morgan WF, Wolff S: Induction of sister chromatid exchange by 3-aminobenzamide is independent of bromodeoxyuridine. Cytogenet Cell Genet 38: 34-38, 1984
18. Natarajan AT, Csukas I, van Zeeland AA: Contribution of incorporated 5-bromodeoxyuridine in DNA to the frequencies of sister-chromatid exchanges induced by inhibitors of poly(ADP-ribose) polymerase. Mutation Res 84: 125-132, 1981
19. Morgan WF, Bodycote J, Doida, Fero ML, Hahn P, Kapp IN: Spontaneous and 3-aminobenzamide-induced sister-chromatid exchange frequencies estimated by ring chromosome analysis. Mutagenesis. 1: 453-459, 1986
20. Morgan WF, Cleaver JE: 3-Aminobenzamide synergistically increases sister-chromatid exchanges in cells exposed to methyl methanesulfonate but not to ultraviolet light. Mutation Res 104: 361-366, 1982
21. Park SD, Kim CG, Kim MG: Inhibitors of poly(ADP-ribose) polymerase enhance DNA strand breaks, excision repair, and sister chromatid exchanges induced by alkylating agents. Environ Mutagen 5: 515-525, 1983
22. Schwartz JL, Morgan WF, Weichselbaum RR: Different efficiencies of interaction between 3-aminobenzamide and various monofunctional alkylating agents in the induction of sister chromatid exchanges. Carcinogenesis 6: 699-704, 1985
23. Shall S: ADP-ribose in DNA repair: A new component of DNA excision repair. Adv Radiat Biol 11: 1-69, 1984
24. Berger NA, Sikorski GW: Nicotinamide stimulates repair of DNA damage in human lymphocytes. Biochem Biophys Res Commun 95: 67-72, 1980
25. Berger NA, Sikorski GW, Petzold SJ, Kurohara KK: Association of poly-adenosine diphosphoribose) synthesis with DNA damage and repair in normal human lymphocytes. J Clin Invest 63: 1164-1171, 1979
26. Cleaver JE, Morgan WF: Poly(ADP-ribose) polymerase: a perplexing participant in cellular responses to DNA breakage. Mutation Res 257: 1-18, 1991
27. Sims JL, Sikorski GW, Catino DM, Berger SJ, Berger NA: Poly-Adenosine diphosphoribose) polymerase inhibitors stimulate unscheduled deoxyribonucleic acid synthesis in normal human lymphocytes. Biochemistry 21: 1813-1821, 1982
28. Sims JL, Berger SJ, Berger NA: Poly(ADP-ribose) polymerase inhibitors preserve nicotinamide adenine dinucleotide and adenosine 5'-triphosphate pools in DNA-damaged cells: Mechanism of stimulation of unscheduled DNA synthesis. Biochemistry 22: 5188-5194, 1983
29. Ben-Hur E: Involvement of poly(ADP-ribose) in the radiation response of mammalian cells. Int J Radiat Biol 46: 659-671, 1984
30. Hunting DJ, Gowans BJ, Henderson JF: Specificity of inhibitors of poly(ADP-ribose) synthesis. Effects of nucleotide metabolism in cultured cells. Mol Pharmacol 28: 200-206, 1985
31. Milam KM, Thomas GH, Cleaver JE: Disturbances in DNA precursor metabolism associated with exposure to an inhibitor of poly(ADP-ribose) synthetase. Exp Cell Res 165: 260-268, 1986
32. Cleaver JE, Morgan WF: Poly-ADP-ribose synthesis is involved in the toxic effects of alkylating agents but does not regulate DNA repair. Mutation Res 150: 69-76, 1985
33. Cleaver JE, Milam KM, Morgan WF: Do inhibitor studies demonstrate a role for poly(ADP-ribose) in DNA repair? Radiation Res 101: 16-28, 1985
34. Chatterjee S, Petzold SJ, Berger SJ, Berger NA: Strategy for selection of cell variants deficient in poly(ADP-ribose) polymerase. Exp Cell Res 172: 245-257, 1987
35. Chatterjee S, Hirschler NV, Petzold SJ, Berger SJ, Berger NA: Mutant

cells defective in poly(ADP-ribose) synthesis due to stable alterations in enzyme activity or substrate availability. Exp Cell Res 184: 1-15, 1989

36. Chan JYH, Becker FF, German J, Ray JH: Altered DNA ligase I activity in Bloom's syndrome cells. Nature 325: 357-359, 1987
37. Creissen D, Shall S: Regulation of DNA ligase activity by poly(ADP-ribose). Nature (London) 296: 271-272, 1982
38. Jacobson EL, Meadows R, Measel J: Cell cycle perturbations following DNA damage in the presence of ADP-ribosylation inhibitors. Carcinogeneses 6: 711-714, 1985
39. Jacobson EL, Smith JY, Wielckens K, Hilz H, Jacobson, MK: Cellular recovery of dividing and confluent C3H10T1/2 cells from *N*-methyl-*N'*-nitrosoguanidine in the presence of ADP-ribosylation inhibitors. Carcinogenesis 6: 715-718, 1985
40. Boorstein RJ, pardee AB: 3-Aminobenzamide is lethal to MMS-damaged human fibroblasts primarily during S phase. J Cell Physiol 120: 345-353, 1984
41. Wiencke JK, Morgan WF: Cell cycle-dependent potentiation of x-ray-induced chromosomal aberration by 3-aminobenzamide. Biochem Biophys Res Commun 143: 372-376, 1987
42. Chatterjee S, Trivedi D, Petzold SJ, Berger NA: Mechanism of Epipodophyllotoxin-induced cell death in poly(adenosine diphosphate-ribose) synthesis-deficient V79 Chinese hamster cell lines. Cancer Res 50: 2713-2718, 1990
43. Chatterjee S, Cheng MF, Berger NA: Hypersensitivity to clinically useful alkylating agents and radiation in poly(ADP-ribose) polymerase-deficient cell lines. Cancer Commun 2: 401-407, 1990
44. Chatterjee S, Cheng MF, Berger SJ, Berger NA: Alkylating agent hypersensitivity in poly(adenosine diphosphate-ribose) polymerase deficient cell lines. Cancer Commun 3: 71-75, 1991
45. Spirodonidis CA, Chatterjee S, Petzold SJ, Berger NA: Topoisomerase II-dependent and -independent mechanisms of etoposide resistance in Chinese hamster cell lines. Cancer Res 49: 644-650,, 1989
46. Anhström G, Erixon K: Measurement of strand breaks by alkaline denaturation and hydroxyapatite chromatography. In: Friedberg EC, Hanawalt PC (eds). DNA Repair, A laboratory Manual of Research Procedures, Vol 1, Part B. Marcel Dekker Inc, New York and Basel. 1981, pp 403-418

Molecular and Cellular Biochemistry **138:** 71–75, 1994.

The function of poly (ADP-ribosylation) in DNA breakage and rejoining

Sydney Shall

Cell and Molecular Biology Laboratory, Biology Building, University of Sussex, Brighton, East Sussex, BN1 9QG, England

Abstract

Poly (ADP-ribose) polymerase has an obligatory requirement for DNA strand-breaks in order to show full enzyme activity. Exposure of cells to DNA damaging agents activates this enzyme presumably through the production of DNA strand-breaks, either directly or via cellular enzymes. Recent evidence from manipulations of the cloned cDNA of this enzyme confirm the earlier evidence, obtained using enzyme inhibitors, that this enzyme is involved in DNA excision repair, probably at or near the ligation step. A very unusual human genetic disease has provided direct evidence for a link between the enzyme activities of poly (ADP-ribose) polymerase and of DNA ligase I. There is also some evidence that this enzyme may be involved in other cases of DNA breakage and rejoining, such as homologous and non-homologous DNA recombination, for example, in sister chromatid exchanges, in DNA transfection, in the integration of retroviral proviral DNA and in variable antigen switching in African trypanosomes. (Mol Cell Biochem **138:** 71–75, 1994)

Key words: NAD^+, poly (ADP-ribose) polymerase, 3-aminobenzamide, DNA ligase, DNA repair, DNA breakage and rejoining, DNA transfection, DNA recombination, retroviral infection, sister chromatid exchange, trypanosome

Introduction

Over the last decade it has become firmly established that ADP-ribosylation of chromatin protein(s) is a component of efficient DNA excision repair [1–6]. This conclusion derives primarily from the observations that poly (ADP-ribose) polymerase is sensitive to DNA breakage and that it seems to have a function in the DNA rejoining process. I will review the information that leads to these statements, and in particular, I will describe recent evidence from molecular biology experiments and from experiments with sub-cellular (*in vitro*) systems, in support of these conclusions.

The first and very important observation is that the enzyme poly (ADP-ribose) polymerase is dependent on the presence of DNA nicks (strand-breaks) or double-strand breaks for its full enzyme activity. In the absence of DNA (or of activating forms of DNA) the amount of enzyme activity remaining seems to be about 0.2% of the maximum possible enzyme activity. Thus it seems that this enzyme is crucially dependent on the presence of DNA strand-breaks for its activity. So far, there has not been any report of the activation of this enzyme in the clear absence of DNA strand-breaks. Consequently, we must presume that the physiological function of this enzyme is related to DNA strand-breaks.

Response of poly (ADP-ribose) to DNA damage

When cells or tissues are exposed to high concentrations of alkylating agents, or to large doses of ionizing radiation there is an inhibition of glycolysis, which is due to a loss of NAD^+. The drop in NAD^+ levels is due to activation of nuclear poly (ADP-ribose) polymerase, and not to inhibition of the synthesis of NAD^+[7]. Moreover, it has been directly shown that the loss of NAD^+ is associated with synthesis of the polymer, poly (ADP-ribose). There is also a good correlation between the kinetics of loss of NAD^+ and the appearance of poly (ADP-ribose). All the inhibitors of poly (ADP-ribose) polymerase are effective in inhibiting the loss of NAD^+ after DNA damage, while none of the non-inhibitory analogues have this property.

Address for offprint: S. Shall, Cell and Molecular Biology Laboratory, Biology Building, University of Sussex, Brighton, East Sussex, BN1 9QG, England

It was demonstrated a long time ago that these inhibitors of poly (ADP-ribose) polymerase are also able to inhibit DNA excision repair, while none of the non-inhibitory analogues are able to do this [1]. It is significant that inhibition of DNA repair by the enzyme inhibitors can be observed even when the damage is so small that there is no drop in the level of cellular NAD^+. Thus a drop in the level of cellular NAD^+ is not a necessary condition of the activity of poly (ADP-ribose) polymerase in DNA repair. Moreover, it is noted that enzyme inhibitors which prevent the drop in cellular NAD^+ levels after DNA damage, also increase the level of cell killing. This implies that it is the direct effect of the DNA damaging agent on the DNA that is responsible for cell killing and not the secondary effect on the levels of cellular NAD^+. At very high levels of DNA damage it is very clear that there may be a very substantial loss of cellular NAD^+, even to a complete loss of all the coenzyme; whether this contributes to cell death in these cases is unclear. It is also unclear whether loss of NAD^+ is a significant contributory element in the cell death that follows massive DNA degradation in apoptosis.

Further independent evidence for the involvement of poly (ADP-ribose) polymerase in DNA excision repair comes from experiments in which the level of NAD^+ in the cells is experimentally manipulated. Mammalian cells can be grown in media with very low concentrations of nicotinamide. Under these special conditions the steady-state level of intracellular NAD^+ is much lowered. It is possible by careful control to achieve cell growth with 10% to 20% of the normal NAD^+ level. Such cells are viable and can grow almost as fast as the control cells. However, it turns out that such cells are markedly deficient in DNA excision repair [1]. This association between DNA excision repair and the level of cellular NAD^+ is further evidence for the involvement of poly (ADP-ribose) polymerase in DNA excision repair.

Further useful information may be derived from a consideration of which step in DNA repair requires the presence of an active poly (ADP-ribose) polymerase. The appearance of DNA strand-breaks early in the process of DNA repair does not require an active enzyme; the strand-breaks occur even when the enzyme is inhibited. Further, removal of methylated bases also does not seem to require an active enzyme, because here too the rate of removal of lesions from the DNA is not affected by inhibition of the enzyme. The next step is DNA repair synthesis; the data regarding this step have led to some confusion although the data are reasonably clear. At moderate doses of damage there is neither an increase nor a decrease in the level of repair replication as assessed by two different laboratories using quite different assays of repair replication [8,9]. At very high doses of DNA damaging agents, the enzyme inhibitors inhibit DNA repair to the extent that extensive degradation of the DNA occurs, generating many new breaks. These cells are able to execute these enzymic steps although they will never again divide because their DNA becomes extensively degraded. It seems that there are attempts at repair at these sites, whose precise origins are unknown but which only occur in cells with very extensive damage and an inability to properly achieve repair. Thus the evidence indicates that the site of action of the inhibitors of poly (ADP-ribose) polymerase in DNA repair are at some stage after the DNA repair synthesis step; this means at or near the ligation step.

Manipulations with cloned genes

More recent molecular biology experiments have consistently confirmed the older experiments described above. HeLa cell lines containing antisense poly (ADP-ribose) polymerase genes driven by a mouse mammary tumour virus promoter have been established [10]. Without activation of the promoter the transfected cells seemed to be healthy. Anti-sense RNA appeared in the cells when the viral promoter was activated with dexamethasone. After 72 hours there was about 10 to 20% of the original enzyme protein or enzyme activity remaining in the cells. In these enzyme-depleting cells, the ability to rejoin DNA strand-breaks was sharply decreased, although the DNA strand-breaks were eventually rejoined; this response is identical to that previously reported when enzyme inhibitors were used to inhibit poly (ADP-ribose) polymerase. In these enzyme-depleted cells there were also other alterations which are interesting; cell morphology was altered, with multinucleated syncytia evident after 72 hours. This work has also directly confirmed the previous observations that the increase in enzyme activity as a consequence of DNA damage is solely due to activation of preformed enzyme protein and not at all due to induction of new transcription.

In a related approach [11], the isolated DNA-binding domain of poly (ADP-ribose) polymerase was transiently over-expressed in mammalian cells (see Chapter IIA, de Murcia *et al*). This isolated DNA-binding domain was shown to inhibit the endogenous enzyme in the host cells. A similar inhibition was observed when these cells were treated with an alkylating agent to induce DNA strand-breaks. In an extension of this work, these two groups have overproduced the 46kDa DNA-binding domain of the human poly (ADP-ribose) polymerase in E. coli, in CV-1 monkey cells or in human fibroblasts. They found a positive correlation between the *in vitro* DNA-binding capacity of the recombinant polypeptides and their inhibitory effect on the enzyme activity which was stimulated by an alkylating agent. In addition, when the wild-type DNA-binding region of the enzyme was overproduced in living cells, it blocked unscheduled DNA synthesis (this is repair synthesis) induced by alkylating agents but not by ultra-violet irradiation. Their experiments with inactivating mutations in the various zinc fingers, has implicated the second zinc finger of the enzyme in the binding to single-strand breaks in DNA. Their results have also emphasised the involvement of poly (ADP-ribose) polymerase in DNA excision repair.

DNA repair in cell extracts

Further support for the conclusion that poly (ADP-ribose) polymerase plays some part in DNA repair has come from experiments with cell extracts [12,13]. A human cell extract system which is able to repair some of the DNA damage induced by ionizing radiation has been shown to be dependent on the presence of NAD^+. When the extract was passed over a DNA column to remove the poly (ADP-ribose) polymerase, then the reaction was no longer dependent on NAD^+. The re-addition of purified enzyme protein re-established the NAD^+ dependence. An interesting observation from this work was that the repair in the absence of both NAD^+ and the enzyme was the same as in the presence of both components. Predictably, the presence of the enzyme protein alone was inhibitory; this is analogous to the experiment with the isolated DNA-binding domain of the enzyme. Both procedures show inhibition presumably because in both cases the DNA strand breaks are bound to poly (ADP-ribose) polymerase molecules which cannot be easily removed.

DNA ligase I and 3-aminobenzamide

Perhaps the most convincing evidence connecting poly (ADP-ribose) polymerase and DNA repair has come from a detailed study of a rare human genetic disease; this work has also produced compelling evidence associating poly (ADP-ribose) polymerase and the DNA ligation reaction. The disease was identified in a single patient who presented with symptoms of hypogammaglobulinaemia; this patient had a slightly dwarfed appearance with venous dilations and a selective IgA deficiency. The patient had recurring pulmonary infections, several features suggesting premature ageing and a possible slow-growing splenic lymphoma. There were no neurological abnormalities.

A cell strain was established from this patient and this cell strain was called 46BR [14]. This cell strain is unique in that it shows hypersensitivity to the lethal effects of a wide range of DNA-damaging agents. Thus, it is hypersensitive to both gamma and UV irradiation, as well as to dimethyl sulphate, methyl methane sulphonate, ethyl methyl sulphonate, N-methyl-N'-nitro-N-nitroso-guanidine and N-methyl-N-nitrosourea, but very significantly it is not sensitive to N-ethyl-N-nitrosourea [14]. It is very unusual to find a human DNA repair defect that is hypersensitive to a large variety of DNA-damaging agents, because there are a variety of particular repair pathways for different types of DNA lesion. This unusual feature immediately suggested that a late step in the DNA repair pathway was involved. This was quickly confirmed when it was shown that in this cell line there is a DNA ligation defect [15]. Finally, and very remarkably, this cell line was found to be hypersensitive to 3-aminobenzamide in the absence of any DNA-damaging agent. Thus, this cell line shows three important features; it is hypersensitive to a broad range of DNA damaging agents, it has a DNA ligation defect and it is excessively sensitive to the presence of inhibitors of poly (ADP-ribose) polymerase. The explanation for this phenotype was discovered when the DNA ligase I genes in this cell line were sequenced. It was observed that both alleles of the DNA ligase I gene in this individual carry independent, different mutations with the result that one allele is completely inactive and the other allele seems to have only about 5% of the wild-type DNA ligase activity. A mutant DNA ligase activity in this cell line seems to be the explanation for its unusual phenotype [16]. In particular, it seems that the hypersensitivity of this cell line is due to the mutations in the DNA ligase I gene. This inference has recently been formally proven by re-introducing a wild-type DNA ligase I gene into this cell line [17]. This manipulation resulted in a substantial recovery of resistance to alkylating agents and to a complete loss of the sensitivity to 3-aminobenzamide. This experiment clearly formally associates 3-aminobenzamide with a defect in DNA ligase I. It is now extremely difficult to avoid the conclusion that 3-aminobenzamide by its inhibition of the activity of poly (ADP-ribose) polymerase in some way modulates the activity of DNA ligases. In our original proposal that poly (ADP-ribose) polymerase may work in this way, we associated the polymerase with DNA ligase II, because it was generally believed in those days that the latter enzyme was the DNA ligase involved in DNA repair. Newer data make clear that both DNA ligase I and DNA ligase II are able to participate in DNA ligation reactions during DNA repair. It remains to be established exactly what the functions and contributions of the two ligases are in DNA repair.

Other possible functions of poly (ADP-ribose) polymerase in DNA breakage and rejoining

The conclusion, which now seems to be compelling, is that poly (ADP-ribose) polymerase plays some part in DNA repair, at or near the ligation step. This conclusion implying that the enzyme participates by recognizing DNA strand-breaks and then by modulating in some way the rejoining process is also consistent with further data suggesting that this enzyme may have some role in other situations in which DNA strand-breaks are generated and are then rejoined. For example, it is well established that interference with the enzyme generates sister chromatid exchanges in undamaged cells. Sister chromatid exchanges are examples of homologous recombination. Furthermore, if the same experiment is performed after DNA damage, then, in addition to an increased frequency of sister chromatid exchanges, there is also observed a substantial frequency of chromosomal abberations, which are of course examples of non-homologous recombination. A further example of non-homologous recombination in which the enzyme seems to have a part is in the process of DNA transfection. It has been observed that inhibitors of the enzyme are able to interfere significantly with the process of DNA transfection. In some cell types there is observed an almost complete inhibition of DNA transfection, and this was shown to be due to inhibition of the step at which the transfected DNA is integrated into the chromosomal DNA, this means at a step at which there is DNA strand-breakage and rejoining [18]. Furthermore, it has been observed that integration of retroviral proviral DNA into the chromosome may be prevented by the inhibitors of poly (ADP-ribose) polymerase, as well as by the standard molecular biology procedures of generating anti-sense RNA. Gene amplification is modulated by poly (ADP-ribose) polymerase. Finally, as another example of homologous recombination I can mention the interference of Variable Antigen Type switching which occurs in African trypanosome spp. This process has been demonstrated to be due in some cases, to gene conversion events; in these cases inhibition of poly (ADP-ribose) polymerase substantially decreases the frequency of successful surface glycoprotein switching in these organisms. I would comment in an aside that these African and South American trypanosome spp. which belong to the class of kinetoplastidae are possibly the most ancient (or youngest?) organisms in which poly (ADP-ribose) polymerase has been unequivocally demonstrated.

Conclusion

In conclusion then, I find that there is compelling evidence for a role of poly (ADP-ribose) polymerase in DNA repair, probably at the ligation step. Further, there is suggestive evidence that this enzyme might also have a role to play in other situations where there are DNA strand-breaks and subsequent rejoining.

It remains to be established exactly how this enzyme participates in these DNA reactions. Perhaps this information will only be forthcoming from a combination of reverse genetic experiments with animals in which both alleles of this enzyme have been deleted and from cell extracts in which a detailed biochemical analysis can be made of the components necessary for the execution of proper DNA repair reaction. There is still much to be learned about the turnover of the polymer, about the kinetics of the reaction and in particular concerning the identification of the significant proteins which are modified by polymer in the course of the DNA repair reaction. And ultimately it will be gratifying to understand the significance of the presence of the polymer poly (ADP-ribose) since such a highly charged poly-anion in the vicinity of chromatin can only be expected to have a dramatic effect on the structure and function of chromatin.

Acknowledgements

The work in my Laboratory has been supported over many years by the Cancer Research Campaign, the Medical Research Council, the Science and Engineering Research Council and the Royal Society of London.

References

1. Durkacz BW, Omidiji O, Gray DA, Shall S: (ADP-ribose)$_n$ participates in DNA excision repair. Nature (Lond) 283: 593–596, 1980
2. Shall S: ADP-ribosylation, DNA repair, cell differentiation and cancer. In: M. Miwa, S. Shall, T. Sugimura. (eds). ADP-ribosylation, DNA repair and cancer. International Symposium of the Princess Takamatsu Cancer Research Fund. Japan Scientific Societies Press, Tokyo, 13: 3–25, 1983
3. Shall S: ADP-ribose in DNA repair: a new component of DNA excision repair. Adv Rad Biol 11: 1–69. 1984
4. Shall S: Inhibition of DNA repair by inhibitors of nuclear ADP-ribosyl transferase. Nucleic Acids Symp Ser 13: 143–191, 1984
5. Althaus FR, Hilz H, Shall S: (eds) ADP-ribosylation of proteins. Springer-Verlag, Berlin. 1985
6. Shall S: ADP-ribosylation of proteins: a ubiquitous cellular control mechanism. Biochem Soc Trans 17: 317–322, 1989
7. Brightwell MD, Leech CE, O'Farrell MK, Whish WJD, Shall S: Poly (adenosine diphosphate ribose) polymerase in Physarum polycephalum. Biochem J 147: 119–129, 1975
8. Durkacz BW, Irwin J, Shall S: Inhibition of (ADP-ribose)n biosynthesis retards DNA repair but does not inhibit DNA repair synthesis. Biochem Biophys Res Commun 101: 1433–1441, 1981
9. James MR, Lehmann AR: Role of poly(adenosine diphosphate ribose) in deoxyribonucleic acid repair in human fibroblasts. Biochemistry 21: 4007–4013, 1982
10. Ding R, Pommier Y, Kang VH, Smulson M: Depletion of poly(ADP-ribose) polymerase by antisense RNA expression results in a delay in DNA strand break rejoining. J Biol Chem 267: 12804–12812, 1992
11. Molinete M, Vermeulen W, Burkle A, Menissier-de Murcia J, Kupper JH, Hoeijmakers JH, de Murcia G: Overproduction of the poly(ADP-ribose) polymerase DNA-binding domain blocks alkylation-induced

DNA repair synthesis in mammalian cells. EMBO J 12: 2109–2117, 1993
12. Satoh MS, Lindahl T: Role of poly(ADP-ribose) formation in DNA repair. Nature 356: 356–358, 1992
13. Satoh MS, Poirier GG, Lindahl T: NAD(+)-dependent repair of damaged DNA by human cell extracts. J Biol Chem 268: 5480–5487, 1993
14. A biochemical defect in the repair of alkylated DNA in cells from an immunodeficient patient (46BR). Carcinogenesis 4: 559–564, 1983
15. Lehmann AR, Willis AE, Broughton BC, James MR, Steingrimsdottir H, Harcourt SA, Arlett CF, Lindahl T: Relation between the human fibroblast strain 46BR and cell lines representative of Bloom's syndrome. Cancer Res 48: 6343–6347, 1988
16. Barnes DE, Tomkinson AE, Lehmann AR, Webster ADB, Lindahl T: Mutations in the DNA ligase I gene of an individual with immunodeficiencies and cellular hypersensitivity to DNA-damaging agents. Cell 69: 495–503, 1992
17. Somia NV, Jessop JK, Melton DW: Phenotypic correction of a human cell line (46BR) with aberrant DNA ligase I activity. Mutat Res 294: 51–58, 1993
18. Farzaneh F, Panayatou GN, Bowler LD, Hardas BD, Broom T, Walther C, Shall S: ADP-ribosylation is involved in the integration of foreign DNA into the mammalian cell genome. Nucl Acids Res 16: 11319–11326, 1988

Molecular and Cellular Biochemistry **138:** 77–84,1994.

Poly(ADP-ribose) polymerase gene on chromosome 1q: Early role in differentiation linked replication; gene on human chromosome 13q: Marker of carcinogenesis

Mark E. Smulson
Georgetown University Medical School, Dept. of Biochemistry and Molecular Biology, Washington, DC 20007, USA

Abstract

We have focused in part on utilizing molecular aspects of the PADPRP gene sequences located on chromosome 1q as well as on 13 to help clarify the biological role of these genes. With respect to chromosome 1q, stably transfected cell lines have been established in-both HeLa as well as 3T3-L1 preadipocyte cells, which express a reverse complementary transcription product *in vivo* to the exons coded for on the chromosome 1 locus. Also, in the case of the PADPRP-like gene on chromosome 13q, we have fine mapped this sequence and have observed in a linkage of deletion of this region with predisposition to various forms of cancer, especially in the African-American population. (Mol Cell Biochem **138:** 77–84, 1994).

Key words: cancer, African Americans, carcinogenesis, poly(ADP-Ribose) polymerase gene, chromosome 1q, differentiation, chromosome 13q, DNA Repair, cytotoxicity, cell survival

Introduction

In a study of PADPRP deregulation, we established earlier a stably transfected cell line of human HeLa-S3 cells (PADPRP-as[7]) that express PADPRP antisense transcripts driven by a mouse mammary tumor virus LTR promoter (MMTV) in response to glucocorticoid hormone. In spite of the fact that the biological half-life of PADPRP in cells is relatively long (15 hours), we found that induction of the PADPRP antisense transcripts could be adjusted to cause a progressive decrease in both nuclear PDPRP activity as well as content [1].

Results

Time course of lowering of PADPRP activity and nuclear content

The specific activity of PADPRP remained constant in the control cells during 72 hr of culture; Dex *per se* had no effect on the PADPRP activity of the control cells transfected with the MMTV vector alone (Table 1). More significantly, no marked change in PADPRP activity in antisense cells incubated in the absence of Dex for the 72 hr culture period was noted. In contrast, Dex incubation caused a significant inhibition of PADPRP specific activity in the antisense cell culture. The specific activity in PADPRP-as[7] was reduced by 55% and 83% after 48 and 72 h induction of Dex, respectively.

Depletion in nuclear PADPRP content with antisense induction

In control cells no apparent changes in PADPRP content were observed for as long as 72 hr due to incubation with Dex. Additionally, the PADPRP concentration, in the absence of induction, was approximately the same in antisense

Address for offprints: M. E. Smulson, Georgetown University Medical School, Dept. of Biochemistry and Molecular Biology, Washington, DC 20007, USA

and control cells [data not shown; taken from Ref. 1]. In contrast to these data, the enzyme content was markedly reduced by induction of complementary PADPRP mRNA. After 48 hr of induction, only a minute amount of immuno

Table 1. Kinetics of antisense RNA-induced loss of PADPRP activity in HeLa cells incubated with dexamethasone.

Incubation time (h)	PADPRP activity pmol min^{-1}	mg^{-1} protein	Inhibition (%)
	–Dex	+Dex	
Control cells (Vector alone)			
48	440 ± 50	430 ± 26	–
72	460 ± 38	450 ± 23	–
Antisense cells (PADPRP-as[7])			
48	430 ± 43	190 ± 9	57
72	390 ± 30	70 ± 10	83

Cells (2 × 10^6 to 3 × 10^6/plate) were incubated in the presence or absence of dexamethasone (1 μM) for the indicated times. Subsequently, cells were washed, detached, sonicated in 0.25 M sucrose buffer containing 50 mM Tris (pH 8.0), 2 mM $MgCl_2$, 1 mM dithiothreitol, and 0.1 mM phenylmethylsulfonyl fluoride as described in Methods. Initial velocity (1 min) assays of [^{32}P]NAD incorporation were performed in triplicate as previously described. The data is expressed as the average ± the standard deviation. Taken from Ref. 1.

logically PADPRP could be detected, and by 72 hr essentially no PADPRP band was visible, consistent with the 83% inhibitory level of activity noted in Table 1.

A marked difference in DNase I sensitivity was observed when comparing nuclear DNA derived from cells with depleted PADPRP content [1]. During the linear portion of digestion (2 min) 30% of DNA was digested from nuclei derived antisense compared to 10% in control nuclei. These observations suggest that chromatin of the two cell lines may be organized differently; future studies are being directed at exploiting the ability to experimentally alter nuclear content of this enzyme to better clarify whether PADPRP represents a significant structural component of chromatin fibers.

Antisense-induced reduction in poly ADP-ribosylation and DNA repair.

At 2.0 mM MMS, DNA damage representing 600 rad-equivalents was observed of which over 90% was repaired by 5 hr. No difference in the *extent* of SSB-repair in the antisense cells, whether induced or uninduced for antisense transcription was observed. Thus, only minute nuclear amounts of PADPRP seem sufficient for SSB repair. The induced PADPRP-as[7] cells showed a significant reduction in the SSB repair rate (Fig. 1). After 10 and 20 min., about 460 rad-equivalents of SSB remained, equal to the initial damage, indicating that no SSB rejoining had occurred during this time period. Even after 45 min of post-damage incubation, 330 rad-equivalents of breaks remained in the antisense cells, representing only 25% of SSB rejoining. However, repair resumed very rapidly from 45 to 90 min. In agreement with the results of others, when endogeneous PADPRP activity was completely inhibited by benzamide (5 mM), no SSB rejoining was observed for up to 45 min [2]. Taken together, our results suggest that PADPRP makes an important contribution to early stages of DNA strand break resolution.

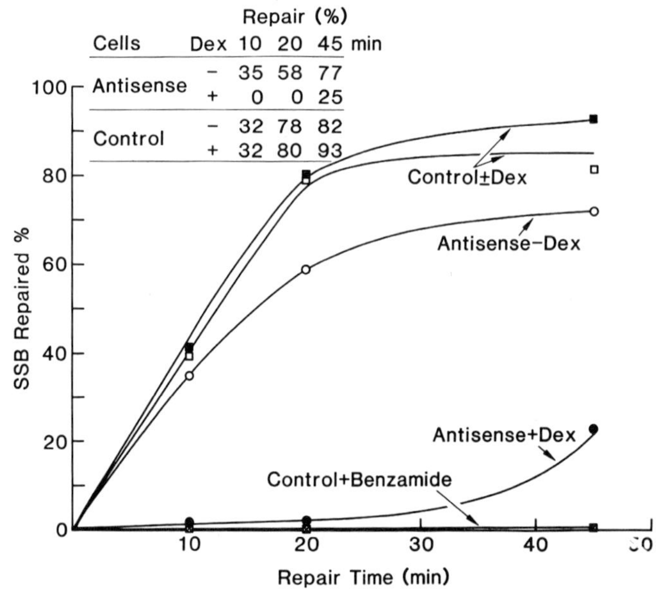

Fig. 1. The Short Time Repair of MMS-Induced SSB in Control and Antisense Cells. Control and antisense cells were treated with MMS following Dex induction. The treated cells were allowed to repair for 10, 20, and 45 min. MMS-induced SSB was measured by alkaline elution. (taken from Ref. 1).

Effect of reduced nuclear PADPRP content on cell survival

To further extend these observations on MMS-induced strand break rejoining, the survival of control and antisense HeLa cells after exposure to various concentration of MMS and their ability to form colonies was determined (data not shown). The data indicated that at a dose of 4 mM MMS, the control cells and uninduced antisense cells showed one log of kill. But, induced antisense cells showed a 10-fold sensitivity to the same dose of MMS. This biological assay further confirms earlier data that reduced PADPRP inhibits DNA repair [3].

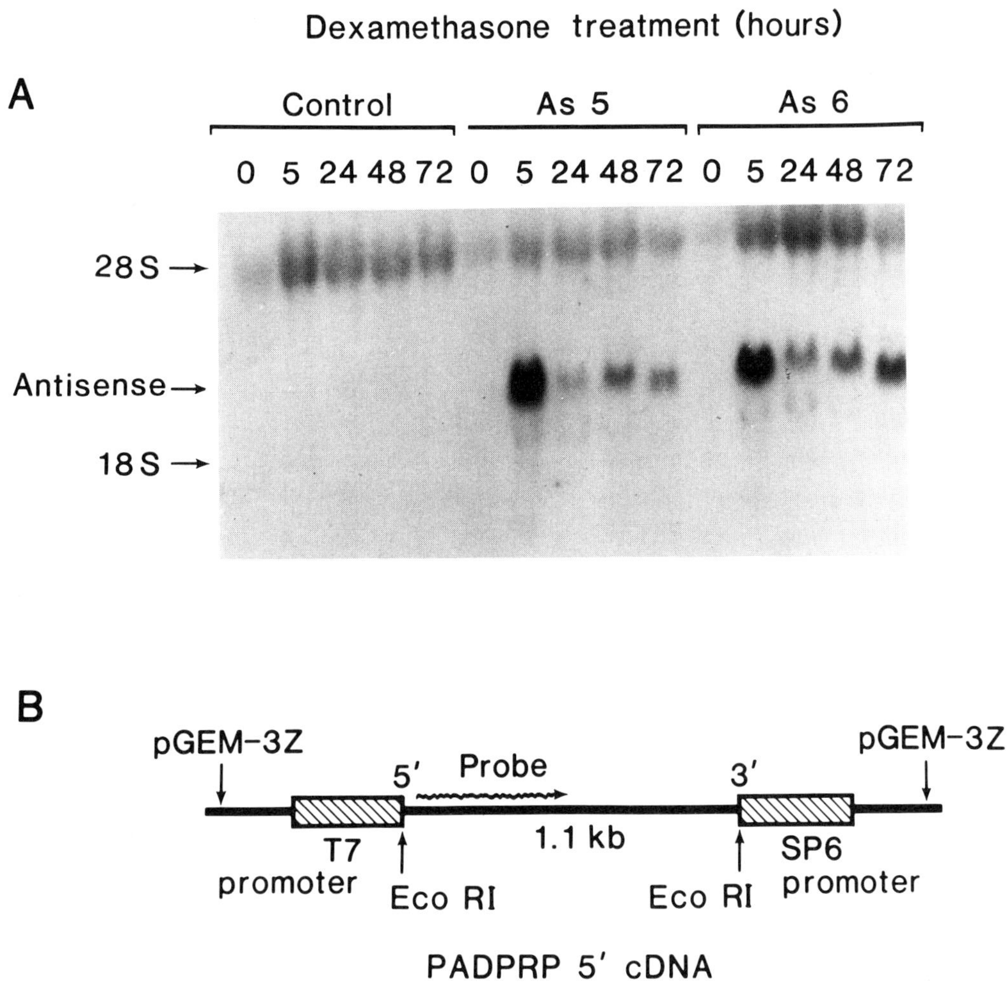

Fig. 2. Expression of PADPRP antisense transcript in 3T3-L1 cells stably transfected with antisense plasmid pMAM-As after induction by dexamethasone (manuscript submitted; taken from Ref. 8).

Studies on 3T3-L1 preadipocytes, differentiation and cytotoxicity

In more recent studies, a 1.1-kb fragment including the DNA binding domain of the murine cDNA as well as encompassing the 5′ portion of the automodification region, was subcloned into the expression vector pMAM-neo under the control of the MMTV promoter [8]. To characterize and quantify the kinetics of antisense PADPRP RNA expression, Northern analysis was performed with control and two clones of transfected 3T3-L1 cells over a broad range of Dex incubation periods. Fig. 2 demonstrates that an antisense PADPRP transcript of 2.4 kb was detected in clones As 5 and As 6 within 5 hours of induction by Dex and continued for more than 72 hours. The induction of this antisense RNA in these cells caused a depletion of the endogenous PADPRP transcripts and PADPRP protein content (not shown). Activity measurements in these cells also indicated greatly reduced activity for the enzyme after 72 hours.

Changes in PADPRP activity and nuclear content during differentiation of 3T3-L1 preadipocytes

Control 3T3-L1 cells were grown to confluence and differentiation was initiated by introduction of insulin, Dex, and MIX. Cells were collected daily and were examined for both morphological and biochemical alterations which are characteristic of adipocytes. The cells which are initially at post-confluency, nevertheless progressed through two additional rounds of cell division, a pre-requisite for differentiation. The differentiating cells become morphologically rounded, and synthesized triglyceride droplets by day 3 of differentiation. Accompanying these changes, an immediate and transient 2–3 fold increase in PADPRP activity was observed during the first 2 days of differentiation (Fig. 3). We confirmed the increase in activity of PADPRP by performing a Western analysis. It was of interest that the enhancement of PADPRP expression occurred at the same time as increased cellular replication, which appears to be re-

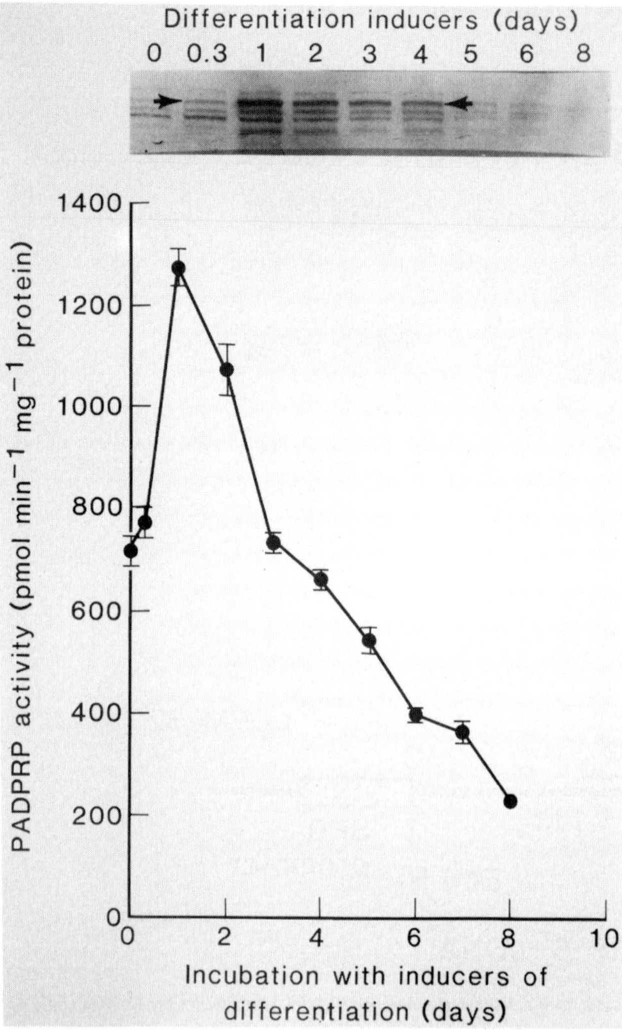

Fig. 3. Poly(ADP-ribose) polymerase activity and enzyme content of control 3T3-L1 preadipocytes during differentiation (manuscript submitted; taken from Ref. 8).

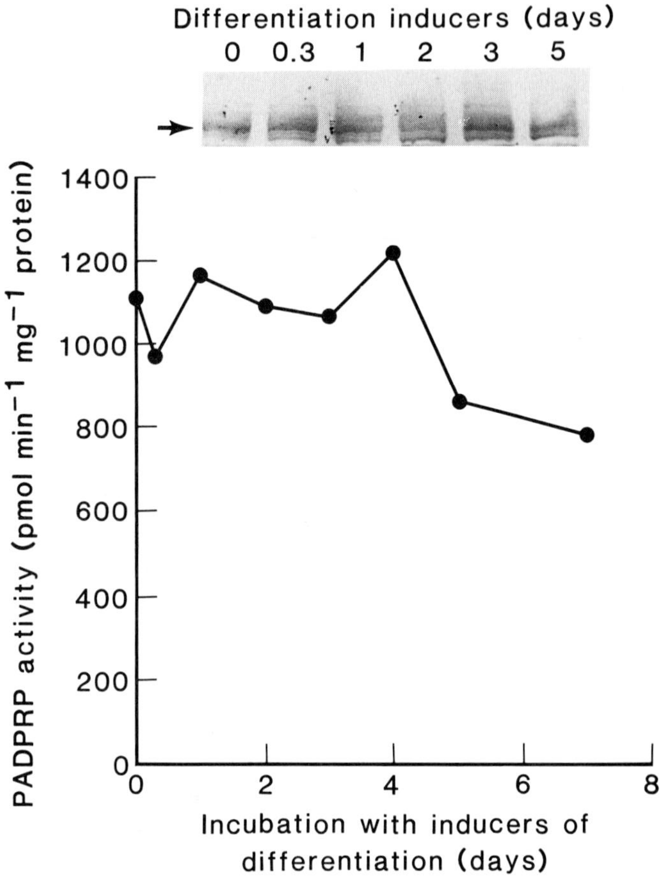

Fig. 4. PADPRP activity and enzyme content of antisense 3T3-L1 clones during differentiation (manuscript submitted; taken from Ref. 8).

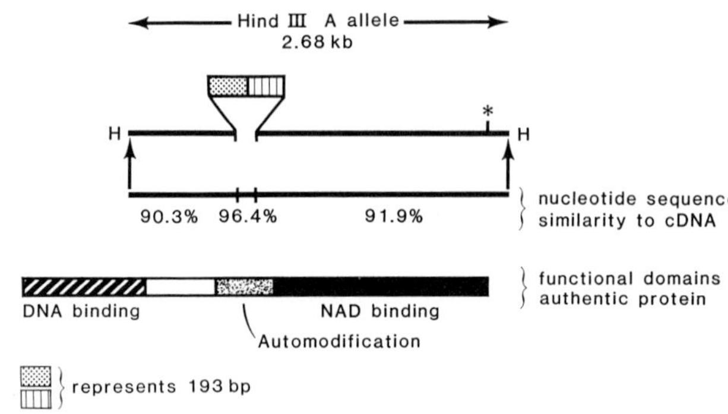

Fig. 5. Schematic representation of the relationship of the isolated *Hin*dIII A allele to the functional domains of the authentic PADPRP protein. The analogous termination codon of the authentic PADPRP protein on the A allele is indicated by an asterisk (*) (taken from Ref. 5).

quired for differentiation.

Effects of reduced PADPRP on preadipocyte growth, morphology and differentiation

The marked increase in both PADPRP content and activity observed during the first two days of differentiation (Fig. 3), did not occur and activity observed during the first two days of differentiation (Fig. 3), did not occur in the antisense cells, and in fact the activity level remained constant (Fig. 4). Additionally, the antisense cells did not exhibit detectable cytoplasmic triglyceride accumulation. These cells with significantly reduced capacity for poly(ADP-ribosyl)ation compared to control did not undergo the normal 2–3 rounds of cell division, required during the initial 1–2 days of differentiation induction (Fig. 5). In fact, cells with reduced PADPRP became easy detached from plates and eventually lost viability. Alternatively, in several instances, the antisense cells progressed through one round of cell division; however, this resulted in even more extensive cytotoxicity and the cells eventually died (Fig. 5). We feel that these latter effects are the result of Okazaki fragments which were not able to be ligated and accumulated in cells due to the low

Table 2. Increased frequency of B allele in Burkitt DNA from different population groups. Distribution of genotypes based upon RFLP banding patterns obtained following restriction of DNA with HindIII

Population	DNA from normal cells with genotype distribution of *AA*	*AB*	*BB*	Total	*B* allele frequency	*BB* genotypes over noncancer control (fold increase)
Black American	15	18	4	37	0.35	
African	23	18	9	50	0.36	
Caucasian American	45	12	2	59	0.14	
Chinese	18	3	0	21	0.07	
	Tumor DNA					
Black	0	11	8	19	0.70	2.8
Caucasian	12	9	2	23	0.28	2.5

Taken from Ref. 4.

Table 3. Allelic zygosity of the chromosome 13 PADPRP sequences in genomic DNA from matched normal and tumor samples

DNA	Genotype *AA*	*AB*	*BB*	Total	*B*-allele frequency	P[1a]	P[2]
Tumor							
B-cell lymphoma	12	7	4	23	0.33	0.0001	
Lung carcinoma	11	9	4	24[b]	0.35	0.001	
Breast carcinoma	34	16	3	53	0.21	0.003[c]	
Myeloid leukemia	11	2	0	13	0.08	0.45	
Colorectal carcinoma	41	15	7	63	0.22	0.0006[c]	
From normal tissue of above individuals with cancer							
Lung carcinoma	4	4	1	9	0.33	0.012	0.88
Breast carcinoma	13	16	2	31	0.32	0.0006	0.395
Colorectal carcinoma	38	20	4	62	0.23	0.023	0.70
From normal tissue of black individuals with various cancers[d]	4	15	9	28	0.59		

[a]P[1] calculations compare *B* allele frequencies of respective tumor or normal DNA from cancer patients with DNA from noncancer individuals (0.12). P[2] calculations compare *B* allele frequencies of matched normal DNA respective tumor DNA. [b]Of the 24 lung tumors, 20 were diagnosed as small cell lung carcinoma while 4 others were classified as non-small lung tumors. [c]Although the results show a statistical significance, we cannot determine the significance of the increase in the *B* allele frequency since although most of these samples were from Caucasians, we do not know the exact racial distribution. [d]Normal DNA obtained from black cancer patients were genotypes for the tumor DNA was not determined.
Taken from Ref. 4.

Table 4. Frequency of the *B* allele in germline DNA from patients with various cancers

Cancer type	Population	Genotype *AA*	*AB*	*BB*	Total	*B* allele Frequency	*p**
Multiple	Black	5	11	15	31	0.66	<0.003
Myeloma	White	28	7	2	37	0.15	0.968
Prostate	Black	1	3	5	9	0.72	0.010
	White	10	6	0	16	0.19	0.653
Lung	Black	1	5	3	9	0.61	0.080
	White	4	4	1	9	0.33	0.075
Colon	Black	0	8	3	11	0.64	0.033
Breast	Black	12	7	2	21	0.26	0.430
Control:	Black	15	18	4	37	0.35	
noncancer population	White	45	12	2	59	0.14	

Germline DNA was restricted with *Hin*dIII and hybridized to full-length PADPRP cDNA after Southern blotting. **P* calculations compare the *B* allele frequency from cancer patients with noncancer population of the same racial group.
Taken from Ref. 5.

PADPRP content. Accordingly, this cell system should prove to be a good model system to better ascertain the in influence of PADPRP on DNA strand break rejoining reactions [8].

Strategies to investigate the PADPRP-like sequences on chromosome 13q

In the course of analyzing potential rearrangements within the PADPRP gene in tumor cells, a simple two allele (A/B) polymorphism localized to chromosome 13q33-qtr was observed . A two-fold increase in the frequency of the B allele was also noted in tumor DNA from patients with Burkitts lymphoma, B-cell follicular lymphoma and lung carcinoma when compared to germline DNA in a noncancer population (4). RFLP was identified as a 2.7 kb or 2.5 kb *Hind*III fragment after hybridization to full-length human PADPRP cDNA. These fragments were thought to originate from either a processed PADPRP pseudogene or a gene with extensive identity to PADPRP. Our studies on the PADPRP-like exon sequences on chromosome 13q have centered mainly on correlations with a deletion in this region (i.e. 13 34qtr) and predisposition to cancer [4]. The initial analysis has revealed the following: 1) as shown by the composite data of Tables 2 and 3, the frequency of the deletion was high in multiple types of cancers, and also in the normal tissue of the same individuals; 2) instances of loss of heterozygosity were observed (data not shown), suggesting a linkage to a new tumor suppressor (4); and 3) the frequency of the deletion, detected in normal tissue, is 3-fold higher in non-cancerous African–Americans compared to Caucasians as demonstrated by the data in Table 2.

Analysis of the PADPRP sequences on chromosome 13

An important aspect of the earlier study as noted above was the observation that at least one copy of the B allele was always present in tumor DNA from patients with endemic Burkitts lymphoma, and the frequency of the B allele was at least two-fold higher than in the black, noncancer populaton. Since the incidence of multiple myeloma, prostate and lung cancer is higher in the US black population, we have analyzed the B allele frequency in germline DNA to determine whether the PADPRP gene correlates with a polymorphic susceptibility to these diseases (Table 4). For multiple myeloma and prostate cancer, an increased frequency of the B allele appears to be striking only in black patients. In contrast, the distribution of the B allele in germline DNA does not differ among white patients with these diseases when compared to the control group (Table 4). These observations suggest that the PADPRP polymorphism may provide a valid marker for a predisposition to these cancers in black individuals. To determine the genomic structure of the polymorphic PADPRP sequences, a 2.68 kb *Hind*III clone was isolated and sequenced from a chromosome 13-enriched library [5]. Sequence analysis of this clone (A allele) revealed a close sequence similarity (91.8%) to PADPRP cDNA (1q42) and an absence of introns, suggesting that the gene on 13q exists as a processed pseudogene.

PADPRP gene on chromosome 13 has a duplicated 193-bp region

A 193 bp conserved duplicated region within the A allele [5] has now been identified as the source of the polymorphism (Fig. 6). The nucleotide differences between the PADPRP gene on chromosome 13 and related PADPRP genes were exploited to develop oligonucleotides which can detect the difference between the A/B genotypes in a PCR).

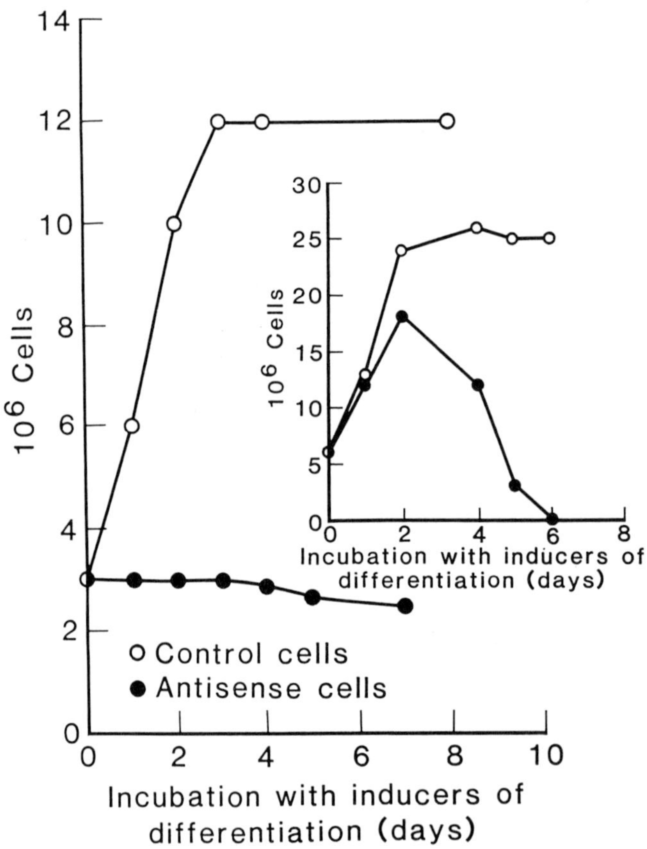

Fig. 6. Cell proliferation in control and pMAM-As cells after induction of differentiation (manuscript submitted; taken from Ref. 8).

This PCR assay offers the opportunity for anlyzing additional black cancer patients to determine how the PADPRP processed pseudogene, or an unidentified gene which

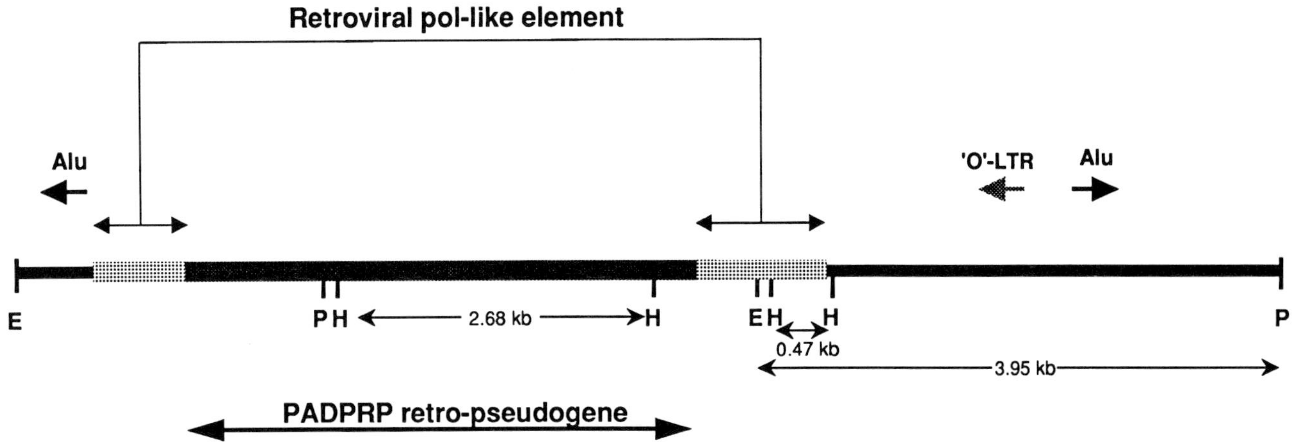

Fig. 7. Schematic representation of the *Eco*RI-*Pst*I genomic region. The complete nucleotide sequence was determined from two overlapping clones (*Eco*RI 5.3 kb and *Pst*I 7.4 kb). Restriction enzyme sites are denoted as E, *Eco*RI; H, *Hin*dIII; P, *Pst*I. Alu and 'O'-like repetitive elements are shown with the arrowhead indicating the orientation of each sequence. The fragments used as hybridization probes (2.7 kb *Hin*dII, 0.47 kb *Hin*dIII, 3.95 kb *Eco*RI-PstI) are indicated (manuscript submitted; taken from Ref. 7).

cosegregates with the PADPRP gene might be involved with the development of malignancy. Taken together, our sequence data suggest that the integration of the PADPRP (cDNA-like) sequences at the 13q33-qter locus occurred before the human/great ape divergence, and we recent data suggest that a PADPRP processed pseudogene similar to that found in man is present in gorillas; the sequences on other primates such as chimps are not as strongly similar to the human gene [6].

The polymorphic poly(ADP-ribose) polymerase gene in humans interrupts an endogenous reverse transcriptase-like element on 13q34

A recently constructed chromosome 13 enriched library was recently used to isolate the entire poly(ADP-ribose) polymerase (PADPRP) retro-pseudogene on 13q34 [7]. The complete PADPRP sequences 3.77 kb (3808 bases) were determined and found to resemble an intronless cDNA including 137 bases of the 5' untranslated region and a short poly (A) tail. An endogenous *pol*-like retroviral element (1.53 kb) was identified in which the DNA sequences were interrupted by the polymorphic PADPRP retro-pseudogene (Fig. 7).

Concluding remarks

The 13q34 locus appears to be a recipient genomic site for retroposons. We infer that this region may be prone to recombination, as retroposition requires a chromosome break followed by integration of the retroposed fragment. The PADPRP retro-pseudogene represents a polymorphic marker associated with a predisposition to cancer. It is possible that the PADPRP sequences may co-segregate with an unknown gene which is involved in the development of malignancy.

It is also interesting that data described earlier in this review show that consistent with the work of a number of other colleagues in this field, PADPRP has a major functional role in DNA strand break rejoining events, which might be associated with such above-mentioned transpositions. Whether these two interesting facets of PADPRP genes on chromosome 1 and 13 are related is as yet unknown.

Acknowledgements

This work was supported in part by the following grants: CA 13195, NIH; F49620-92-J-0242, U.S. Air Force Office of Scientific Research; CA 25344, NIH; and DAMD 17-90-C-0053), U.S. Army Medical Research and Development Command.

The author would like to acknowledge the work of Kishor Bhatia, Ruchuang Ding, Veronica Kang, and Deborah Lyn for major contributions to the data in this review.

The author would also like to thank the following colleagues who participated in many of the experiments outlined above: Francisco Barriga, James Berenson, George Bonney, Barry Cherney, Jeffrey Cossman, Maria DeBernardi, Larry Deaven, Barry Gause, Konrad Huppi, Nancy Istock, Bruce Johnson, Marc Lalande, Alan Lichtenstein, Shelley Lupold, Ian Magrath, Janet Neequayi, James Ntambi, Yves Pommier, Dean Rosenthal, and Edward Sausville.

References

1. Ding R, Pommier Y, Kang V, Smulson ME: Depletion of poly (ADP-ribose)polymerase by antisense RNA expression results in a delay in

DNA strand break rejoining. J Biol Chem 267: 12804–12, 1992
2. Boorstein RJ, Pardee AB: Factors modifying 3-aminobenzamide cytotoxicity in normal and repair-deficient human fibroblasts. J Cell Phys 120: 335–344, 1984
3. Ding R, Smulson ME: Depletion of Nuclear Poly(ADP-ribose) Polymerase by Antisense RNA Expression: Influences on Genomic Stability and Chromatin Organization. In press, *Cancer Research*, 1994
4. Bhatia K, Cherney B, Huppi K, Magrath I, Cossman J, Sausville E, Barriga F, Johnson B, Gause B, Bonney G, Needquayi J, DeBernardi M, Smulson ME: A deletion linked to a poly(ADP-ribose) polymerase gene on chromosome 13q33-qter occurs frequently in the normal black population as well as in multiple tumor DNA. Cancer Research 50: 5406–13, 1990
5. Lyn D, Cherney B, Lalande M, Berenson J, Lichtenstein A, Lupold S, Bhatia K, Smulson ME: A duplicated region is responsible for the poly(ADP-ribose) polymerase polymorphism, on chromosome 13, associated with a predisposition to cancer. Am J Hum Genet 52: 124–134, 1993
6. Lyn D, Istock N, Smulson ME: Resolution of the hominoid trichotomy by use of the ADP-ribosyltransferase (NAD^+) pseudogene 1. Manuscript submitted, 1993
7. Lyn D, Deaven L, Istock N, Smulson ME: Conservation of sequences between human and gorilla lineages: the ADP-ribosyltransferase (NAD^+) pseudogene 1 and the neighboring retroposons. Genomics 18: 206–211, 1993
8. Kang K, Ntambi J, Rosenthal D, Smulson ME: Expression of poly(ADP-ribose) polymerase antisense prevents increases in enzyme expression and DNA replication during differentiation of 3T3-L1 preadipocytes. manuscript submitted, 1994

Molecular and Cellular Biochemistry **138**: 85–90, 1994.

Poly(ADP-ribose) polymerase activity in intact or permeabilized leukocytes from mammalian species of different longevity

Alexander Bürkle, Marcus Müller, Ivo Wolf and Jan-Heiner Küpper
Deutsches Krebsforschungszentrum, Forschungsschwerpunkt "Angewandte Tumorvirologie", D-69009 Heidelberg, Germany

Abstract

Poly(ADP-ribosyl)ation is a eukaryotic posttranslational protein modification catalyzed by poly(ADP-ribose) polymerase (PARP), a highly conserved nuclear enzyme which uses NAD as substrate. We have previously tested PARP activity in permeabilized mononuclear blood cells (MNC) from 13 mammalian species as a function of the species-specific life span. A direct and maximal stimulus of PARP activation was provided by including saturating amounts of a double-stranded oligonucleotide in the PARP-reaction buffer. The data yielded a strong positive correlation between PARP activities and the species' maximal life spans ($r = 0.84$; $p << 0.001$). Here, we investigated the formation of poly(ADP-ribose) in *living* MNC from two mammalian species with widely differing longevity (rat and man) by immunofluorescence detection of poly(ADP-ribose). The fraction of positive cells was recorded, following γ-irradiation of intact MNC, as a semiquantitative estimation of poly-(ADP-ribose) formation. Human samples displayed a significantly higher percentage of positivity than did those from rats, consistent with our previous results on permeabilized cells. While rat MNC had a higher NAD content than human MNC, the number of radiation-induced DNA strand breaks was not significantly different in the two species. Since poly(ADP-ribosyl)ation is apparently involved in DNA repair and the cellular recovery from DNA damage, we speculate that the higher poly(ADP-ribosyl)ation capacity of long-lived species might more efficiently help to slow down the accumulation of unrepaired DNA damage and of genetic alterations, as compared with short-lived species. (Mol Cell Biochem **138**: 85–90, 1994)

Key words: Poly(ADP-ribose) polymerase, DNA repair, life span, mononuclear leukocytes

Introduction

Poly(ADP-ribosyl)ation is a eukaryotic posttranslational protein modification catalyzed by poly(ADP-ribose) polymerase (PARP; EC 2.4.2.30), a highly conserved nuclear enzyme which uses NAD as substrate [for review, see Refs. 1,2; and the article by T. Sugimura, this volume]. The DNA-binding domain of PARP specifically binds to DNA single- or double-strand breaks, mediated by two zinc fingers [3–5], resulting in a drastic activation of enzyme catalysis. The catalytic center is located in the NAD-binding domain which is separated from the DNA-binding domain by an automodification domain. A large number of studies done in a variety of experimental systems led to the view that poly-(ADP-ribosyl)ation should play a role in DNA repair and other cellular responses to DNA damage. In fact, it was recently shown that overexpression of the isolated PARP DNA-binding domain, known to inhibit polymer synthesis in a *trans*-dominant manner [6], strongly inhibits DNA repair synthesis of living human fibroblasts treated with an alkylating agent, highlighting the potential of PARP to act as critical regulatory component in DNA repair [5].

In the past two decades, evidence has accumulated that the DNA repair capacity of mammalian species is positively correlated with life span [7–9]. Taking this as a base line, we addressed the question, initially tackled by R. Pero's

Address for offprints: A. Bürkle, Deutsches Krebsforschungszentrum, Forschungsschwerpunkt 'Angewandte Tumorvirologie', Abt. 0610, Postfach 10 19 49, D-69009 Heidelberg, Germany

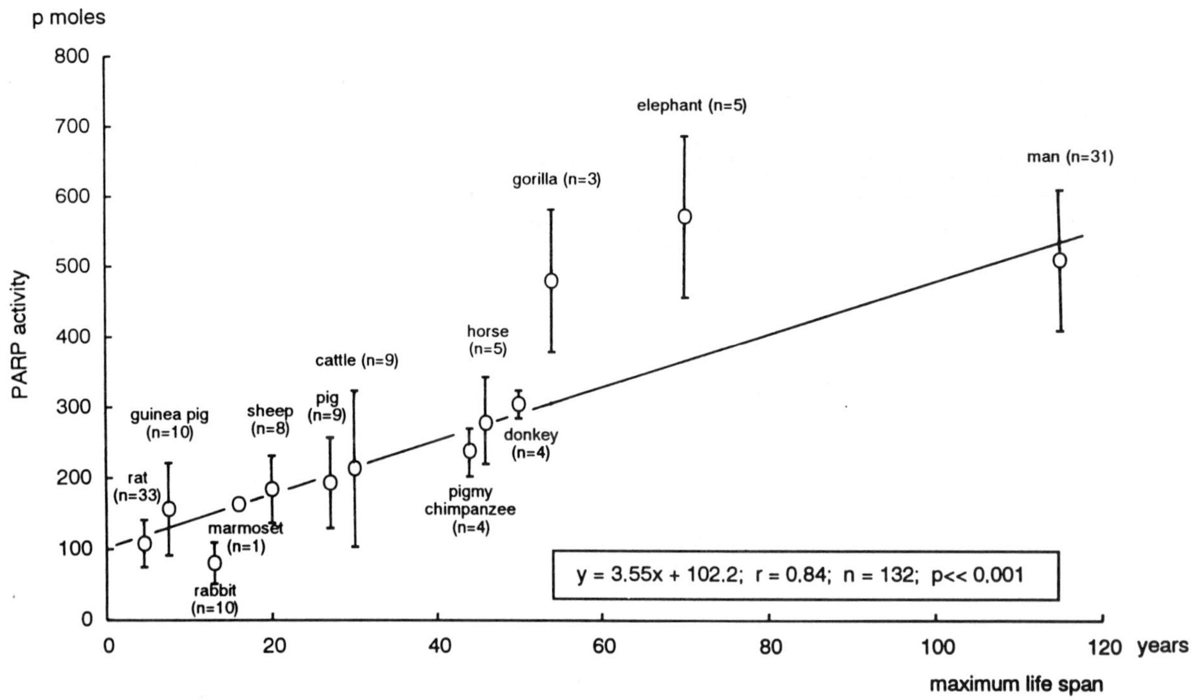

Fig. 1. Correlation between PARP activity and maximal life span of 13 different mammalian species. Percoll gradient-purified, permeabilized MNC were assayed for maximal oligonucleotide-stimulated PARP activity per 2×10^6 cells as described in detail in Ref. 11. Results of 10-min assays are given as mean values of the respective species (n = numbers of individuals) ± 1 SD. Reprinted from Ref. 11.

group [10], of a relationship between poly(ADP-ribosyl)ation and the species-specific life spans of mammals. We tested PARP activity in permeabilized mononuclear blood cells (MNC) of 132 individuals, representing 13 mammalian species, at high NAD concentrations (333 μM) as a function of species-specific life span [11]. A direct and maximal stimulus for PARP activation was provided by including saturating amounts of a double-stranded oligonucleotide in the PARP-reaction buffer [12]. As is shown in Fig. 1, regression analysis of the data yielded a strong positive correlation between PARP activities and the species' maximal life spans (rank correlation coefficient r = 0.84; p<< 0.001). Plotting the data against average life spans resulted in a similar correlation (r = 0.82). Intraspecies comparisons with both rat and human cells from donors of all ages revealed some decline of PARP activity with age in humans (57% decrease projected over maximal life span) and in rats (39% decrease over maximal life span), but these correlations were relatively weak (r = –0.54 and r = –0.34, respectively) [11]. For the species survey shown in Fig. 1, yielding a five-fold difference in PARP activity between man and rat, we had recruited only young blood donors; *i.e.*, no more than 37% of the respective species-specific maximal life span had elapsed. That activity difference, therefore, could not be explained by age effects.

To compare the relative amounts of PARP protein, Western blot comparisons were done on whole-cell lysates of equal numbers of MNC from different species, using a polyclonal antiserum against the PARP NAD-binding domain, whose primary sequence has been extremely well conserved during evolution [13, 14]. In contrast to the enzyme activity data, no obvious correlation was detected between the amount of PARP protein from different species and life span [11]. The combined results suggested that differences in the specific enzyme activity are likely.

Results

Since the data shown in Fig. 1 had been obtained in a subcellular system in which no significant polymer-catabolic activities could be detected [11], it remained to be seen whether in the context of living cells the higher maximal PARP activities of long-lived species would indeed lead to higher levels of poly(ADP-ribose) under conditions of DNA breakage. Alternatively, the rate of poly(ADP-ribose) turnover might be increased due to an enhanced activation of poly(ADP-ribose) glycohydrolase *in vivo* [15]. We therefore set out to study poly(ADP-ribosyl)ation in *living* MNC, based on a polymer-specific immunofluorescence method we had

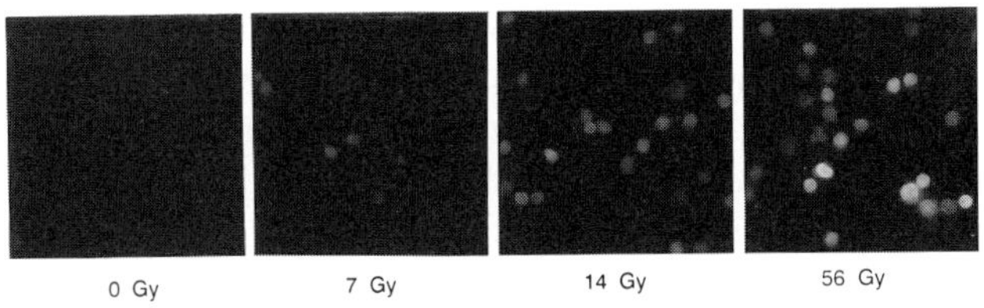

Fig. 2. Indirect immunofluorescence of poly(ADP-ribose) formed in γ-irradiated, living human peripheral blood MNC. Percoll gradient-purified human MNC, adherent to poly-L-lysine-coated coverslips [28], were exposed to γ-irradiation (as indicated) in PBS at ambient temperature at a dose rate of 14 Gy/min. Five minutes after the start of irradiation, cells were fixed in ice-cold TCA and processed for polymer immunofluorescence as described recently [16]. Prior to its use as the first antibody, hybridoma supernatant containing anti-poly(ADP-ribose) monoclonal antibody [17] was concentrated about 10-fold by ultrafiltration through Centriprep cartridges (Amicon, Witten, Germany). 160-fold magnification.

developed recently for the analysis of monolayer cell cultures [16]. Very typically, the fluorescence signals consist of individual, very fine dots, scattered throughout the cell nucleus, that become confluent when present in large number, yielding a 'homogeneous' picture at lower microscopic magnification [16]. As is shown in Fig. 2, this *in-situ* technique may also be applied to cells in suspension, e.g., to MNC. When Percoll-purified human MNC were allowed to settle on poly-L-lysine-coated coverslips, irradiated with γ-rays to generate PARP-activating DNA breaks, fixed in TCA, and processed for immunofluorescence analysis, using monoclonal antibody 10H [17], a dose-dependent increase in nuclear poly(ADP-ribose) signals was clearly visible (Fig. 2); negative cells displayed only a very weak, nonspecific cytoplasmic background staining (invisible on Fig. 2, due to the short photographic exposure time). Even at the rather low dose of 2.8 Gy (= 280 rad), a minority of 'homogeneously' positive cells could be identified (not shown). Therefore, so as to avoid significant NAD depletion which could confound the results, we decided to use 2.8 Gy as a standard dose for a species comparison. Furthermore, since in different species γ-irradiation might lead to different amounts of breaks, due to varying activities of radical scavengers or repair endonucleases [18, 8], it was necessary to measure in parallel the relative number of DNA breaks. In a first attempt, two mammalian species with widely differing longevity were selected, *i.e.*, rat and man. As a semiquantitative estimation of polymer formation, the fraction of 'homogeneously' positive cells was recorded. As is shown in Fig. 3a, human cells exhibited a higher percentage of positivity than did rat cells. The relatively low percentages apparently reflect the 'tip of the iceberg', since the scoring of *all* positive cells (*i.e.*, including those with sparse but specific, dot-like nuclear signals, visible at high maginification) yielded a similar difference between rat and human cells in larger fractions of the cell population (Fig. 3b). Interestingly, even a few of the untreated cells from both rat and man did show poly(ADP-ribose) staining, a finding we have occasionally noted in cultured primary rat pancreatic islet cells [B. Heller, A. Bürkle, E. Fengler, J. Radons, M. Müller, V. Burkart, and H. Kolb, submitted for publication] but not in several

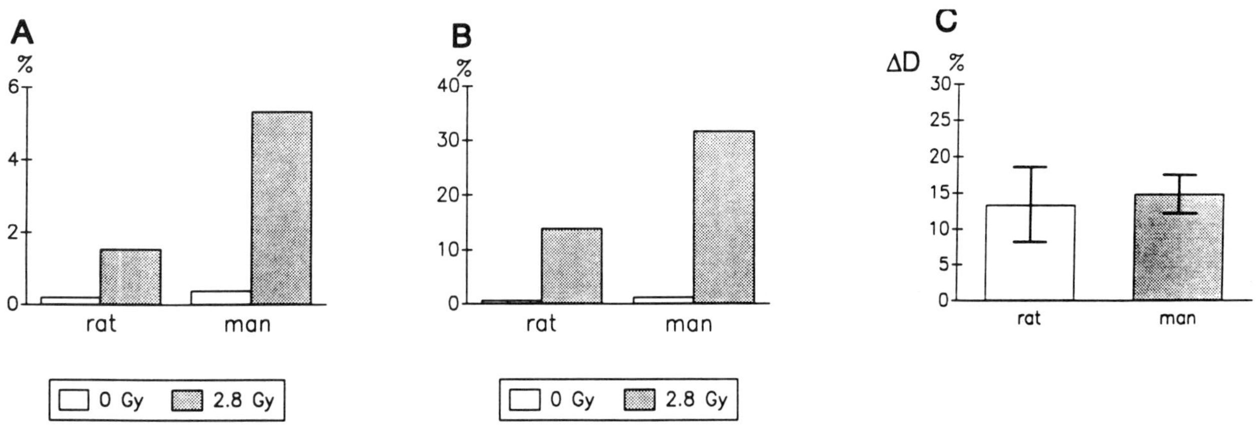

Fig. 3. Percentage of rat and human MNC staining positive for poly(ADP-ribose) after 0 or 2.8 Gy γ-irradiation and determination of DNA strand breaks. Equal numbers of MNC from two Sprague-Dawley rats or two human subjects, respectively, were pooled and processed as in the Legend to Fig. 2. All donors were young males (no more than 37% of the respective species-specific maximal life span elapsed; *cf.* Ref. 11) in apparently good health. The fraction of 'homogeneously' positive (*A*), or of all positive (*B*) cells was determined by examining at least 400 (Panel *A*) or 150 (Panel *B*) cells in randomly selected microscopic fields, the examiner being unaware of the respective experimental conditions. A cell was considered 'homogeneously' positive when the nuclear fluorescence, consisting of many individual dot-like signals, gave a homogeneous appearance at low magnification (*cf.* Fig. 2; [7 Gy]), while 'all positive cells' included those with sparse, but specific, nuclear fluorescence dots visible only at high magnification [*cf.* Ref. 16].

C. In the same MNC pools, DNA strand breaks were determined by the 'alkaline unwinding' technique in triplicates immediately after irradiation, exactly as described by Birnboim and Jevcak [19]. Strandbreaks were determined in irradiated and control samples, respectively, as the percentage of remaining double-stranded DNA in the cell lysates at the end of the exposure to alkali. Results are given as difference between irradiated and control samples ('ΔD' [mean ± SD]), which is a measure of the radiation-induced DNA breaks.

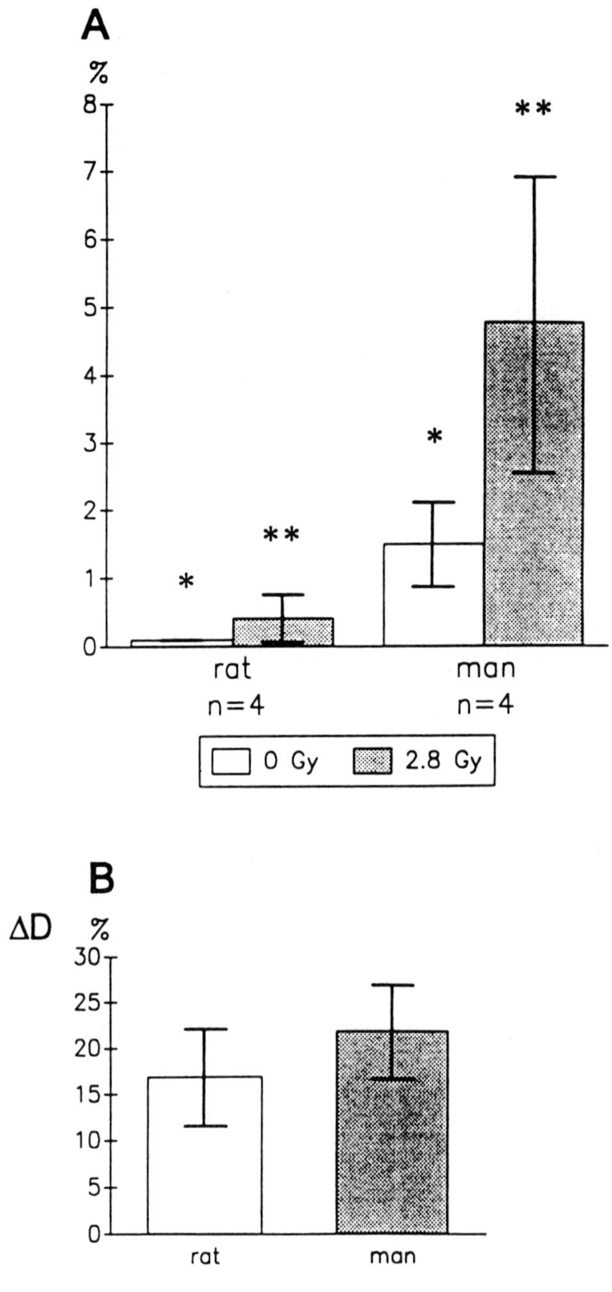

Fig. 4. Percentage of 'homogeneously positive' rat and human MNC after 0 or 2.8 Gy γ-irradiation and determination of DNA strand breaks. *A*. MNC from 4 rats (female) and 4 human donors (3 females, 1 male) were processed separately and evaluated as in Fig. 3. All donors were young (*cf.* Legend to Fig. 3) and apparently healthy. Results are given as mean ± SD or rat or human cells, respectively. The values labelled with single or double asterisks, respectively, yield differences that are statistically significant at $p<0.05$, as determined by t-Test. *B*. DNA strand breaks were determined on pools of rat or human MNC (the same samples as in *A*) in triplicates, as described in the Legend to Fig. 3*C*. Difference is not significant.

different cell lines analyzed so far, using this antibody preparation [16, and unpublished observations]. The different extent of radiation-induced poly(ADP-ribosyl)ation in the two species was apparently not due to different numbers of radiation-induced DNA strand breaks (Fig. 3c), as determined by alkaline unwinding [19].

An independent set of experiments (Fig. 4) confirmed the results shown in Fig. 3 in samples from four other individuals, respectively. Finally, NAD determinations were made with a NAD cycling technique [20] on MNC samples from several individuals. Rat MNC had a significantly higher NAD content than human MNC (mean ± SD from 10 rats: 75.8 ± 38.1 pmoles NAD/10^6 MNC; mean ± SD from 12 humans: 41.0 ± 9.3 pmoles/10^6 MNC; $p<0.05$, as determined by t-Test), making it unlikely that a differential availability of substrate in living cells could be held responsible for the differences in polymer formation.

Discussion

So far, only a few reports have dealt with possible relations between aging or lifespan and poly(ADP-ribosyl)ation: Jackowski and Kun [21] reported that PARP activity was lower in cardiocytes from 90 d-old rats than in those from 5 d-old, despite an increase in the number of DNA breaks. On the other hand, Bizec *et al.* [22] examined PARP activity in bovine eye lens epithelial cells and reported an increase in basal activity with age, in parallel with an increase in DNA strand breakage. Quesada *et al.* [23] investigated basal PARP activity in cell nuclei from rat ventral prostate during aging: A large decline was noted with advancing age, along with a changing pattern of modified acceptor proteins. However, in those studies, maximally stimulated activity was not compared. Dell'Orco and Anderson [24] found a decline of basal PARP activity in aging human fibroblast cultures, but surprisingly not after DNAse I stimulation. Our results are in agreement with and extend the work of Pero and colleagues who described a positive correlation between PARP activities in nucleotide-permeable leukocytes of different mammalian species after high-dose gamma-irradiation of the cells and the species-specific life spans [10]. Due to the experimental design of that study, however, it was not clear whether the reported correlation was direct (*i.e.*, due to a higher enzyme content or a greater enzyme activity) or indirect (*i.e.*, due to other cellular functions).

The preliminary data on poly(ADP-ribose) formation in living MNC shown here (Figs. 3 and 4) are consistent with the results obtained previously in permeabilized cells from species of different life spans, under conditions of oligonucleotide saturation [11]. Remarkably, even at a rather low level of DNA strand breakage, a significant difference in the fraction of polymer-positive cells is detectable between the two species, underscoring the potential biological relevance of this finding. It is obvious that the polymer-fluorescence data need to be corroborated by precise biochemical quantitations of polymer levels under different conditions of DNA breakage [25], and this should be conducted both in a larger number of individual blood donors and in a series of different species. Likewise, it remains to be studied whether or not the same differences between species exist in other tissues. Furthermore, the mechanism(s) underlying the differences in PARP activity between species remain(s) to be investigated. Several possibilities are conceivable: Subtle, but functionally important differences in the primary structure of PARP or in its posttranslational modifications, or putative different accessory chromatin factors could result in different specific PARP activities. Finally, it will be interesting to see whether or not the overexpression of human PARP or, in parallel, of murine PARP in stably transfected rodent cell lines would lead to increased polymer levels in living cells, and if so, what the biological consequences would be in terms of DNA repair, cell survival, and genetic instability under conditions of DNA breakage.

Aging is associated with genetic instability in a variety of organisms studied [26, 27]. It appears very likely that DNA damages generated by endogenous and exogenous agents that continuously attack the genomes of living organisms, (*e.g.*, oxygen free radicals, reducing sugars and other physiological cell metabolites, environmental carcinogens, or irradiations) play an important role in the induction of genetic instability. This view is fully consistent with the correlation between mammalian life span and DNA repair [7–9]: DNA repair would antagonize the accumulation of damages more efficiently in longer-lived species, and thus, genome integrity and stability could be maintained over a longer time scale. This might also be a determinant for the apparent delay in tumor formation in long-lived species as compared with short-lived ones. Since poly(ADP-ribosyl)ation is apparently involved in DNA repair and the cellular recovery from DNA damage, we speculate that the higher poly(ADP-ribosyl)ation capacity of long-lived species might be a factor contributing to (if not controlling) their more efficient defence against the accumulation of unrepaird DNA damage and of genetic alterations.

Acknowledgements

We wish to thank Dr. H. zur Hausen for his continuous interest, encouragement and support; Drs. M. Miwa (Tsukuba, Japan) and T. Sugimura (Tokyo, Japan) for the generous gift of 10H hybridoma cells; Dr.G. de Murcia (Strasbourg, France) for the antiserum directed against the human PARP NAD-binding domain; and Drs. H. zur Hausen and M. Pawlita for critical reading of the manuscript. This work was supported by the Deutsche Forschungsgemeinschaft (grant Bu 698/2-2).

References

1. Althaus FR, Richter C: ADP-Ribosylation of Proteins. Enzymology and Biological Significance. Molecular Biology, Biochemistry and Biophysics, Vol 37. Springer Verlag, Berlin, 1987, pp 1–126
2. Boulikas T: Relation between carcinogenesis, chromatin structure and poly(ADP-ribosylation)*(Review)*. Anticancer Res 11: 489–528, 1991
3. Gradwohl G, Menissier de Murcia J, Molinete M, Simonin F, Koken M, Hoeijmakers JHJ, de Murcia G: The second zinc-finger domain of poly(ADP-ribose) polymerase determines specificity for single-stranded breaks in DNA. Proc Natl Acad Sci USA 87: 2990–2994, 1990
4. Ikejima M, Noguchi S, Yamashita R, Ogura T, Sugimura T, Gill M, Miwa M: The zinc fingers of human poly(ADP-ribose) polymerase are differentially required for the recognition of DNA breaks and nicks and the consequent enzyme activation. J Biol Chem 265: 21907–21913, 1990
5. Molinete M, Vermeulen W, Bürkle A, Ménissier-de Murcia J, Küpper J-H, Hoeijmakers JHJ, de Murcia G: The poly(ADP-ribose) polymerase DNA-binding domain blocks alkylation-incuded DNA repair synthesis in living cells. EMBO J 12: 2109–2117, 1993
6. Küpper JH, de Murcia G, Bürkle A: Inhibition of poly(ADP-ribosyl)ation by overexpressing the poly(ADP-ribose) polymerase DNA-binding domain in mammalian cells. J Biol Chem 265: 18721–18724, 1990
7. Hart RW, Setlow RB: Correlation between deoxyribonucleic acid excision-repair and life-span in a number of mammalian species. Proc Natl Acad Sci USA 71: 2169–2173, 1974
8. Hart RW, Sacher GA, Hoskins TL: DNA repair in a short- and a long-lived rodent species. J Gerontol 34: 808–817, 1979
9. Bernstein C, Bernstein H: Aging, Sex, and DNA Repair. Academic Press, New York, NY, 1991, pp 108–116
10. Pero RW, Holmgren K, Persson L: Gamma-radiation induced ADP-ribosyl transferase activity and mammalian longevity. Mutat Res 142: 69–73, 1985
11. Grube K, Bürkle A: Poly(ADP-ribose) polymerase activity in mononuclear leukocytes of 13 mammalian species correlates with species-specific life span. Proc Natl Acad Sci USA 89: 11759–11763, 1992
12. Grube K, Küpper JH, Bürkle A: Direct stimulation of poly(ADP-ribose) polymerase in permeabilized cells by double-stranded DNA oligomers. Anal Biochem 193: 236–239, 1991
13. Ittel M-E, Garnier J-M, Jeltsch J-M, Niedergang CP: Chicken poly(ADP-ribose) synthetase: complete deduced amino acid sequence and comparison with mammalian enzyme sequences. Gene 102: 157–164, 1991
14. Uchida K, Hanai S, Ishikawa K-I, Ozawa Y-I, Uchida M, Sugimura T, Miwa M: Cloning of cDNA encoding *Drosophila* poly(ADP-ribose) polymerase: Leucine zipper in the auto-modification domain. Proc Natl Acad Sci USA 90: 3481–3485, 1993
15. Alvarez-Gonzales R, Althaus RF: Poly(ADP-ribose) catabolism in mammalian cells exposed to DNA-damaging agents. Mutat Res 218: 67–74, 1989
16. Bürkle A, Chen G, Küpper J-H, Grube K, Zeller WJ: Increased poly(ADP-ribosyl)ation in intact cells by cisplatin treatment. Carcinogenesis (London) 14: 559–561, 1993
17. Kawamitsu H, Hoshino H, Okada H, Miwa M, Mimoi H, Sugimura T: Monoclonal antibodies to poly(adenosine diphosphate ribose) recognize different structures. Biochemistry 23: 3771–2777, 1984
18. Tolmasoff JM, Ono T, Cutler RG: Superoxide dismutase: Correlation with life-span and specific metabolic rate in primate species. Proc Natl Acad

Sci USA 77: 2777–2781, 1980

19. Birnboim HC, Jevcak JJ: Fluorometric method for rapid detection of DNA strand breaks in human white blood cells produced by low doses of radiation. Cancer Res 41: 1889–1892, 1981
20. Jacobson EL, Jacobson MK: Pyridine nucleotide levels as a function of growth in normal and transformed 3T3 cells. Arch Biochem Biophys 175: 627–634, 1976
21. Jackowski G, Kun E: Age-dependent variation of rates of polyadenosinediphosphoribose synthesis by cardiocyte nuclei and the lack of correlation of enzymatic activity with macromolecular size distribution of DNA. J Biol Chem 256: 3667–3670, 1981
22. Bizec JC, Klethi J, Mandel P: Regulation of protein adenosine diphosphate ribosylation in bovine lens during ageing. Ophthalmic Res 21: 175–183, 1989
23. Quesada P, Faraone-Mennella MR, Jones R, Malanga M, Farina B: ADP-ribosylation of nuclear proteins in rat ventral prostate during ageing. Biochem Biophys Res Com 170: 900–907, 1990
24. Dell'Orco RT, Anderson LE: Decline of poly(ADP-ribosyl)ation during *in vitro* senescence in human diploid fibroblasts. J Cell Physiol 146: 216–221, 1991
25. Jacobson MK, Payne DM, Alvarez-Gonzales R, Juarez-Salinas H, Sims JL, Jacobson EL: Determination of *in vivo* levels of polymeric and monomeric ADP-ribose by fluorescence methods. Meth Enzymol 106: 483–494, 1984
26. Slagboom PE, Vijg J: Genetic instability and aging: theories, facts and future perspectives. Genome 31: 373–385, 1989
27. Osiewacz HD: Molecular analysis of aging processes in fungi. Mutat Res 237: 1–8, 1990
28. Harlow E, Lane D: Antibodies. A laboratory manual. Cold Spring Harbor Laboratory, Cold Spring Harbor, NY, 1988, p. 372

Molecular and Cellular Biochemistry **138**: 91–98, 1994.

ADP-ribosylation and gene expression

Peter Zahradka and Lorraine Yau

Division of Cardiovascular Sciences, Department of Physiology, St. Boniface General Hospital Research Centre, University of Manitoba, Winnipeg, Manitoba, R2H 2A6, Canada

Abstract

Gene expression can be defined as the conversion of information existing in a molecule of DNA into a mature RNA or protein product and each step in the process, which requires the concerted action of several macromolecules for completion, may be perturbed by the post-translational modification of specific proteins with ADP-ribose. The participation of poly(ADP-ribose) in the regulation of transcription initiation was examined using cell-free systems for both ribosomal RNA and ribosomal proteins. The presence or absence of poly(ADP-ribose) polymerase did not influence the transcription process. Similarly, under conditions optimal for poly(ADP-ribose) polymerase activity, no change in transcription was observed. A direct contribution of poly(ADP-ribosyl)ation to gene transcription thus could not be detected. In contrast, the addition of 3-aminobenzamide to quiescent hepatoma cells treated with insulin inhibited the stimulation of rRNA synthesis. The high concentrations necessary for this effect suggest that a mono(ADP-ribosyl)ation event participates in the cellular action of insulin. A role in the signal transduction pathway leading to activation of rRNA gene expression has been proposed. (Mol Cell Biochem **138**: 91–98, 1994)

Key words: poly(ADP-ribose), transcription, insulin, 3-aminobenzamide, ribosomal RNA

Introduction

The regulation of gene expression in response to various developmental and physiological signals may involve any one of several distinct steps necessary for synthesis of a gene product. Among the possible events subject to control are (i) decondensation of chromatin, (ii) initiation of transcription, (iii) processing and splicing to form the mature messenger RNA, (iv) translation of the message into protein and (v) stability of either the mRNA or the protein product [1, 2]. In general, selective modulation of either an enzymatic activity or a nucleic acid-protein association can alter the parameters of one of the processes listed in this scheme and thus block generation of the gene product. Alternatively, the signal transduction system that transmits extracellular messages from the cell surface into the nucleus may be dormant or absent as designated by cell type, growth state or physiological environment. This signalling mechanism is vital under rapidly changing conditions where an immediate cellular response is required and for which the generally slower gene activation processes are inadequate.

Phosphorylation of specific proteins can regulate certain reactions in all of the scenarios mentioned above, suggesting that the contribution of post-translational modification to gene activity is highly significant [3, 4]. As with phosphorylation, the covalent attachment of ADP-ribose to proteins, either as single moieties or as a long polymer, has also been shown to regulate a variety of cellular processes [5, 6]. Attention has generally been focused on defining the role of nuclear poly(ADP-ribosyl)ation and cytoplasmic mono-ADP-ribosyltransferases, and this dichotomy of monomeric and polymeric ADP-ribose reactions is also reflected by the mechanism of action affiliated with each process. More specifically, poly(ADP-ribosyl)ation modulates the binding of proteins to nucleic acids while mono(ADP-ribosyl)ation controls enzymatic activity directly [7].

The association between poly(ADP-ribose) and the nucleus provides a possible link between ADP-ribosylation and gene expression, and several investigations using inhibitors of poly(ADP-ribose) polymerase have indicated that

Address for offprints: P. Zahradka, Division of Cardiovascular Sciences, St. Boniface General Hospital Research Centre, 351 Tache Avenue, Winnipeg, MB, R2H 2A6, Canada

ADP-ribosylation reactions may regulate gene activity [8, 9]. Additionally, it has been shown that poly(ADP-ribose) polymerase prevents spurious gene transcription *in vitro*, can be associated with the ribonucleoprotein particles involved in RNA splicing, and modifies RNA polymerases I and II leading to an inhibition of their activity [10, 13]. Based on this information, it may be concluded that poly(ADP-ribosyl)ation participates at several points in the progression from gene to protein; however, the specific targets for this modification have yet to be identified. On the other hand, the alternative possibility that ADP-ribosylation can influence the signalling systems responsible for gene activation has not been considered in great detail. Typical difficulties that may be encountered include the need for intact cell systems and the questionable specificity of the various inhibitors that are available. It must be noted, however, that the regulatory effects of mono-ADP-ribosyltransferases on G-protein function [14] makes it extremely likely that some connection to gene expression does exist.

Gene activation

The assembly of ribosomes in response to a cell growth activation signal requires the synthesis of 4 distinct ribosomal RNAs and over 80 different proteins, and involves the combined action of all three RNA polymerase species [15]. This is, therefore, an ideal system for investigating all of the possible activities contributing to gene expression. In addition, the close connection between cell proliferation and ribosome biogenesis allows the study of hormonal action on gene activity. By using a variety of methods available for studies of ribosome formation, it has been possible to examine some of the possible functions for ADP-ribosylation during gene activation.

Ribosomal RNA

RNA polymerase I synthesizes a single RNA molecule of 45S that is subsequently processed into 3 distinct RNA species of 28S, 18S and 5.8S. There is only one gene which codes for this rRNA precursor, but it is amplified into clusters of many tandem repeats that are physically contained in the nucleolus. Although the synthesis and assimilation of rRNA into the ribosome consists of numerous steps, it is generally transcriptional initiation that acts as the rate limiting step in this process [15].

The *cis* and *trans*-acting factors which control rRNA gene transcription have been identified using specific cell-free

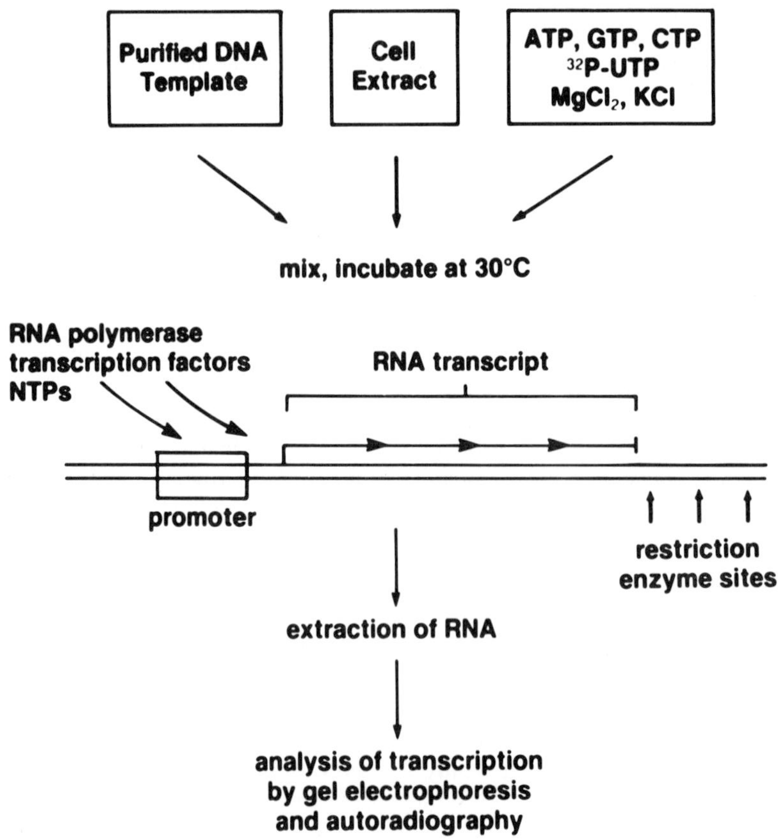

Fig. 1. A schematic representation of a cell-free transcription assay.

systems consisting of a cell extract and truncated plasmids that contain the 5′ region of the 45S rRNA gene. In this assay, specific transcription of the plasmid template results in the synthesis of an RNA molecule of defined size that can be monitored by gel electrophoresis (Fig. 1). When extracts were prepared from proliferating L6 myoblasts which have a high rate of rRNA gene transcription, or differentiated myotubes in which the rate of rRNA gene activity is reduced by approximately 80%, the cell-free assay yielded comparable values to those obtained with intact cells [16]. When the contribution of poly(ADP-ribose) polymerase was assessed, no difference was noted between myoblasts and myotubes [16]. A direct involvement in the regulation of 45S rRNA gene transcription is therefore unlikely. Nevertheless, the polymerase has been demonstrated to act as a suppressor of non-specific transcription [17], presumably by binding to nicks present in the template.

Ribosomal proteins

The coordinate production of r-proteins in response to an increased cellular demand for ribosomes is typically regulated post-transcriptionally [15]. While this information would suggest ADP-ribosylation does not participate in processes which control gene activity at the level of transcription, the availability of a cell-free system for transcription of these genes [18] provides a means to determine whether modification of a component(s) with poly(ADP-ribose) could influence gene activity.

Phosphocellulose chromatography has proven a useful resin for separating factors essential for gene transcription by RNA polymerase II. A transcriptionally active myoblast S100 extract was partitioned by phosphocellulose chromatography with 0.1 M, 0.4 M, 0.6 M and 1.0 M KC1 into four fractions (PI, PII, PIII, PIV, respectively). Measuring

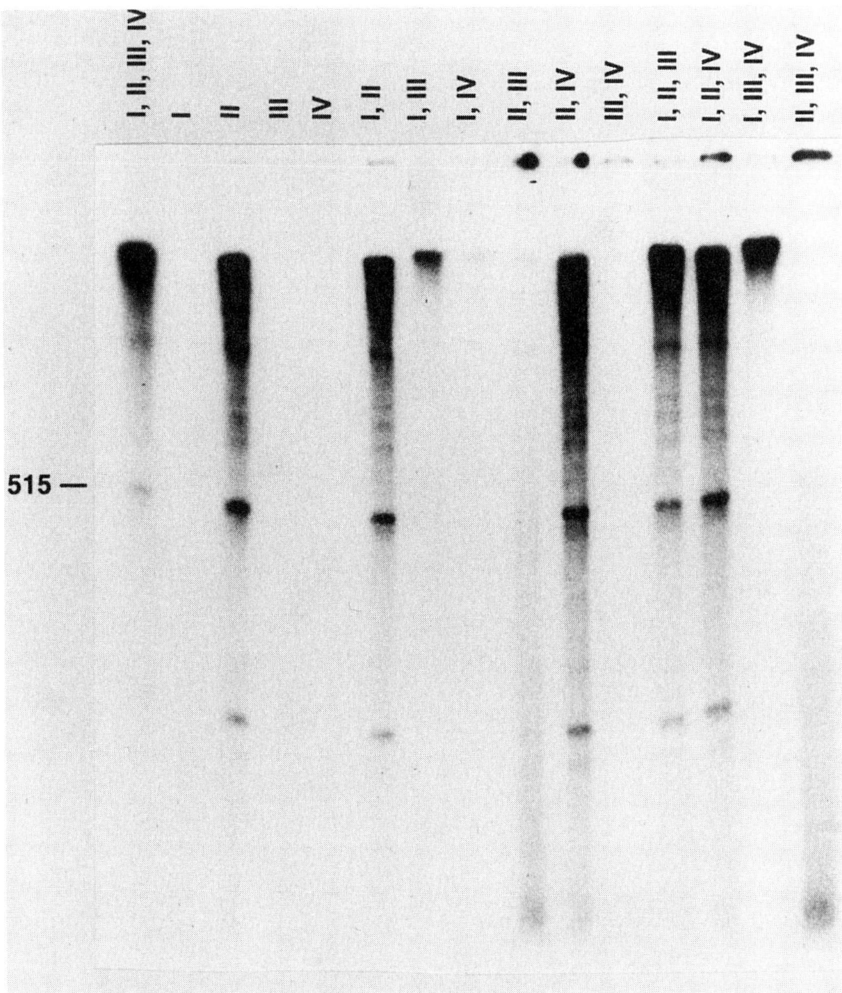

Fig. 2. Transcription *in vitro* of a ribosomal protein plasmid template. An S100 extract was fractionated by phosphocellulose chromatography and the four fractions tested individually and in combination for transcriptional activity using the template p3AR2.8 (truncated with *Hind*III) which produces a specific transcript of 515 nucleotides [16].

the capacity of each column fraction to transcribe the r-protein gene *in vitro* demonstrated that only PII was required (Fig. 2). This fraction also contained all of the RNA polymerase II activity present in the extract. In contrast, the poly(ADP-ribose) polymerase activity was present in PI. Mixing experiments demonstrated that two additional components present in the extract were capable of influencing the transcription rate. When fraction PIII was incubated with PII, no transcript was obtained unless PI was also included (Fig. 2). These results suggest that PIII contained an inhibitor of transcription while a component of PI suppressed the activity present in PIII. Since the unfractionated extract efficiently transcribes the r-protein gene template, sufficient suppressor must be available to mask the inhibitor activity.

To determine whether the PI component was poly(ADP-ribose) polymerase, several experiments were conducted to define the properties of both the suppressor and inhibitor. The effect of PIII was similar whether it was included throughout the incubation period (Fig. 3a) or added for 10 minutes following a 60 minute incubation in its absence (Fig. 3b). The latter result clearly indicated that the RNA transcript rather than the plasmid template was degraded by the PIII activity. Previously, both RNase [19, 20] and DNase [21] activities which affect transcription *in vitro* have

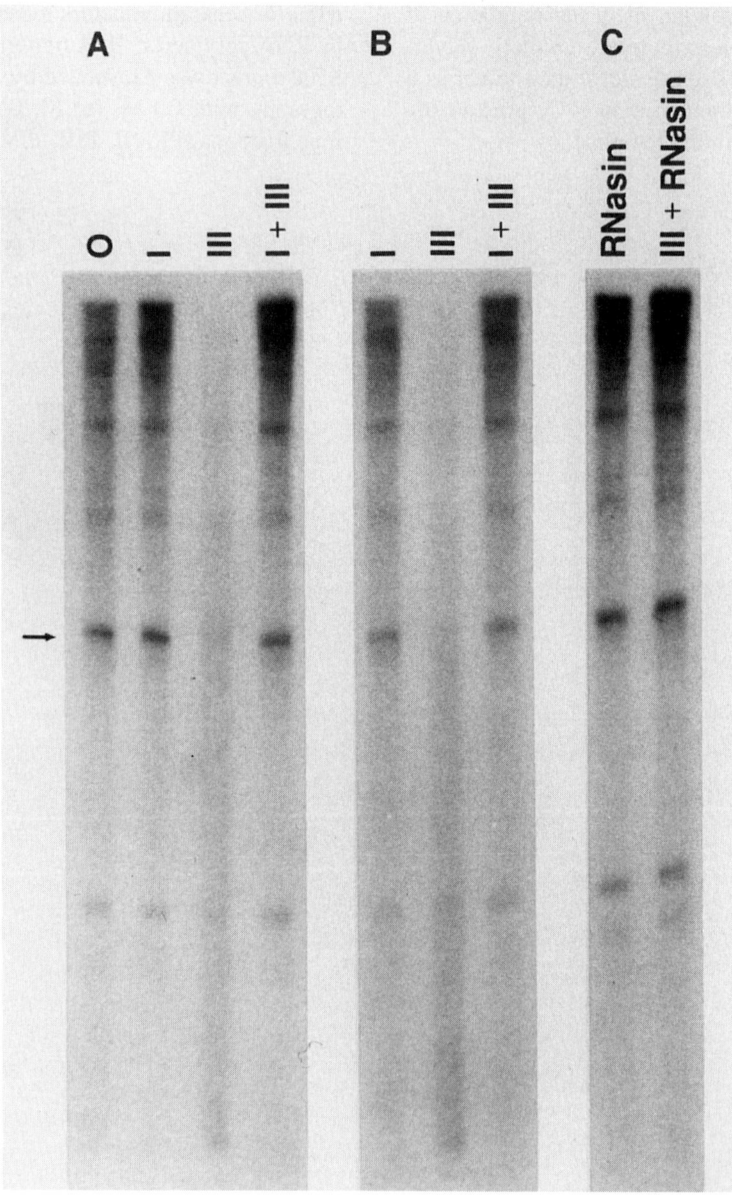

Fig. 3. Analysis of the activities present in phosphocellulose fractions PI and PIII. The r-protein gene template was incubated (A) with PII for 60 min in the absence or presence of PIII, (B) with PII for 60 min and continued for another 10 min following addition of PI and PIII, and (C) with PII for 60 min in the presence of RNase inhibitor or both RNase inhibitor and PIII.

been observed following chromatographic separation of cell extracts. To verify this inference, placental RNase inhibitor (RNAguard, Pharmacia), which alone failed to inhibit transcription (data not shown), arrested the action of PIII (Fig. 3c). Since purified poly(ADP-ribose) polymerase could not achieve the same result (data not shown), no apparent function can be suggested for this enzyme in the cell-free assay. Nevertheless, it may prove a useful peripheral factor for suppressing non-specific transcription events [10].

While a direct involvement in transcription may not exist for poly(ADP-ribose) polymerase, modification of either RNA polymerase II or its associated transcription factors could affect the transcription process [13]. To assess this possibility, cell extracts containing poly(ADP-ribose) polymerase were incubated for 30 minutes in the presence of NAD^+, 3-aminobenzamide or a combination of both. R-protein gene transcription by the extracts was subsequently measured. The data obtained from this experiment (Table 1) suggest there was no significant change in transcription efficiency following treatment with NAD^+. A repeat of the same experiment in which the Mg^{2+} concentration was reduced to minimize polymerase automodification [22] produced similar results (data not shown). Since a similar protocol inhibited RNA polymerase I transcription *in vitro* [23], it is unlikely that poly(ADP-ribosyl)ation has a significant influence on RNA polymerase II mediated gene transcription.

Table 1. Relative r-protein transcription *in vitro* following treatment with NAD^+

Treatment	Transcription Rate (%)
none	100.0
NAD^+	93.9 ± 4.5
3AB	97.9 ± 3.2
NAD^+ + 3AB	97.0 ± 4.0

Myoblast S100 extracts were incubated for 30 minutes at 25°C with either no addition, 0.1 mM NAD^+, 2 mM 3-aminobenzamide or both compounds. Truncated plasmid templates (0.25 μg) were included during the incubation period and these activated poly(ADP-ribose) polymerase to 75% of maximal activity. Transcription assays were conducted by subsequent addition of labelled nucleotide triphosphates and a second incubation for 60 minutes at 37°C. Formation of specific transcript was monitored by polyacrylamide gel electrophoresis and quantitated by scanning densitometry. The values represent averages of triplicate experiments.

Signalling processes

rRNA gene activation

The correlation between cell proliferation and RNA synthesis has been established by numerous investigators [15, 24]. It is well known that mitogens rapidly stimulate proto-oncogene expression through induction of phosphorylation cascades that modulate specific transcription factors. The gene products, in turn, activate the cellular machinery responsible for cell proliferation. Eventually, the cells progress from the quiescent state into S phase to commence DNA replication and subsequently divide when mitosis is reached. These events increase the cellular demand for ribosomes and an increased rate of rRNA production is essential for continuous cell growth. For this reason, activation of rRNA gene transcription is an immediate early response of cells following stimulation with a mitogen [24].

Insulin has been linked to increased ribosome biogenesis in numerous studies and is the sole agent necessary for inducing the proliferation of H4IIE hepatoma cells [25]. To determine whether ADP-ribosylation was coupled to the signalling systems which activated the rRNA genes, quiescent H4IIE hepatomas were stimulated with insulin in the presence of 3-aminobenzamide. Although these cells are not actively growing after a 72 hour period in serum-reduced medium, the typical housekeeping processes remain operating. Thus RNA synthesis continues at a significant rate. Nevertheless, re-entry into the cell cycle following addition of insulin results in an enhanced rate of synthesis which approaches 200% of that observed in quiescent cells. Addition of 3-aminobenzamide simultaneously with the insulin decreased the effectiveness of insulin in stimulating RNA synthesis, resulting in a decline to 135% (Fig 4). This result is not due to the action of 3-aminobenzamide on the cells, since 3-aminobenzamide alone does not significantly affect the RNA synthesis rate of quiescent cells in the absence of insulin. It is therefore apparent that 3-aminobenzamide interferes with the signal connecting RNA synthesis and insulin.

The large excess of ribosomal RNA in a cell (>80% of the total RNA) suggests that the changes in RNA synthesis documented in Fig. 4 represent fluctuations specifically in rRNA gene activity rather than mRNA synthesis. However, since this assumption could be incorrect, the 28S rRNA molecule was isolated from cells incubated with radiolabelled uridine following separation by agarose gel electrophoresis. The comparison of label present in the rRNA and mRNA fractions following stimulation with insulin (Table 2) clearly demonstrates that any increase recorded in general RNA synthesis reflects an increased rate of rRNA, not mRNA, production. Therefore, the data presented in Fig. 4 provide the first evidence that ADP-ribosylation may be necessary for rRNA gene activity by insulin.

Prospects

The connection between poly(ADP-ribosyl)ation and various nuclear processes has been under debate for many years. Various methods have defined some of the biochemical

Table 2. Incorporation of uridine into mRNA and rRNA pools following stimulation with insulin

	rRNA	mRNA
untreated	104.8 ± 7.7	64.8 ± 4.7
insulin	183.3 ± 9.0	67.5 ± 1.2

H4IIE hepatoma cells, cultured in 39 mm diameter multi-well plates for 72 hours in serum-free α-MEM, were incubated for 6 hours with 2 μCi/ml ^{3}H-uridine (Dupont-NEN) in the absence or presence of insulin (10^{-6} M). Total RNA was harvested by the SDS lysis method of Salvatori *et al.* [29] and separated by electrophoresis in 1.2% low melting agarose. The 28S rRNA bands were visualized by staining with ethidium bromide, excised and extracted [30]. The regions of the gel flanking the 28S rRNA bands were also extracted to control for changes in labelling of mRNA. The recovered RNA was dissolved in 0.1 M NaOH and the radioactivity present in each sample was measured by liquid scintillation counting. The individual data were normalized to the amount of RNA present in each lane as determined by densitometric scanning of a photograph of the stained gel. The final values represent averages of triplicate experiments and are given in arbitrary units.

parameters of this post-translational modification process, thereby providing clues concerning the cellular processes connected to its mobilization. In particular, the production of long polymers of ADP-ribose can influence the interaction between specific proteins and DNA [7]. Based on this information plus the results of inhibitor studies, a connection to gene transcription was proposed, and several enticing pieces of evidence to support this contention have been reported. In particular, it was demonstrated that expression of the genes coding for phosphoenolpyruvate carboxykinase, procollagen and prolyl hydroxylase are activated following the addition of 3-aminobenzamide to the cell medium [8, 9]. While it was inferred that these results argued for a regulatory role for poly(ADP-ribosyl)ation in these systems, definitive proof for such a conclusion was not available.

The use of poly(ADP-ribose) polymerase inhibitors has flourished because of the difficulties encountered in distin-

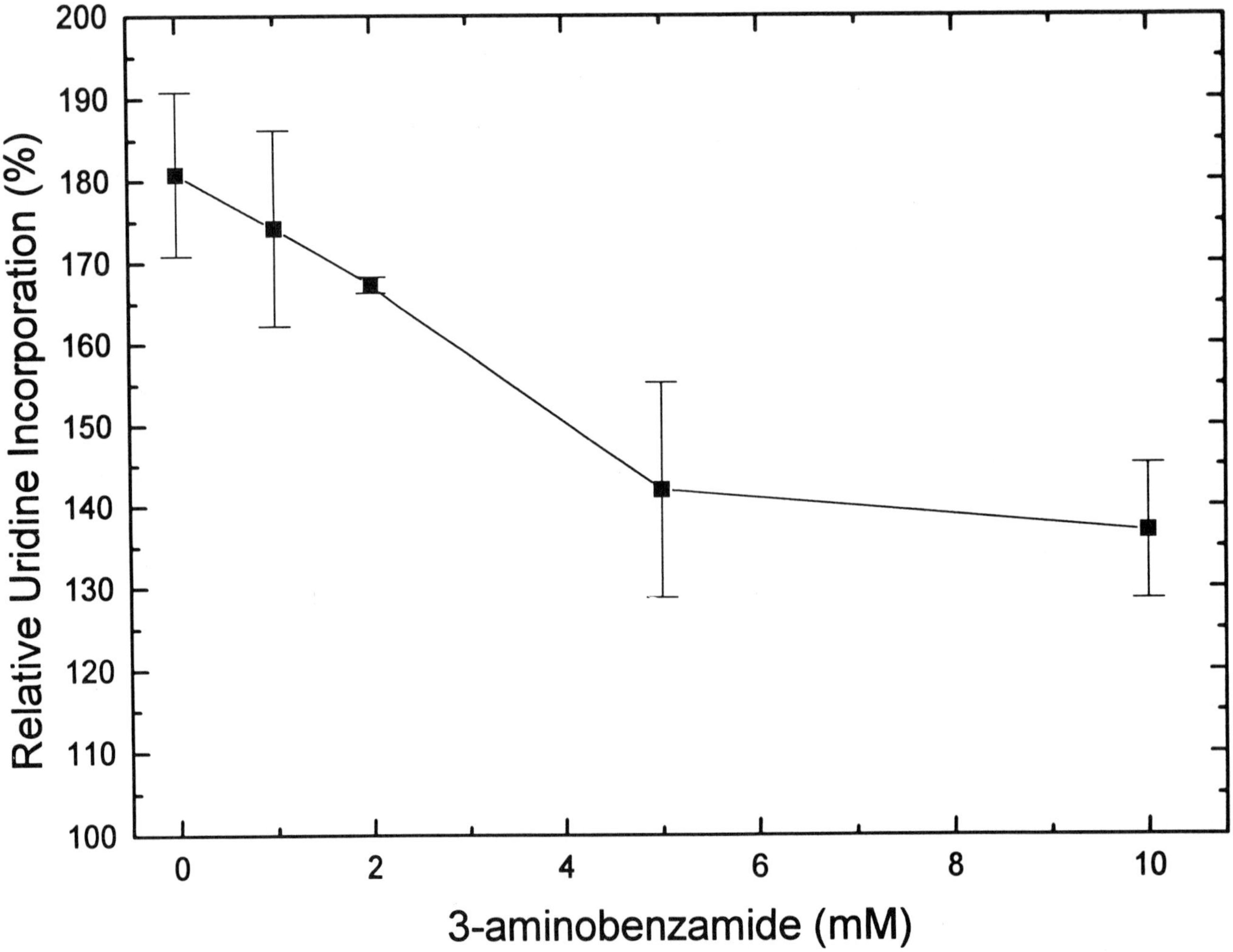

Fig. 4. Action of 3-aminobenzamide on the stimulation of RNA synthesis by insulin. H4IIE hepatoma cells were placed in serum-free α-MEM for 72 hours. RNA synthesis was measured by incorporation of ^{3}H-uridine over a 6 hour period into trichloroacetic acid insoluble material following the addition of insulin (10^{-6} M). Increasing amounts of 3-aminobenzamide were included in the medium over the course of the assay.

guishing specific from non-specific ADP-ribosylated proteins in whole cells. It has been frequently noted, however, that high concentrations of these inhibitors are lethal. Based on the information already published [8], it could be maintained that the concentration of 3-aminobenzamide necessary to elicit a response from the PEPCK gene was at toxic levels. Thus, the observed changes in gene expression may reflect one facet of a general stress response rather than an event specifically associated with transcriptional regulation. Alternatively, the high amounts of 3-aminobenzamide are needed to inhibit a process that is remarkably resistant to this compound. To bypass the problems associated with inhibitor studies, specific cell-free systems were used to assess the contribution of poly(ADP-ribose) polymerase to gene expression. No direct interaction between modification by poly(ADP-ribose) and gene transcription *in vitro* was noted (Figs. 2, 3). These data potentially conflict with recent results observed with a cell-free RNA polymerase I transcription system [23], however, the ability to stimulate an inactive cell extract was not tested in the course of our studies. A direct function for ADP-ribosylation in gene transcription by RNA polymerase II is therefore unlikely.

The activation of rRNA gene transcription by insulin is an essential step in the re-establishment of cell proliferation [15, 24]. When 3-aminobenzamide was added to quiescent hepatoma cells, no changes in either cell appearance or RNA synthesis were observed (Fig. 4), suggesting the concentrations used in the experiment (up to 10 mM) did not have a toxic effect over a 6 hour time period. In contrast, 3-aminobenzamide prevented the stimulation of rRNA synthesis by insulin (Fig. 4). Although it is impossible to suggest a mechanism responsible for this effect, the independent observation that an insulin-regulated gene such as phosphoenolpyruvate carboxykinase is induced by 3-aminobenzamide [8] provides support for the existence of a connection between the actions of insulin and ADP-ribose. The difficulty in identifying a suitable interpretation for these distinct findings is related to the concentration of inhibitor (>5 mM) required to produce an effect when the K_i of 3-aminobenzamide is approximately 5 µM in a standard poly(ADP-ribose) polymerase assay (Zahradka and Ebisuzaki, unpublished). Although the penetration of 3-aminobenzamide into a cell is probably impeded to some degree, the difference in these values is highly significant. Recent studies by Ueda and colleagues [26] have indicated that 3-aminobenzamide is capable of inhibiting mono-ADP-ribosyltransferase at high concentration (K_i for benzamide is 4.5 mM). When coupled with the data presented in Table 1, it may be proposed that 3-aminobenzamide functions by inhibiting an ADP-ribosyltransferase essential for cell signalling by insulin. This proposal does not exclude a role for poly(ADP-ribosyl)ation in gene expression, since the experiments presented above were limited solely to an examination of transcription initiation. Considering the close association of poly(ADP-ribose) with chromatin structure [27], a role in gene activation would appear evident. Similarly, RNA splicing/processing reactions and translation would be legitimate targets for regulation by ADP-ribosylation [11, 28]. However, each process must be examined independently before a final conclusion can be reached.

Summary

Poly(ADP-ribose) polymerase has been considered a secondary factor in gene transcription, since it will suppress non-specific initiation events *in vitro*. However, this investigation has shown that the enzyme makes no direct contribution to transcription by either RNA polymerase I or II in a cell-free system. Nevertheless, an indirect regulatory function remains possible, especially when the recent report by Mishima *et al.* [23] is considered. The most notable observation made in the course of these studies related to the inhibition of insulin-stimulated RNA synthesis by 3-aminobenzamide. Although a role for poly(ADP-ribosyl)-ation in the activation of rRNA gene transcription may be inferred on the basis of this finding, it is more likely that 3-aminobenzamide blocks an essential step in signal transduction associated with mono(ADP-ribosyl)ation.

Acknowledgements

This research was supported by a grant from the Natural Sciences and Engineering Research Council of Canada. Additional support was provided by a graduate studentship to L.Y. from the Medical Research Council of Canada.

References

1. Goldman MA: The chromatin domain as a unit of gene regulation. BioEssays 9: 50–56, 1988
2. Darnell JE Jr: Variety in the level of gene control in eukaryotic cells. Nature 297: 365–371, 1982
3. Edelman AM, Blumenthal DK, Krebs EG: Protein serine/threonine kinases. Annu Rev Biochem 56: 567–613, 1987
4. Hunter T, Karin M: The regulation of transcription by phosphorylation. Cell 70: 375–387, 1992
5. Williamson KC, Moss J: Mono-ADP-ribosyltransferases and ADP-ribosylarginine hydrolases: a mono-ADP-ribosylation cycle in animal cells. In: J Moss, M Vaughan (eds) ADP-ribosylating toxins and G proteins: Insights into signal transduction. American Society for Microbiology, Washington, DC, pp 493–510, 1990
6. Lautier D, Lagueux J, Thibodeau J, Menard L, Poirier GG: Molecular and biochemical features of poly(ADP-ribose) metabolism. Mol Cell Biochem 122: 171–193, 1993
7. Ueda K, Hayaishi O: ADP-ribosylation. Annu Rev Biochem 54: 73–100,

1985
8. Leverence RR, Beale EG, Granner DK: 3-aminobenzamide inhibits poly(ADP-ribose) synthetase activity and induces phosphoenolpyruvate carboxykinase (GTP) in H4IIE hepatoma cells. Arch Biochem Biophys 260: 667–673, 1988
9. Ghani QP, Hussain MZ, Zhang J, Hunt TK: Control of procollagen gene transcription and prolyl hydroxylase activity by poly(ADP-ribose). In: G Poirier, P Moreau (eds) ADP-ribosylation Reactions. Springer-Verlag, New York, pp 211–217, 1992
10. Slattery E, Dignam JD, Matsui T, Roeder RG: Purification and analysis of a factor which suppresses nick-induced transcription by RNA polymerase II and its identity with poly(ADP-ribose) polymerase. J Biol Chem 258: 5955–5959, 1983
11. Thomassin H, Niedergang C, Mandel P: Characterization of the poly(ADP-ribose) polymerase associated with free cytoplasmic mRNA protein particles. Biochem Biophys Res Comm 133: 654–661, 1985
12. Muller WEG, Zahn RW: Poly ADP-ribosylation of DNA-dependent RNA polymerase I from quail oviduct. Mol Cell Biochem 12: 147–159, 1976
13. Taniguchi T, Suzuki S, Shizuta Y: Poly(ADP-ribosyl)ation of RNA polymerase II from wheat germ. Biochem Biophys Res Comm 127: 525–532, 1985
14. Jacquemin C, Thibout H, Lambert B, Correze C: Endogenous ADP-ribosylation of G_s subunit and autonomous regulation of adenylate cyclase. Nature 323: 182–184, 1986
15. Larson DE, Zahradka P, Sells BH: Control points in eucaryotic ribosome biogenesis. Biochem Cell Biol 69: 5–22, 1991
16. Zahradka P, Sells BH: Transcription factors mediate rRNA transcription during myogenesis. Eur J Biochem 171: 37–43, 1988
17. Kurl RN, Jacob ST: Characterization of a factor that can prevent random transcription of cloned rDNA and its probable relationship to poly(ADP-ribose) polymerase. Nucl Acids Res 13: 89–101, 1985
18. Zahradka P, Larson DE, Sells BH: Characterization of a mammalian ribosomal protein gene promoter. Biochem Cell Biol 68: 949–956, 1990
19. Samuels M, Fire F, Sharp PA: Separation and characterization of factors mediating accurate transcription by RNA polymerase II. J Biol Chem 257: 14419–14427, 1982
20. Sawadogo M, Roeder RG: Interaction of gene-specific transcription factor with the adenovirus major late promoter upstream of the TATA box region. Cell 43: 165–175, 1985
21. Sluder AE, Price DH, Greenleaf AL: An activity necessary for *in vitro* transcription is a DNase inhibitor. Biochimie 69: 1199–1205, 1987
22. Tanaka Y, Hashida T, Yoshihara H, Yoshihara K: Bovine thymus poly(ADP-ribose) polymerase histone-dependent and Mg^{2+}-dependent reaction. J Biol Chem 254: 12433–12438, 1979
23. Mishima Y, Nishimura T, Muramatsu M, Kominami R: Transcription of mouse ribosomal RNA gene with inactive extracts is activate by NAD^+ *in vitro*. J Biochem 113: 36–42, 1993
24. Mauck C, Green H: Regulation of RNA synthesis in fibroblasts during transition from resting to growing state. Proc Natl Acad Sci USA 70: 2819–2822, 1973
25. Lauris V, Crettaz M, Kahn CR: Coordinate roles of insulin and glucose on the growth of hepatoma cells in culture. Endocrinology 118: 2519–2524, 1986
26. Banasik M, Komura H, Shimoyama M, Ueda K: Specific inhibitors of poly(ADP-ribose) synthetase and mono(ADP-ribosyl)transferase. J Biol Chem 267: 1569–1575, 1992
27. Poirier GG, de Murcia G, Niedergang C, Jongstra-Billen J, Mandel P: Poly(ADP-ribosyl)ation of polynucleosomes causes relaxation of the chromatin structure. Proc Natl Acad Sci USA 79: 3423–3427, 1982
28. Iglewski WJ, Dewhurst S: Cellular mon(ADP-ribosyl) transferase inhibits protein synthesis. FEBS Lett 283: 235–238, 1991
29. Salvatori R, Bockman RS, Guidon PT Jr: A simple modification of the Peppel/Baglioni method for RNA isolation from cell culture. BioTechniques 13: 510–512, 1992
30. Favre D: Improved phenol-based method for the isolation of DNA fragments from low melting temperature agarose gels. BioTechniques 13: 22–26, 1992

Molecular and Cellular Biochemistry **138**: 99–104, 1994.

Transcriptional regulation and autoregulation of the human gene for ADP-ribosyltransferase

Shiao Li Oei[1], Herbert Herzog[2]*, Monica Hirsch-Kauffmann[3], Rainer Schneider[2], Bernhard Auer[2] and Manfred Schweiger[1]

[1]*Inst. für Biochemie, Freie Universität, Fachbereich Chemie, Thielallee 63, D-14195-Berlin, Germany;* [2]*Institut für Biochemie (Nat. Fak.), Universität Innsbruck, Peter Mayr Str. 1a, A-6020 Innsbruck, Austria (*Present address: Garvan Institute for Medical Research, Darlinghurst, NSW 2010, Sydney, Australia);* [3]*Inst. f. Med. Biol. und Humangenetik. Schöpfstr. 41, A-6020 Innsbruck, Austria)*

Abstract

Human nuclear poly(ADP-ribosyl)transferase (ADPRT) modifies proteins with branched ADP-ribose-polymers. Various proteins, including ADPRT itself, serve as acceptors for polyADP-ribose. Target proteins include those controlling basic cellular processes such as DNA repair, differentiation and proliferation. Because of the outstanding features of this enzyme: automodification, several functional domains and central role in physiology of the cell, the molecular biology of ADPRT gained wide interest. The promoter structure contains several CCAAT/TATA boxes and SP1 sites. However, there is no CCAAT/TATA box in the neighbourhood of an SP1 site and, thus no obvious site for initiation of transcription. Within this region there are several noteworthy inverted repeats, which by internal basepairing could form two types of cruciform structures. Deletion analysis revealed that these cruciform structures have functional significance. Removal of one type increases the promoter activity, whereas removal of the other diminishes the promoter function.

Overexpression of ADPRT from heterologous promoters (MMTV, SV40) leads to repression of the activity of the ADPRT promoter. Indeed, ADPRT was shown to bind specifically to one type of cruciform structure. This specific interaction indicates autorepression of the ADPRT gene: the enzyme ADPRT acts directly as a negative modulator of the activity of its own promoter. (Mol Cell Biochem **138**: 99–104, 1994)

Key words: poly(ADP-ribosyl) transferase (human), autoregulation, gene expression, promoter structure, cruciform structure

Introduction

Nuclear NAD$^+$: protein ADP-ribosyltransferase, polymerizing [ADPRT; EC 2.4.2.30] modifies proteins by forming branched poly(ADP-ribose) chains and seems to be ubiquitous and highly conserved in eukaryotic cells. Target proteins for ADP-ribosylation are, besides ADPRT itself, histones, high mobility group proteins (HMG), DNA ligase, topoisomerases and other nuclear proteins. The function of the ADP-ribosylation modification is as yet unclear, but the very short half-life of the ADP-ribosyl polymers implies an involvement in nuclear control mechanisms. Indeed, the enzyme appears to participate in central processes of the cell such as DNA repair, differentiation, tumorigenic cell transformation and others (for a review see [1]).

In accordance with its central biochemical role, the amino acid sequence and structure of this enzyme are highly conserved among phylogenetically distant species as indicated by the crossreaction of antibodies against the human enzyme with purified trout and snail ADPRT [2, 3].

The enzyme ADPRT consists of several functionally distinguishable domains: an aminoterminal DNA-binding domain containing two 'zinc-fingers', a nuclear localization signal, a site for automodification, and the NAD$^+$-binding domain located in the carboxy terminal region.

The DNA-binding domain of ADPRT exhibits a very high

Address for offprints: Shiao Li Oei, Institut für Biochemie, Fachbereich Chemie, Thielallee 63, D-14195, Berlin, Germany

affinity towards DNA, especially to single strand DNA breaks, which act as coenzyme for catalytic function of ADPRT. During DNA repair, the presence of DNA single-strand breaks is required to activate ADPRT. The second 'zinc-finger' plays a fundamental role in specific single strand DNA-binding [4]. Furthermore, it has been reported, that ADPRT binds and stabilizes DNA-loops [5].

During DNA repair, the appearance of DNA breaks activates ADPRT function, which utilizes NAD^+ for ADP-ribosylation, thereby reducing the concentration of NAD^+.

Specific inhibition of ADP-ribosylation increases the lethality of noxious chemicals. The actual role of ADPRT in DNA repair, differentiation and proliferation is not understood. Besides this there is interestingly enough post-transcriptional regulation of ADPRT expression [6, 7]. Within the 3′ end of ADPRT messenger RNA there are two independent polyadenylation sites, which lead to alternative mRNAs with different stabilities. In several untransformed cells the poly A tail starts 186 nucleotides downstream from a polyadenylation site which is used in Hela cells and in SV40 transformed fibroblasts [6]. The proximal polyadenylation signal, which is utilized in Hela cells leads to a mRNA with a considerably longer half-life than that found in primary human fibroblasts. The stability of ADPRT messenger RNA was also found to be significantly higher in replicative cells than in confluent cells [8].

In a proliferating system, i.e., in activated human lymphocytes, it has been shown that the transcription of ADPRT was induced at an early stage in the cell cycle before the onset of DNA synthesis, whereas in quiescent lymphocytes the levels of ADPRT mRNA were almost undetectable [9]. In addition to the posttranscriptional regulation, ADPRT seems to be controlled at the transcriptional level [10]. The cellular concentrations of the enzyme are rigidly regulated and maintained, most likely, by means of direct regulation of promoter activity.

To understand the function and regulation of both the ADPRT enzyme and its expression, the human gene for ADPRT was sequenced and characterized and its potential functions studied.

The gene structure of human ADPRT

Access to this most interesting enzyme was opened by the development of an efficient affinity chromatography procedure [11]. cDNA for ADPRT was isolated and characterized simultaneously in several laboratories [12–15]. The deduced amino acid sequence and protein structure prediction revealed two 'zinc fingers' for DNA binding [15]. The human gene for ADPRT was cloned, mapped, and its genomic localization determined [16]. Remarkably, the ADPRT gene maps to the same region of chromosome 1 (q41–q42) as the genes for protooncogene *trk* and for transforming growth factor β.

The ADPRT gene consists of 23 exons ranging in size from 62–553 bp, spread over a region of 43 kb. All the exons contain canonical splice consensus sequences. The two 'zinc-fingers' of the DNA-binding domain are encoded by two exons each.

The human ADPRT promoter

A clone isolated from a human liver cDNA library containing the first exon and surrounding sequences, served for sequencing and characterization of the ADPRT promoter region. Multiple transcription initiation sites were found by primer extension analysis. The major initiation site was determined to be 160 bp upstream from the translational start site. The use of multiple initiation sites for transcription has also been found in other housekeeping genes [17]. The sequence of the ADPRT promoter contains several interesting structures (see Fig. 1): Polymerase II-specific promoter structures like CCAAT- and TATA-boxes, SP1 binding sites and two sets of inverted repeats. Most interestingly, none of the CCAAT/TATA boxes is located sufficiently close to a SP1 site to form an active polymerase II transcription initiation complex. The closest CCAAT/TATA-box to a SP1 site resides 580 bp upstream to the transcription initiation site.

Four inverted repeats, 90 bp in length with about 90% internal basepairing sequence identity, allow the formation of two potential DNA cruciform structures with long base-paired stems (see Fig. 1). The formation of a loop structure could bring the essential promoter structures CCAAT/TATA and SP1 closer together. Since the formation of an active polymerase II transcription initiation complex requires a much shorter distance between the TATA-box and the transcription start site, it was tempting to assume that the loops may be involved in the formation of a transcription initiation complex.

The first putative DNA-loop B (Fig. 1b) of the ADPRT promoter does not affect the distance between the most proximal SP1 site and the nearest upstream CCAAT/TATA-box. However the second type of loop structure C (Fig. 1c) would shorten this distance considerably and thereby bring the CCAAT/TATA sequence into the vicinity of the SP1 site near the transcription start site, suggesting loop C forms the active promoter structure.

This assumption was supported by experiments with deletion mutants in the promoter sequence, eliminating either loop type B or loop type C. Promoter activities of the deletion mutants were measured by connecting them to the reporter gene chloramphenicol acetyltransferase (CAT) and transfection of cells with these constructs (Fig. 2).

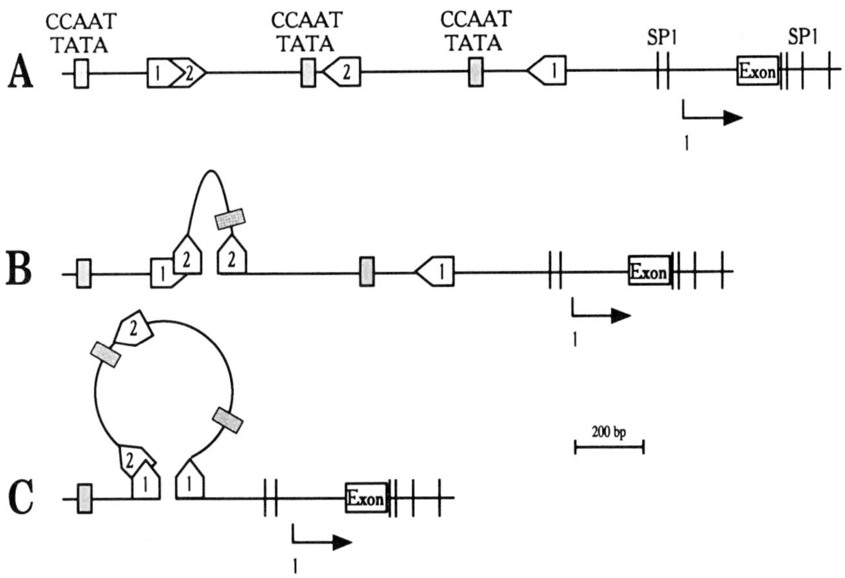

Fig. 1. Schematic representation of the predicted secondary structures of the ADPRT promoter. Localization and direction of the inverted repeats are given by arrows [1, 2]. SP1 binding site clusters, CCAAT-TATA boxes and open reading frames are marked.

The complete ADPRT promoter produced the highest amount of CAT activity. Deletion of the proximal parts of the inverted repeats (mutant 1), preventing formation of both types of DNA loops, reduced the promoter activity considerably. However, elimination of DNA loop B, which still allowed the DNA loop C to bring the CCAAT/TATA-box into the vicinity of the transcription initiation site, did not affect promoter activity. When formation of loop C was rendered impossible the promoter activity was low. This led to the assumption that loop B formation inhibited the initiation of transcription and that promoter activity was dependent on DNA loop C formation.

The alternative formation of loop B (inactive) and of loop C (active) could be a perfect tool to control the promoter activity. Stabilization of loop B would inactivate the promoter, whereas the alternative stabilization of loop C would activate the promoter. The loop structures could be stabilized by binding of specific proteins. Indeed, it has been shown that the ADPRT enzyme binds specifically to DNA cruciform structures [3]. Thus, it could be assumed, that poly(ADP-ribosyl) transferase binds specifically to cruciform structures of its own promoter and thereby regulates its own expression via stabilization of certain DNA-loop structures.

Autoregulation of ADPRT expression

To support the notion of autoregulation of expression of ADPRT, promoter activities were studied in the presence of overexpressed ADPRT. Activity of ADPRT promoter was measured by fusion with the reporter gene CAT. Cotransfection with ADPRT-cDNA under the control of the SV40 promoter led to overexpression of ADPRT and resulted in inhibition of the ADPRT promoter. Induction of ADPRT cDNA (controlled by the MMTV promoter) by dexamethasone from a transfected plasmid also diminished ADPRT promoter-dependent expression of CAT. There was neither influence of ADPRT on the SV40 or MMTV promoters, nor was the activity of the ADPRT promoter repressed by cotransfection of the empty SV40 promoter plasmid. Thus the decrease of ADPRT promoter activity turns out to be specifically dependent on ADPRT expression.

These experiments indicated a causal relationship between the overexpression of ADPRT enzyme and the down-regulation of the ADPRT gene. Moreover, repression of the ADPRT promoter appears to be exerted by the DNA binding domain of ADPRT enzyme.

Specific binding of ADPRT to its promoter

Further evidence for ADPRT as a regulator of its own promoter arose from binding experiments. Poly(ADP-ribosyl) transferase coupled to a solid support was able to bind fragments of the ADPRT promoter. Similarly, the DNA binding domain of the enzyme also bound these DNA fragments. The specificity of this binding was shown by competition experiments. Nonspecific DNA fragments did not compete successfully with the promoter DNA or ADPRT. Importantly,

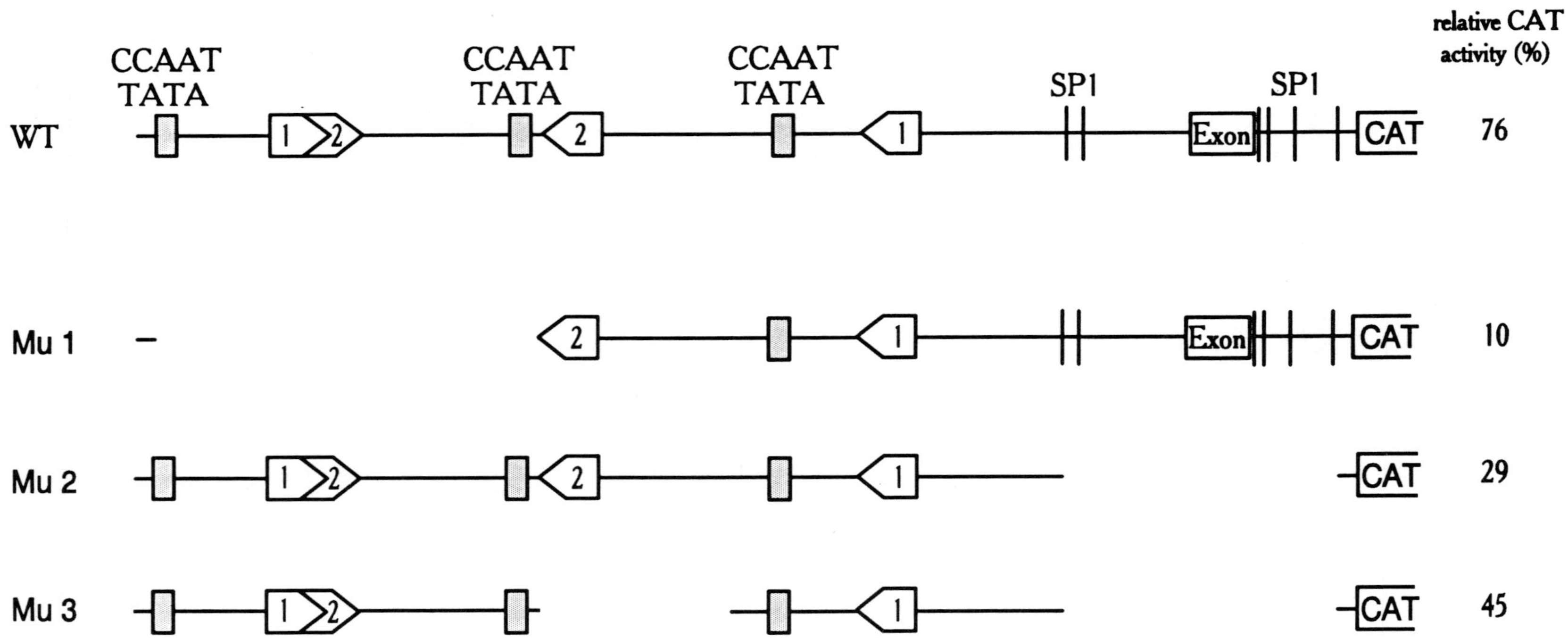

Fig. 2. Promoter activity of various deletion mutants of the human ADPRT promoter linked to the CAT reporter gene. Localization and direction of the inverted repeats [1, 2]. SP1 binding cluster, CCAAT-TATA boxes, and open reading frame are indicated.

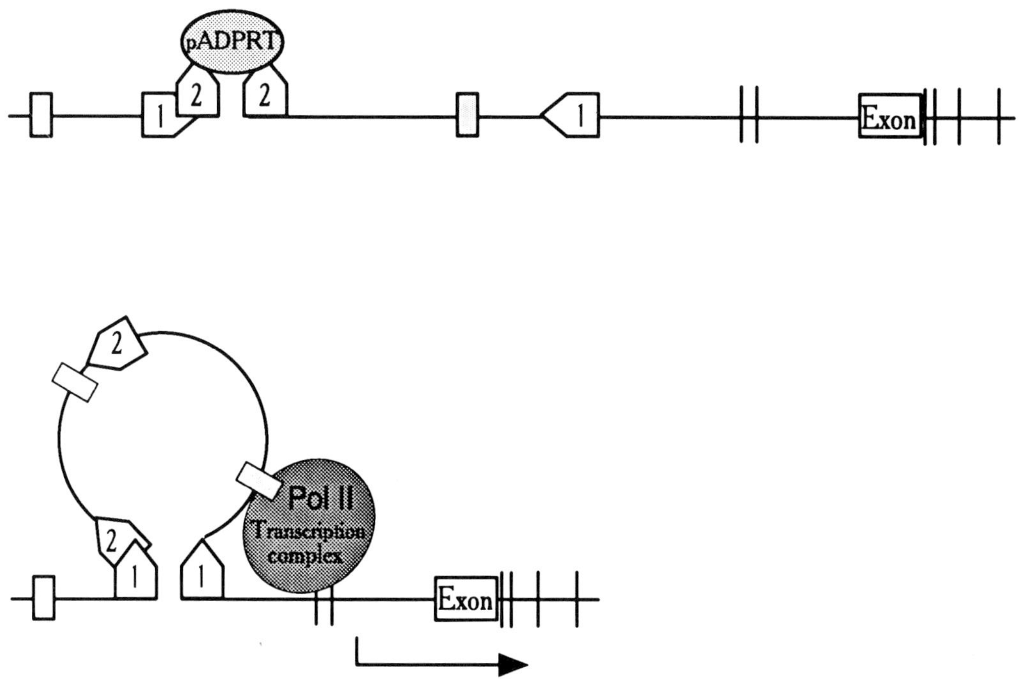

Fig. 3. Autoregulation of ADPRT. Model of interaction between ADPRT and its own promoter resulting in inhibition of transcription.

binding of the promoter fragment to ADPRT could be inhibited by a DNA fragment which was able to form loop B. Since ADPRT was shown to have the ability to stabilize DNA loops [3], these results document a direct interaction between the ADPRT enzyme and the potential cruciform structure on its own promoter. Thus, ADPRT enzyme is able to repress its own transcription by stabilization of the small loop B and thereby preventing the formation of loop C, the active form of the promoter. In Fig. 3, we depict a model for the autoregulation of ADPRT by its binding to loop B.

Discussion

Autoregulation of transcription was described previously in prokaryotes for the expression of ribosomal constituents [16]. RNA polymerase in *E. coli* is also autoregulated. In eukaryotes, transcription of homeobox genes is regulated by binding of the gene products to their promoter structures [19].

The mechanism for transcriptional regulation of housekeeping genes is still unknown. The mechanism of autoregulation shown here for the ADPRT gene expression could serve as a model for the regulation of housekeeping genes encoding DNA-binding proteins.

Considering the complex mechanism of autoregulation of glutamine synthetase in bacteria [20], which represents the most effective way to control a whole group of operons, it should be worthwhile to investigate the interaction of ADPRT with the promoters of DNA-repair genes in the future.

References

1. Althaus FR, Richter C: ADP-ribosylation of proteins: Enzymology and biological significance. Molecular Biol Biochem and Biophys 37: 1–226, 1987
2. Burtscher HJ, Klocker H, Schneider R, Auer B, Hirsch-Kauffmann M, Schweiger M: ADP-ribosyltransferase from Helix pomatia. Purification and characterization. Biochem J 248: 859–864, 1987
3. Burtscher HJ, Schneider R, Klocker H, Auer B, Hirsch-Kauffmann M, Schweiger M: ADP-ribosyltransferase is highly conserved: Purification and characterization of ADP-ribosyltransferase from a fish and its comparison with the human enzyme. J Comp Physiol Biol 157: 567–572, 1987
4. Gradwohl G, Ménissier-De Murcia JM, Molinete M, Simonin F, Koken M, Hoeijmakers JHJ, De Murcia G: The second zinc-finger domain of Poly (ADP-ribose) polymerase determines specificity for single-stranded breaks in DNA. Proc Natl Sci USA 87: 2990–2994, 1990
5. Sastry SS, Kun E: The interaction of adenosine diphosphoribosyl transferase (ADPRT) with a cruciform DNA. Biochem Biophys Res Commun 167: 842–847, 1990
6. Herzog H, Zabel BU, Schneider R, Auer B, Hirsch-Kauffmann M, Schweiger M: Human nuclear NAD⁺ ADP-ribosyltranferase: Localization of the gene on chromosome 1q41–q42 and expression of an active human enzyme in *Escherichia coli*. Proc Natl Acad Sci USA 86: 3514–3518, 1989
7. Schweiger M, Auer B, Herzog H, Hirsch-Kauffmann M, Kaiser P, Flick K, Nagl U, Schneider R: Molecular biology of human Nuclear NAD⁺: ADP-

Ribosyl-transferase (Polymerizing). In G.G. Poirier and P. Moreau (eds) ADP-Ribosylation Reactions. Springer-Verlag, New York, 1992, pp 20–30
8. Bhatia K, Pommier Y, Giri C, Fornance AJ, Imaizumi M, Breitmann TR, Cherney B, Smulson M: Expression of the poly(ADP-ribose) polymerase gene following natural and induced DNA strand breakage and effect of hyperexpression on DNA repair. Carcinogenesis 11: 123–128, 1990
9. Negroni M, Bertazzoni U: Differential expression and stability of poly (ADP-ribose) polymerase mRNA in human cells. Biochimica et Biophysica Acta 1173: 133–140, 1993
10. Wein KH, Netzker R, Brand K: Cell cycle-related expression of poly(ADP-ribosyl)transferase in proliferating rat thymocytes. Biochimica et Biophysica Acta 1176: 69–76, 1993
11. Burtscher HJ, Auer B, Klocker H, Schweiger M, Hirsch-Kauffmann M: Isolation of ADP-ribosyltransferase by affinity chromatography. Anal Biochem 152: 285–290, 1986
12. Schneider R, Auer B, Kühne C, Herzog H, Klocker H, Burtscher HJ, Hirsch-Kauffmann M, Wintersberger U, Schweiger M: Isolation of a cDNA clone for human NAD^+: protein ADP-ribosyltransferase. Eur J Cell Biol 44: 302–307, 1987
13. Alkhatib HM, Chen D, Cherney B, Bhatia K, Notario V, Giri C, Stein G, Slattery E, Roeder RG, Smulson M: Cloning and expression of cDNA for human poly (ADP-ribose) polymerase. Proc Natl Acad Sci USA 84: 1224–1228, 1987
14. Kurosaki T, Ushiro H, Mitsuchi Y, Suzuki S, Matsuda M, Matsuda Y, Katanuma N, Kanugawa K, Matsuo H, Hirose T, Inayama S, Shizuta Y: Primary structure of human poly (ADP-ribose) synthetase deduced from cDNA sequence. J Biol Chem 262: 15990–15997, 1987
15. Uchida K, Morita T, Sato T, Ogura T, Yamashita R, Noguchi S, Suzuki H, Nyunoya H, Miwa M, Sugimura T: Nucleotide sequence of a full-length DNA for human fibroblast poly (ADP-ribose)-polymerase. Biochem Biophys Res Comm 148: 617–622, 1987
16. Auer B, Nagl U, Herzog H, Schneider R, Schweiger M: Human nuclear NAD^+ ADP-ribosyltransferase(polymerizing): Organization of the gene. DNA 8: 575–580, 1989
17. Struve I, Rausch T, Bernasconi P, Taiz L: Structure and function of the promoter of the Carrot V-type H^+-ATPase catalytic subunit gene. J Biol Chem 265: 7927–7932, 1990
18. Gourse RL, Sharrock RA, Nomura M: Control of Ribosome Synthesis in *Escherichia coli*. In B. Hardesty and G. Kramer (eds). Structure and Function, and Genetics of Ribosomes. Springer-Verlag, New York, 1985, pp 766–788
19. Fainsod A, Leonhard DB, Ruusala T, Lubin M, Crothers D, Ruddle FH: The homeo domain of a murine protein binds 5′ to its own homeo box. Proc Natl Acad Sci USA 83: 9532–9536, 1986
20. Magasanik B: Reversible phosphorylation of an enhancer binding protein regulates the transcription of bacterial nitrogene utilization. TIBS 13: 475–479, 1988

PART III

MONO(ADP-RIBOSYLATION)

Molecular and Cellular Biochemistry **138**: 107–112, 1994.

Vertebrate mono-ADP-ribosyltransferases

Anna Zolkiewska, Ian J. Okazaki and Joel Moss

Laboratory of Cellular Metabolism, National Heart, Lung, and Blood Institute, National Institutes of Health, Bethesda, Maryland 20892, USA

Abstract

Mono-ADP-ribosylation appears to be a reversible modification of proteins, which occurs in many eukaryotic and prokaryotic organisms. Multiple forms of arginine-specific ADP-ribosyltransferases have been purified and characterized from avian erythrocytes, chicken polymorphonuclear leukocytes and mammalian skeletal muscle. The avian transferases have similar molecular weights of ~28 kDa, but differ in physical, regulatory and kinetic properties and subcellular localization. Recently, a 38-kDa rabbit skeletal muscle ADP-ribosyltransferase was purified and cloned. The deduced amino acid sequence contained hydrophobic amino and carboxy termini, consistent with known signal sequences of glycosylphosphatidylinositol (GPI)-anchored proteins. This arginine-specific transferase was present on the surface of mouse myotubes and of NMU cells transfected with the cDNA and was released with phosphatidylinositol-specific phospholipase C. Arginine-specific ADP-ribosyltransferases thus appear to exhibit considerable diversity in their structure, cellular localization, regulation and physiological role. (Mol Cell Biochem **138**: 107–112, 1994)

Key words: ADP-ribosylarginine hydrolase, glycosylphosphatidylinositol anchor, skeletal muscle, cholera toxin, ADP-ribosyltransferases

Introduction

Mono-ADP-ribosylation of proteins has been detected in many eukaryotic organisms. Arginine, cysteine, diphthamide, histidine and asparagine residues have been identified as acceptors for amino acid-specific ADP-ribosyltransferases. However, very few enzymes that catalyze ADP-ribosylation reactions have been purified and characterized. A recent review [1] summarizes most of the reports in which the presence of mono-ADP-ribosylation was only suggested. Here, we focus on arginine-specific mono-ADP-ribosyltransferases, which are the best characterized transferases thus far. This group includes several avian transferases [2–9, 12], as well as a transferase from skeletal muscle. The latter enzyme was recently purified [10, 11] and cloned [11], and shown to be glycosylphosphatidylinositol (GPI)-anchored in the muscle cell membrane.

Avian ADP-ribosyltransferases

Turkey erythrocyte ADP-ribosyltransferases

Four distinct NAD:arginine ADP-ribosyltransferases were purified from turkey erythrocytes [2–9]. Transferases A and B were isolated from the erythrocyte cytosol, transferase C from membrane fraction and transferase A′ from the nucleus. Transferase A was purified 500,000-fold by sequential chromatography on phenyl-Sepharose, carboxymethyl cellulose, NAD-agarose and concanavalin A-agarose and had an apparent size of 28.3 kDa by sodium dodecyl sulfate-polyacrylamide gel electrophoresis (SDS-PAGE) [2]. In addition to proteins such as casein and ovalbumin, ADP-ribose acceptors included simple guanidino compounds such as arginine and agmatine. The specific activity of the purified transferase was 350 $\mu mol \cdot min^{-1} mg^{-1}$ with arginine methyl ester as the ADP-ribose acceptor. The presence of positively

Address for offprints: A. Zolkiewska, Building 10, Room 5N-307, National Institutes of Health, 9000 Rockville Pike, Bethesda, MD 20892, USA

charged groups near the guanidino moiety made ADP-ribose acceptors more effective: arginine methyl ester > agmatine ≥ arginine > guanidinopropionate ≥ guanidine[1].

In the absence of salt, transferase A existed in a relatively inactive, rapidly sedimenting form, consistent with the formation of multimers, which eluted in the void volume of a G-200 gel permeation column [3]. In the presence of salt, histones, lysophosphatidylcholine or some detergents, the transferase dissociated and eluted as a slowly sedimenting, active monomer. Stimulation of ADP-ribosylagmatine formation by histones in the absence of salt was observed at histone concentrations of 10–20 μg/ml, less than 1% of those necessary to serve as ADP-ribose acceptors [4, 5]. Histones H1 and H4 were slightly more active than were H2a, H1b and H3 in stimulating transferase activity. Stimulation by histones was independent of DNA [4]. The effectiveness of salts on transferase activity appeared related to the anionic rather than cationic moiety, with $SCN^- > Br^- > Cl^- > F^- > PO_4^{3-}$ [3].

ADP-ribosyltransferase A activity was enhanced ~6 fold by lysophosphatidylcholine but not lysophosphatidylglycerol, lysophosphatidylserine, lysophosphatidylethanolamine, lysophosphatidic acid, glycerophosphatidylcholine, or choline [6]. Overall, lysophosphatidylcholine derivatives containing long chain fatty acids were more effective than those with short chains. Lysolecithin caused a rapid and reversible activation of the transferase A and stabilized the enzyme against thermal denaturation. The zwitterionic detergent 3-[(cholamidopropyl) dimethylammonio]-1-propanesulfonate (CHAPS) and certain nonionic detergents such as Triton X-100, Triton X-114, Tween-20 and Triton X-305 also activated the transferase, but were less effective than lysolecithin, histone or NaCl [6].

The K_m values for NAD and agmatine were 7 μM and 260 μM, respectively. Kinetic studies of transferase A were consistent with a rapid-equilibrium random sequential reaction mechanism in which binding of the first substrate (NAD or agmatine) alters affinity of enzyme for the second substrate [7]. The mechanism was similar to that observed for cholera toxin, a bacterial ADP-ribosyltransferase.

ADP-ribosyltransferase B was purified 270,000-fold by sequential chromatography on phenyl-Sepharose, concanavalin A-Sepharose, carboxymethyl cellulose, Procion red-agarose, and Ultrogel AcA-54 [8]. The purified enzyme had an estimated size of 32.7 kDa by gel filtration and apparent K_m values for NAD and arginine methyl ester of 36 μM and 3.0 mM, respectively. Unlike transferase A, transferase B appeared not to self-associate in the absence of salt or histones and was inhibited ~50% by chaotropic salts [8].

ADP-ribosyltransferases C and A′ were purified from turkey erythrocyte particulate preparations by chromatography on phenyl-Sepharose, carboxymethyl cellulose, concanavalin A-Sepharose, NAD-agarose, and DNA-agarose [9]. Transferase C was enriched in the membrane fraction and most of transferase A′ was found in the nucleus. Transferase A′ was similar in properties to its cytosolic counterpart. The apparent sizes of transferase C and A′ by gel permeation chromatography were 26 kDa and 25.5 kDa, respectively. Transferase C was relatively insensitive to salt or histones, whereas the activity of transferase A′ was increased 5–20 fold by 300 mM NaCl or histones. Apparent K_m values for transferase C were 15 μM and 2 mM for NAD and agmatine, respectively, similar to those for transferases A and B.

Hen liver nuclear ADP-ribosyltransferase

A 27.5-kDa ADP-ribosyltransferase was purified 620-fold from hen liver nuclei by sequential chromatography through Sephadex G-200, hydroxyapatite, phenyl-Sepharose, carboxymethylcellulose and DNA-Sepharose [12]. The hen liver transferase used the same ADP-ribose acceptors as the turkey erythrocyte enzyme. The effectiveness of histones as ADP-ribose acceptors was H3 > H2a > H4 > H2b > H1. The K_m values for NAD were 0.5 mM with H1 as ADP-ribose acceptor, 0.29 mM with H2a, 0.18 mM with agmatine and 0.07 mM with arginine methyl ester, all of which were higher than the values for the turkey enzyme. The K_m values for arginine methyl ester and agmatine were 24 and 19 mM, respectively.

The pH optimum for the enzyme was 9.0 with histones as acceptor proteins. Inhibition of the ADP-ribosylation reaction by small amounts of DNA was observed, and could be reversed by histones. The hen liver nuclear transferase was stimulated by dithiothreitol and inhibited by increasing concentration of NaCl up to 200 mM.

The ADP-ribosyltransferase from hen liver nuclei served as an initiator of poly(ADP-ribose) synthesis *in vitro* for a small fraction of poly(ADP-ribosylated) histones [12]. *In vitro* ADP-ribosylation of calf thymus whole histones by the transferase inhibited subsequent phosphorylation by cAMP-dependent protein kinase [13, 14]. Calf thymus histone H1 was ADP-ribosylated at arginine 34 by the hen liver nuclear transferase, which inhibited phosphorylation of H1 at serine 38 [15]. In addition, purified rabbit skeletal muscle phosphorylase kinase was an *in vitro* substrate for the hen liver transferase and ADP-ribosylation of the α and β subunits of the kinase blocked subsequent cAMP-dependent phosphorylation, resulting in a decrease in phosphorylase kinase activity [16]. Similarly, pig liver L-type pyruvate kinase was ADP-ribosylated by the nuclear transferase, which resulted in inhibition of phosphorylation-induced inactivation of L-type pyruvate kinase [17].

Chicken heterophile ADP-ribosyltransferase

An ADP-ribosyltransferase was isolated from chicken polymorphonuclear leukocytes (heterophiles) [18]. The protein was localized to the granule fraction on Percoll density gradient centrifugation. The physical and kinetic properties of this transferase and the hen liver enzyme were very similar.

Arginine-specific mono-ADP-ribosyltransferase from skeletal and cardiac muscle

Characterization and purification of the enzyme

A mono-ADP-ribosyltransferase specific for guanidino group of low molecular weight model substrates [19] or arginine residues in proteins [20–22] was present at a relatively high level in skeletal and cardiac muscle tissue and in cultured skeletal muscle cells [23]. Most of the enzyme activity was detected in a cell membrane fraction, with specific activities of 0.005–2 $nmol{\cdot}min^{-1}mg^{-1}$ in preparations from rabbit, canine or porcine muscles with simple guanidino compounds as acceptors of ADP-ribose [19].

The transferase was rapidly inactivated by dithiothreitol, β-mercaptoethanol, reduced glutathione or cysteine [20]. Divalent cations Mg^{2+} and Ca^{2+} had no effect on the ADP-ribosyltransferase activity [20]. Triton X-100 at low concentration (0.1%) slightly activated the enzyme, but higher concentrations, as well as deoxycholate above 0.01% were inhibitory [20]. Other detergents (e.g. Lubrol (up to 0.5%) [10], CHAPS (up to 1%) [11]) had no effect on enzyme activity.

The enzyme was highly specific for NAD; NADP and NADH were very poor substrates [20]. Using partially purified transferase from rabbit skeletal muscle, K_m values of 0.56 mM and 1.2 mM were determined for NAD and L-arginine methyl ester as substrates, respectively [24]. The kinetic mechanism was defined as random sequential [24].

Arginine-specific ADP-ribosyltransferase was extracted from rabbit skeletal muscle membrane fraction with 0.3% deoxycholate and purified by sequential chromatography on DE-52 cellulose and concanavalin A agarose [10, 11] and then on high resolution anion exchange (DEAE) matrix and by gel-filtration HPLC [11]. The enzyme was purified 215,000-fold with respect to the muscle homogenate and 16,000-fold with respect to membrane fraction, with a yield of 4% [11]. Specific activity was 14 $mmol{\cdot}min^{-1}mg^{-1}$ when assayed with 0.1 mM NAD and 20 mM agmatine and 68 $mmol{\cdot}min^{-1}mg^{-1}$ with 2 mM NAD [11]. Analysis of the purified transferase by SDS-PAGE revealed the presence of 38-kDa protein [10, 11]. However, the size of the transferase determined by gel filtration-HPLC was 61 kDa [11], consistent either with dimerization of the enzyme or strong interaction with the detergent (CHAPS) used at this step of purification. An alternative method of purification of the rabbit skeletal muscle ADP-ribosyltransferase has been proposed recently [25], which results in an enzyme of similar purity.

Cloning and expression of the rabbit skeletal muscle ADP-ribosyltransferase

cDNAs for an ADP-ribosyltransferase were cloned from a rabbit skeletal muscle library by screening with oligonucleotide probes based on partial amino acid sequence of the purified ADP-ribosyltransferase, and by a polymerase chain reaction (PCR)-based procedure [11]. The composite cDNA contained a 981-base-pair open reading frame, encoding a polypeptide of 36,134 Da. The deduced amino acid sequence did not have extended homology to any of the bacterial ADP-ribosylating toxins, the ADP-ribosyltransferase from *Rhodospirillum rubrum*, or poly(ADP-ribose)polymerase. It contained two potential sites for N-linked glycosylation.

The deduced amino acid sequence of the transferase had very hydrophobic amino and carboxy termini (stretches of approximately 20 amino acids), with a hydrophilic center [26]. This hydrophilicity profile resembled the profiles of glycosylphosphatidylinositol (GPI)-anchored membrane proteins, which are synthesized as nascent polypeptides, containing hydrophobic signal sequences at amino and carboxy termini [27, 28]. For GPI-linked proteins, the amino terminal sequence serves as a signal for translocation into endoplasmic reticulum and the carboxy terminal sequence is a signal for GPI-attachment. Both signal sequences are absent from the mature form of a GPI-anchored protein.

A transferase-specific oligonucleotide probe hybridized with a 4-kb mRNA expressed primarily in rabbit skeletal and heart muscle. It did not hybridize with mRNA from smooth muscle, brain, lung, kidney, spleen, or liver [11]. Polyclonal antibodies to the rabbit transferase cross-reacted with partially purified proteins from other mammalian species [26], consistent with partial conservation of the structure.

Based on the assumption that the native rabbit transferase has a GPI modification and given the fact that mammalian signal sequences are not recognized in *E. coli*, a truncated form of the transferase cDNA, lacking 5′ and 3′ ends encoding the putative signal peptides, was expressed, yielding an active and soluble enzyme [11]. Thus, the amino and carboxy terminal amino acids were not required for enzyme activity.

More convincing evidence that ADP-ribosyltransferase is a GPI-anchored protein was obtained from expression experiments in mammalian cells. Expression of the complete coding region of the rabbit transferase cDNA in rat mammary adenocarcinoma (NMU) cells resulted in an active, membrane-bound enzyme [26]. All GPI-anchored proteins found so far are located on the exterior of the plasma membrane or, to a lesser extent, on the luminal face of intracellular vesicles [27, 28]. Treatment of the intact transfected cells with phosphatidylinositol-specific phospholipase C from *Bacillus thuringiensis* caused almost complete release of the transferase to the medium [26], consistent with the extracellular localization of a GPI-anchored transferase. Expression of the mutant protein, lacking the C-terminal hydrophobic sequence, resulted in the production of a soluble, secreted form of the transferase [26]. This is consistent with the view that the C-terminal sequence is required for GPI-attachment and, in its absence, the transferase is transported from the endoplasmic reticulum to the plasma membrane, but it is not membrane-anchored. Based on the similarity between ADP-ribosyltransferase amino and carboxy terminal hydrophobic sequences and signal sequences of other GPI-anchored proteins, as well as amino acid requirements adjacent to the sites of cleavage of signal peptides [29], structures of the nascent and the mature forms of the arginine-specific ADP-ribosyltransferase are proposed (Fig. 1).

The possibility that ADP-ribosyltransferase might be located on the surface of muscle cells is unexpected and intriguing. Although it was reported that after fractionation of rat cardiac muscle most of the arginine-specific ADP-ribosyltransferase activity was present in the plasma membrane [22, 30], the external orientation of the enzyme was not suspected. There was a report [31] of labeling of several proteins during incubation of intact cells (lymphocytes, T cell lymphomas, hybridomas and myoblasts) with [^{32}P]NAD (extracellular) and it was suggested that the transferase activity might be associated with a cell surface protein. On the other hand, numerous papers referred to the muscle ADP-ribosyltransferase as an enzyme located in the sarcoplasmic reticulum (SR) [21, 25, 32, 33]. It remains to be determined whether the skeletal muscle SR contains a distinct ADP-ribosyltransferase or whether the SR used in those studies contained plasma membrane, bearing the GPI-linked form of the enzyme. It should be stressed that even after extensive separation of cardiac sarcoplasmic reticulum and plasma membrane vesicles, using marker enzymes for both types of membranes, these two fractions were found to be considerably cross-contaminated [34].

Other mammalian ADP-ribosyltransferases

The family of known mammalian ADP-ribosyltransferases specific for amino acids other than arginine includes asparagine-specific transferase from bovine brain [35] and diphthamide- and histidine-specific transferase modifying elongation factor 2 in hamster kidney cells [36–39]. Moreover, four ADP-ribosyltransferases (each ~66 kDa) were purified from rat brain [40]. Though non-muscle actin was used routinely in the ADP-ribosylation assay, each of these transferases could label purified alpha subunits of guanine

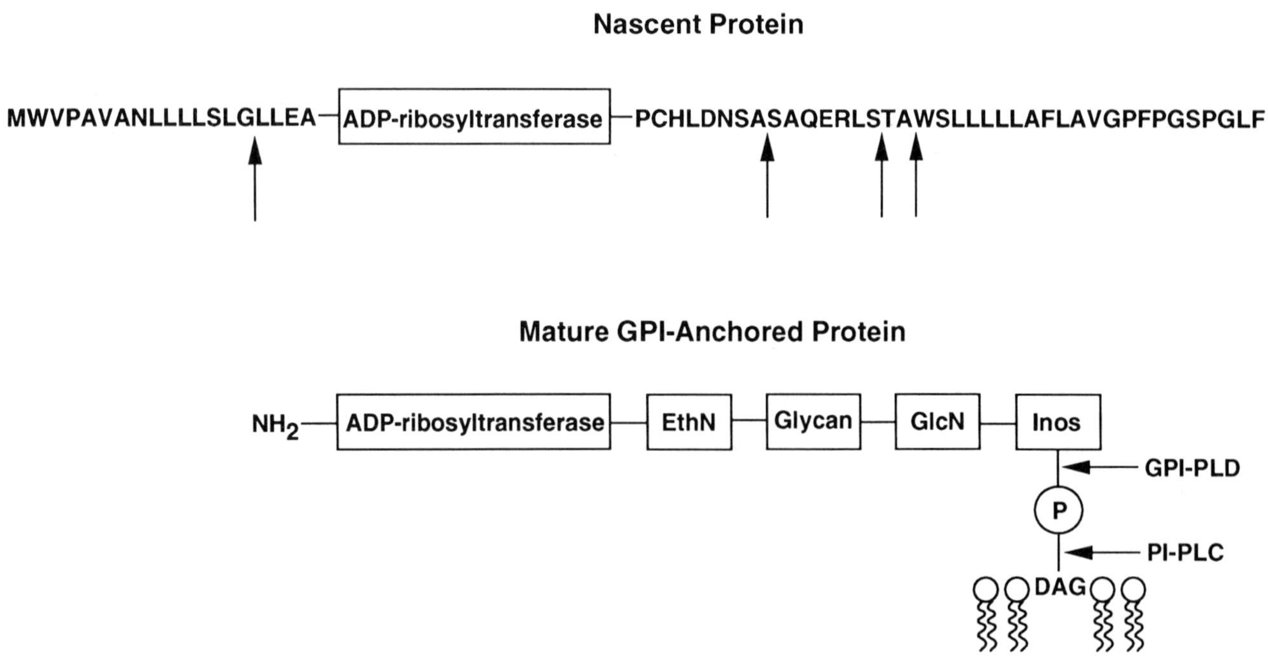

Fig. 1. Proposed structures of nascent and mature forms of GPI-anchored rabbit skeletal muscle ADP-ribosyltransferase. Top: Possible cleavage sites for amino and carboxy terminal signal peptides in the nascent form of the transferase are indicated with arrows. Bottom: Structure of the mature form of the enzyme is proposed, showing the anchoring into membrane via diacylglycerol moiety (EthN: ethanolamine, GlcN: glucosamine, Inos: inositol, DAG: diacylglycerol). The sites of the cleavage by glycosylphosphatidylinositol-specific phospholipase D (GPI-PLD) and phosphatidylinositol-specific phospholipase C (PI-PLC) are indicated.

nucleotide-binding (G) proteins G_s and G_o. Endogenous ADP-ribosylation of G proteins was proposed in other reports [41–47], but the putative transferases involved in these reactions have not been purified.

The immunoglobin heavy chain-binding protein (BiP, also known as 78-kDa glucose-regulated protein) has been postulated to be ADP-ribosylated inside the endoplasmic reticulum of mouse hepatoma cells [48–52]. ADP-ribosyltransferases were also identified in human platelets [53] and rat renal brush border membranes [54–56]. In less direct studies, agents that are inhibitors of mono-ADP-ribosylation were found to cause various impairments of cellular functions, suggesting an involvement of ADP-ribosyltransferases in many vital processes [57–60]. Many of the inhibitors used are, however, not completely specific and might have effects other than those related to the inhibition of ADP-ribosylation.

References

1. Williamson KC, Moss J: Mono-ADP-ribosyltransferases and ADP-ribosylarginine hydrolases: a Mono-ADP-ribosylation cycle in animal cells. In: J. Moss, M. Vaughan (eds) ADP-ribosylating toxins and G proteins: Insights into signal transduction. American Society for Microbiology, Washington, DC, pp 493–510, 1990
2. Moss J, Stanley SJ, Watkins PA: Isolation and properties of an NAD- and guanidine-dependent ADP-ribosyltransferase from turkey erythrocytes. J Biol Chem 255: 5838–5840, 1980
3. Moss J, Stanley SJ, Osborne JC Jr: Effect of self-association on activity of an ADP-ribosyltransferase from turkey erythrocytes. J Biol Chem 256: 11452–11456, 1981
4. Moss J, Stanley SJ: Histone-dependent and histone-independent forms of an ADP-ribosyltransferase from human and turkey erythrocytes. Proc Natl Acad Sci USA 78: 4809–4812, 1981
5. Moss J, Stanley SJ, Osborne JC Jr: Activation of an NAD:arginine ADP-ribosyltransferase by histone. J Biol Chem 257: 1660–1663, 1982
6. Moss J, Osborne CJ Jr, Stanley SJ: Activation of an erythrocyte NAD:arginine ADP-ribosyltransferase by lysolecithin and nonionic and zwitterionic detergents. Biochemistry 23: 1353–1357, 1984
7. Osborne JC Jr, Stanley SJ, Moss J: Kinetic mechanisms of two NAD:arginine ADP-ribosyltransferases: the soluble, salt-stimulated transferase from turkey erythrocytes and choleragen, a toxin from *Vibrio cholerae*. Biochemistry 24: 5235–5240, 1985
8. Yost DA, Moss J: Amino acid-specific ADP-ribosylation. J Biol Chem 258: 4926–4929, 1983
9. West RE Jr, Moss J: Amino acid specific ADP-ribosylation: specific NAD:arginine mono-ADP-ribosyltransferases associated with turkey erythrocyte nuclei and plasma membranes. Biochemistry 25: 8057–8062, 1986
10. Peterson JE, Larew JS-A, Graves DJ: Purification and partial characterization of arginine specific ADP-ribosyltransferase from skeletal muscle microsomal membranes. J Biol Chem 265: 17062–17069, 1990
11. Zolkiewska A, Nightingale MS, Moss J: Molecular characterization of NAD:arginine ADP-ribosyltransferase from rabbit skeletal muscle. Proc Natl Acad Sci USA 89: 11352–11356, 1992
12. Tanigawa Y, Tsuchiya M, Imai Y, Shimoyama M: ADP-ribosyltransferase from hen liver nuclei. J Biol Chem 259: 2022–2029, 1984
13. Tanigawa Y, Tsuchiya M, Imai Y, Shimoyama M: Mono-(ADP-ribosyl)ation of hen liver nuclear proteins suppresses phosphorylation. Biochem Biophys Res Commun 113: 135–141, 1983
14. Tanigawa Y, Tsuchiya M, Imai Y, Shimoyama M: ADP-ribosylation regulates the phosphorylation of histones by the catalytic subunit of cyclic AMP-dependent protein kinase. FEBS Lett 160: 217–220, 1983
15. Ushiroyama T, Tanigawa Y, Tsuchiya M, Matsuura R, Ueke M, Sugimoto O, Shimoyama M: Amino acid sequence of histone H1 at the ADP-ribose-accepting site and ADP-ribose histone-H1 adduct as an inhibitor of cyclic-AMP-dependent phosphorylation. Eur J Biochem 151: 173–177, 1985
16. Tsuchiya M, Tanigawa Y, Ushiroyama T, Matsuura R, Shimoyama M: ADP-ribosylation of phosphorylase kinase and block of phosphate incorporation into the enzyme. Eur J Biochem 147: 33–40, 1985
17. Matsuura R, Tanigawa Y, Tsuchiya M, Mishima K, Yoshimura Y, Shimoyama M: ADP-ribosylation suppresses phosphorylation of the L-type pyruvate kinase. Biochim Biophys Acta 969: 57–65, 1988
18. Mishima K, Terashima M, Obara S, Yamada K, Imai K, Shimoyama M: Arginine-specific ADP-ribosyltransferase and its acceptor protein p33 in chicken polymorphonuclear cells: co-localization in the cell granules, partial characterization, and *in situ* mono(ADP-ribosyl)ation. J Biochem 110: 388–394, 1991
19. Soman G, Mickelson JR, Louis CF, Graves DJ: NAD:guanidino group specific mono ADP-ribosyltransferase activity in skeletal muscle. Biochem Biophys Res Commun 120: 973–980, 1984
20. Soman G, Graves DJ: Endogenous ADP-ribosylation in skeletal muscle membranes. Arch Biochem Biophys 260: 56–66, 1988
21. Hara N, Mishima K, Tsuchiya M, Tanigawa Y, Shimoyama M: Mono(ADP-ribosyl)ation of Ca^{2+}-dependent ATPase in rabbit skeletal muscle sarcoplasmic reticulum and the effect of poly L-lysine. Biochem Biophys Res Commun 144: 856–862, 1987
22. Piron KJ, McMahon KK: Localization and partial characterization of ADP-ribosylation products in hearts from adult and neonatal rats. Biochem J 270: 591–597, 1990
23. Kharadia SV, Huiatt TW, Huang H-Y, Peterson JE, Graves DJ: Effect of an arginine-specific ADP-ribosyltransferase inhibitor on differentiation of embryonic chick skeletal muscle cells in culture. Exp Cell Res 201: 33–42, 1992
24. Larew JS-A, Peterson JE, Graves DJ: Determination of the kinetic mechanism of arginine-specific ADP-ribosyltransferase using a high performance liquid chromatographic assay. J Biol Chem 266: 52–57, 1991
25. Taniguchi M, Tsuchiya M, Shimoyama M: Comparison of acceptor protein specificities on the formation of ADP-ribose-acceptor adducts by arginine-specific ADP-ribosyltransferase from rabbit skeletal muscle sarcoplasmic reticulum with those of the enzyme from chicken peripheral polymorphonuclear cells. Biochim Biophys Acta 1161: 265–271, 1993
26. Okazaki IJ, Zolkiewska A, Nightingale MS, Moss J: Immunological and molecular conservation of mammalian glycosylphosphatidylinositol-linked ADP-ribosyltransferase from skeletal muscle. Submitted, 1994
27. Ferguson MAJ, Williams AF: Cell-surface anchoring of proteins via glycosylphosphatidylinositol structures. Annu Rev Biochem 57: 285–320, 1988
28. Low MG: Glycosylphosphatidylinositol: a versatile anchor for cell surface proteins. FASEB J 3: 1600–1608, 1989
29. Gerber LD, Kodukula K, Udenfriend S: Phosphatidylinositol glycan (PI-G) anchored membrane proteins. J Biol Chem 267: 12168–12173, 1992
30. McMahon KK, Piron KJ: The 52 kDa ADP-ribosylated protein in the rat heart plasma membrane: Is it G_{sa}? In: G.G. Poirier, P. Moreau (eds) ADP-ribosylation reactions. Springer Verlag, NY, pp 377–379, 1992
31. Soman G, Haregewoin A, Hom RC, Finberg RW: Guanidine group specific ADP-ribosyltransferase in murine cells. Biochem Biophys Res Commun 176: 301–308, 1991
32. Hara N, Tsuchiya M, Mishima K, Tanigawa Y, Shimoyama M: ADP-ribosylation of Ca^{2+}-dependent ATPase *in vitro* suppresses the enzyme activity. Biochem Biophys Res Commun 148: 989–994, 1987
33. Taniguchi M, Tanigawa Y, Tsuchiya M, Mishima K, Obara S, Yamada K,

Shimoyama M: Arginine-specific ADP-ribosyltransferase from rabbit skeletal muscle sarcoplasmic reticulum is solubilized as the active form with trypsin. Biochem Biophys Res Commun 164: 128–133, 1988

34. Jones LR, Besch HR, Fleming JW Jr, McConnaughey MM, Watanabe AM: Separation of vesicles of cardiac sarcolemma from vesicles of cardiac sarcoplasmic reticulum. J Biol Chem 254: 530–539, 1979
35. Maehama T, Takahashi K, Ohoka Y, Ohtsuka T, Ui M, Katada T: Identification of a botulinum C_3-like enzyme in bovine brain that catalyzes ADP-ribosylation of GTP-binding proteins. J Biol Chem 266: 10062–10065, 1991
36. Lee H, Iglewski WJ: Cellular ADP-ribosyltransferase with the same mechanism of action as diphtheria toxin and *Pseudomonas* toxin A. Proc Natl Acad Sci USA 81: 2703–2707, 1984
37. Fendrick JL, Iglewski WJ: Endogenous ADP-ribosylation of elongation factor 2 in polyoma virus-transformed baby hamster kidney cells. Proc Natl Acad Sci USA 86: 554–557, 1989
38. Iglewski WJ, Dewhurst S: Cellular mono(ADP-ribosyl)transferase inhibits protein synthesis. FEBS Lett 283: 235–238, 1991
39. Fendrick JL, Iglewski WJ, Moehring JM, Moehring TJ: Characterization of the endogenous ADP-ribosylation of wild-type and mutant elongation factor 2 in eukaryotic cells. Eur J Biochem 205: 25–31, 1992
40. Matsuyama S, Tsuyama S: Mono-ADP-ribosylation in brain: purification and characterization of ADP-ribosyltransferases affecting actin from rat brain. J Neurochem 57: 1380–1387, 1991
41. Feldman AM, Levine MA, Baughman KL, Van Dop C: NAD^+-mediated stimulation of adenylate cyclase in cardiac membranes. Biochem Biophys Res Commun 142: 631–637, 1987
42. Abramowitz J, Jena BP: Evidence for a rabbit luteal ADP-ribosyltransferase activity which appears to be capable of activating adenylyl cyclase. Int J Biochem 23: 549–559, 1991
43. Obara S, Yamada K, Yoshimura Y, Shimoyama M: Evidence for the endogenous GTP-dependent ADP-ribosylation of the α-subunit of the stimulatory guanyl-nucleotide-binding protein concomitant with an increase in basal adenylyl cyclase activity in chicken spleen cell membrane. Eur J Biochem 200: 75–80, 1991
44. Duman RS, Terwilliger RZ, Nestler EJ: Endogenous ADP-ribosylation in brain: initial characterization of substrate proteins. J Neurochem 57: 2124–2132, 1991
45. Donnelly LE, Boyd RS, MacDermont J: $G_s\alpha$ is a substrate for mono(ADP-ribosyl)transferase of NG108-15 cells. Biochem J 288: 331–336, 1992
46. Jacquemin C, Thibout H, Lambert B, Correze C: Endogenous ADP-ribosylation of G_s subunit and autonomous regulation of adenylate cyclase. Nature 323: 182–184, 1986
47. Molina y Vedia L, Nolan RD, Lapetina EG: The effect of iloprost on the ADP-ribosylation of $G_s\alpha$ (the α-subunit of G_s). Biochem J 261: 841–845, 1989
48. Carlsson L, Lazarides E: ADP-ribosylation of the M_r 83,000 stress-inducible and glucose-regulated protein in avian and mammalian cells: Modulation by heat shock and glucose starvation. Proc Natl Acad Sci USA 80: 4664–4668, 1983
49. Ledford BE, Jacobs DF: Translational control of ADP-ribosylation in eukaryotic cells. Eur J Biochem 161: 661–667, 1986
50. Leno GH, Ledford BE: ADP-ribosylation of the 78-kDa glucose-regulated protein during nutritional stress. Eur J Biochem 186: 205–211, 1989
51. Leno GH, Ledford BE: Reversible ADP-ribosylation of the 78 kDa glucose-regulated protein. FEBS Lett 276: 29–33, 1990
52. Staddon JM, Bouzyk MM, Rozengurt E: Interconversion of GRP78/BiP. J Biol Chem 267: 25239–25245, 1992
53. Brune B, Molina y Vedia L, Lapetina EG: Agonist-induced ADP-ribosylation of a cytosolic protein in human platelets. Proc Natl Acad Sci USA 87: 3304–3308, 1990
54. Hammerman MR, Hansen VA, Morrissey JJ: ADP-ribosylation of canine renal brush border membrane vesicle proteins is associated with decreased phosphate transport. J Biol Chem 257: 12380–12386, 1982
55. Kempson SA, Curthoys NP: NAD^+-dependent ADP-ribosyltransferase in renal brush-border membranes. Am J Physiol 245: C449–C456, 1983
56. Kempson SA: Mechanism of stimulation of ADP-ribosyltransferase in the renal brush-border membrane by EDTA. Biochim Biophys Acta 770: 101–104, 1989
57. Duncan MR, Rankin PR, King RL, Jacobson MK, Dell'Orco RT: Stimulation of mono (ADP-ribosyl)ation by reduced extracellular calcium levels in human fibroblasts. J Cell Physiol 134: 161–165, 1988
58. Halldorsson H, Bodvarsdottir T, Kjeld M, Thorgeirsson G: Role of ADP-ribosylation in endothelial signal transduction and prostacyclin production. FEBS Lett 314: 322–326, 1992
59. Juedes MJ, Kass GEN, Orrenius S: m-iodobenzylguanidine increases the mitochondrial Ca^{2+} pool in isolated hepatocytes. FEBS Lett 313: 39–42, 1992
60. Mishima Y, Nishimura T, Muramatsu M, Kominami R: Transcription of mouse ribosomal RNA gene with inactive extracts is activated by NAD^+ *in vitro*. J Biochem 113: 36–42, 1993

Molecular and Cellular Biochemistry **138:** 113–118, 1994.

Target protein for eucaryotic arginine-specific ADP-ribosyltransferase

Mikako Tsuchiya and Makoto Shimoyama
Department of Biochemistry, Shimane Medical University, Izumo 693, Japan

Abstract

Among ADP-ribosyltransferases reported in eucaryotes, arginine-specific transferases from turkey erythrocytes, chicken heterophils and rabbit skeletal muscle have been purified and extensively studied. They were reported to modify a number of proteins *in vitro*. ADP-ribosylation of Ha-*ras*-p21 and transducin by the turkey erythrocyte transferase inhibits their GTPase and GTP-binding activities. Chicken heterophil enzyme modifies several substrate proteins for protein kinases and decreases the phosphate-acceptor activity. Rabbit skeletal muscle Ca^{2+}-ATPase is inhibited by ADP-ribosylation catalyzed by the muscle transferase. Three transferases all ADP-ribosylate small molecular weight guanidino compounds such as arginine, arginine methylester and agmatine and poly-L-arginine and nuclear histones. However, the observation that muscle transferase did not ADP-ribosylate casein or actin, both of which can be modified by the heterophil transferase under the same conditions indicates that substrate specificity of these two enzymes are different. Substrate-dependent effects were observed with polyions of nucleotides such that polyanions stimulate the ADP-ribosylation of possible target protein, p33 by chicken heterophil transferase but has no effect on the modification of casein by the same enzyme. (Mol Cell Biochem **138**: 113–118, 1994)

Key words: ADP-ribosyltransferase, ADP-ribosylation, target protein, arginine, post-translational modification

Introduction

For procaryotes and eucaryotes, there are enzymes which catalyze the transfer of the ADP-ribose moiety from NAD to an arginine residue of specific proteins. Cholera toxin, an arginine-specific ADP-ribosyltransferase [1, 2], ADP-ribosylates α subunit of stimulatory guanine nucleotide binding protein (Gs) to irreversibly inhibit its GTPase activity [3–5]. In photosynthetic bacteria *Rhodospirillum rubrum*, endogenous ADP-ribosyltransferase modifies arginine-101 on the target protein dinitrogenase reductase, resulting in inactivation of the reductase [6, 7]. Recovery from the state of inactivation is achieved through cleavage of the arginine-ADP-ribose linkage by dinitrogenase reductase-activating glycohydrolase [8, 9]. This is the only well-defined example of a regulatory role for reversible ADP-ribosylation on an endogenous functional protein.

For eucaryotic cells, several lines of evidence suggest the presence of an ADP-ribosylation-deADP-ribosylation cycle functioning as a regulatory mechanism of cellular protein activity: (1) arginine-specific ADP-ribosyltransferase activities have been noted in variety of cells and tissues, including turkey and human erythrocytes [10–13], rat liver [14], *Xenopus* oocytes [15], rabbit and pig skeletal muscle, [16, 17], chicken heterophils [18] , human leukocytes [19] and variety of murine cell lines, including lymphocytes [20]; (2) ubiquitous distribution of ADP-ribosylarginine hydrolase together with purification, characterization and molecular cloning of the enzyme have been reported [21–26]; (3) ADP-ribose bound to the arginine residue of proteins *in vivo* can be identified on the chemical stability of the bond [27–30]. To define the role of arginine-specific reversible ADP-ribosylation, the target protein has to be identified. The purpose of this review is to summarize the ADP-ribosylation of protein substrates catalyzed by eucaryotic ADP-ribosyltransferases and the effect of modification on the activity of the substrate proteins. Review articles describing target proteins for the reaction have been published [31, 32].

Address for offprints: M. Tsuchiya, Department of Biochemistry, Simane Medical University, Izumo 693, Japan

Eucaryotic arginine-specific ADP-ribosyltransferase

From turkey erythrocytes, four different arginine-specific ADP-ribosyltransferases have been purified: transferase A [33] and B [12] from soluble fraction, C from membrane fraction and A' from nuclear fraction [13]. Although all four enzymes catalyze the transfer of ADP-ribose from NAD to arginine, they have different properties. Among them transferase A, a 28 kDa single polypeptide activated by salts has been extensively studied. Chicken arginine-specific ADP-ribosyltransferase was first detected in liver nuclear fraction [34]. Further investigation including tissue distribution studies led us to conclude that ADP-ribosyltransferase primarily originated from granules of polymorphonuclear leukocytes (heterophils) located in hepatic interlobular spaces [19, 35]. Although both enzymes are from avian blood cells and have a similar molecular weight (27.5 kDa), heterophil transferase clearly differs from transferase A with regard to effects of salt on the activity. A membrane-associated arginine-specific ADP-ribosyltransferase was purified from rabbit skeletal muscle [36, 37]. Whereas these three enzymes differ both in properties and characteristics, they all ADP-ribosylate simple guanidino-compounds such as arginine, agmatine or arginine methyl ester and poly-L-arginine and histones [10, 34, 38–44]. Other than those general acceptor substrates, a number of proteins were reported to be ADP-ribosylated and their activity modified *in vitro* (Table 1), as described below.

Turkey erythrocyte transferase A

Ovine brain glutamine synthetase, with a critical arginine residue at the catalytic site, was inactivated after modification by transferase A with NAD [45]. ADP-ribosylation of recombinant Ha-*ras*-p21, a proto-oncogene product, decreased both guanine nucleotide binding and GTPase activities [46]. Transferase A also modified α and β subunits of transducin and this ADP-ribosylation decreased both GTPase and guanine nucleotide binding activities [47]. The ADP-ribosylation of chicken erythrocyte tubulin by transferase A inhibited the formation of microtubule assembly [48]. A similar result was obtained with bovine brain microtuble protein [49]. In addition, turkey transferase A was reported to modify several other proteins, including poly(ADP-ribose) polymerase [50] and serum proteins [51].

Chicken heterophil transferase

The enzyme can modify a number of proteins including histones, protamine, phosvitin, casein [34], human Ha-Ras [52] and rabbit skeletal muscle Ca^{2+}-ATPase [44], *in vitro*. Substrate proteins for cAMP-dependent protein kinase (A-kinase) including nuclear histones [53], rabbit phosphorylase kinase [54], and porcine liver pyruvate kinase [55], were ADP-ribosylated by the transferase and the subsequent phosphorylation by A kinase was inhibited. Studies using calf thymus histone H1 [56] and kemptide, the synthetic peptide Leu-Arg-Arg-Ser-Ala-Leu-Gly, representing the phosphorylation site of pyruvate kinase [57] by A-kinase revealed that ADP-ribosylation of arginine(s) immediately N-terminal to the phosphorylatable serine residue inhibited A-kinase-catalyzed phosphorylation. Heterophil transferase ADP-ribosylated the arginine residue(s) of the vasodilator peptide, bradykinin, resulting in a decrease in binding and contractile activities [58].

p33 is a 33 kDa heterophil granule protein coexisting with heterophil transferase [18]. ADP-ribosylation of this protein by the enzyme *in vitro* was extensively studied [59, 18]. The incorporation of ADP-ribose into the protein was

Table 1. ADP-ribosylation catalyzed by arginine-specific transferases and effects on the activity of substrate proteins

	Substrate protein	Effect	References
Turkey erythrocyte transferase A	Glutamine synthetase	Inhibition of activity	[45]
	Ha-ras p21	Inhibition of GTPase and GTP-binding activities	[46]
	Transducin α and β	Inhibition of GTPase and GTP-binding activities	[47]
	Tubulin	Inhibition of microtubule assembly	[48, 49]
Chicken heterophil transferase	Whole histones	Suppression of acceptor-activity for A-kinase	[53]
	Histone H1		[56]
	Phospholylase kinase		[54]
	Pyruvate kinase		[55]
	Kemptide		[57]
	Bradykinin	Inhibition of binding and contractile activities	[58]
	Actins	Inhibition of polymerization	[62]
Rabbit skeletal muscle transferase	Ca^{2+}-ATPase	Inhibition of activity	[41]

strikingly increased by the addition of polyanions such as DNA and poly-L-aspartate [59, 18]. A protein database search revealed that except for one amino acid, the partial amino acid sequence of p33 matched that of *myb*-induced myeloid specific protein-1, Mim-1, the expression of which is regulated by the nuclear oncogene product Myb [60]. Although the expression of the *mim-1* correlates with granulocyte differentiation [61], the function of the protein is not clear. In permeabilized heterophils, the 33 kDa and to a lesser extent 43 kDa proteins, corresponding to p33 and cytosolic actin, respectively, were labelled after incubation with ^{32}P-NAD, thereby indicating that these two proteins are endogenous targets for chicken transferase [18, 62]. β/γ and actins were ADP-ribosylated *in vitro* and the modification inhibited the polymerization of actin monomers, thereby suggesting that ADP-ribosylation of actin catalyzed by the heterophil enzyme may affect the morphology or movement of the cells by altering the state of actin polymerization [62].

Rabbit skeletal muscle ADP-ribosyltransferase

In rabbit skeletal muscle, transferase activity was detected in the sarcoplasmic reticulum (SR), sarcolemma and glycogen pellet fractions [17, 63]. To search for the substrates, incubation of each fraction with ^{32}P-NAD was conducted, followed by SDS-PAGE and autoradiography [40, 63]. Among the several bands detected, Ca^{2+}-ATPase, a major protein in the SR, was studied further. Modification of the ATPase was enhanced by adding poly-L-lysine [40]. In the system reconstituted with partially purified proteins, ADP-ribosylation of the ATPase inhibited the ATP hydrolyzing activity, thereby implicating an effect on Ca^{2+} transport in the SR [41].

Table 2. Substrate specificities of arginine-specific ADP-ribosyltrasferase.

Substrate	ADP-ribosyltransferase	
	Heterophil[a]	Muscle[b]
Poly-L-arginine	+	+
Whole histones	+	+
Whole histones + DNA	++	–
p33	+	–
p33 + DNA	++	–
Casein	+	–
α-actin	+	–
Ca^{2+}-ATPase	+	–
Ca^{2+}-ATPase + poly-L-lysine	+	+

[a]Transferase purified from chicken heterophil granules; [b]transferase purified from rabbit skeletal muscle sarcoplasmic reticulum.

Comparison of substrate specificity

Substrate specificity of ADP-ribosyltransferases purified from chicken heterophil and rabbit skeletal muscle SR was compared using the same assay conditions except for the presence of an SH reagent for the heterophil transferase activity [44]. The results are summarized in Table 2. Although both transferases modified poly-L-arginine and histones, SR transferase did not ADP-ribosylate the substrates for heterophil enzyme, p33, casein or actins. The addition of DNA inhibited the muscle enzyme-catalyzed ADP-ribosylation of histones but it increased the modification of histones or p33 by heterophil transferase. ADP-ribosylation of Ca^{2+}-ATPase, a possible endogenous target, by SR transferase was observed only when poly-L-lysine was present in the reaction mixture, yet the heterophil enzyme was able to modify the protein without polyanion-dependency. Therefore the heterophil transferase has a broader specificity for its substrate proteins compared to the SR enzyme, and the effects of polyions are completely different for reactions catalyzed by the two enzymes.

Acceptor-dependent effect

Acceptor-dependent effects on the arginine-specific ADP-ribosylation *in vitro* were observed with polyions [40, 59] or nucleotides [51]. Although DNA enhanced the modification of p33 or histones by heterophil transferase, the polyanion did not increase the ADP-ribosylation of casein [59]. Poly-L-lysine, required for the ADP-ribosylation of Ca^{2+}-ATPase by SR transferase, did not affect the modification of arginine catalyzed by the same enzyme [40]. The effect of ATP on ADP-ribosylation by turkey transferase A differed, depending on the substrate proteins [51]. Automodification of heterophil transferase decreased or increased the rate of ADP-ribosylation, depending on the substrate proteins [64]. These substances or an automodified domain of the enzyme seem to interact with the substrate protein and change its conformation to a more accessible form rather than to exert a direct effect on the catalytic activity.

Endogenous modification of Gsα

Several studies [10, 65–67] suggested a link between the endogenous arginine-specific ADP-ribosylation and adenylyl cyclase activity, as summarized in an earlier review [31]. Recently, additional data from chicken spleen membrane [68] and human platelet [69] were reported, thereby supporting the hypothesis that endogenous transferase modifies the arginine residue of Gsα, the acceptor site for the cholera toxin-catalyzed ADP-ribosylation, resulting in the activation of adenylyl cyclase.

The eucaryotic arginine-specific ADP-ribosyltransferases seem to have broad substrate specificities *in vitro*: even the SR transferase, which has the highest specificity among the three purified eucaryotic transferases, can modify a greater variety of substrates than the transferase from *Rhodospirillum rubrum* which modifies only the dinitrogenase reductases but never utilizes simple guanidino-compounds such as arginine or poly-L-arginine and other polypeptides as ADP-ribose-acceptors [7].

To ADP-ribosylate a specific target protein *in vivo*, there are probably several mechanisms restricting the ADP-ribose-accepting protein. The observation that permeabilized heterophils incubated with ^{32}P-NAD yielded a 33 kDa labelled protein (probably p33), while many additional proteins were labelled with the cell homogenate [18] implies that the subcellular compartmentalization of the enzyme and acceptor substrate is important for directing the modification of specific proteins. Recently, a novel class of protein, termed a "targetting subunit", has been proposed as a part of protein kinase or phosphatase [70]. The targetting subunit reversibly binds to the catalytic subunit and thereby directs it to the target locus and modulates catalytic and regulatory properties. The possibility that such a protein component might also function with arginine-specific ADP-ribosyltransferases *in vivo*, to exert their physiological roles, merits exploration.

References

1. Moss J, Manganiello VC, Vaughan M: Hydrolysis of nicotinamide adenine dinucleotide by choleragen and its A protomer: Possible role in the activation of adenylate cyclase. Proc Natl Acad Sci USA 73: 4424–4427, 1976
2. Moss J, Vaughan M: Mechanism of action of choleragen: Evidence for ADP-ribosyltransferase acitivity with arginine as an acceptor. J Biol Chem 252: 2455–2457, 1977
3. Cassel D, Pfeuffer T: Mechanism of cholera toxin action: Covalent modification of the guanyl nucleotide-binding protein of the adenylate cyclase system. Proc Natl Acad Sci USA 75: 2669–2673, 1978
4. Gill DM, Meren R: ADP-ribosylation of membrane proteins catalyzed by cholera toxin: Basis of the activation of adenylate cyclase. Proc Natl Acad Sci USA 75: 3050–3054, 1978
5. Northup JK, Sternweis PC, Smigel MD, Schleifer LS, Ross EM, Gilman AG: Purification of the regulatory component of adenylate cyclase. Proc Natl Acad Sci USA 77: 6516–6520, 1980
6. Pope MR, Murrell SA, Ludden PW: Covalent modification of the iron protein of nitrogenase from *Rhodospirillum rubrum* by adenosine diphosphoribosylation of a specific arginine residue. Proc Natl Acad Sci USA 82: 3173–3177, 1985
7. Lowery RG, Ludden PW: Purification and properties of dinitrogenase reductase ADP-ribosyltransferase from the photosynthetic bacterium *Rhodospirillum rubrum*. J Biol Chem 263: 16714–16719, 1988
8. Saari LL, Pope MR, Murrell SA, Ludden PW: Studies on the activating enzyme for iron protein of nitrogenase from *Rhodospirillum rubrum*. J Biol Chem 261: 4973–4977, 1986
9. Pope MR, Saari LL, Ludden PW: N-Glycohydrolysis of adenosine diphosphoribosyl arginine linkages by dinitrogenase reductase activating glycohydrolase (activating enzyme) from *Rhodospirillum rubrum*. J Biol Chem 261: 10104–10111, 1986
10. Moss J, Vaughan M: Isolation of an avian erythrocyte protein possessing ADP-ribosyltransferase activity and capable of activating adenylate cyclase. Proc Natl Acad Sci USA 75: 3621–3624, 1978
11. Moss J, Stanley SJ: Histone-dependent and histone-independent forms of an ADP-ribosyltransferase from human and turkey erythrocytes. Proc Natl Acad Sci USA 78: 4809–4812, 1981
12. Yost DA, Moss J: Amino acid-specific ADP-ribosylation: Evidence for two distinct NAD:arginine ADP-ribosyltransferases in turkey erythrocytes. J Biol Chem 258: 4962–4929, 1983
13. West RE, Jr, Moss J: Amino acid specific ADP-ribosylation: Specific NAD:arginine mono-ADP-ribosyltransferases associated with turkey erythrocyte nuclei and plasma membranes. Biochemistry 25: 8057–8062, 1986
14. Moss J, Stanley SJ: Amino acid-specific ADP-ribosylation: Identification of an arginine-dependent ADP-ribosyltransferase in rat liver. J Biol Chem 256: 7830–7833, 1981
15. Godeau F, Belin D, Koide SS: Mono(adenosine diphosphate ribosyl) transferase in *Xenopus* tissues. Direct demonstration by a zymographic localization in sodium dodecyl sulfate-polyacrylamide gels. Anal Biochem 137: 287–296, 1984
16. Soman G, Tomer KB, Graves DJ: Assay of mono ADP-ribosyltransferase activity by using guanylhydrazones. Anal Biochem 134: 101–110, 1983
17. Soman G, Mickelson JR, Louis CF, Graves DJ: NAD: guanidino group specific mono ADP-ribosyltransferase activity in skeletal muscle. Biochem Biophys Res Commun 120: 973–980, 1984
18. Mishima K, Terashima M, Obara S, Yamada K, Imai K, Shimoyama M: Arginine-specific ADP-ribosyltransferase and its acceptor protein p33 in chicken polymorphonuclear cells: Co-localization in the cell granules, partial characterization and *in situ* mono(ADP-ribosyl)ation. J Biochem 110: 388–394, 1991
19. Obara S, Mishima K, Yamada K, Taniguchi M, Shimoyama, M: DNA-regulated arginine-specific mono(ADP-ribosyl)ation and de-ADP-ribosylation of endogenous acceptor proteins in human neutrophils. Biochem Biophys Res Commun 163: 452–457, 1989
20. Soman G, Haregewoin A, Hom RC, Finberg RW: Guanidine group specific ADP-ribosyltransferase in murine cells. Biochem Biophys Res Commun 176: 301–308, 1991
21. Moss J, Jacobson MK, Stanley SJ: Reversibility of arginine-specific mono(ADP-ribosyl)ation: Identification in erythrocytes of an ADP-ribose-L-arginine cleavage enzyme. Proc Natl Acad Sci USA 82: 5603–5607, 1985
22. Smith KP, Benjamin RC, Moss J, Jacobson MK: Identification of enzymatic activities which process protein bound mono(ADP-ribose). Biochem Biophys Res Commun 126: 136–142, 1985
23. Chang Y-C, Soman G, Graves DJ: Identification of an enzymatic activity that hydrolyzes protein-bound ADP-ribose in skeletal muscle. Biochem Biphys Res Commun 139: 932–939, 1986
24. Moss J, Oppenheimer NJ, West RE, Jr, Stanley SJ: Amino acid specific ADP-ribosylation: Substrate specificity of an ADP-ribosylarginine hydrolase from turkey erythrocytes. Biochemistry 25: 5408–5414, 1986
25. Moss J, Tsai S-C, Adamik R, Chen H-C, Stanley SJ: Purification and characterization of ADP-ribosylarginine hydrolase from turkey erythrocytes. Biochemistry 27: 5819–5823, 1988
26. Moss J, Stanley SJ, Nightingale MS, Murtagh JJ, Jr, Monaco L, Mishima K, Chen H-C, Williamson KC, Tsai S-C: Molecular and immunological

characterization of ADP-ribosylarginine hydrolases. J Biol Chem 267: 10481–10488, 1992

27. Adamietz P, Wielckens K, Bredehorst R, Lengyel H, Hilz H: Subcellular distribution of mono(ADP-ribose) protein conjugates in rat liver. Biochem Biophys Res Commun 101: 96–103, 1981
28. Moss J, Yost DA, Stanley SJ: Amino acid-specific ADP-ribosylation: Stability of the reaction products of an NAD:arginine ADP-ribosyltransferase to hydroxylamine and hydroxide. J Biol Chem 258: 6466–6470, 1983
29. Hsia JA, Tsai S-C, Adamik R, Yost DA, Hewlett EL, Moss J: Amino acid-specific ADP-ribosylation: Sensitivity to hydroxylamine of [cysteine(ADP-ribose)] protein [arginine(ADP-ribose)] protein linkages. J Biol Chem 260: 16187–16191, 1985
30. Payne DM, Jacobson EL, Moss J, Jacobson MK: Modification of proteins by mono(ADP-ribosylation) *in vivo*. Biochemistry 24: 7540–7549, 1985
31. Williamson KC, Moss J: Mono-ADP-ribosyltransferase and ADP-ribosylarginine hydrolases: a mono-ADP-ribosylation cycle in animal cells. In: J. Moss and M. Vaughan (eds), ADP-ribosylating toxins and G proteins: Insights into signal transduction. American Society for Microbiology, Washington, DC, pp 493–510, 1990
32. Shimoyama M, Tsuchiya M, Mishima K: Endogenous arginine-specific ADP-ribosyltransferases and their target proteins. In: S. Tuboi, N. Taniguchi and N. Katunuma (eds). The post-translational modification of proteins: roles on molecular and cellular biology. Japan scientific societies press, Tokyo and CRC press, Boca Raton, pp 183–192, 1992
33. Moss J, Stanley SJ, Watkins PA: Isolation and properties of an NAD- and guanidine-dependent ADP-ribosyltransferase from turkey erythrocytes. J Biol Chem 255: 5838–5840, 1980
34. Tanigawa Y, Tsuchiya M, Imai Y, Shimoyama M: ADP-ribosyltransferase from hen liver nuclei. Purification and characterization. J Biol Chem 259: 2022–2029, 1984
35. Osagao H, Mishima K, Tsuchiya M, Tanigawa Y, Umeno T, Shimoyama M: Localization of an endogenous ADP-ribose acceptor, p33, in polymorphonuclear cell granules in chicken liver interlobular connective tissue. Biochem Biophys Res Commun 180: 64–68, 1991
36. Peterson JE, Larew JS-A, Graves DJ: Purification and partial characterization of arginine-specific ADP-ribosyltransferase from skeletal muscle microsomal membranes. J Biol Chem 265: 17062–17069, 1990
37. Zolkiewska A, Nightingale MS, Moss J: Molecular characterization of NAD:arginine ADP-ribosyltransferase from rabbit skeletal muscle. Proc Natl Acad Sci USA 89: 11352–11356, 1992
38. Moss J, Stanley SJ, Oppenheimer NJ: Substrate specificity and partial purification of a stereospecific NAD- and guanidine-dependent ADP-ribosyltransferase from avian erythrocytes. J Biol Chem 254: 8891–8894, 1979
39. Tsuchiya M, Tanigawa Y, Mishima K, Shimoyama M: Determination of ADP-ribosyl arginine anomers by reverse-phase high-performance liquid chromatography. Anal Biochem 157: 381–384, 1986
40. Hara N, Mishima K, Tsuchiya M, Tanigawa Y, Shimoyama M: Mono(ADP-ribosyl)ation of Ca^{2+}-dependent ATPase in rabbit skeletal muscle sarcoplasmic reticulum and the effect of poly L-lysine. Biochem Biophys Res Commun 144: 856–862, 1987
41. Hara N, Tsuchiya M, Mishima K, Tanigawa Y, Shimoyama M: ADP-ribosylation of Ca^{2+}-dependent ATPase *in vitro* suppress the enzyme activity. Biochem Biophys Res Commun 148: 989–994, 1987
42. Larew JS-A, Peterson JE, Graves DJ: Determination of the kinetic mechanism of arginine-specific ADP-ribosyltransferase using a high performance liquid chromatographic assay. J Biol Chem 266: 52–57, 1991
43. Taniguchi M, Tanigawa Y, Tsuchiya M, Mishima K, Obara S, Yamada K, Shimoyama M: Arginine-specific ADP-ribosyltransferase from rabbit skeletal muscle sarcoplasmic reticulum is solubilized as the active form with trypsin: Partial purification and characterization. Biochem Biophys Res Commun 164: 128–133, 1989
44. Taniguchi M, Tsuchiya M, Shimoyama M: Comparison of acceptor protein specificities on the formation of ADP-ribose acceptor adducts by arginine-specific ADP-ribosyltransferase from rabbit skeletal muscle sarcoplasmic reticulum with those of the enzyme from chicken peripheral polymorphonuclear cells. Biochim Biophys Acta 1161: 265–271, 1993
45. Moss J, Watkins PA, Stanley SJ, Purnell MR, and Kidwell WR: Inactivation of glutamine synthetase by an NAD:arginine ADP-ribosyltransferase. J Biol Chem 259: 5100–5104, 1984
46. Tsai S-C, Adamic R, Moss J, Vaughan M, Manne V, Kung H-F: Effects of phospholipid and ADP-ribosylation on GTP hydrolysis by *Escherichia coli*-synthesized Ha-*ras*-encoded p21. Proc Natl Acad Sci USA 82: 8310–8314, 1985
47. Watkins PA, Kanaho Y, Moss J: Inhibition of the GTPase activity of transducin by an NAD^+:arginine ADP-ribosyltransferase from turkey erythrocytes. Biochem J 248: 749–754, 1987
48. Raffaeilli N, Scaife RM, Purich DL: ADP-ribosylation of chicken red cell tubulin and inhibition of microtubule self-assembly *in vitro* by the NAD^+-dependent avian ADP-ribosyl transferase. Biochem Biophys Res Commun 184: 414–418, 1992
49. Scaife RM, Wilson L, Purich DL: Microtubule protein ADP-ribosylation *in vitro* leads to assembly inhibition and rapid depolymerization. Biochemistry 31: 310–316, 1992
50. Martinez M, Price SR, Moss J, Alvarez-Gonzalez R: Mono(ADP-ribosyl)ation of poly(ADP-ribose)polymerase by cholera toxin. Biochem Biophys Res Commun 181: 1412–1418, 1991
51. Watkins PA, Moss J: Effects of nucleotides on activity of a purified ADP-ribosyltransferase from turkey erythrocytes. Arch Biochem Biophys 216: 74–80, 1982
52. Kawamitsu H, Miwa M, Tanigawa Y, Shimoyama M, Noguchi S, Nisimura S, Ohtsuka E, Sugimura T: A hen enzyme ADP-ribosylates normal human and mutated c-Ha-*ras* oncogene products synthesized in *Escherichia coli.* Proc Japan Acad 62 (B): 102–104, 1986
53. Tanigawa Y, Tsuchiya M, Imai Y, Shimoyama M: ADP-ribosylation regulates the phosphorylation of histones by the catalytic subunit of cyclic AMP-dependent protein kinase. FEBS Lett 160: 217–220, 1983
54. Tsuchiya M, Tanigawa Y, Ushiroyama T, Matsuura R, Shimoyama M: ADP-ribosylation of phosphorylase kinase and block of phosphate incorporation into the enzyme. Eur J Biochem 147: 33–40, 1985
55. Matsuura R, Tanigawa Y, Tsuchiya M, Mishima K, Yoshimura Y, Shimoyama M: ADP-ribosylation suppresses phosphorylation of the L-type pyruvate kinase. Biochim Biophys Acta 969: 57–65, 1988
56. Ushiroyama T, Tanigawa Y, Tsuchiya M, Matsuura R, Ueki M, Sugimoto O, Shimoyama M: Amino acid sequence of histone H1 at the ADP-ribose-accepting site and ADP-ribose histone-H1 adduct as an inhibitor of cyclic-AMP-dependent phosphorylation. Eur J Biochem 151: 173–177, 1985
57. Matsuura R, Tanigawa Y, Tsuchiya M, Mishima K, Yoshimura Y, Shimoyama M: Preferential ADP-ribosylation of arginine-3 in synthetic heptapeptide Leu-Arg-Arg-Ala-Ser-Leu-Gly. Biochem J 253: 923–926, 1988
58. Mishima K, Tanigawa Y, Hara N, Tsuchiya M, Ushiroyama T, Yoshimura Y, Shimoyama M: ADP-ribosylation of bradykinin and effects on its biological activities. J Biochem 103: 342–347, 1988
59. Mishima K, Tsuchiya M, Tanigawa Y, Yoshimura Y, Shimoyama M: DNA-dependent mono(ADP-ribosyl)ation of p33, an acceptor protein in hen liver nuclei. Eur J Biochem 179: 267–273, 1989
60. Yamada K, Tsuchiya M, Mishima K, Shimoyama M: p33, an endogenous target protein for arginine-specific ADP-ribosyltransferase in chicken polymorphonuclear leukocytes, is highly homologous to

mim-1 protein (*myb*-induced myeloid protein-1). FEBS Lett 311: 203–205, 1992

61. Ness SA, Marknell A, Graf T: The v-*myb* oncogene product binds to and activates the promyelocyte-specific *mim*-1 gene. Cell 59: 1115–1125, 1989
62. Terashima M, Mishima K, Yamada K, Tsuchiya M, Wakutani T, Shimoyama M: ADP-ribosylation of actins by arginine-specific ADP-ribosyltransferase purified from chicken heterophils. Eur J Biochem 204; 305–311, 1992
63. Soman G, Graves DJ: Endogenous ADP-ribosylation in skeletal muscle membranes. Arch Biochem Biophys 260: 56–66, 1988
64. Yamada K, Tsuchiya M, Nishikori Y, Shimoyama M: Automodification of arginine-specific ADP-ribosytransferase purified from chicken peripheral heterophils and alteration of the transferase activity. Arch Biochem Biophys 308: 31–36, 1994
65. De Wolf MJS, Vitti P, Ambesi-Impiombato FS, Kohn LD: Thyroid membrane ADP-ribosyltransferase activity. Stimulation by thyrotropin and activity in functioning and nonfunctioning rat thyroid cells in culture. J Biol Chem 256: 12287–12296, 1981
66. Jacquemin C, Thibout H, Lamgert B, Correze C: Endogenous ADP-ribosylation of Gs subunit and autonomous regulation of adenylate cyclase. Nature 323: 182–184, 1986
67. Feldman AM, Levine MA, Baughman KL, Van Dop C: NAD$^+$-mediated stimulation of adenylate cyclase in cardiac membranes. Biochem Biophys Res Commun 142: 631–637, 1987
68. Obara S, Yamada K, Yoshimura Y, Shimoyama M: Evidence for the endogenous GTP-dependent ADP-ribosylation of the α-subunit of the stimulatory guanyl-nucleotide-binding protein concomitant with an increase in basal adenylyl cyclase activity in chicken spleen cell membrane. Eur J Biochem 200: 75–80, 1991
69. Inageda K, Nishina H, Tanuma S: Mono-ADP-ribosylation of Gs by an eukaryotic arginine-specific ADP-ribosyltransferase stimulates the adenylate cyclase system. Biochem Biophys Res Commun 176: 1014–1019, 1991
70. Hubbard MJ, Cohen P: On target with a new mechanism for the regulation of protein phosphorylation. Trends Biochem Sci 18: 172–177, 1993

Molecular and Cellular Biochemistry **138**: 119–122, 1994.

ADP-ribosylarginine hydrolases

Tatsuyuki Takada, Ian J. Okazaki and Joel Moss
Laboratory of Cellular Metabolism, National Heart, Lung, and Blood Institute, National Institutes of Health, Bethesda, MD 20892, USA

Abstract

ADP-ribosylation is a reversible post-translational modification of proteins involving the addition of the ADP-ribose moiety of NAD to an acceptor protein or amino acid. NAD:arginine ADP-ribosyltransferase, purified from numerous animal tissues, catalyzes the transfer of ADP-ribose to an arginine residue in proteins. The reverse reaction, catalyzed by ADP-ribosylarginine hydrolase, removes ADP-ribose, regenerating free arginine. An ADP-ribosylarginine hydrolase, purified extensively from turkey erythrocytes, was a 39-kDa monomeric protein under denaturing and non-denaturing conditions, and was activated by Mg^{2+} and dithiothreitol. The ADP-ribose moiety was critical for substrate recognition; the enzyme hydrolyzed ADP-ribosylarginine and (2-phospho-ADP-ribosyl)arginine but not phosphoribosylarginine or ribosylarginine. The hydrolase cDNA was cloned from rat and subsequently from mouse and human brain. The rat hydrolase gene contained a 1086-base pair open reading frame, with deduced amino acid sequences identical to those obtained by amino terminal sequencing of the protein or of HPLC-purified tryptic peptides. Deduced amino acid sequences from the mouse and human hydrolase cDNAs were 94% and 83% identical, respectively to the rat. Anti-rat brain hydrolase polyclonal antibodies reacted with turkey erythrocyte, mouse and bovine brain hydrolase. The rat hydrolase, expressed in *E. coli*, demonstrated enhanced activity in the presence of Mg^{2+} and thiol, whereas the recombinant human hydrolase was stimulated by Mg^{2+} but was thiol-independent. In the rat and mouse enzymes, there are five cysteines in identical positions; four of the cysteines are conserved in the human hydrolase. Replacement of cysteine 108 in the rat hydrolase (not present in the human enzyme) resulted in a thiol-independent hydrolase without altering specific activity. Rabbit anti-rat brain hydrolase antibodies reacted on immunoblot with the wild-type rat hydrolase and only weakly with the mutant hydrolase. There was no immunoreactivity with either the wild-type or mutant human enzyme. Cysteine 108 in the rat and mouse hydrolase may be responsible in part for thiol-dependence as well as antibody recognition. Based on these studies, the mammalian and avian ADP-ribosylarginine hydrolases exhibit considerable conservation in structure and function. (Mol Cell Biochem **138**: 119–122, 1994)

Key words: ADP-ribosylarginine hydrolase, mono-ADP-ribosylation, ADP-ribosyltransferase, cholera toxin

Introduction

Mono-ADP-ribosylation is a post-translational modification of proteins consisting of the transfer of the ADP-ribose moiety of NAD to an acceptor amino acid or protein catalyzed by an ADP-ribosyltransferase [1]. The function of ADP-ribosylation has been well characterized for bacterial toxins where ADP-ribosylation of proteins alters critical metabolic pathways [2–6]. Cholera toxin ADP-ribosylates an arginine in the stimulatory guanine nucleotide-binding protein of the adenylyl cyclase system leading to an increase in intracellular cyclic AMP [2]. Pertussis toxin modifies a specific cysteine and thus inactivates the inhibitory component of the adenylyl cyclase system [3, 4]. Diphtheria toxin ADP-ribosylates a modified histidine in elongation factor 2, thus inhibiting protein synthesis [5].

It has been postulated that in mammalian and avian cells, as well as bacteria, ADP-ribosylation is a reversible post-translational modification of proteins [1]. NAD:arginine and NAD:cysteine ADP-ribosyltransferases, which have been identified in a variety of animal species, catalyze the forward reaction [1, 7–9]. ADP-ribosylarginine hydrolases remove the ADP-ribose group, regenerating free arginine and completing an ADP-ribosylation cycle that can reversibly

Address for offprints: T. Takada, National Institutes of Health, 9000 Rockville Pike, Building 10, Room 5N307, Bethesda, MD 20892, USA

regulate the function of substrate proteins (Fig. 1). The existence of this cycle was established in the bacterium *Rhodospirillum rubrum* where it regulates dinitrogenase reductase, a key enzyme in nitrogen fixation [10, 11].

Biochemical characterization of ADP-ribosylarginine hydrolase

Among the ADP-ribosyltransferases in animal cells, the arginine-specific transferases have been isolated from turkey erythrocytes [12–14], chicken liver [15], rabbit skeletal muscle [16] and other tissues (see vertebrate mono-ADP-ribosyltransferases). ADP-ribosylarginine hydrolase activity was initially detected in mouse fibroblasts where the ribosylarginine bond was cleaved when ADP-ribosylated histones were used as substrate [17]. An ADP-ribosylarginine hydrolase was identified in, and partially purified from turkey erythrocytes [18]; enzyme activity was stimulated by dithiothreitol (DTT) and further enhanced by Mg^{2+}. Hydrolase activity was also detected in the soluble fraction of rat skeletal muscle, where ADP-ribosylated lysozyme served as substrate [19].

An ADP-ribosylarginine hydrolase was purified ~200,000-fold from the soluble fraction of turkey erythrocytes [20]. The enzyme is a monomeric protein with an apparent size of 39 kDa by sodium dodecylsulfate-polyacrylamide gel electrophoresis and gel permeation chromatography. The ADP-ribosylarginine hydrolase preferentially hydrolyzed the α-anomer of ADP-ribosylarginine [21] which is the stereospecific product of the NAD:arginine ADP-ribosyltransferase and other transferases including the bacterial toxins choleragen and *E. coli* heat-labile enterotoxin [22–25]. The K_m for ADP-ribosylarginine was 65 μM whereas that for (2-phospho-ADP-ribosyl)arginine was 47 μM; the V_{max} for (2-phospho-ADP-ribosyl)arginine, however, was only 39% that of ADP-ribosylarginine [21]. (Phosphoribosyl)arginine and ribosylarginine, products of sequential degradation of ADP-ribosylarginine by phosphodiesterase and phosphatase, were neither substrates nor inhibitors of the hydrolase. Further, hydrolase activity was inhibited by ADP-ribose>ADP>AMP. These data are consistent with the conclusion that the ADP-ribose moiety is important for substrate recognition.

The hydrolase was able to cleave ADP-ribosylguanidine (K_m = 27 μM) and ADP-ribosylagmatine. Thus the arginine moiety was not essential for hydrolysis of the ribosyl-guanidine bond. The hydrolase cleavage product was arginine, and able to serve as substrate for ADP-ribosyltransferase which, in the presence of NAD, regenerated ADP-ribosylarginine [18].

Mg^{2+} (5–10 mM) or Mg^{2+} plus DTT (5–10 mM) activated the hydrolase [18]. Incubation in 20 mM DTT alone at 37°C resulted in rapid loss of enzyme activity but addition of Mg^{2+} stabilized the hydrolase against thermal inactivation [21]. ADP-ribosylarginine hydrolase activity was also inhibited by incubation with NAD and the turkey erythrocyte ADP-ribosyltransferase presumably due to ADP-ribosylation of the hydrolase [20]. In contrast, when incubating the hydrolase in the presence of NAD, transferase and Mg^{2+} or Mg^{2+}

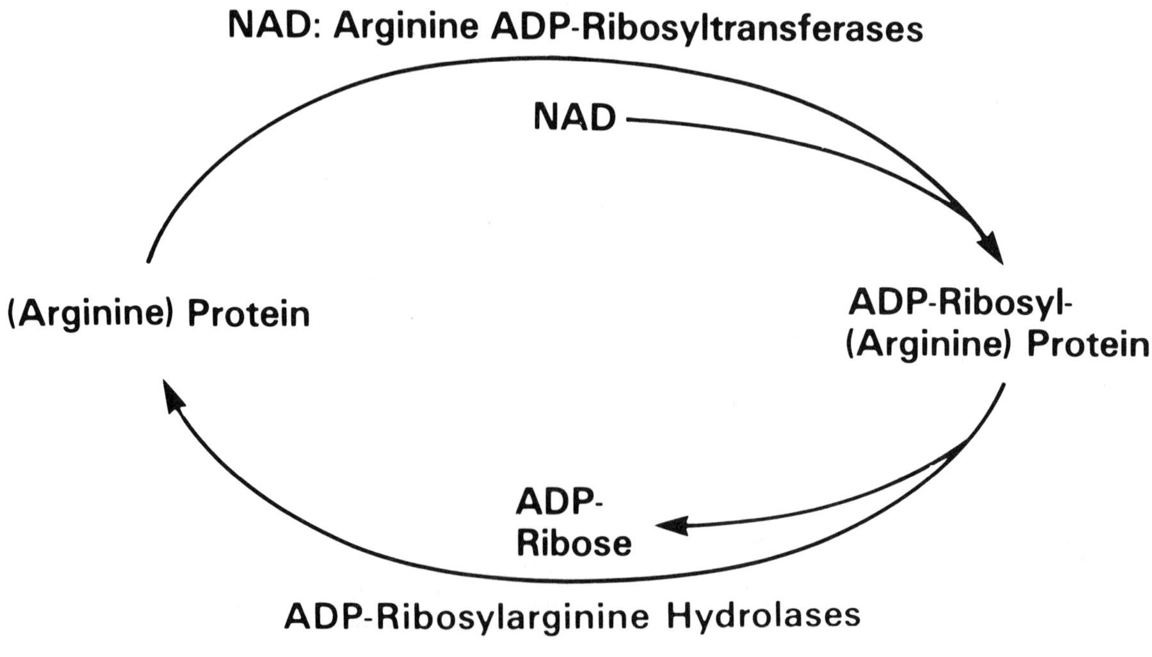

Fig. 1. Reversible mono-ADP-ribosylation of proteins in eucaryotic cells. Proteins are ADP-ribosylated by NAD:arginine ADP-ribosyltransferases and free arginine (protein) is generated by ADP-ribosylarginine hydrolases which remove the ADP-ribose moiety.

plus DTT, there was a decrease in ADP-ribosylation of the hydrolase and no hydrolase inactivation. The protection by Mg^{2+} and DTT from inactivation by transferase was not due to removal of ADP-ribose from an ADP-ribosylated hydrolase but rather, resulted from the ability of Mg^{2+} and DTT to promote an active conformation of the hydrolase, masking critical arginine residues that could otherwise serve as sites for ADP-ribosylation. The hydrolase was also inhibited >80% by 5 mM NaF and 200 mM NaCl; histone had no effect [21]. Both NaCl and histone activated an erythrocyte ADP-ribosyltransferase.

During isolation of the turkey erythrocyte ADP-ribosylarginine hydrolase, both thiol-resistant and thiol-sensitive species were observed depending on the purification procedure [20]. The thiol-resistant enzyme was obtained by sequential chromatography on DE52, phenyl-Sepharose, hydroxyapatite, Sephadex G-25, DE52, Ultrogel AcA54, and mono Q HR 5/5 HPLC columns. The thiol-sensitive hydrolase was purified on DE52, phenyl-Sepharose, hydroxyapatite, Affigel 501, DE52, Sephadex G-25, DE52, Ultrogel AcA54 and mono Q HR 5/5 columns. The hydrolase activity in the supernatant of the erythrocyte homogenate was thiol-dependent possibly due to the presence of endogenous thiols. Partial purification through the hydroxyapatite step yielded a thiol-resistant hydrolase. The thiol-resistant form, purified on organomercurial agarose (Affigel 501), followed by removal of DTT used for elution during purification, generated a thiol-sensitive hydrolase. Similarly, incubation of the thiol-resistant species with $HgCl_2$ followed by addition of thiol, and subsequent gel permeation chromatography, resulted in the isolation of a thiol-sensitive species. When the thiol-sensitive and thiol-resistant forms were incubated with DTT at 4°C prior to N-ethylmaleamide (NEM), enzyme inactivation occurred only with the thiol-sensitive species. After exposure to DTT at 30°C followed by NEM at 4°C, both forms were inactivated. Thiol activation of the hydrolase thus appeared to be a reversible process.

Molecular and immunologic characterization of mammalian ADP-ribosylarginine hydrolase

ADP-ribosylarginine hydrolase was purified ~20,000-fold from rat brain and partially purified, for determination of kinetic and immunologic properties, from mouse, guinea pig, rabbit, pig, sheep and calf brains and numerous rat tissues including brain, heart, lung, liver, kidney, spleen, testis and muscle [26]. ADP-ribosylarginine hydrolase activity in rat and mouse brains was higher than that in guinea pig, rabbit, sheep, pig or calf brains. In addition, rat and mouse hydrolase was stimulated by Mg^{2+} and DTT, whereas guinea pig and calf hydrolase was stimulated by Mg^{2+} but not thiol. The turkey erythrocyte enzyme was enhanced by DTT in the absence of additional Mg^{2+}, although Mg^{2+} caused a further increase in activity. Despite differences in enzymatic properties among species, 39-kDa proteins in partially purified hydrolase preparations from turkey erythrocytes and mouse, rat and bovine brains reacted on immunoblots with rabbit anti-rat brain hydrolase polyclonal antibodies. These data are consistent with partial cross-species conservation of hydrolase sequences.

The rat brain ADP-ribosylarginine hydrolase was cloned from a rat brain cDNA library (Lambda ZAP) [26]. The deduced amino acid sequence obtained from the 1086-base pair open reading frame, was identical to that determined by direct sequencing of the intact enzyme and HPLC-purified tryptic peptides. The hydrolase, synthesized as a glutathione S-transferase fusion protein in *E. coli*, exhibited Mg^{2+} and DTT-dependent activity similar to the native enzyme and reacted with anti-hydrolase antibodies. On Northern analysis, the hydrolase cDNA hybridized to a 1.7=kb band in poly $(A)^+$ RNA from rat and mouse brains but not chicken, rabbit or bovine brains, or HL60 or IMR32 cells. A hydrolase-specific oligonucleotide probe hybridized to a 1.7-kb band in total RNA from various rat tissues but was most abundant in brain, spleen, testis and lung. In general, there was a correlation between hydrolase activity and levels of the 1.7-kb mRNA in the rat tissues.

The role of cysteines in hydrolase activation

The mouse brain ADP-ribosylarginine hydrolase cDNA was amplified by polymerase chain reaction from mouse Quick-clone cDNA and the human hydrolase cDNA was cloned from a human fetal brain Lambda Zap II library [27]. The deduced amino acid sequences from the mouse and human hydrolase cDNA were 94% and 83% identical, respectively, to that of the rat sequence. There were five cysteines in identical positions in the rat and mouse hydrolases, four of which were conserved in human. Cysteine 108 in the rat and mouse sequence was replaced by a serine at position 103 in the native human hydrolase. Whereas the rat and mouse hydrolases were activated by Mg^{2+} and DTT, the human enzyme required only Mg^{2+} for activity and was thiol-independent. Replacing serine 103 in the human hydrolase with cysteine by site-directed mutagenesis converted the hydrolase to a thiol-dependent enzyme. The specific activity did not differ between the wild-type and mutated hydrolase. The K_m values for ADP-ribosylarginine of the wild-type and mutated hydrolase were similar at 26 μM and 46 μM, respectively. Alternatively, substitution of a serine for the cognate cysteine 108 in the DTT-dependent rat hydrolase, resulted in a DTT-independent enzyme, without altering specific activity. In the absence of DTT, cysteine 108 in the rat and mouse hydrolases may form reversible disulfide bonds with other cysteine residues or otherwise be susceptible to oxidation, altering enzyme conformation and activity. The mutated rat hydrolase, synthesized as a fusion protein in *E. coli*, reacted poorly with anti-rat brain hydrolase antibodies on immunoblot suggesting that the single amino acid substitution caused a significant change in immunoreactivity without affecting specific activity.

References

1. Williamson KC, Moss J: Mono-ADP-ribosyltransferases and ADP-ribosylarginine hydrolases: a MonoADP-ribosylation cycle in animal cells. In: J. Moss, M. Vaughan (eds). ADP-ribosylating toxins and G proteins: Insights into signal transduction. American Society for Microbiology, Washington DC, pp 493–510, 1990
2. Moss J, Vaughan M: ADP-ribosylation of guanyl nucleotide-binding proteins by bacterial toxins. Adv Enzymol 61: 303–379, 1988
3. Katada T, Ui M: ADP-ribosylation of the specific membrane protein of C6 cells by islet-activating protein associated with modification of adenylate cyclase activity. J Biol Chem 257: 7210–7216, 1982
4. Ui M: Pertussis toxin as a valuable probe for G-protein involvement in signal transduction. In: J. Moss, M. Vaughan (eds). ADP-ribosylating toxins and G proteins: Insights into signal transduction. American Society for Microbiology, Washington DC, pp 45–77, 1990
5. Collier RJ: Diphtheria toxin: Structure and function of a cytocidal protein. In: J. Moss, M. Vaughan (eds). ADP-ribosylating toxins and G proteins: Insights into signal transduction. American Society for Microbiology, Washington DC, pp 3–9, 1990
6. Aktories K, Just I: Botulinum C2 toxin. In: J. Moss, M. Vaughan (eds). ADP-ribosylating toxins and G proteins: Insights into signal transduction. American Society for Microbiology, Washington DC, pp 79–95, 1990
7. Tanuma S-I, Kawashima K, Endo H: An NAD:cysteine ADP-ribosyltransferase is present in human erythrocytes. J Biochem 101: 821–824, 1987
8. Tanuma S-I, Kawashima K, Endo H: Eukaryotic mono(ADP-ribosyl)-transferase that ADP-ribosylates GTP-binding regulatory G_i protein. J Biol Chem 263: 5485–5489, 1988
9. Tanuma S-I, Endo H: Mono(ADP-ribosyl)ation of G_i by eukaryotic cysteine-specific mono(ADP-ribosyl)transferase attenuates inhibition of adenylate cyclase by epinephrine. Biochim Biophys Acta 1010: 246–249, 1989
10. Pope MR, Murrell SA, Ludden PW: Covalent modification of the iron protein of nitrogenase from *Rhodospirillum rubrum* by adenosine diphosphoribosylation of a specific arginine residue. Proc Natl Acad Sci USA 82: 3173–3177, 1985
11. Lowery RG, Ludden PW: Endogenous ADP-ribosylation in procaryotes. In: J. Moss, M. Vaughan (eds). ADP-ribosylating toxins and G proteins: Insights into signal transduction. American Society for Microbiology, Washington DC, pp 459–477, 1990
12. Moss J, Stanley SJ, Watkins PA: Isolation and properties of an NAD- and guanidine-dependent ADP-ribosyltransferase from turkey erythrocytes. J Biol Chem 255: 5838–5840, 1980
13. Yost DA, Moss J: Amino acid-specific ADP-ribosylation. J Biol Chem 258: 4926–4929, 1983
14. West RE Jr, Moss J: Amino acid-specific ADP-ribosylation: specific NAD:arginine mono-ADP-ribosyltransferases associated with turkey erythrocyte nuclei and plasma membranes. Biochem 25: 8057–8062, 1986
15. Tanigawa Y, Tsuchiya M, Imai Y, Shimoyama M: ADP-ribosyltransferase from hen liver nuclei. J Biol Chem 259: 2022–2029, 1984
16. Soman G, Mickelson JR, Louis CF, Graves DJ: NAD:guanidino group specific mono-ADP-ribosyltransferase activity in skeletal muscle. Biochem Biophys Res Commun 120: 973–980, 1984
17. Smith KP, Benjamin RC, Moss J, Jacobson MK: Identification of enzymatic activities which process protein bound mono(ADP-ribose). Biochem Biophys Res Commun 126: 136–142, 1985
18. Moss J, Jacobson MK, Stanley SJ: Reversibility of arginine-specific mono(ADP-ribosyl)ation: Identification in erythrocytes of an ADP-ribose-L-arginine cleavage enzyme. Proc Natl Acad Sci USA 82: 5603–5607, 1985
19. Chang Y-C, Soman G, Graves DJ: Identification of an enzymatic activity that hydrolyzes protein-bound ADP-ribose in skeletal muscle. Biochem Biophys Res Commun 139: 932–939, 1986
20. Moss J, Tsai S-C, Adamik R, Chen H-C, Stanley SJ: Purification and Characterization of ADP-ribosylarginine hydrolase from turkey erythrocytes. Biochem 27: 5819–5823, 1988
21. Moss J, Oppenheimer NJ, West RE Jr, Stanley SJ: Amino Acid specific ADP-ribosylation: Substrate specificity of an ADP-ribosylarginine hydrolase from turkey erythrocytes. Biochem 25: 5408–5414, 1986
22. Moss J, Stanley SJ, Oppenheimer NJ: Substrate specificity and partial purification of a stereospecific NAD- and guanidine-dependent ADP-ribosyltransferase from avian erythrocytes. J Biol Chem 254: 8891–8894, 1979
23. Oppenheimer NJ: Structural determination and stereospecificity of the choleragen-catalyzed reaction of NAD^+ with guanidines. J Biol Chem 253: 4907–4910, 1978
24. Oppenheimer NJ: ADP-ribosylarginine. Methods Enzymol 106: 399–403, 1984
25. Moss J, Garrison S, Oppenheimer NJ, Richardson SH: NAD-dependent ADP-ribosylation of arginine and proteins by *Escherichia coli* heat-labile enterotoxin. J Biol Chem 254: 6270–6272, 1979
26. Moss J, Stanley SJ, Nightingale MS, Murtagh JJ Jr, Monaco L, Mishima K, Chen H-C, Williamson KC, Tsai S-C: Molecular and immunological characterization of ADP-ribosylarginine hydrolases. J Biol Chem 267: 10481–10488, 1992
27. Takada T, Iida K, Moss J: Cloning and site-directed mutagenesis of human ADP-ribosylarginine hydrolase. J Biol Chem, 268: 17837–17843, 1993

Molecular and Cellular Biochemistry **138**: 123–129, 1994.

Reversible ADP-ribosylation as a mechanism of enzyme regulation in procaryotes

Paul W. Ludden

Department of Biochemistry, University of Wisconsin, Madison, WI 53706, USA

Abstract

Several cases of ADP-ribosylation of endogenous proteins in procaryotes have been discovered and investigated. The most thoroughly studied example is the reversible ADP-ribosylation of the dinitrogenase reductase from the photosynthetic bacterium *Rhodospirillum rubrum* and related bacteria. A dinitrogenase reductase ADP-ribosyltransferase (DRAT) and a dinitrogenase reductase ADP-ribose glycohydrolase (DRAG) from *R. rubrum* have been isolated and characterized. The genes for these proteins have been isolated and sequences and show little similarity to the ADP-ribosylating toxins. Other targets for endogenous ADP-ribosylation by procaryotes include glutamine synthetase in *R. rubrum* and *Rhizobium meliloti* and undefined proteins in *Streptomyces griseus* and *Pseudomonas maltophila*. (Mol Cell Biochem **138**: 123–129, 1994)

Key words: ADP-ribosylation, nitrogen fixation, nitrogenase, dinitrogenase reductase, photosynthetic bacteria, *nif*, *dra*, glutamine synthetase, signal transduction

Introduction

Following the discovery of ADP-ribosylation of target proteins in eucarytoes by bacterial toxins [1–3] it was reasonable to suspect that bacteria might use ADP-ribosylation as a mechanism to control their own metabolism. The first report of an endogenous target for ADP-ribosylation was that of the modification of RNA polymerase in virus-infected *Escherichia coli cells* [4]. This discovery was followed by the description of the ADP-ribosylation of the dinitrogenase reductase protein in *Rhodospirillum rubrum* [5, 6] and, more recently, by the finding that some bacterial glutamine synthetases appear to be targets for endogenous ADP-ribosylation [7, 8]. A few other proteins are also thought to be targets for endogenous ADP-ribosylation in procaryotes including undefined proteins in *Streptomyces griseus* [9] and *Pseudomonas maltophila* [10]. This review will discuss the evidence for the ADP-ribosylation of proteins in bacteria. Most attention will be paid to the well-studied case of ADP-ribosylation of dinitrogenase reductases in *R. rubrum* and related organisms with analysis of the biochemical and genetic aspects of the signal transduction pathway leading to enzyme modification in this system. Glutamine synthetase is an interesting case, because it appears to be concurrently subject to two systems of covalent modification in some bacteria: the well-studied adenylylation pathway and the more recently studied ADP-ribosylation pathway.

Regulation of nitrogenase activity by reversible ADP-ribosylation

In order to discuss the effects and significance of ADP-ribosylation in *R. rubrum* it is necessary to briefly describe the activities of this complex enzyme (see review by Burris [11]. Nitrogenase carries out the reduction of N_2 to ammonium and the enzyme system consists of two proteins. Dinitrogenase reductase (also referred to as the Fe protein) is an α_2 dimer of the *nifH* gene product and contains a single Fe_4S_4 cluster that bridges the two subunits of the protein at cysteine 98 and cystein 132 [12]. Dinitrogenase reductase binds two MgATP molecules [13] and these are hydrolized when an electron is passed from dinitrogenase reductase to the dinitrogenase component of the enzyme [14]. It is the dinitrogenase reductase component of nitrogenase that is the target for ADP-ribosylation. The other component of nitro-

Address for offprints: P.W. Ludden, Department of Biochemistry, University of Wisconsin, Madison, WI 53706, USA

genase is the dinitrogenase protein, which is an $\alpha_2\beta_2$ tetramer of the *nifK* and *D* gene products, respectively [15]. This protein contains two Fe_8S_8 clusters [16] and two molecules of the iron-molybdenum cofactor per protein molecule [17]. The dinitrogenase protein contains the substrate reduction site of the enzyme and it not a target for ADP-ribosylation. Both nitrogenase proteins are extremely oxygen-labile and thus their purifications and all manipulations are performed under anaerobic conditions. Nitrogenase reduce a variety of substrates, including N_2 to ammonium, protons to H_2, and acetylene to ethylene. This last reaction provides a sensitive and easy assay for the enzyme, as the product ethylene is readily measured by a simple gas chromatograph [18].

In vivo, flavodoxins or ferredoxins serve as electron donors; but during *in vitro* studies, dithionite is usually employed as the donor. Dithionite serves both to reduce the dinitrogenase reductase protein and to scavenge any oxygen that enters the system, thus protecting the oxygen-labile nitrogenase proteins. 2 MgATP are hydrolyzed for each electron transferred to dinitrogenase and each reduction of N_2 requires 8 electrons (6 for the reduction of N_2 and 2 for the reduction of 2 protons of H_2 as an obligate side reaction of N_2 reduction) [19].

Dinitrogenase reductase is ADP-ribosylated on arginine 101 by DRAT (Dinitrogenase Reductase ADP-ribosyl Transferase) which uses NAD as the ADP-ribose donor. ADP-ribosylation of dinitrogenase reductase occurs in response to fixed nitrogen sources such as ammonium or glutamine or in response to darkness. The protein is ADP-ribosylated on only one of the two identical subunits of the protein and is completely inactivated by the modification. *In vitro*, modification reaction requires the presence of ADP. Dinitrogenase reductase is activated by removal of the ADP-ribose group by DRAG (Dinitrogenase Reductase Activating Glycohydrolase) in a MgATP- and Mn^{2+}-dependent reaction. The current model for the regulation of nitrogenase by reversible ADP-ribosylation is shown in Fig. 1.

The modifying group was isolated both as the free nucleotide and as a peptide-bound nucleotide [6]. The free nucleotide was obtained by heating the modified protein at pH>8.0 [20] and the peptide-bound form was obtained by treating the protein with subtilisin [6]. In each case, the nucleotide was then purified by chromatography on boronate resin, which binds the hydroxyl groups of the riboses, and finally, by HPLC [21]. The chemical structure of the nucleotide modification of dinitrogenase reductase was determined by mass spectrometric and NMR analysis of the isolated nucleotide [6]. The linkage of the ADP-ribose molecule to the protein was established as an Nα-glycosidic bond to the guanidinium N of arginine, and the position of the arginine residue in the chain was deduced to be arginine 101 (or arg 100, depending on the numbering system used) by comparison of the amino acid sequnce of the hexapeptide isolated with ADP-ribose to the amino acid sequences known for other dinitrogenase reductase proteins. The unique sequence of gly-agr-gly-val-ile-thr is conserved in all dinitrogenase reductases with the exception of a couple containing an isoleucine residue in place of the valine [22, 23]. Of obvious interest is the fact that the target arginine is only 3 residues removed from cysteine98 which serves as one of the sites of ligitation of the FeS cluster to the protein [12]. The stereochemistry of the linkage was established by comparison of the NMR of the anomeric protons of known α and β N-glycoside bonds [25].

It was the discovery of the DRAG enzyme, then called Activating Factor [26] or Membrane Component [27], that led to the elucidation of the nitrogenase ADP-ribosylation system. DRAG removes the ADP-ribose from dinitrogenase reductase, thus activating the enzyme and allowing its assay by the acetylene reduction technique [25, 28, 29]. DRAG is found associated with the chromatophore membranes in extracts of *R. rubrum* (chromatophores are membrane vesicles that contain the photosynthetic apparatus of phototrophic bacteria) and can be eluted with 0.5 M NaCl [26, 27]. DRAG is active in both its membrane-associated and soluble forms. DRAG was initially purified by a combination of ion exchange, hydroxyl apatite, dye affinity and gel filtration chromatography after elution from the chromatophore membranes [29, 30]. It is a very low abundance protein and a 12,000 fold purification is required. Recently a strain of *R. rubrum* that overexpresses DRAG by 150 fold has been constructed, thus making the isolation of DRAG much simpler.

The substrate speciality of DRAG has been studied in detail. DRAG specifically cleaves the α anomer of the N-glycosidic bond between ADP-ribose and the guanidinium N of arginine [25, 31]. DRAG is equally effective in the cleavage of ADP-ribose from native dinitrogenase reductase, oxygen denatured dinitrogenase reductase and proteolytically-cleaved dinitrogenase reductase [32]. It will not cleave the N-glycosidic bond of ribosyl arginine or phosphoribosyl arginine, bit it will cleave guanosine-dansylarginine methyl ester. DRAG cleaves ADPR-arginine (with the amino group of arginine unblocked) very poorly but cleaves the amino-dansylated derivative very effectively [25, 32]. Blocking of the carboxyl group of arginine had no significant effect on activity. The Km for the ADP-ribosylated dinitrogenase reductase was reported as 74 μM while the Kms for the ADP-rybosylated hexapeptide obtained by proteolysis of dinitrogenase reductase and ADP-ribose-dansylarginine methyl ester were 11–12 μM with Vmax at 500–600 pmole ADPR released per min per mg DRAG [25, 32]. These kinetic parameters are consistent with DRAG operating *in vivo* with dinitrogenase reductase as its substrate.

DRAG is inhibited by free ADP-ribose and by α-NAD, but not by β-NAD [32]. The *in vivo* significance of inhibition

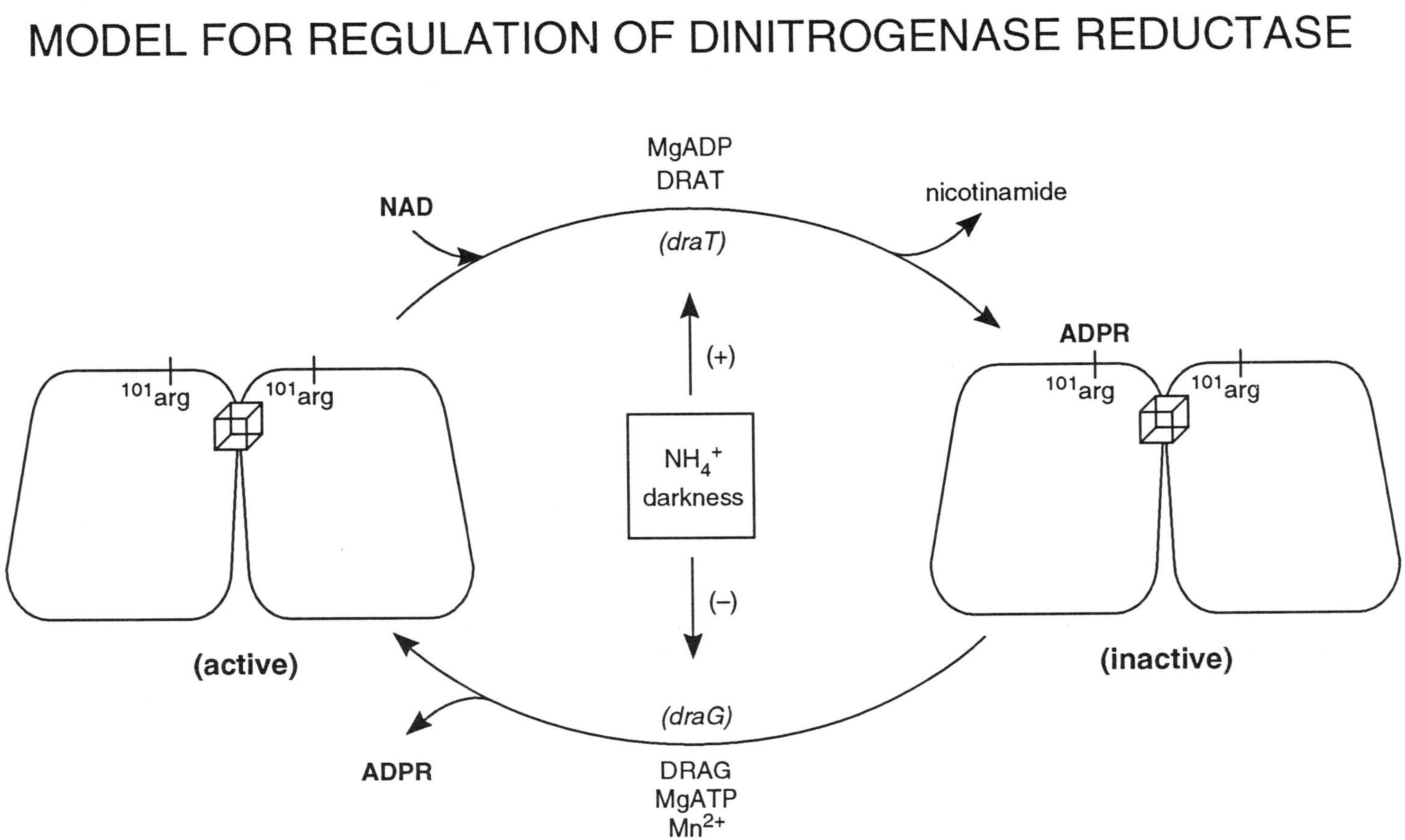

Fig. 1. Model for regulation of dinitrogenase reductase

by ADP-ribose is not known because the fate of ADP-ribose released by the action of DRAG *in vivo* is not known. Hg^{2+}, Cu^{2+}, and Zn^{2+} inhibit the enzyme, but it is not particularly sensitive to iodoacetamide [25]. Sodium borate inhibits DRAG, presumably by binding to the cis-diols of the ribose groups [33]. Neither ammonium ion, the product of nitrogenase, nor glutamine, the first product of NH_4^+ assimilation, inhibits DRAG *in vitro* [25].

DRAG requires MgATP and free divalent metal for its activity with ADP-ribosylated dinitrogenase reductase as product [25, 27, 29]. MgATP is not required when DRAG cleaves low molecular weight compounds [25, 31] and thus it is proposed that the role of MgATP is to bind to the substrate protein, dinitrogenase reductase, causing a conformational change that makes the ADP-ribose linkage accessible to DRAG. No other nucleotide substitutes for ATP, and ADP is a potent inhibitor of the activation of dinitrogenase reductase by DRAG [32]. The free divalent metal is required for activity against all substrates of DRAG and thus the binding site for the divalent metal is thought to be on the DRAG protein [25, 31]. Divalent metals are used in the following order of effectiveness: $Mn^{2+}>Fe^{2+}>Mg^{2+}$ [34].

DRAG exhibits extreme oxygen-lability ($t_{1/2}$ in the air of 2 min) when isolated by the originally published protocols, both of which called for dithionite in the extraction medium to scavenge oxygen [26, 27]. Recently, it has been found that the extreme oxygen-lability arises from the reaction of DRAG with H_2O_2 produced when dithionite is exposed to oxygen (Gary Nielsen, unpublished results). This lability is greatly increased when Mn^{2+} is present, suggesting a Fenton-type reaction on the enzyme. When purified in the absence of dithionite, DRAG is much more resistant to denaturation by oxygen and is stable for hours in air, although activity is still slowly lost. Interestingly, Fe^{2+}, the normal catalyst of Fenton chemistry, does not accelerate denaturation of DRAG, even though it is known to substitute for Mn^{2+} in the DRAG reaction. DRAG can be purified to homogeneity in the absence of dithionite in degassed buffers. The enzyme requires a reducing agent such as dithiothreitol for long-term stability and is stable indefinitely as 4°C under N_2 in solution containing 1 mM dithiothreitol. No tightly bound metals or other prosthetic groups have been detected on DRAG.

The demonstration that dinitrogenase reductase was covalently modified by ADP-ribose suggested that an NAD-dependent modification, similar to that performed by cholera toxin, occurred when the enzyme was inactivated. This hypothesis was confirmed when an NAD-dependent activity was isolated in extracts of *R. rubrum* [35]. The enzyme was named dinitrogenase reductase ADP-ribysol transferase (DRAT) and shown to be distinct from DRAG. In contrast to DRAG, DRAT is very specific in its modification of native dinitrogenase reductase and will not recognise other proteins, denatured dinitrogenase reductase or small molecule substrates [36]. All wild-type dinitrogenase reductases sequenced to date contain an arginine at position 101 or its equivalent and they can all be ADP-ribosylated by DRAT. The dinitrogenase reductase from *Azospirillum amazonense* has been purified but not sequenced, and this protein cannot be modified by DRAT *in vitro*. Several mutant proteins with changes at arg-101 have been constructed and investigated for their ability to be modified [37]. Proteins with lys, his, val, ile, leu, tyr and met at position 101 are unable to be modified; several of these including the arg–> tyr and the arg–> leu mutants, exhibit considerable electron transfer activity, so the protein is in its native conformation but is unable to be modified. DRAT is specific for β-NAD and will not use NADH or NADP as the donor [36]. Etheno-NAD and the NAD analogs in which guanine or hypoxanthine replace the adenine moiety of NAD are effective as ADP-ribose donors.

Like DRAG, DRAT is a non-abundant protein and is purified approximately 20,000 fold to homogeneity by a combination of ion exchange, dye affinity, gel filtration and HPLC steps [36]. The protein is not oxygen-labile and is stabilized in the presence of ADP. DRAT is soluble in crude extracts and is a monomer of 30 kD. It bears no significant sequence similarity to any of the ADP-ribosylating toxins that have been studied and, unlike the toxins, it has no detectable NADase activity in the absebce of its acceptor.

For optimal activity DRAT requires ADP, and the current model holds that the ADP binds to the target protein to put it in a conformation favoring ADP-ribosylation [38]. In the absence of ADP, the dinitrogenase reductase protein can still be ADP-ribosylated, but the observed Km for NAD increases greater than 30 fold in the presence of saturating dinitrogenase reductase without ADP. It is interesting that the effect of ADP on the ADP-ribosylation of dinitrogenase reductases from different organisms varies dramatically, with the target from *Azotobacter vinelandii* being modified at 50% of its maximal rate in the absence of any ADP whereas the target from *R. rubrum* is scarely modified unless ADP is present [38]. ATP does not support ADP-ribosylation of dinitrogenase reductase and serves as an inhibitor of DRAT [36].

ADP-ribosylation of dinitrogenase reductase on arg-101 occurs on only one of the two possible sites and results in complete inactivation of the enzyme. While a trace of activity is always observed with *in vivo*-modified enzyme, ADP-ribosylation of dinitrogenase reductase can be driven to completion *in vitro* and the resulting enzyme exhibits no detectable activity. The accepted hypothesis as to why only one of two identical subunits is modified is that the two arg-101 residues are very close to each other at a point of interaction of the two subunits. The crystal structure of the dinitrogenase reductase protein from *A. vinelandii* has been solved by Georgiadis et al. [39] and the two arg-101 residues occupy a site at the end of the helix near the surface of the protein. These residues are very near the cys-98 residues which serve as two of the 4 ligands to the Fe_4S_4 cluster that bridges the two subunits. ADP-ribosylation appears to block the ability of dinitrogenase reductase to bind to dinitrogenase and transfer electrons. This conclusion is based on the observation that ADP-ribosylation of the dinitrogenase reductase from *Clostridium pasteurianum* blocks its ability to bind tightly to (and inhibit) the dinitrogenase of *A. vinelandii* [40]. The importance of arg-101 to electron transfer between the nitrogenase proteins was made clear by the studies of Lowery et al., who showed that a mutant dinitrogenase reductase with an arg-101 to histidine mutation was able to perform ATP hydrolysis in the presence of the electron-accepting dinitrogenase protein, but that it was unable to transfer electrons [41].

The genes for both DRAT and DRAG (*dra*T and *dra*G, respectively) have been identified, isolated and, in some cases, sequenced in *R. rubrum* [42], *Azospirillum lipoferum* [43], *Azospirillum brasilense* [44] and *Rhodobacter capsulatus* [45]. In the well-studied *R. rubrum* system, the *dra* genes begin 400 bases from the *nifH* gene and read in the opposite direction. At least one open reading frame downstream of *draG* has been identified and sequenced [42]. The *dra* genes are not under the same tight control as the *nif* genes and appear to be expressed at a low level under all conditions of growth [30]. Recently, the vectors containing the *dra* genes under the powerful *nif* promoter have been constructed and inserted into *R. rubrum*. These strains produce 150 to 200 fold higher levels of the gene products without deleterious effects on the organism (Doug Lies and Gary Roberts, unpublished results). The increase in activity coincides with the increase in DRAT and DRAG proteins accumulated, as measured by antibodies; thus neither DRAT nor DRAG is limited by some other gene product or cofactor in the cell.

Constructs with the *draTG* genes under their own promoter have also been inserted into *Klebsiella pneumoniae* [46]. This nitrogen-fixing organism does not possess the post-translational ADP-ribosylation system, but upon transformation with the *draTG* genes is capable of ADP-ribosylating dinitrogenase reductase in response to annomium in the medium, thus demonstrating that the *draT* and *G* gene products are sufficient for ADP-ribosylation *in vivo*.

Mutations have been made in the *draT* and *draG* genes of *R. rubrum* and investigated for their effects on the nitrogenase activity [47]. Both polar and non-polar mutations in *draT* result in the loss of ability to ADP-ribosylate dinitrogenase reductase in response to stimuli. Mutations in *draG* result in the inability of cells to recover nitrogenase activity following ADP-ribosylation. Strains with mutations in *draG* accumulate dinitrogenase reductase in the active form *in vivo* and inactivate the enzyme only when signalled. Thus DRAT activity is regulated *in vivo*; otherwise, dinitrogenase reductase would be modified as it was synthesized and would accumulate in the ADP-ribosylated form [47].

The role of ADP-ribosylation in the regulation of nitrogenase activity *in vivo* has been studied in several bacteria, most thoroughly in the phototrophic organism *Rhodospirillum rubrum* [48]. In *R. rubrum*, dinitrogenase reductase is ADP-ribosylated in response to several physiological signals including darkness and ammonium. The acetylene reduction assay, in which acetylene is reduced to ethylene by nitrogenase [18], allows the nitrogenase to be assayed *in vivo*. A correlation between the modification state of dinitrogenase reductase and the loss or recovery of nitrogenase activity has been made and this data, along with the analysis of mutants, provides proof that the loss of activity observed upon treatment of cells with darkness or ammonium is due to the ADP-ribosylation of dinitrogenase reductase.

Evidence that DRAG is regulated *in vivo* was obtained by a pulse chase experiment in which dinitrogenase reductase was labelled with ^{32}P-ADPR *in vivo* by treating cells with ammonium ion [48]. The cells were then "chased" with an excess of $KH_2{}^{31}PO_4$ and the amount of label associated with dinitrogenase reductase was monitored. While the total cell ^{32}P chased out with a half-life of about 4 hrs, the ^{32}P associated with dinitrogenase reductase chased out with a half-life of over 13 hrs, leading to the conclusion that, under the conditions tested, DRAG was not active *in vivo* and thus DRAG activity is regulated *in vivo*.

The mechanisms by which DRAT and DRAG and the entire ADP-ribosylation cascade are regulated have not been elucidated, but a number of compounds have been tested as potential regulators. ATP, ADP, AMP, NAD(H) and NADP(H) pools have been investigated, but no correlation between modification and pools of these compounds has been established [49–51]. A correlation between the glutamine pool and nitrogenase activity was reported [50], but further investigation showed that the glutamine pool could vary independently of the modification status of the nitrogenase [52]. At this point, the study of nitrogenase modification is an interesting study in signal transduction.

Modification of glutamine synthetase

Glutamine synthetase (GS) from *R. rubrum* has been shown to be modified in an NAD-dependent reaction [7]. This finding was surprising, as GS from enteric bacteria is known to be tightly regulated by a complex cascade that results in the adenylylation of tyrosine397 of GS [53, 54]. Thus an NAD-dependent modification, presumed to be ADP-ribosylation, would mean that GS was the subject of two cascades that result in different chemical modifications of the enzyme. The modification of GS in *R. rubrum* has been demonstrated *in vivo* and *in vitro* and the site of modification has been shown to be near the amino terminus of the protein, far away from tyr397 which is near the carboxy terminus of the protein [7]. *R. rubrum* has been shown to have the adenylylation system and the adenylyl trasferase has been partially purified. The NAD-dependent modification of GS in *R.rubrum* is not performed by DRAT, as the modification is observed in *draT* mutants. Tests of chemical stability of the modification of GS using the protocols of Jacobson et al. indicate that the modification is not an arginyl N-glycosidic bond [55]. The GS3 from *Rhizobium meliloti* has been isolated and shown to be a target for ADP-ribosylation as well, and in this case ADP-ribosylation results in loss of activity of the enzyme [8].

Summary. ADP-ribosylation reactions in procaryotes have not been found as extensively as in eucaryotic organisms, but the demonostration of several well-documented cases and the suggestion of others leads to the conclusion that NAD-dependent ADP-ribosylations may play many more important roles in the endogenous regulation of microbial metabolism.

Acknowledgements

Work from the author's laboratory was supported by grants from the USDA Competitive Grants Program and the National Science Foundation.

References

1. Honjo T, Nishizuka Y, Hayaishi O, and Kato I: Diptheria toxin-dependent adenosine diphosphate ribosylation of aminoacyl transferase II and inhibition of protein synthesis. J Biol Chem 243: 3553–3555, 1968
2. Gill DM, Pappenheimer AM, Brown R, and Kurnick JT: Studies on the mode of action of diptheria toxin. VII. Toxin-stimulated hydrolysis of nicotinamide adenine dinucleotide in mammalian cell extracts. J Exp Med 129: 1–21, 1969
3. Moss J, and Vaughan M: Mechanism of action of choleragen – evidence for ADP-ribosyltransferase activity with arginine as an acceptor, J Biol Chem 252: 2455–2457, 1977
4. Goff CG: Chemical structure of a modification of the E. coli ribonucleic acid polymerase polypeptides induced by bacteriophage T4 infection. J Biol Chem 249: 6181–6190, 1974
5. Ludden PW, Roberts GP: Regulation of nitrogenase activity by reversible ADP-ribosylation, In B. Horecker, et al. (ed) Current Topics of Cellular Regulation, 989, Academic Press Inc.: Orlando. p. 23–55
6. Pope MR, Murrell SA, Ludden PW: Covalent modification of the iron protein of nitrogenase from *Rhodospirillum rubrum* by adenosine diphosphoribosylation of a specific arginyl residue. Proc Natl Acad Sci USA 82: 3173–3177, 1985
7. Woehle DL, Lueddecke BA, Ludden PW: ATP-dependent and NAD-dependent modification of glutamine synthetase from *Rhodospirillum rubrum in vitro*. J Biol Chem 265: 13741–13749, 1990
8. Shatters RG, Liu Y, Kahn ML: Isolation and characterization of a novel glutamine synthetase from *Rhizobium meliloti*. J Biol Chem 268: 469–475, 1993
9. Penyige A, Barabas G, Szabo I, Ensign JC: ADP-ribosylation of membrane proteins of Streptomyces griseus strain 52–1. FEMS Microbiol Lett 69: 293–298, 1990
10. Edmonds C, Griffin GE, Johnstone AP: Demonstration and partial characterization of ADP-ribosylation in *Pseudomonas maltophilia*. Biochem J. 261: 113–118, 1989
11. Burris RH: Nitrogenases. J Biol Chem 266: 9339–9342, 1991
12. Hausinger RP, Howard J: Thiol reactivcity of the nitrogenase Fe-protein from *Azotobacter vinelandii*. J Biol Chem 258: 13486–13492, 1983
13. Tso MYW, Burris RH: The binding of ATP and ADP by nitrogenased components from *Clostridium pasteurianum*. Biochim Biophys Acta 309: 263–70, 1973
14. Ljones T, Burris RH: ATP hydrolysis and electron transfer in the nitrogenase reaction with different combinations of the iron protein and rhe molybdenum-iron protein. Biochim Biophys Acta 275: 93–101, 1972
15. Roberts GP, Brill WJ: Gene-product relationships of the nif regulon of *Klebsiella pneumoniae*. J. Bact. 144: 210–216, 1980
16. Bolin JT, Ronco AE, Lorgan TV, Mortenson LE, Xuong N-H: The unusual metal clusters of nitrogenase: Structural features revealed by x-ray anomolous diffraction studies of the MoFe protein from *Clostridium pasteurianum*. Proc Natl Acad Sci USA 90: 1078–1082, 1993
17. Shah VK, Brill WJ: Isolation of an iron-molybdenum cofactor (FeMo-co) from nitrgenase. Proc Natl Acad Sci USA 74: 3249–3253, 1977
18. Stewart WPD, Fitzgerald GP, Burris RH: *In situ* studies on N_2 fixation using the acetylene reduction technique. Proc Natl Acad Sci USA 58: 2071–2078, 1967
19. Simpson FB, Burris RH: A nitrogen pressue of 50 atmospheres does not prevent evolution of hydrogen by nitrogenase. Science 224: 1095–1097, 1984
20. Dowling TE, Preston GG, Ludden PW: Heat activation of the Fe protein of nitrogenase from *Rhodospirillum rubrum*. J Biol Chem 257: 13987–13992, 1982
21. Pope MR, Murrell SA, Ludden PW: Purification and properties of the heat released nucleotide modifying group from the inactive Fe protein of nitrogenase from *Rhodospirillum rubrum*. Biochemistry 24: 2374–2380, 1985
22. Hausinger RP, Howard JB: The amino acid sequence of the nitrogenase iron protein from *Azotobacter vinelandii*. J Biol Chem 257: 2483–2490, 1982
23. Pretorius I-M, Rawlins DE, O'Neill EG, Jones WA, Kirby R, Woods DR: Nucleotide sequence of the gene encoding the nitrogenase ion protein of *Thiobacillus ferrooxidans*. J Bacteriol, 169: 367–270, 1987
24. Lehman LJ, Fitzmaurice WP, Roberts GP: The cloning and functional characterization of the *nifH* gene of *Rhodospirillum rubrum*. Gene 95: 143–147, 1990
25. Pope MR, Saari LL, Ludden PW: N-glycohydrolysis of adenosine diphosphoribosyl argine linkages by dinitrogenase reductase activating glycohydrolase (activating enzyme) from *Rhodospirillum rubrum*. J Biol Chem 261: 10104–10111, 1986
26. Ludden PW, Burris RH: Activating factor for the iron protein of nitrogenase from *Rhodospirillum rubrum*. Science 194: 424–426, 1976
27. Nordlund S, Eriksson U, Blatscheffsky H: Necessity of a membrane component for nitrogenase activity in *Rhodospirillum rubrum*. Biochem Biophys Acta 462: 187–195, 1977
28. Ludden PW, Burris RH: Removal of an adenine-loke molecule during activation of dinitrogenase reductase from *Rhodospirillum rubrum*. Proc Natl Acad Sci USA 76: 6201–6205, 1979
29. Saari LL, Triplett EW, Ludden PW: Purification and properties of the activating enzyme for iron protein of nitrogenase from the photosynthetic bacterium *Rhodospirillum rubrum*. J Biol Chem 259: 15502–15508, 1984
30. Triplett EW, Wall JD, Ludden PW: Expression of the activating enzyme and Fe protein of nitrogenase from *Rhodospirillum rubrum*. J Bacteriol 152: 786–791, 1982
31. Pope MR, Saari LL, Ludden PW: Fluorometric assay for ADP-ribosylarginine cleavage enzymes. Anal Biochem 160: 68–77, 1987
32. Saari LL, Pope MR, Murrell SA, Ludden PW: Studies on the activating enzyme for iron protein of nitrogenase from *Rhodospirillum rubrum*. J Biol

Chem 261: 4973–4977, 1986

33. Ludden PW: Borate inhibits activation of inactive dinitrogenase reductase from *Rhodospirillum rubrum*. Biochem J 197: 503–505, 1981
34. Nordlund S, Noren A: Dependence on divalent cations of the activation for inactive Fe protein of nitrogenase from *Rhodospirillum rubrum*. Biochem Biophys Acta 791: 21–27, 1984
35. Lowery RG, Saari LL, Ludden PW: Reversible regulation of the iron protein of nitrogenase from *Rhodospirillum rubrum* by ADP-ribosylation *in vitro*. J Bacteriol 166: 513–518, 1986
36. Lowery RG, Ludden PW: Purification and properties of the dinitrogenase reductase inactivating ADP-ribosyltransferase from *Rhodospirillum rubrum*. J Biol Chem 263: 16714–16719, 1988
37. Pierrard J, Ludden PW, Roberts GP: Posttranslational regulation of nitrogenase in *Rhodobacter capsulatus*: Existence of two independent regulatory effects of ammonium. J Bacteriol 175: 1358–1366, 1993
38. Lowery RG, Ludden PW: Effect of Nucleotides on the Activity of Dinitrogenase Reductase ADP-ribosyltransferase from *Rhodospirillum rubrum*. Biochemistry 28: 4956–4961, 1989
39. Georgiardis MM, Komiya P, Chakrabarti P, Woo D, Kornuc JJ, Rees DC: Crystallographic structure of the nitrogenase iron protein from *Azotobacter vinelandii*. Science 257: 1653–1659, 1992
40. Murrell SA, Lowery RG, Ludden PW: ADP-ribosylation of dinitrogenase reductase from *Clostridium pasteurianum* prevents its inhibition of nitrogenase from *Azotobacter vinelandii*. Biochem J 251: 609–612, 1988
41. Lowery RG, Chang CL, Davis LC, McKenna MC, Stephens PJ, Ludden PW: Substitution of Histidine for Arginine-101 of Dinitrogenase reductase Disrupts Electron Transfer to Dinitrogenase. Biochemistry 28: 1206–1212, 1989
42. Fitzmaurice WP, Saari LL, Lowery RG, Ludden PW, Roberts GP: Genes coding for the reversible ADP-ribosylation system of dinitrogenase reductase from *Rhodospirillum rubrum*. Mol Gen Genet 218: 340–347, 1989
43. Fu H-A, Fitzmaurice WP, Roberts GP, Burris RH: Cloning and expression of *draTG* genes from *Azospirillum lipoferum*. Gene 86: 95–98, 1990
44. Zhang Y, Burris RH, Roberts GP: Cloning, sequencing, mutagenesis, and functional characterization of *draT* and *draG* genes from *Azospirillum brasilense*. J Bacteriol 174: 3364–3369, 1992
45. Masephol B, Krey R, Klipp W: The *draTG* region of *Rhodobacter capsulatus* is required for posttranslational regulation of both the molybdenum and the alternative nitrogenase. J Gen Microbiol 1993 (in press)
46. Fu H, Burris RH, Roberts GP: Reversible ADP-ribosylation is demonstrated to be a regulatory mechanism in prokaryotes by heterologous expression. Proc Natl Acad Sci USA 87: 1720–1724, 1990
47. Liang J, Nielsen GM, Lies DP, Burris RH, Roberts GP, Ludden PW: Mutations in the *draT* and *draG* genes of *Rhodospirillum rubrum* result in loss of regulation of nitrogenase by reversible ADP-ribosylation. J Bacteriology 173: 6903–6909, 1991
48. Kanemoto RH, Ludden PW: Effect of ammonia, darkness, and phenazine methosulfate on whole-cell nitrogenase activity and Fe protein modification in *Rhodospirillum rubrum*. J Bacteriol 158: 713–720, 1984
49. Paul TD, Ludden PW: Adenine nucleotide levels in *Rhodospirillum rubrum* during switch-off of whole-cell nitrogenase activity. Biochem J 224: 961–969, 1984
50. Li J, Hu C, Yoch DC: Changes in amino acid and nucleotide pools of *Rhodospirillum rubrum* during switch-off of nitrogenase activity initiated by NH_4+ or darkness. J Bacteriol 169: 231–237, 1987
51. Nordlund S, Hogland L: Studies of the adenylate and pyridine nucleotide pools during nitrogenase 'switch-off' in *Rhodospirillum rubrum*. Plant Soil 90: 203–209, 1986
52. Kanemoto RH, Ludden PW: Amino acid concentrations in *Rhodospirillum rubrum* during expression and switch-off of nitrogenase activity. J Bacteriol 169: 3035–3043, 1987
53. Shapiro BM, Stadtman ER: 5'-adenylyl-)-tyrosine. The novel phosphodiester residue of adenylylated glutamine synthetase from *Escherichia coli*. J Biol Chem 243: 3769–3771, 1968
54. Rhee SG, Chock PB, Stadtman ER: Glutamine synthetase from *Escherichia coli*. Methods Enzymol 113: 213–241, 1985
55. Cervantes-Laurean D, Minter DE, Jacobson EL, Jacobson MK: Protein glycation by ADP-ribose: Studies of model conjugates. Biochemistry 32: 1528–1534, 1993

Molecular and Cellular Biochemistry **138:** 131–133, 1994.

Cellular ADP-ribosylation of Elongation Factor 2

Wallace J. Iglewski

Department of Microbiology and Immunology, University of Rochester Medical Center, 601 Elmwood Avenue, Box 672 Rochester, New York 14642, USA

Abstract

A cellular ADP-ribosyltransferase activity has been found in a variety of animals and tissues. The enzyme transfers ADP-ribose from NAD to elongation factor 2, inactivating the factor and thus inhibiting *in vitro* protein synthesis. Although, the mechanism of action of the cellular enzyme appears similar to diphtheria toxin and *Pseudomonas exotoxin* A, it differs from the toxins in that only a fraction of the EF-2 pool is modified. The endogenously ADP-ribosylated EF-2 has been detected by a variety of methods including two-dimensional electrophoresis and immunoprecipitation with elongation factor 2 antibody. The nature of the cellular ADP-ribosyltransferase and its physiological significance are unknown. (Mol Cell Biochem **138:** 131–133, 1994)

Key words: ADP-ribosyltransferase, elongation factor 2, protein synthesis, bacterial toxins

Introduction

Elongation factor 2 (EF-2) is a single polypeptide of 857 amino acids (96kDa) which catalyzes the last step of the elongation cycle, translocation, during protein synthesis. EF-2 is the only known protein which is ADP-ribosylated and inactivated by diphtheria toxin or *Pseudomonas* toxin in the presence of NAD [1, 2]. ADP-ribosylation occurs at the unique post-translationally modified histidine residue (diphthamide 715) only found in EF-2, which is conserved throughout eukaryotic evolution [3, 4]. The strict conservation of this residue in EF-2 and the identification of a cellular ADP-ribosyltransferase with the same mechanism of action as the toxins [5] suggests that residue 715 is most likely a site of regulatory modification of the factor.

Properties of the cellular ADP-ribosyltransferase

The cellular ADP-ribosyltransferase was initially found in preparations of EF-2 from polyomavirus-transformed baby hamster kidney (pyBHK) cells [5]. Like the bacterial toxins, the cellular enzyme transfers [^{14}C]ADP-ribose from NAD to EF-2. The cellular transferase, however, is immunologically distinct from the bacterial toxins and can also be distinguished from them by its inhibition by cytoplasmic extracts and histamine [5]. Snake venom phosphodiesterase digestion of the [^{14}C]adenosine – labelled EF-2 product of the cellular enzyme yields [^{14}C]AMP, indicating the enzyme is a mono-ADP-ribosyltransferase. Complete digestion of EF-2 with trypsin demonstrates that the same tryptic peptide of EF-2 is ADP-ribosylated by either fragment A of diphtheria toxin or the cellular transferase, suggesting that similar products are formed by the two enzymes. Furthermore, the forward ADP-ribosylation reaction catalyzed by the cellular enzyme is reversed by fragment A of diphtheria toxin, suggesting that the products of the two ADP-ribosylation reactions are identical in stereochemistry [5]. Modification of EF-2 by the cellular transferase inactivates the ability of EF-2 to function in protein synthesis [6], as does the similar modification catalyzed by diphtheria toxin and *Pseudomonas* toxin A [7, 8].

Similar cellular ADP-ribosyltransferase activity has been isolated from beef liver [9], rabbit reticulocytes [10] and rat liver [11]. The transferase activity co-purifies with EF-2 from liver and kidney but cannot be detected in EF-2 preparations from reticulocytes. In rabbit recitulocytes, the cellular transferase activity is among the proteins associated with polyribosomes. The identity of the cellular transferase remains unknown.

Because of the co-purification of the cellular transferase activity with EF-2, the amino acid sequence of the enzymatic domains of diphtheria toxin and *Pseudomonas* toxin were compared to the sequence of mammalian EF-2. Interestingly,

Address for offprints: W. J. Iglewski, Department of Microbiology and Immunology, University of Rochester Medical Center, 601 Elmwood Avenue, Box 672 Rochester, New York 14642, USA

there is a significant similarity between the toxins and the carboxyl terminal third of human EF-2 [6]. This EF-2 subfragment was cloned and expressed in the TNT transcription-translation system (Promega). A 35kDa protein was expressed and detected by Western Blot analysis with EF-2 antibody. The 35kDa fragment of EF-2 lacked transferase activity and residue 715 could not posttranslationally modified to diphthamide for ADP-ribosylation by diphtheria toxin. Presumably, the 35kDa fragment of EF-2 had not folded into the proper native configuration, resulting in an inactive form (unpublished data).

Detection of ADP-ribosylated elongation factor 2

ADP-ribosylated EF-2 has been detected in EF-2 immunoprecipitated directly from cell lysates with EF-2 antibody [12, 13], in polyribosomes from rabbit reticulocytes [10] and in EF-2 fractions purified from polyomavirus transformed baby hamster kidney cells [13] and from rat liver [14]. Three distinct methods have been employed to identify the ADP-ribosylated EF-2 in the fractions. When immunoprecipitated EF-2 was analyzed on polyacrylamide gels, eluted and digested with snake venom phosphodiesterase, AMP, the expected product from a mono-ADP-ribosylated protein, was identified [12, 13]. Alternatively, a purified EF-2 preparation was incubated with excess [^{14}C] nicotinamide under conditions favoring reversal of the ADP-ribosylation reaction and the resulting NAD was identified, indicating that ADP-ribosylated EF-2 was a component of the EF-2 preparation [13]. Finally, EF-2 purified from rat liver was separated into its various isoforms using O'Farrell two-dimensional electrophoresis [14]. Both ADP-ribosylated EF-2 and phosphorylated EF-2 were identified in the EF-2 preparations following electrophoresis. A native form of EF-2 containing both the ADP-ribosylated and phosphorylated modification was not detected. We have also detected ADP-ribosylated EF-2 in EF-2 preparations from polyomavirus transformed baby hamster kidney cells by two-dimensional electrophoresis (unpublished data). The phosphorylated form of EF-2 arises by phosphorylation of threonine residues by a Ca^{2+}/calmodulin-dependent protein kinase III. This protein modification has been shown to inactivate the factor, while dephosphorylation restores its activity [15, 16]. Thus phosphorylation of EF-2 is a mechanism for the regulation of protein synthesis and appears to operate during the cell cycle [17, 18]. The calmodulin kinase III actively is detectable in proliferating cells but is absent in nonproliferating cells [19].

Hypothetical models for the proposed function of ADP-ribosylated EF-2

EF-2 functions in translation at the level of elongation of nascent polypeptide chains. The effect of the cellular transferase is to ADP-ribosylate EF-2, inactivating the enzyme and inhibiting protein synthesis [5, 6]. However, only a fraction of the total EF-2 pool is ADP-ribosylated in cells by the cellular transferase and pulse-chase experiments demonstrate a complete loss of ^{32}P-label from EF-2 within 30 min, suggesting that ^{32}P-labelled ADP-ribose may be removed from EF-2 leading to reactivation of the enzyme [12]. These observations are consistent with ADP-ribosylation of EF-2 being a regulatory modification of the factor.

One model for the proposed function of ADP-ribosylated EF-2 focuses on the arrest of peptide elongation during translocation of secreted proteins across the endoplasmic reticulum membrane [20]. Presumably, ADP-ribosylation of EF-2 may result in the arrest of elongation following synthesis of the signal peptide of the secreted protein. Following translocation of the signal peptide into the lumen of the endoplasmic reticulum, ADP-ribose would be removed from the EF-2, allowing peptide chain elongation to continue. Thus far an enzyme catalyzing cleavage of ADP-ribose from EF-2 has not been identified.

Posttranslational modification of EF-2 may operate in a general fashion to fine-tune the activity of the EF-2 pool without *de novo* synthesis of the factor [5]. The fraction of phosphorylated EF-2 changes in parallel with the total protein synthetic activity during the cell cycle [18, 19]. Similarly, the amount of ADP-ribosylatable EF-2 varies two-fold during the cell cycle of normal and cancerous cell [21] and during ageing of cells in culture [22]. If an enzyme for removal of ADP-ribose from EF-2 is found, it is possible that ADP-ribosylation of EF-2 may act in a manner similar to phosphorylation to regulate the amount of active EF-2 in the cellular pool.

References

1. Collier RJ: Effect of diphtheria toxin on protein synthesis: Inactivation of one of the transfer factors. J Mol Biol 25: 83–89, 1967
2. Iglewski BH, Kabat D: NAD-dependent inhibition of protein synthesis by *Pseudomonas aeruginosa* toxin. Proc Natl Acad Sci USA 72: 2284–2288, 1975
3. Oppenheimer NJ, Bodley JW: Diphtheria toxin. Site and configuration of ADP-ribosylation of diphthamide in elongation factor-2. J Biol Chem 256: 8579–8581, 1981
4. Pappenheimer AM, Jr: Diphtheria toxin. Ann Rev Biochem 46: 69–94, 1977
5. Lee H, Iglewski WJ: Cellular ADP-ribosyltransferase with the same mechanism of action as diphtheria toxin and *Pseudomonas* toxin A. Proc Natl Acad Sci USA 81: 2703–2707, 1984

6. Iglewski WJ, Dewhurst S: Cellular mono(ADP-ribosyl) transferase inhibits protein synthesis. FEBS Lett 173: 113–118, 1991
7. Collier RJ: Diphtheria toxin: Mode of action and structure. Bacteriol Rev 39: 58–85, 1975
8. Iglewski BH, Liu P, Kabat, D: Mechanism of action of *Pseudomonas aeruginosa exotoxin* A: Adenosine diphosphate-ribosylation of mammalian elongation factor 2 *in vitro* and *in vivo.* Infect Immun 15: 138–144, 1977
9. Iglewski WJ, Lee H, Muller P: ADP-ribosyltransferase from beef liver which ADP-ribosylates elongation factor 2. FEBS Lett 173: 113–118, 1984
10. Sitikov AS, Davydova EK, Ovchinnikov LP: Endogenous ADP-ribosylation of elongation factor 2 in polyribosome fraction of rabbit reticulocytes. FEBS Let 176: 261–263, 1984
11. Sayhan O, Ozdermirle M, Nurten R, Bermek E: On the nature of the cellular ADP-ribosyltransferase from rat liver specific for elongation factor 2. Biochem Biophys Res Commun 139: 1210–1214, 1986
12. Fendrick JL, Iglewski WJ: Endogenous ADP-ribosylation of elongation factor 2 in polyomavirus-transformed baby hamster kidney cells. Proc Natl Acad Sci USA 86: 554–557, 1989
13. Fendrick JL, Iglewski WJ, Moehring JM, Moehring TJ: Characterization of the endogenous ADP-ribosylation of wild-type and mutant elongation factor 2 in eukaryotic cells. Eur J Biochem 205: 25–31, 1992
14. Marzouki A, Lavergne JP, Reboud AM: Heterogeneity of native rat liver elongation factor 2. FEBS Lett 255: 72–76, 1989
15. Nairn AC, Palfrey HC: Identification of the major Mr 100,000 substrate for calmodulin-dependent protein kinase III in mammalian cells as elongation factor 2. J Biol Chem 262: 17299–17303, 1987
16. Ryazanov AG, Shestakova EA, Natapov PG: Phosphorylation of elongation factor 2 by EF-2 kinase affects rate of translation. Nature (London) 334: 170–173, 1988
17. Palfrey HC, Nairn AC, Muldoon LL, Villered ML: Rapid activation of calmodulin-dependent protein kinase III in mitogen-stimulated human fibroblasts. J Biol Chem 262: 9785–9792, 1987
18. Celis JE, Madsen P, Ryazanov AG: Increased phosphorylation of elongation factor 2 during mitosis in transformed human amnion cells correlates with decreased rate of protein synthesis. Proc Natl Acad Sci USA 87: 4231–4235, 1990
19. Bagaglio DM, Chen EH, Gorelicks FS, Nairn AC, Hart WN: Phosphorylation of elongation factor 2 in normal and malignant rat glial cells. Cancer Res 53: 2260–2264, 1993
20. Iglewski WJ, Fendrick JL: ADP-ribosylation of elongation factor 2 in animal cells. In: J Moss and M Vaughan (eds). ADP-ribosylating toxins and G proteins: Insights into signal transduction. American Society for Microbiology, Washington, D.C., 1990, pp 511–524
21. Riis B, Rattan SIS, Cavallius J, Clark BC: ADP-ribosylatable content of elongation factor 2 changes during cell cycle of normal and cancerous cells. Biochem Biophys Res Comm 159: 1141–1146, 1989
22. Riis B, Rattan SIS, Derventzi A, Clark BC: Reduced levels of ADP-ribosylatable elongation factor-2 in aged and SV40-transformed human cell cultures. FEBS Lett 226: 45–47, 1990

Molecular and Cellular Biochemistry **138:** 135–140, 1994.

Characterization of botulinum C3-catalyzed ADP-ribosylation of *rho* proteins and identification of mammalian C3-like ADP-ribosyltransferase

Tomohiko Maehama, Nobuyuki Sekine, Hiroshi Nishina, Katsunobu Takahashi, and Toshiaki Katada
Department of Life Science, Tokyo Institute of Technology, Yokohama 227, Japan

Abstract

The exoenzyme C3 produced by *Clostridium botulinum* catalyzes ADP-ribosylation of *rho* gene products which belong to a family of small molecular-weight GTP-binding proteins. The C3 enzyme-catalyzed ADP-ribosylation of *rho* proteins partially purified from bovine brain was markedly activated by certain types of detergents or phospholipids and by endogenous factors present in the brain cytosol. *Rho* A protein that had been expressed in *E. coli* and subsequential purified was readily ADP-ribosylated by the C3 enzyme even in the absence of the activating factors. These results suggest that partially purified *rho* proteins contain an inhibitor, probably *rho* GDI (GDP-dissociation inhibitor for *rho* p21), of C3-catalyzed ADP-ribosylation. The activity of an endogenous enzyme, having the same substrate as botulinum C3 enzyme, was also found in brain cytosol. The enzyme activity was partially purified and characterized. The enzyme appeared to have a molecular mass of approximately 20,000 on a gel filtration and displayed unique properties similar to those observed with the botulinum C3 enzyme. The α-subunits of αβγ-trimeric G proteins which served as the substrates of cholera or pertussis toxin were not ADP-ribosylated by the brain enzyme. (Mol Cell Biochem **138:** 135–140, 1994)

Key words: ADP-ribosylation, botulinum ADP-ribosyltransferase C3, GTP-binding protein, *rho*-gene product

Introduction

ADP-ribosylation, which was first identified in an enzyme reaction catalyzed by diphtheria toxin, is one of the post-translational modifications of cellular proteins. It was later found that other bacterial toxins or enzymes also catalyzed ADP-ribosylation of cellular proteins, such as GTP-binding proteins and nonmuscle actin. The ADP-ribosylating toxins can be divided into at least four types, defined by the amino acid used as an ADP-ribose acceptor. Diphtheria toxin utilizes diphthamide which is an amino acid derived from histidine and present study in eukaryotic elongation factor 2. Pertussis toxin uses cysteine as an ADP-ribose acceptor, but this toxin rather selectively modifies a specific cysteine residue in the α-subunits of G proteins. The toxins that use arginine as their ADP-ribose acceptors may be subdivided on the basis of their substrate specificity, *i.e.*, cholera toxin for G proteins and botulinum C2 toxin for nonmuscle actin. The fourth amino acid utilized for an ADP-ribose acceptor is asparagine; the botulinum C3 exoenzyme produced by *Clostridium botulinum* type C strain [1] ADP-ribosylates a specific asparagine residue located in the putative effector domain of *rho* gene products (*rho* p21s), a family of the small molecular-weight GTP-binding proteins [2]. Although the definitive functions of *rho* proteins have not been fully established, recent studies have revealed that the GTP-binding proteins are responsible for the regulation of cell proliferation and differentiation as a result of their interactions with the cytoskeletal system [3–11].

Several ADP-ribosylating enzymes similar to the bacterial toxins have been recently identified and characterized in eukaryotic cells. NAD:arginine ADP-ribosyltransferases were purified from turkey erythrocytes [12] and rabbit skeletal muscle [13]. A cDNA encoding the enzyme was cloned from the latter tissue [14]. NAD:cysteine ADP-ribosyltransferase was also identified and purified from hu-

Address for offprints: T. Katada, Department of Physiological Chemistry, Faculty of Pharmaceutical Sciences, University of Tokyo, Hongo, Tokyo 113, Japan

man erythrocyte cytosol [15]. However, there has been no report of an endogenous ADP-ribosylating enzyme similar to botulinum C3 enzyme. In the present study, we describe the characteristics of botulinum C3-induced ADP-ribosylation of *rho* proteins and a novel C3-like endogenous enzyme partially purified from bovine brain. The properties of the endogenous enzyme have been investigated and compared with botulinum C3 ADP-ribosyltransferase [16, 17].

Characteristics of botulinum C3-catalyzed ADP-ribosylation of rho *proteins*

In previous studies, we reported that botulinum C3-catalyzed ADP-ribosylation of 22/25-kDa *rho* proteins was markedly stimulated by the addition of factors present in bovine brain cytosol when substrate proteins, partially purified from the brain cytosol, were used [18]. The factor-induced stimulation was profoundly affected by the ionic composition of the reaction mixture [18]. In addition to the endogenous factors, the botulinum C3-catalyzed ADP-ribosylation of *rho* proteins was also activated by certain types of detergents and phospholipids [16]. The major findings obtained in these studies are summarized as follows. (a) Botulinum C3-catalyzed ADP-ribosylation of the cytoplasmic 22/25-kDa proteins was stimulated by anionic detergents, such as cholate or deoxycholate, at their optimum concentrations, but inhibited by CHAPS or Lubrol-PX (Fig. 1). (b) NAD glycohydrolase activity of botulinum C3 enzyme, observed in the absence of substrate proteins, was stimulated by any of the detergents tested (Fig. 1). (c) Certain types of phospholipids stimulated, while other phospholipids either did not activate or inhibit, the botulinum C3-catalyzed ADP-ribosylation. (d) Botulinum C3 enzyme-catalyzed ADP-ribosylation appeared to be enhanced synergistically by the endogenous factors and detergents or phospholipids. Thus, it is very likely that detergents act not only on botulinum C3 enzyme but also on the substrate proteins in a concentration-dependent manner since CHAPS and Lubrol-PX exerted their stimulatory effect only on the NAD glycohydrolase activity of botulinum C3 enzyme despite their inhibitory action on the ADP-ribosylation of the added substrate protein (Fig. 1).

Recent reports have revealed that *rho* proteins form a tight complex with *rho* GDI (GDP-dissociation inhibitor for *rho* p21) in cells and that the *rho* GDI inhibits ADP-ribosylation catalyzed by the botulinum C3 enzyme probably through a masking of the modification site on *rho* proteins [19–22]. *Rho* proteins contain a C-A-A-X consensus sequence in the carboxyl termini, and the post-translational addition of a

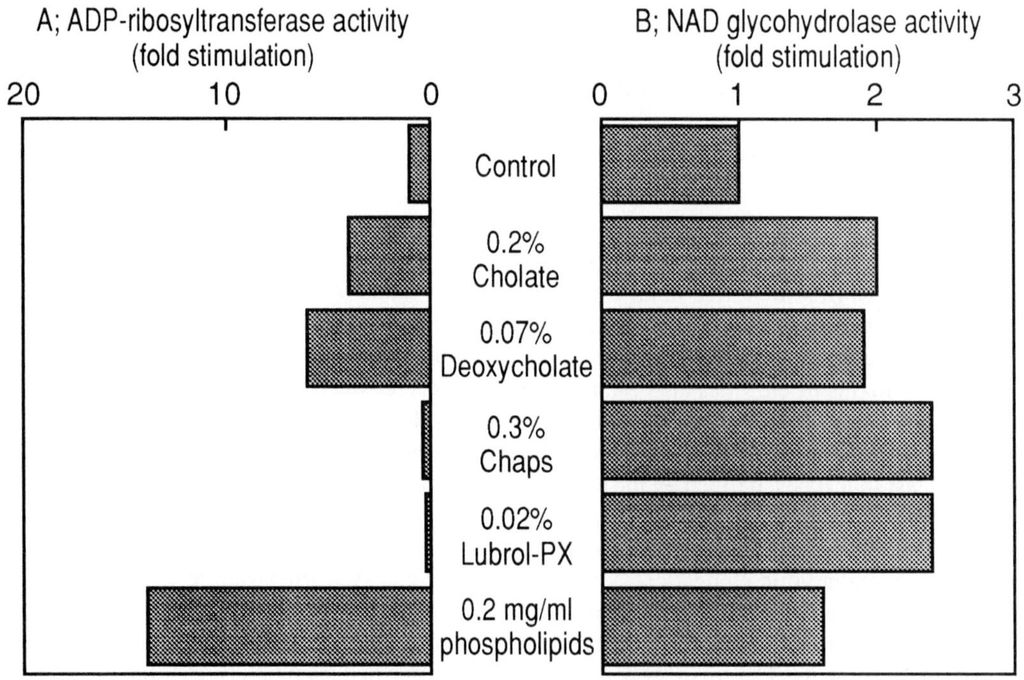

Fig. 1. Effects of various detergents and phospholipids on botulinum C3-catalyzed reactions. A, the partially purified cytoplasmic substrates (4 μg; approximately 3 pmol of the substrate activity) were incubated at 37°C for 1 min in 25 μl of a reaction mixture containing botulinum C3 enzyme (10 ng) and [^{32}P]NAD in the presence of the indicated concentration of detergents or crude phospholipids (0.2 mg/ml). The initial rate of ADP-ribosylation was measured and the results are expressed as fold stimulation of the control activity. B, [4-^{3}H-*nicotinamide*]NAD (20 μM) was incubated at 37°C for 3 h botulinum C3 enzyme in the presence of the additions indicated, and the activities are expressed as fold stimulation of the control activity.

polyisoprenoid (geranylgeranyl group) to the cysteine through a thioesther linkage appears to be essential for the association of the GTP-binding proteins with *rho* GDI [20]. Thus, the post-translationally processed form of *rho* proteins was, but the unprocessed form was not, capable of interacting with *rho* GDI [20, 23].

Based on these findings, we further investigated the effect of the endogenous factors on botulinum C3-catalyzed ADP-ribosylation of recombinant *rho*A protein which had been expressed and purified from *E. coli.* As shown in Fig. 2, the rate of ADP-ribosylation was extremely low when substrate proteins partially purified from bovine bran cytosol were incubated with botulinum C3 enzyme (Fig. 2a). However, this reaction was markedly enhanced by the addition of the endogenous factors. On the other hand, the ADP-ribosylation of purified recombinant *rho*A protein, which lack an isoprenylated carboxy terminal cysteine, occurred very rapidly even in the absence of the endogenous factors (Fig. 2b). These results suggest that partially purified *rho* proteins contain a significant amount of *rho* GDI. Indeed, if the substrate proteins were further purified, the rate of botulinum C3-catalyzed ADP-ribosylation was increased even without addition of the endogenous factor (data not shown). Moreover, the activation of ADP-ribosylation by certain types of detergents or phospholipids may be due to the dissociation of *rho* protein-*rho* GDI complex. However, this does not exclude the possibility that the action of the endogenous factors results from their interaction directly with botulinum C3 enzyme rather than substrate proteins, as has been reported for ADP-ribosylation factor, which interacts and stimulates the ADP-ribosyltransferase activity of cholera toxin [24].

Materials and methods

Identification of botulinum C3-like endogenous ADP-ribosyltransferase in bovine brain cytosol

As described above, certain types of detergents and phospholipids could stimulate the enzyme activity of botulinum C3 ADP-ribosyltransferase. Based on these findings, we sought a novel enzyme activity similar to the botulinum

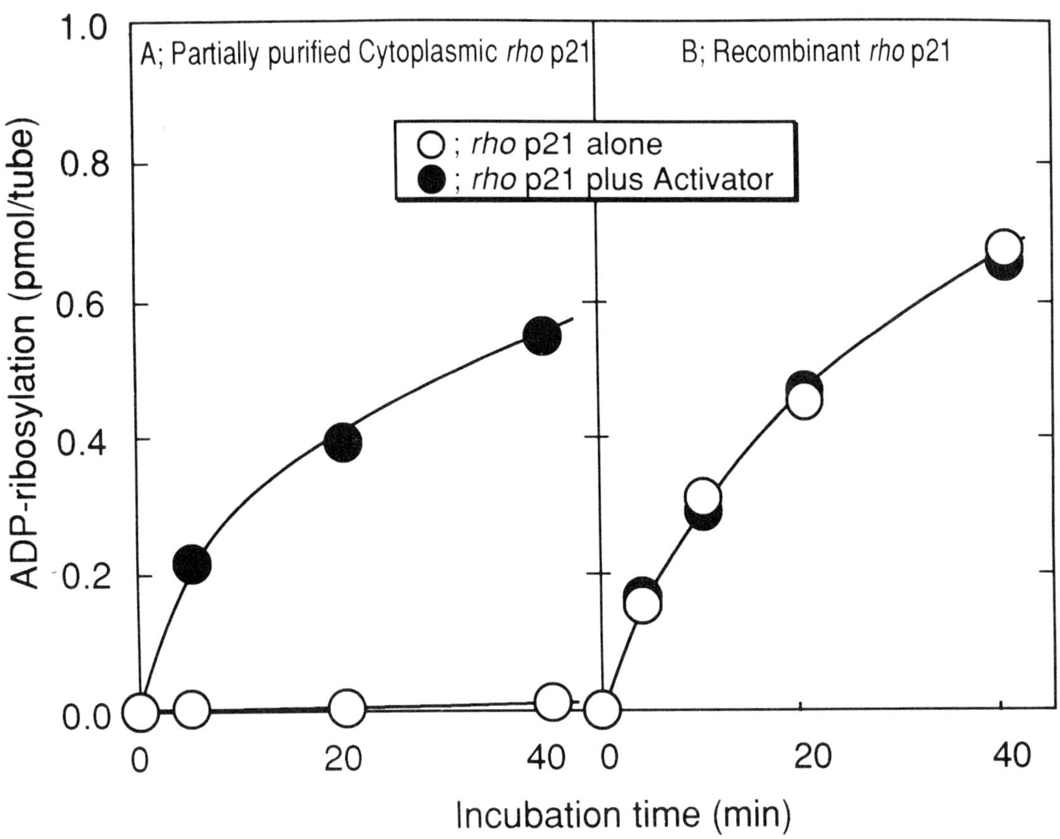

Fig. 2. Effects of endogenous factors on botulinum C3-catalyzed ADP-ribosylation of purified recombinant *rho* p21 and partially cytoplasmic substrate. The partially purified cytoplasmic substrate (A) or the purified *rho* p21 expressed in *E. coli* (B) were incubated at 37°C with botulinum C3 enzyme and [^{32}P]NAD in the presence (●) or the absence (○) of the endogenous factors which were partially purified from bovine brain cytosol. At the indicated times, aliquots were withdrawn, and [^{32}P]ADP-ribosylated protein was then measured.

C3 enzyme and found its activity in bovine brain cytosol. The endogenous enzyme activity present in the cytosol was partially purified by sequential column chromatography on DEAE-Toyopearl, AF-blue-Toyopearl, and Superose 12 [17]. Although the enzyme activity was extremely low in the early purification steps, the AF-blue affinity chromatography was highly effective for enzyme purification. After the AF-blue column, the activity of ADP-ribosyltransferase became feasible to be measured by a filtration assay. As shown in Fig. 3, the molecular weight of the C3-like enzyme appeared to be approximately 20,000, based on the elution profile from the Soperose 12 gel filtration column. The partially purified enzyme thus obtained had a specific activity of 10–20 pmol/min/mg of protein under the present conditions which was less than one-tenth of the purified botulinum C3 enzyme.

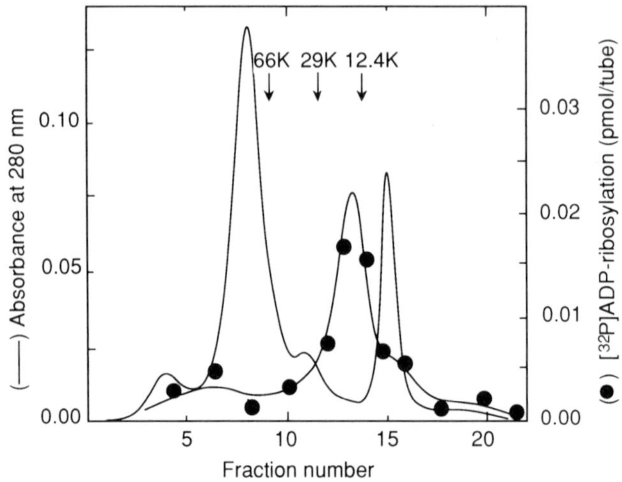

Fig. 3. Superose 12 column chromatography of C3-like endogenous ADP-ribosyltransferase. An endogenous ADP-ribosyltransferase eluted from the AF-blue Toyopearl was fractionated on a column of Superose 12, and the enzyme activity (●) was assayed. The absorbance at 280 nm of the eluted protein was also monitored (——). The *Arrows* on the top indicate the eluted positions of molecular weight markers; *66 K, 29 K,* and *12.4 K* were bovine serum albumin, carbonic anhydrase, and cytochrome c, respectively.

ADP-ribosylation sites of *rho* proteins catalyzed by botulinum C3- and the C3-like endogenous enzymes were next compared by means of tryptic digestion of ^{32}P-radiolabeled substrates (Fig. 4). Digestion with trypsin resulted in conversion of the radiolabeled 25/22-kDa proteins to a 6-kDa fragment via an intermediate of 11-kDa fragment regardless of whether bovine brain membrane C3 substrates had been ADP-ribosylated by the brain (*lanes 2, 4,* and *6*) or botulinum C3 (*lanes 1, 3,* and *5*) enzyme. Both the 22/25-kDa proteins partially purified from brain membrane and those obtained from the cytosol served as substrate for the C3-like enzyme. When the cytoplasmic substrates that had been [^{32}P]ADP-ribosylated by the two enzymes were subjected to

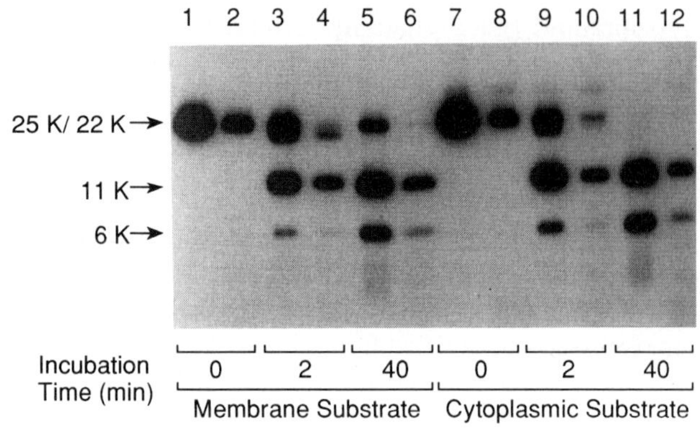

Fig. 4. Limited tryptic digestion of substrate proteins [^{32}P]ADP-ribosylated by botulinum C3 or the C3-like endogenous enzyme. The C3 substrates (approximately 2 pmol of GTPγS-binding activity) that had been purified from the brain membranes (*lanes 1–6*) or cytosol (*lanes 7–12*) were incubated with botulinum C3 (24 ng; *lanes 1, 3, 5, 7, 9,* and *11*) or the C3-like endogenous enzyme (2.4 μg; *lanes 2, 4, 6, 8, 10,* and *12*) at 37°C in a reaction mixture containing [^{32}P]NAD. The reaction mixtures were further supplemented with trypsin and then incubated at 37°C. At the indicated times, aliquots were withdrawn and subjected to SDS-polyacrylamide gel (20%) electrophoresis and autoradiography.

analysis by limited tryptic digestion (*lanes 7* to *12*), essentially the same results were obtained. Moreover, pre-treatment of purified *rho* proteins with C3-like enzyme and NAD decreased their ability to serve as a substrate for the botulinum C3 enzyme. Thus, the sites modified by the brain and botulinum C3 enzymes appeared to be closely related, if not identical, to each other.

Results

Characteristics of ADP-ribosylation reaction catalyzed by brain C3-like enzyme

The characteristics of the ADP-ribosylation reaction catalyzed by brain C3-like enzyme were next compared with those of botulinum C3 enzyme. The activity of brain endogenous enzyme was stimulated by low concentrations of deoxycholate and phospholipids as observed with botulinum C3 enzyme. Moreover, the addition of endogenous factors initially identified as activators supporting botulinum C3-catalyzed ADP-ribosylation also stimulated the ADP-ribosylation of *rho* proteins catalyzed by the C3-like enzyme. The C3-like enzyme activity was destroyed by incubation at 90°C for 5 min, whereas the botulinum C3 ADP-ribosyltransferase was heat-stable. However, treatment with trypsin or *N*-ethylmaleimide did not modify the activities of either C3-like or botulinum C3 enzymes. Reducing agents such as dithiothreitol and 2-mercaptoethanol slightly enhanced the activity of brain C3-like enzyme. Substrate

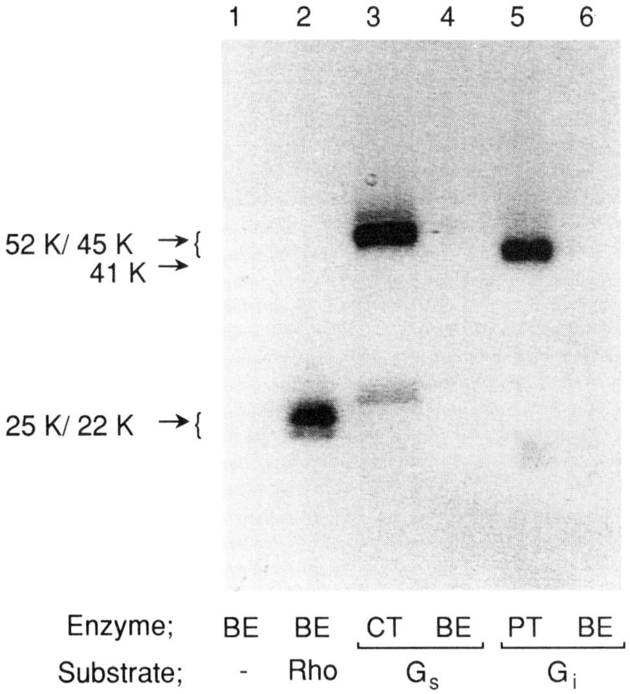

Fig. 5. Substrate specificity for the C3-like endogenous ADP-ribosyltransferase. Botulinum C3 substrates (Rho, approximately 5 pmol of GTPγS-binding activity; *lane 2*), G_s (2 pmol; *lanes 3* and *4*) or G_i (2 pmol; *lanes 5* and *6*) purified from bovine brain membranes were incubated for 2 h with 40 ng of the brain C3-like endogenous enzyme (BE; *lanes 1, 2, 4* and *6*), 5 µg of cholera toxin (CT; *lane 3*) or 0.5 µg of pertussis toxin (PT; *lane 5*) in a reaction mixture containing [^{32}P]NAD. The incubation mixture further contained 25 µg of ADP-ribosylation factor (*lanes 3–4*) partially purified from bovine brain cytosol. The [^{32}P]ADP-ribosylated proteins were then subjected to SDS-polyacrylamide gel (13.5%) electrophoresis and autoradiography.

specificity for the brain C3-like enzyme is illustrated in Fig. 5. The brain C3-like enzyme catalyzed only the ADP-ribosylation of 25/22-kDa *rho* proteins (*lane 2*). The a-subunits of trimeric G proteins, G_s and G_i , substrates for cholera and pertussis toxins, respectively, were not ADP-ribosylated by the brain endogenous enzyme (*lanes 4* and *6*).

Discussion

In the present study, we characterized the properties of botulinum C3-catalyzed ADP-ribosylation of *rho* proteins and identified a C3-like endogenous ADP-ribosyltransferase in bovine brain cytosol. The site (Asn^{41}) of *rho* proteins for botulinum C3-catalyzed ADP-ribosylation appeared to be also modified by an endogenous C3-like enzyme. Interestingly, the activity of the endogenous enzyme was profoundly affected by certain types of phospholipids or endogenous activating factors in a manner similar to botulinum C3 enzyme. If a signal involving phospholipids or the endogenous factor is generated under physiological conditions, the C3-like ADP-ribosyltransferase would be activated to modify the GTP-binding proteins. Thus, it is tempting to speculate that such an ADP-ribosylation of the signal-coupling proteins might occur in intact cells.

References

1. Aktories K, Frevert J: ADP-ribosylation of a 21–24 kDa eukaryotic protein(s) by C3, a novel botulinum ADP-ribosyltransferase, is regulated by guanine nucleotide. Biochem J 247: 363–368, 1987
2. Sekine A, Fujiwara M, Narumiya S: Asparagine residue in the *rho* gene product is the modification site for botulinum ADP-ribosyltransferase. J Biol Chem 264: 8602–8605, 1989
3. Rubin E J, Gill D M, Boquet P, Popoff M R: Functional modification of a 21-kilodalton G protein when ADP-ribosylated by exoenzyme C3 of *Clostridium botulinum.* Mol Cell Biol 8: 418–426, 1988
4. Chardin P, Boquet P, Madaule P, Popoff M R, Rubin E J, Gill D M: The mammalian G protein *rhoC* is ADP-ribosylated by *Clostridium botulinum* exoenzyme C3 and affects actin microfilaments in Vero cells. EMBO J 8: 1087–1092, 1989
5. Paterson H F, Self A J, Garrett M D, Just I, Aktories K, Hall A: Microinjection of recombinant $p21^{rho}$ induces rapid changes in cell morphology. J Cell Biol 111: 1001–1007, 1990
6. Nishiki T, Narumiya S, Morii N, Yamamoto M, Fujiwara M, Kamata Y, Sakaguchi G, Kozaki S: ADP-ribosylation of the *rho/rac* proteins induces growth inhibition, neurite outgrowth and acetylcholine esterase in cultured PC-12 cells. Biochem Biophys Res Commun 167: 265–272, 1990
7. Mohr C, Just I, Hall A, Aktories K: Morphological alterations of *Xenopus* oocytes induced by valine-14 $p21^{rho}$ depend on isoprenylation and are inhibited by *Clostridium botulinum* C3 ADP-ribosyltransferase. FEBS Lett 275: 168–172, 1990
8. Ridley A J, Hall A: The small GTP-binding protein rho regulates the assembly of focal adhesions and actin stress fibers in response to growth factors. Cell 70: 389–399, 1992
9. Morii N, Teru-uchi T, Tominaga T, Kumagai N, Kozaki S, Ushikubi F, Narumiya S: A *rho* gene product in human blood platelets. J Biol Chem 267: 20921–20926, 1992
10. Kishi K, Sasaki T, Kuroda S, Itoh T, Takai Y: Regulation of cytoplasmic division of *Xenopus* embryo by *rho* p21 and its inhibitory GDP/GTP exchange protein (*rho* GDI). J Cell Biol 120: 1113–1121, 1993
11. Toratani S, Katada T, Yokosawa H: Botulinum ADP-ribosyltransferase C3 induces elevation of the vitelline coat of ascidian eggs. Biochem Biophys Res Commun: in press.
12. Yost D A, Moss J: Amino acid-specific ADP-ribosylation. J Biol Chem 258: 4926–4929, 1983
13. Peterson J E, Larew J S-A, Graves D J: Purification and partial characterization of arginine-specific ADP-ribosyltransferase from skeletal microsomal membranes. J Biol Chem 265: 17062–17069, 1990
14. Zolkiewska A, Nightingale M S, Moss J: Molecular characterization of NAD:arginine ADP-ribosyltransferase from rabbit skeletal muscle. Proc Natl Acad USA 89: 11352–11356, 1992
15. Tanuma S, Kawashima K, Endo H: Eukaryotic mono(ADP-ribosyl)transferase that ADP-ribosylates GTP-binding regulatory G_i protein. J Biol Chem 263: 5485–5489, 1988

16. Maehama T, Ohoka Y, Ohtsuka T, Takahashi K, Nagata K, Nozawa Y, Ueno K, Ui M, Katada T: Botulinum ADP-ribosyltransferase activity as affected by detergents and phospholipids. FEBS Lett 263: 376–380, 1990
17. Maehama T, Takahashi K, Ohoka Y, Ohtsuka T, Ui M, Katada T: Identification of a botulinum C_3-like enzyme in bovine brain that catalyzes ADP-ribosylation of GTP-binding proteins. J Biol Chem 266: 10062–10065, 1991
18. Ohtsuka T, Nagata K, Iiri T, Nozawa Y, Ueno K, Ui M, Katada T: Activator protein supporting the botulinum ADP-ribosyltransferase reaction. J Biol Chem 264: 15000–15005, 1989
19. Kuroda S, Kikuchi A, Hirata K, Masuda T, Kishi K, Sasaki T, Takai Y: Cooperative function of *rho* GDS and *rho* GDI to regulate *rho* p21 activation in smooth muscle. Biochem Biophys Res Commun 185: 473–480, 1992
20. Kikuchi A, Kuroda S, Sasaki T, Kotani K, Hirata K, Katayama M, Takai Y: Functional interactions of stimulatory and inhibitory GDP/GTP exchange proteins and their common substrate small GTP-binding protein. J Biol Chem 267: 14611–14615, 1992
21. Regazzi R, Kikuchi A, Takai Y, Wollheim C B: The small GTP-binding proteins in the cytosol of insulin-secreting cells are complexed to GDP-dissociation inhibitor proteins. J Biol Chem 267: 17512–17519, 1992
22. Bourmeyster N, Stasia M-J, Garin J, Boquet P, Vignais P V: Copurification of rho protein and the rho-GDP dissociation inhibitor from bovine neutrophil cytosol. Effect of phosphoinositides on rho ADP-ribosylation by the C3 exoenzyme of *Clostridium botulinum*. Biochemistry 31: 12863–12869, 1992
23. Hori Y, Klkuchi A, Isomura M, Katayama M, Miura Y, Fujioka H, Kaibuchi K, Takai Y: Post-translational modifications of the C-terminal region of the *rho* protein are important for its interaction with membranes and the stimulatory and inhibitory GDP/GTP exchange proteins. Oncogene 6: 515–522, 1991
24. Kahn R A, Gilman A G: Purification of a protein cofactor required for ADP-ribosylation of the stimulatory regulatory component of adenylate cyclase by cholera toxin. J Biol Chem 259: 6228–6234, 1984

Molecular and Cellular Biochemistry **138:** 141–148, 1994.

ADP-ribosylation of the molecular chaperone GRP78/BiP

Barry E. Ledford[1] and Gregory H. Leno[2]
[1]*Department of Biochemistry and Molecular Biology, Medical University of South Carolina, SC 29425, USA* [2]*Department of Biochemistry, University of Mississippi Medical Center, Jackson, MS 39216, USA*

Abstract

Starvation of mouse hepatoma cells for essential amino acids or glucose results in the ADP-ribosylation of the molecular chaperone BiP/GRP78. Addition of the missing nutrient to the medium reverses the reaction. The signal mediating the response to environmental nutrients involves the translational efficiency. An inhibitor of proteins synthesis, cycloheximide, or reduced temperature, both of which reduce translational efficiency, stimulate the ADP-ribosylation of BiP/GRP78. Inhibition of N-linked glycosylation of proteins results in the overproduction of BiP/GRP78. The over produced protein is not ADP-ribosylated suggesting that this is the functional form of BiP/GRP78. The over produced BiP/GRP78 can, however, be ADP-ribosylated if the cells are starved for an essential amino acid. BiP/GRP78 resides in the lumen of the endoplasmic reticulum where it participates in the assembly of secretory and integral membrane proteins. ADP-ribosylation of BiP/GRP78 during starvation is probably part of a nutritional stress response which conserves limited nutrients by slowing flow through the secretory pathway. (Mol Cell Biochem **138:** 141–148, 1994).

Key words: BiP/GRP78, ADP-ribosylation, nutrient starvation, hepatoma secretion

Introduction

The synthesis of albumin by cultured mouse hepatoma cells was shown to be regulated by the availability of essential amino acids [1–3]. These studies followed those of Morgan and Peters [4] which established that serum protein synthesis in normal rat liver was regulated by the availability of dietary protein. The mouse hepatoma cell line, Hepa, was being used in our studies because it secreted serum proteins in amounts comparable to hepatocytes; and because of the ease in controlling the environmental conditions. Limiting the availability of essential amino acids to Hepa cells was shown to reduce the translational efficiency (peptides/message/minute) of the total cellular message pool; and to selectively turn off the synthesis of the secretory proteins albumin, α-fetoprotein, and transferrin [1–3]. Lee and Iglewski [5] had identified an endogenous ADP-ribosyltransferase that modified the same residue on elongation factor 2 as does diphtheria toxin. Furthermore, numerous microbial toxins had been shown to ADP-ribosylate GTP binding proteins; and the protein synthetic apparatus includes a number of essential GTP-binding proteins. It, therefore, seemed reasonable that an ADP-ribosylation event could be responsible for the modulation of protein synthesis during nutritional stress. A series of experiments was designed to test this hypothesis. The results showed that when Hepa cells were starved for essential amino acids, a protein with a relative molecular weight of 80 kDa was ADP-ribosylated. This was subsequently shown to be a consequence, rather than the cause, of the reduced translational efficiency [6]. The protein was isolated and partially sequenced. Amino-terminal sequence analysis identified the protein as the 78-kDa glucose-regulated-protein, GRP78, [7] which is identical to the immunoglobulin heavy chain binding protein, BiP [8–10]. BiP is a molecular chaperone which resides in the lumen of the endoplasmic reticulum and functions there in the folding and assembly of proteins entering the secretory pathway [11–14]. BiP was overproduced in Hepa cells by inhibiting N-linked glycosylation. The overproduced BiP was shown to be the un-ADP-ribosylated form suggesting that this is

Address for offprints: B. Ledford, Dept. of Biochemistry and Molecular Biology, Medical University of South Carolina, 171 Ashley Avenue, Charleston, SC 29425 USA

the functional form of the protein; and that ADP-ribosylation renders BiP nonfunctional [7]. The ADP-ribosylation of BiP was shown to be a freely reversible process indicating that the ratio of modified to unmodified BiP is determined by the relative activities of an ADP-ribosyltransferase and a glycohydrolase [15]. It is this balance of enzymatic activities that is shifted in response to nutritional stress.

This review will describe the identification and characterization of this unique ADP-ribosylation event, and discuss its possible significance.

Results

The initial attempt to determine if translational changes were accompanied by an ADP-ribosylation event involved incubation of Hepa cells with ^{3}H-adenosine during tryptophan starvation. The conditions of starvation had been shown to alter the activity of the protein synthetic appratus [1, 3]. The incorporated adenosine was separated by SDS-PAGE and visualized by fluorography. Total cell homogenates from tryptophan-starved cultures were compared with those from fed cultures. Figure 1 shows the results. The smear of incorporated adenosine is due to nucleic acid synthesis. Lane A shows the pattern from the fed culture; and Lane B shows the result of tryptophan starvation. A distinct band with an apparent molecular weight of 80 kDa (indicated by the arrow) is labeled in lane B [6].

The signal was separated from the background by preparation of a postribosomal supernatant. Figure 2 shows the results of fractionating the Hepa cell homogenate by differential centrifugation. Lane A is the total homogenate; and lane D is the postribosomal supernatant. The adenosine-labeled band was almost completely recovered in the postribosomal supernatant. The protein was clearly too small to be elongation factor 2, which is 97 kDa. The intensely labeled band near the dye front has been shown the vary in intensity from experiment to experiment; and is probably RNA [6].

The incorporated label was shown to be ADP-ribose by cleavage with snake venom phosphodiesterase. The cleavage product co-chromatographed with 5′ AMP on a Dowex AG-1X2 resin. Additionally, the incorporation of adenosine could be blocked by incubation of the cells with nicotinamide during starvation [6].

The ADP-ribosylation of the 80 kDa protein was shown to occur when Hepa cells were deprived of any essential amino acid. Furthermore, the reaction was shown to occur in a variety of cell lines including human and chicken. Later, a variety of mouse tissues were surveyed and shown to ADP-ribosylate the protein [6]. It is thus ubiquitous with no species or tissue restriction.

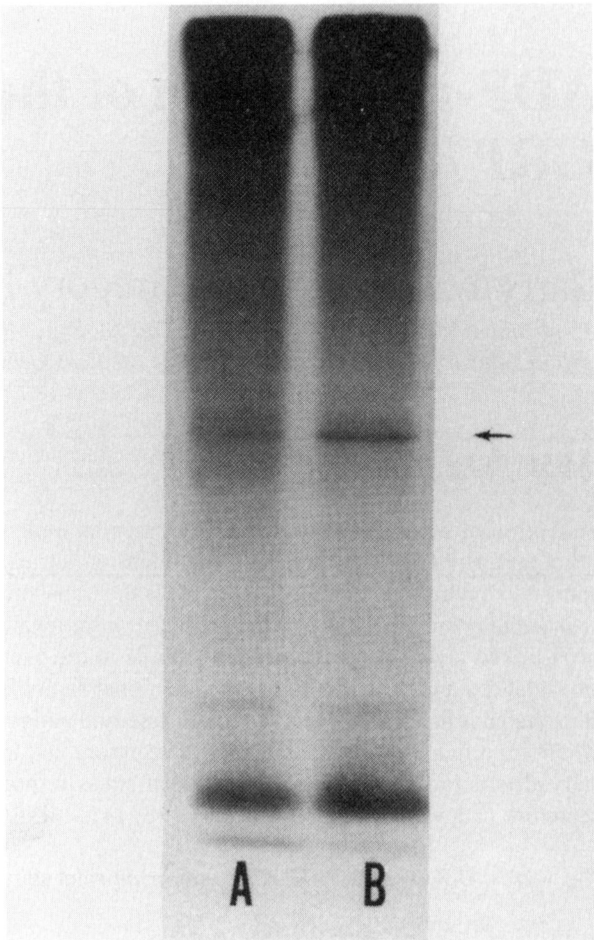

Fig. 1. Effect of tryptophan starvation on adenosine incorporation. The mouse hepatoma cell line, Hepa, was starved for tryptophan for 2 h in the presence of [^{3}H]adenosine. Lane A, fed control; lane B, tryptophan-starved culture. Starvation stimulates adenosine incorporation into an 80-kDa band, P80 (arrow).

The remarkable specificity of the reaction is illustrated in Fig. 3. Postribosomal supernatants were analyzed by 2-D electrophoresis. The top two panels show Coomassie blue stained gels. The top panel is from a fed culture. The middle panel is from a leucine-starved, ^{3}H-adenosine-labeled, culture. The lower panel is a fluorogram of the gel in the middle panel. Overlaying the fluorogram and the gel identified the spot indicated by the arrows on the stained gels to be the ADP-ribosylated protein. Even by overexposing the fluorogram, only a single spot, out of hundreds, was labeled with adenosine.

Addition of ADP-ribose to a protein should make it more acidic and shift its isoelectric point toward the anode. The spot immediately to the cathode side of the labeled spot was, therefore, suspected as the un-ADP-ribosylated form of the protein. Having identified the ADP-ribosylated protein, and its putative precursor, in the 2-D pattern, the laborious task

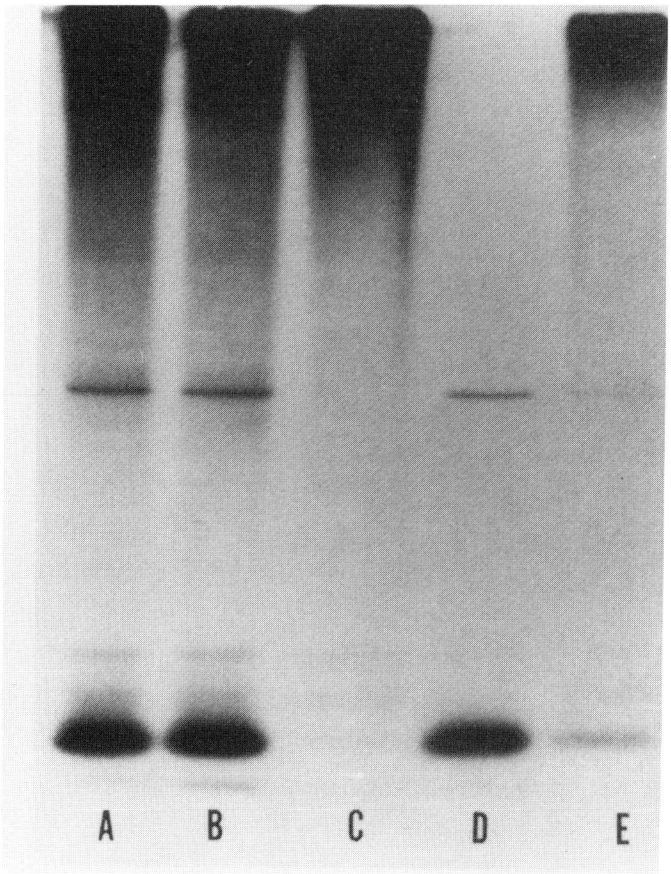

Fig. 2. Fractionation of [³H]adenosine-labeled homogenates by different centrifugation. Cultures of Hepa cells were starved for tryptophan and labeled as before. Nuclei and mitochondria were pelleted by centrifugation at 10,000 × g for 10 minutes. Ribosomes were removed from the resulting supernatant by centrifugation at 200,000 × g for 60 minutes. Lane A, homogenate; lane B, postmitochondrial supernatant; lane C, mitochondrial/nuclear pellet; lane D, postribosomal supernatant; lane E, ribosomal pellet.

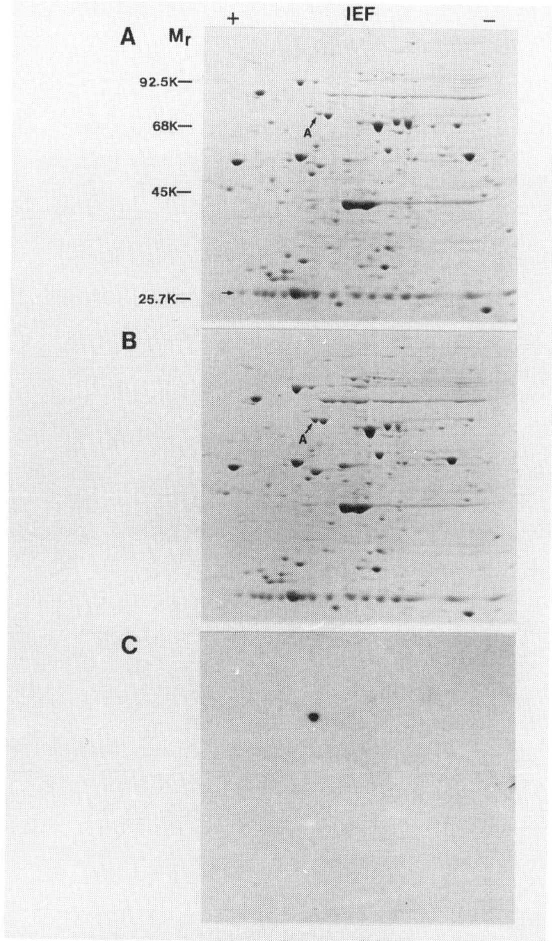

Fig. 3. Localization of the ADP-ribose acceptor protein on two-dimensional PAGE gels. Subconfluent Hepa cultures were incubated with complete or leucine-free medium containing 50 mCi/ml [³H]adenosine for 4 h. Cells were harvested and analyzed on 2-D SDS PAGE. (A) and (B) Coomassie-blue stained gels of proteins from fed and starved Hepa cultures, respectively. (C) A fluorogram of the gel shown in (B) showing [³H]adenosine incorporation. The spot in (B) which aligned with the incorporated adenosine is indicated by an arrow labeled A. The fluorograph of the gel shown in (A) showed no incorporated label. The spot corresponding to the labeled protein in (B) is indicated by an arrow in (A).

of isolating the protein and obtaining a partial sequence was undertaken. Isolating both spots was achieved using 2-D electrophoresis. Figure 4 shows a typical 2-D pattern with the two spots of interest indicated by arrows. The trail of spots at the bottom of the gel are isoelectric focusing standards (carbamylated carbonic anhydrase). Several hundred 2-D gels were run and the individual spots of interest were excised, free of the potentially cross contaminated area, and electroeluted. Panel B shows the cell homogenate in lane 1; and the two isolated proteins in lanes 2 and 3. The ADP-ribosylated form of the protein appears to have a slightly slower mobility than the non-ADP-ribosylated form [7]. N-terminal sequence analysis of the first 19 amino acids of both proteins showed them to be identical to one another [7] and to the N-terminal sequence of the 78 kDa glucose-regulated-protein from rat, [16, 9].

The relationship between the ADP-ribosylation of BiP and the modulation of protein synthesis was examined [6]. The aim was to determine if BiP was ADP-ribosylated before the reduction in translational efficiency; and if the ADP-ribose was removed before the translational efficiency was restored. This latter approach proved to be the most feasible. The protein synthetic activity was examined directly by analysis of polyribosome profiles. Hepa cells were starved for tryptophan for two hours in the presence of ³H-adenosine. Tryptophan was then added back to the medium and cultures were harvested at the indicated times. Polyribosome profiles were developed in sucrose density gradients. A portion of each of the homogenates was used to prepare posttribosomal supernatants for SDS-PAGE analysis. The results are shown in Fig. 5. Panel A shows some of the polyribosome profiles.

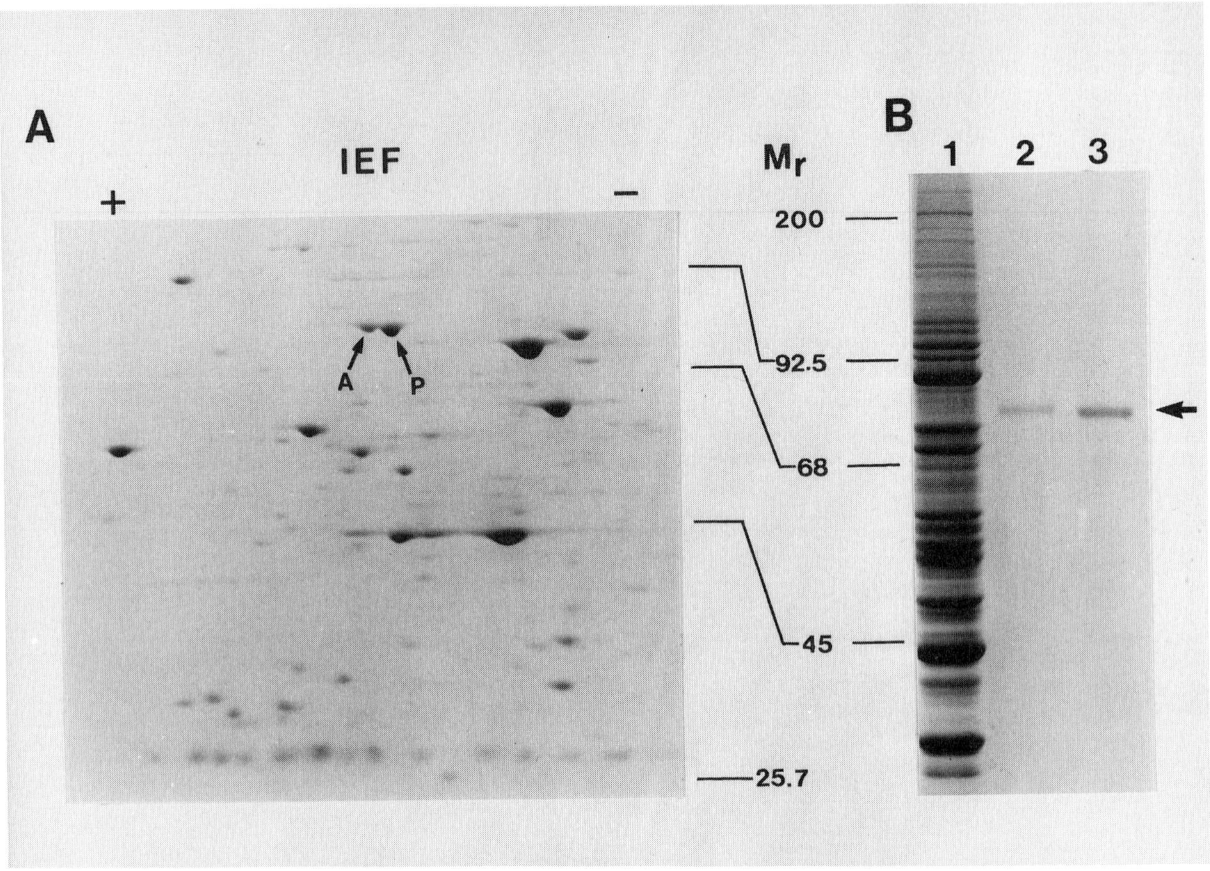

Fig. 4. Isolation of the ADP-ribose acceptor and putative precursor from Hepa cells. The ADP-ribose acceptor, A, and the putative precursor, P, were isolated by electroelution from 2-D SDS/PAGE gels of Hepa cell post-ribosomal supernatants. (A) Representative Coomassie blue-stained gel used for purification; (B) SDS/PAGE gel showing the homogenate (lane 1), and equal volumes of isolated acceptor, A (lane 2), and precursor, P, (lane 3). The positions of the molecular mass standards in kDa (see Fig. 1) are shown.

The areas representing the monomers and subunits (over the bar) were integrated, normalized, and are shown as the filled squares in panel C. Panel B shows the fluorogram of the BiP region of the SDS gel. Densitometric scans of individual lanes are represented as open squares in panel C. Following the addition of tryptophan, monomers and subunits are recruited into polyribosomes well before ADP-ribose is removed from BiP. This result suggested that the ADP-ribosylation of BiP was not the cause of the reduction in translational efficiency, but instead may be a consequence of the reduction in protein synthetic activity. Two approaches were taken to test this hypothesis. First, protein synthesis was inhibited using two concentrations of cycloheximide. One which would inhibit protein synthesis by approximately 10% (1 μM); and one which would block protein synthesis by over 90% (100 μM). The results are shown in Fig. 6. Lane A is the untreated control; lane B is 1 μM cycloheximide; lane C is 100 μM cycloheximide; and lane D is a tryptophan-starved culture. A slight reduction in the rate of protein synthesis resulted in the ADP-ribosylation of BiP to an extent observed during tryptophan-starvation. A second approach was to reduce the protein synthetic activity by incubating Hepa cells at reduced temperatures. The results are shown in Fig. 7. Panel A is a fluorogram of protein labeled with ^{3}H-leucine at the indicated temperatures. The labeling intensity in the lanes was intended to be an index of the overall protein synthetic activity. Panel B is a fluorogram of protein labeled with ^{3}H-adenosine in companion cultures. Reducing the temperature from 37°C to 30° decreased the amount of leucine incorporation and increased the ADP-ribosylation of BiP. Further temperature reduction to 25° completely shut off the incorporation of leucine and caused the ADP-ribosylation of BiP. Thus, both of the approaches taken to reduce translational efficiency resulted in a stimulation of the ADP-ribosylation, supporting the hypothesis that the ADP-ribosylation of BiP is a consequence of the reduction in protein synthesis [6].

In order to determine which form of BiP is the functional form. BiP was overexpressed in cells and the relative abundance of each of the isoforms examined. BiP is known to be overexpressed whenever malfolded or under-glycosylated proteins are present in the secretory pathway [17]. Hepa cells

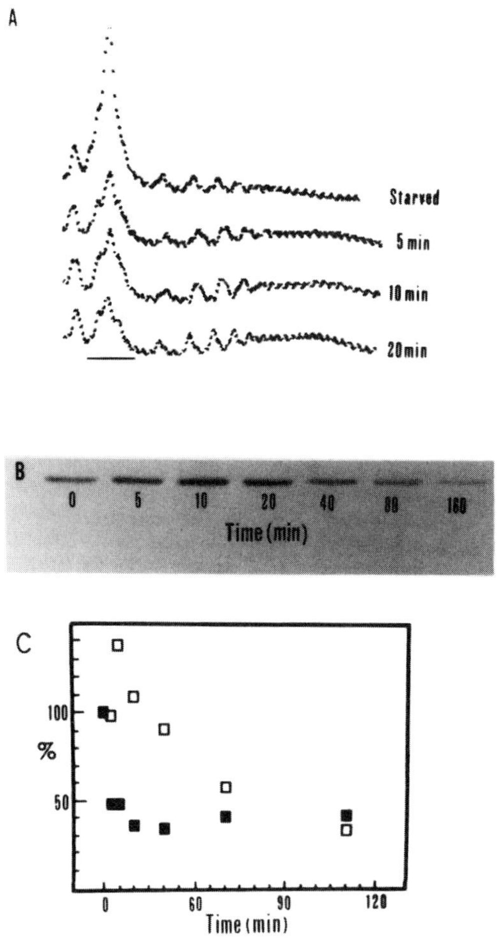

Fig. 5. Temporal comparison of the reformation of polyribosomes and the loss of ADP-ribose from P80. Cultures of Hepa cells were tryptophan-starved for 2 h in the presence of [^{3}H]adenosine. Companion cultures were harvested at the indicated times after addition of tryptophan to the medium. Polyribosome profiles were displayed on linear (0.5M–1.5M) sucrose density gradients centrifuged at 40,000 rpm in a SW41 rotor (Beckman) for 2 h. and are shown in Panel A. The region of each gradient occupied by subunits and monomers (bar) was integrated and the results are shown as solid squares in Panel C. Postibosomal supernatants were resolted by SDS-PAGE and a fluorogram of the gel is shown in Panel B. The densities of each band were determined and are shown as open squares in Panel C.

were treated with a variety of reagents which inhibit N-linked glycosylation. The effect of these treatments on glycosylation was monitored by observing the mobility shift in the secretory glycoprotein, a-fetoprotein (AFP). Figure 8 shows the results. Panel A is an untreated control pattern. Panels D, E, and F are patterns from tunicamycin, 2-deoxy-D-glucose, and glucosamine treated cultures, respectively. Each of these treatments results in the increased electrophoretic mobility of AFP due to the lack of an N-linked oligosaccharide. In each case, there is an overexpression of GRP78 (BiP) which is almost exclusively the un-ADP-ribosylated form. Panels B and C are patterns from tryptophan-starved and glucose-starved cultures, respectively. First, it should be noted that AFP completely disappears from the pattern which is consistent with the previous observations that serum protein synthesis is selectively shut off during nutritional stress [1, 3]. Second, glucose-starvation of Hepa cells increases the ADP-ribosylation of BiP. This latter result seems at odds with the result reported by Carlsson and Lazarides [18] who showed that glucose-starvation of chicken cells caused the rapid decrease in the ADP-ribosylation of SP83 (BiP). However, careful examination of their data shows that the Coomassie Blue staining of the glucose-starved lane is much less than that of the control lane, and could easily account for the observed reduction in the level of ^{3}H-adenosine incorporation in that lane.

The degree of ADP-ribosylation of BiP never exceeded 50%, even under the most stressful of conditions. This observation suggested either that there were two pools of BiP, one subject to ADP-ribosylation, the other not; or that the reaction was reversible, and governed by first order kinetics. This latter possibility was shown to be the case [15]. Figure 9 shows the results of pulse-labeling starved cultures with ^{3}H-adenosine for the last two hours of a period of starvation. The upper portion of the figure shows a fluorogram of the BiP region of the resulting SDS-PAGE gel. The lower portion of the figure shows the normalized densitometric data. The level of labeling was constant across the 12 hour starvation period indicating that a steady-state of ADP-ribosylation was achieved with the two hour labeling period. This was confirmed by demonstrating the metabolic interconvention of the two forms of BiP during starvation and refeeding [15]. Additionally, the overexpressed form of BiP was shown to be a substrate for ADP-ribosylation during starvation [15].

Discussion

The reversible ADP-ribosylation of BiP occurs in response to nutritional stress. Starvation of cells or tissues for either amino acids or glucose results in the modification. The physiological significance of this response is yet to be established. However, reasonable speculations can be advanced based on the data presented here, the known functions of BiP, and the recent descriptions of yeast with mutations in the BiP homolog, Kar2. First, the unmodified form of BiP appears to be the functional one. Frieden, *et al.* [19] have come to the same conclusion based on their data showing that BiP is both phosphorylated and ADP-ribosylated; and that the modified BiP forms oligomers which are not associated with other proteins. BiP has been shown to autophosphorylate [20]. The degree of autophosphorylation is decreased during overexpression [21]. Again, consistent with the unmodified form being the functional one.

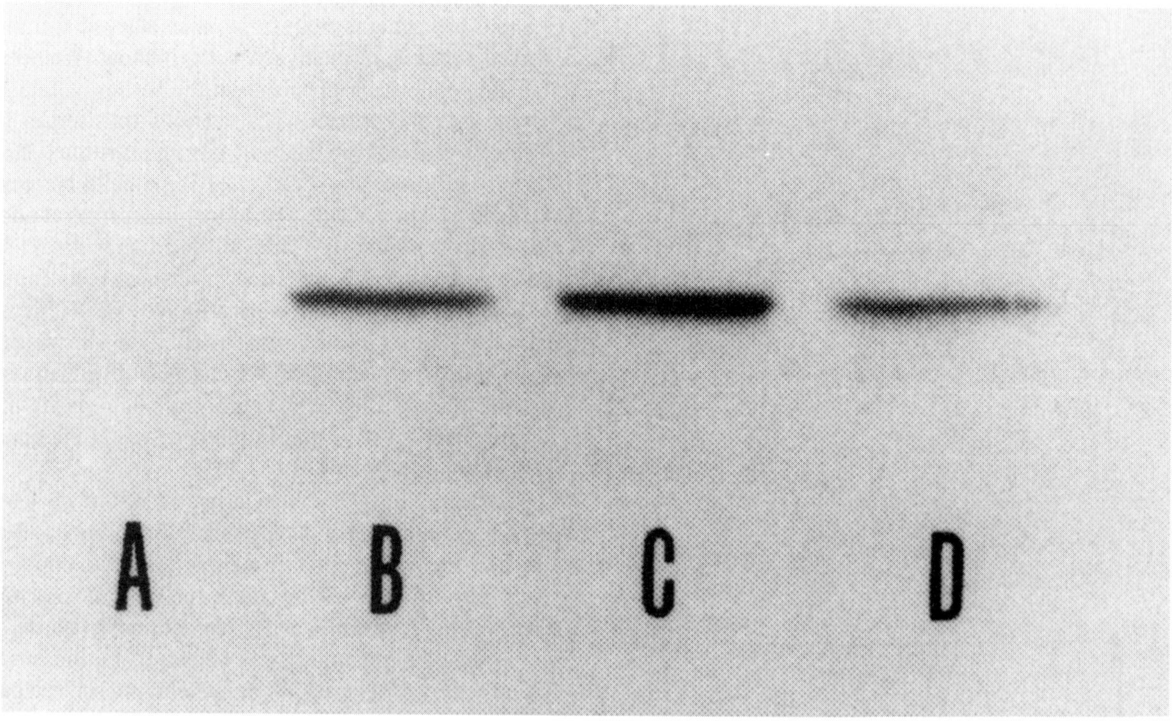

Fig. 6. Effects of cycloheximide on the ADP-ribosylation of P80. Cultures of Hepa cells were treated with either 1 μM cycloheximide or 100 μM cycloheximide for 2 h in the presence of [^{3}H]adenosine. Lane A, untreated control; lane B, 1 μM cycloheximide; lane C, 100 μM cycloheximide; lane D, tryptophan-starved.

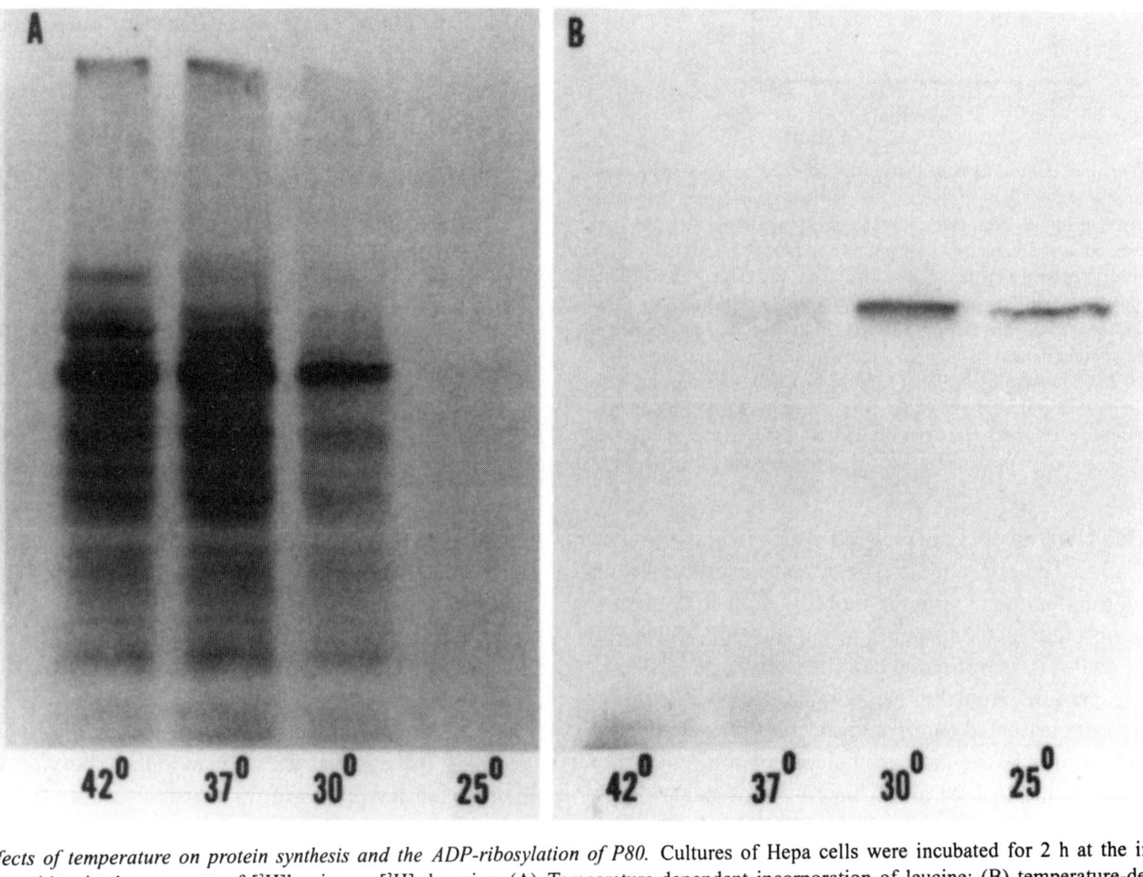

Fig. 7. Effects of temperature on protein synthesis and the ADP-ribosylation of P80. Cultures of Hepa cells were incubated for 2 h at the indicated temperatures either in the presence of [^{3}H]leucine or [^{3}H]adenosine. (A) Temperature-dependent incorporation of leucine; (B) temperature-dependent ADP-ribosylation of P80.

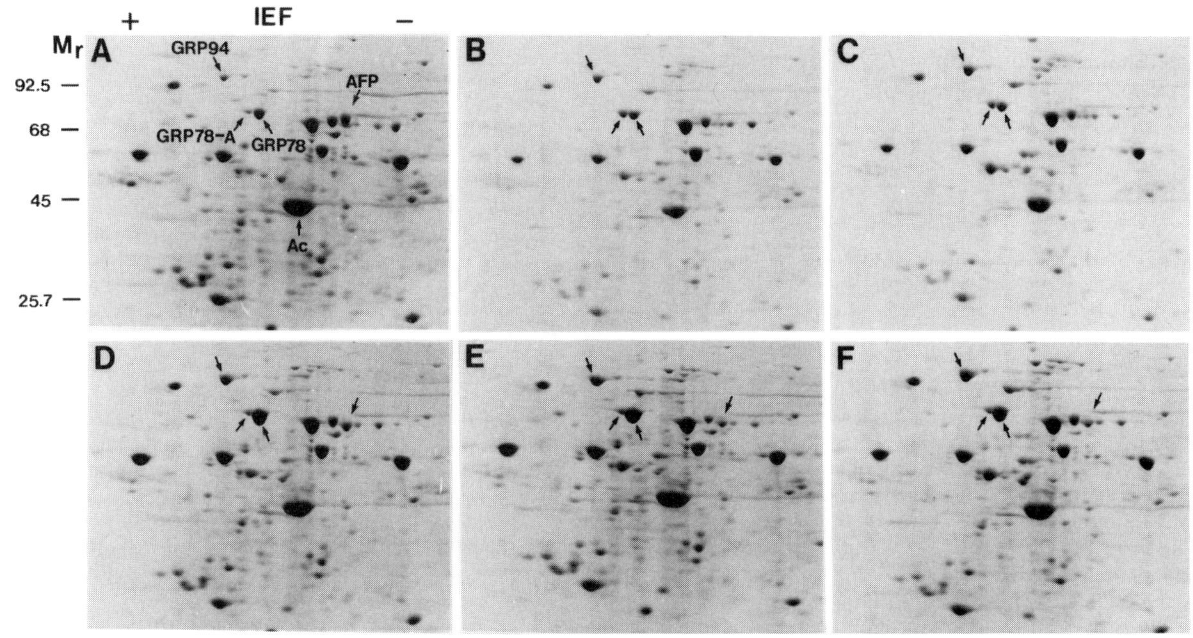

Fig. 8. Induced synthesis and ADP-ribosylation of GRP78 in response to environmental stress. Coomassie blue-stained 2-D PAGE gels from control (A), tryptophan-starved (B), glucose-starved (C), tunicamycin-treated (D), 2-deoxy-D-glucose-treated (E), and glucosamine-treated (F) Hepa cultures. Arrows indicate ADP-ribosylated GRP78 (GRP78-A), non-ADP-ribosylated GRP78 (GRP78), the 94-kDa glucose-regulated protein (GRP94), a-fetoprotein (AFP) and actin (Ac). The positions of the molecular mass standards in kDa (see Fig. 1) are shown.

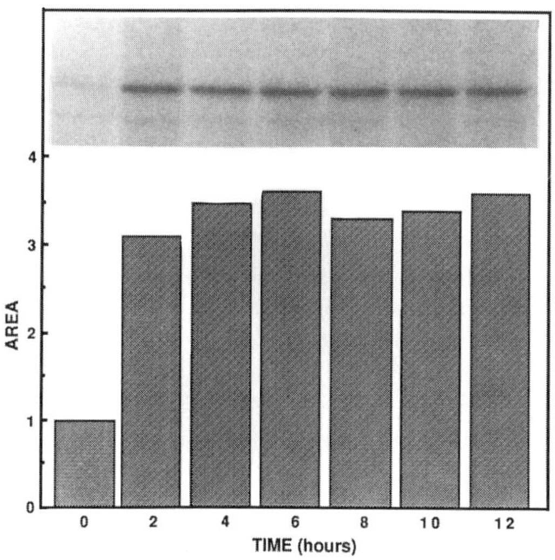

Fig. 9. ADP-ribosylation of GRP78 during amino acid starvation. Hepa cultures were fed complete medium for 2 h prior to addition of complete (control) or tryptophan-free (starvation) medium. The control culture was incubated for 2 h and the remaining cultures

One of the known functions of BiP is the ATP-dependent folding and assembly of proteins in the secretory pathway. BiP binds ATP and can be purified by chromatography on an ATP matrix [11, 12, 22]. BiP purified by chromatography on ATP agarose consists of both the ADP-ribosylated and un-ADP-ribosylated form in relative abundances similar to those in the starting material (unpublished observation). It would, therefore, appear that the ATP-binding activity of BiP is unaltered by ADP-ribosylation.

Yeast with mutations in the BiP homolog, Kar2, are unable to translocate proteins across the membrane of the endoplasmic reticulum [23, 24]. This is a particularly intriguing observation since the data presented here (Figure 8, panels B and C) show that the secretory protein α-fetoprotein disappears from Hepa cells when BiP is ADP-ribosylated. Thus, ADP-ribosylation of BiP may be a mechanism for selectively shutting off the synthesis of membrane and secretory proteins during times of nutritional stress. As such this would conserve nutrients by halting the expansion of the plasma membrane and reduce the rate of proliferation until the nutrient supply was restored. Staddon, *et al.* [25] have reported that incubation of Swiss 3T3 cells with growth factors, which stimulate the proliferation of these cells, resulted in the decreased ADP-ribosylation of BiP, further supporting the premise that the ADP-ribosylation of BiP plays a role in limiting cell proliferation during times of nutritional stringency. If the ADP-ribosylation of BiP proves to be such a stress response, then manipulation of the response could prove to be important in the management of certain hyperproliferative disorders.

Acknowledgements

This work was supported by Grant CA30151 awarded by the National Cancer Institute and by the Medical University of South Carolina Institutional Research Funds. G.H.L was the recipient of a predoctoral fellowship awarded by the College of Graduate Studies, Medical University of South Carolina. We thank Donna F. Jacobs for her helpful comments.

References

1. Ledford BE, Warner RW, Cochran RA: Albumin synthesis in cultured hempatoma cells. Regulation by essential amino acids. Biochim Biophys Acta 475: 90–95, 1977
2. Ledford BE, Davis DF: Analysis of translational parameters in cultured cells. Biochim Biophys Acta 519: 204–212, 1978
3. Ledford BE, Jacobs DF: Translation kinetics in cultured mouse hepatoma cells. Regulation of albumin synthesis by amino acids. Eur J Biochim 152: 611–618, 1985
4. Morgan EH, Peters T Jr.: The biosynthesis of rat serum albumin. V. Effect of protein depletion and refeeding on albumin and transferrin synthesis. J Biol Chem 246: 3500–3507, 1971
5. Lee H, Iglewski WJ: Cellular ADP-ribosyltransferase with the same mechanism of action as diphtheria toxin and Pseudomonas toxin A. Proc Natl Acad Sci USA 81: 2703–2707, 1984
6. Ledford BE, Jacobs DF: Translational control of ADP-ribosylation in eucaryotic cells. Eur J Biochem 161: 661–667, 1986
7. Leno GH, Ledford BE: ADP-ribosylation of the 78-kDa glucose-regulated protein during nutritional stress. Eur J Biochem 186: 205–211, 1989
8. Hendershot LM, Ting J, Lee AS: Identity of the immunoglobulin heavy-chain-binding protein with the 78,000-dalton glucose-regulated protein and the role of posttranslational modifications in its binding function. Mol Cell Biol 8: 4250–4256, 1988
9. Munro S, Pelham HR: An Hsp70-like protein in the ER: identity with the 78 kd glucose-regulated protein and immunoglobulin heavy chain binding protein. Cell 46: 291–300, 1986
10. Kozutsumi Y, Normington K, Press E, Slaughter C, Sambrook J, Gething MJ: Identification of immunoglobulin heavy chain binding protein as glucose-regulated protein 78 on the basis of amino acid sequence, immunological cross-reactivity. J Cell Sci Suppl 11: 115–137, 1989
11. Flynn GC, Chappell TG, Rothman JE: Peptide binding and release by proteins implicated as catalysts of protein assembly. Science 245: 385–390, 1989
12. Flynn GC, Pohl J, Flocco MT, Rothman JE: Peptide-binding specificity of the molecular chaperone BiP. Nature 353: 726–730, 1991
13. Pelham HR: Speculations on the functions of the major heat shock and glucose-regulated proteins. Cell 46: 959–961, 1986
14. Gething MJ, Sambrook J: Protein folding in the cell. Nature 355: 33–45, 1992
15. Leno GH, Ledford BE: Reversible ADP-ribosylation of the 78 kDa glucose-regulated protein. Febs Lett 276: 29–33, 1990
16. Chang SC, Wooden SK, Nakaki T, Kim YK, Lin AY, Kung L, Attenello JW, Lee AS: Rat gene encoding the 78-kDa glucose-regulated protein GRP78: its regulatory sequences and the effect of protein glycosylation on its expression. Proc Natl Acad Sci USA 84: 680–684, 1987
17. Kozutsumi Y, Segal M, Normington K, Gething MJ, Sambrook J: The presence of malfolded proteins in the endoplasmic reticulum signals the induction of glucose-regulated proteins. Nature 332: 462–464, 1988
18. Carlsson L, Lazarides E: ADP-ribosylation of the Mr 83,000 stress-inducible and glucose-regulated protein in avian and mammalian cells: modulation by heat shock and glucose starvation. Proc Natl Acad Sci USA 80: 4664–4668, 1983
19. Freiden PJ, Grant JR, Hendershot LM: Interconversion of three differentially modified and assembled forms of BiP. Embo J 11: 63–70, 1992
20. Leustek T, Toledo H, Brot N, Weissbach H: Calcium-dependent autophosphorylation of the glucose-regulated protein, Grp78. Arch Biochem Biophys 289: 256–261, 1991
21. Leustek T, Amir-Shapira D, Toledo H, Brot N, Weissbach H: Autophosphorylation of 70 kDa heat shock proteins. Cell Mol Biol 38: 1–10, 1992
22. Kassenbrock CK, Kelly RB: Interaction of heavy chain binding protein (BiP/GRP78) with adenine nucleotides. Embo J 8: 1461–1467, 1989
23. Brodsky JL, Hamamoto S, Feldheim D, Schekman R: Reconstitution of protein translocation from solubilized yeast membranes reveals topologically distinct roles for BiP and cytosolic Hsc70. J Cell Biol 120: 95–102, 1993
24. Vogel JP, Misra LM, Rose MD: Loss of BiP/GRP78 functions blocks translocation of secretory proteins in yeast. J Cell Biol 110: 1885–18895, 1990
25. Staddon JM, Bouzyk MM, Roxengurt E: Interconversion of GRP78/BiP. A novel event in the action of Pasteurella multocida toxin, bombesin, and platelet-derived growth factor. J Biol Chem 267: 25239–25245, 1992

PART IV

TOXIN MONO-ADP-RIBOSYLATION

Molecular and Cellular Biochemistry **138**: 151–156, 1994.

Diphtheria toxin-related cytokine fusion proteins: elongation factor 2 as a target for the treatment of neoplastic disease

Johanna vanderSpek[1], Larry Cosenza[1,2], Thasia Woodworth[3], Jean C. Nichols[3] and John R. Murphy[1]

[1]*Evans Department of Clinical Research and Department of Medicine, Boston University Medical Center Hospital, Boston, MA 02118, and* [2]*Department of Pharmacology, Boston University School of Medicine, Boston, MA 02118;* [3]*Seragen Inc., Hopkinton, MA 01748, USA*

Abstract

We have used protein engineering and recombinant DNA methodologies in order to construct a fusion protein in which human interleukin-2 (IL-2) is genetically linked to the catalytic and transmembrane domains of diphtheria toxin. The fusion toxin, DAB_{486}IL-2, is highly cytotoxic for only those cells which display the high affinity interleukin-2 receptor (IL-2R) on their surface. In phase I/II clinical studies the intravenous administration of DAB_{486}IL-2 has been found to be safe, well tolerated and may lead to the induction of durable remissions in patients presenting with a variety of IL-2R positive lymphomas. (Mol Cell Biochem **138**: 151–156, 1994).

Key words: diphtheria toxin, protein engineering, interleukin-2, fusion toxin, phase I/II clinical trials, targeted cytotoxins

Introduction

While there has been some progress in the development of anti-neoplastic agents and the treatment of cancer in the past twenty years, it is clear that additional advances are needed. Current chemotherapeutic regimens may result in long-term remissions and possible cures in patients with Hodgkin's disease, malignant lymphoma, acute leukemias, and several solid tumors, including testicular cancer and early stage breast cancer [1, 2]. In addition, many other malignancies may be initially treated with the use of combination chemotherapeutic regimens. However, when relapses occur following an initial chemotherapeutic induced response, they are usually associated with the emergence of multiple drug resistance [3]. As a result, favorable responses to additional rounds of chemotherapy are less frequent and less durable.

Most traditional chemotherapeutic agents are competitive inhibitors that stoichiometrically act on a variety of cellular targets which include DNA synthesis and repair. Since the majority of these agents passively enter eukaryotic cells, they are non-specific and affect both malignant and normal cells. Consequently, the administration of traditional chemotherapeutic agents is usually associated with many adverse side effects. As a result, the development of new approaches to the treatment of malignant disease should include novel cellular targets, an increase in the specificity of the agent for neoplastic cells versus normal cells, and in the case of biologics, agents which catalytically alter their cellular target.

Before the turn of the century, having observed the localization of dyes as well as antigen antibody reactions, Erlich [4] envisioned the development of 'magic bullets' for the treatment of human disease. In concept, such agents would be composed of a toxic compound coupled to a targeting agent. The conjugate formed would thereby recognize and selectively deliver the therapeutic agent to disease-causing cells. Over the past fifteen years the selective delivery of cytotoxic agents to target cells has become a reality. Two different approaches have been taken toward the development of targeted therapeutics: the immunotoxins and the protein fusion toxins. In general, immunotoxins are composed of monoclonal antibodies to which the toxophore of either plant or microbial toxins have been chemically linked through a

Address for offprints: J. vanderSpek, Evans Department of Clinical Research and Department of Medicine, Boston University Medical Center Hospital, Boston, Massachusetts 02118, USA

disulfide bond (5, 6). Since there are only a few monoclonal antibodies that are cell lineage or tumor cell specific, unexpected cross-reactivity with normal tissue has been a difficult problem with this approach. In contrast, the protein fusion toxins genetically combine the catalytic and transmembrane domains of either diphtheria toxin or *Pseudomonas* exotoxin A with polypeptide hormones or growth factors as targeting ligands. The fusion proteins, or fusion toxins, that are expressed in recombinant strains of *Escherichia coli* retain the cell receptor specificity of the ligand and the cytotoxic potency of the toxophore [7, 8]. Since, in many cases, the biology of the native ligand/receptor complex is well known, the fusion toxin-associated adverse effects can generally be predicted.

Diphtheria toxin-related cytokine fusion proteins

The underlying rationale for the protein fusion toxins is that receptor binding domain substitution will result in the formation of a 'new' and highly selective toxin that will be active only towards those eukaryotic cells which display the targeted cell surface receptor. For example, as shown in Fig. 1a, the ribbon diagram of the crystal structure of native diphtheria toxin clearly shows that the toxin is composed of three distinct domains: the catalytic (C), transmembrane (T), and receptor binding (R) domain (9). In the case of the interleukin-2 receptor-targeted fusion toxin DAB_{389} IL-2, the native diphtheria toxin receptor binding domain has been removed and genetically replaced with amino acids 2–133 of human IL-2 [10]. Figure 1b shows the computer generated ribbon diagram of a model structure of DAB_{389} IL-2 using the relevant portions of the native diphtheria toxin and the corrected IL-2 crystal structures [9, 11, 12]. In a similar fashion, a number of cytokines or growth factors have been used to replace the native diphtheria toxin receptor binding domain.

In 1986, Murphy *et al.* [13] described the genetic construction and properties of a chimeric protein composed of the first 485 amino acids of diphtheria toxin fused to α-melanocyte stimulating hormone (α-MSH), DAB_{486} α-MSH. This first fusion toxin was found to selectively eliminate eukaryotic cells which expressed cell surface α-MSH receptors. Importantly, this fusion toxin was not active against cells which were devoid of the α-MSH receptor. These results suggested that the portion of diphtheria toxin which bound to its cognate receptor was positioned on the extreme C-terminal end of the toxin. While these initial results were encouraging, proteolytic degradation of DAB_{486} α-MSH in recombinant *E. coli* made purification difficult. Analysis of the degradation products by immunoblot suggested the fusion protein was degraded by an endoprotease(s) that recognized sites in the diphtheria toxin portion of the chimera in a region adjacent to the fusion junction with α-MSH sequences. This analysis led to the speculation that the substitution of a more bulky receptor binding domain might give rise to a fusion protein that was resistant to endoproteinase degradation.

Williams *et al.* [14] described the genetic construction, expression, purification, and properties of a fusion toxin in which human IL-2 sequences replaced the native diphtheria toxin receptor binding domain. In contrast to the α-MSH receptor targeted fusion protein, DAB_{486} IL-2 could be readily purified from recombinant *E. coli* in good yield. In these initial genetic constructs, the fusion toxin was expressed from the diphtheria *tox* promoter in precursor form and was secreted into the periplasmic compartment. Following purification by affinity chromatography on either an anti-diphtheria toxin or anti-interleukin-2 matrix and dialysis, DAB_{486} IL-2 was biologically active and selectively inhibited protein synthesis in IL-2 receptor bearing lymphocytes with an IC_{50} of approximately 10^{-10} M. Importantly, cells which were devoid of the IL-2 receptor were markedly resistant to the action of this fusion toxin (e.g., $IC_{50} \approx 10^{-7}$ M). Bacha *et al.* [15] demonstrated that the fusion toxin-mediated inhibition of protein synthesis required binding to the IL-2 receptor. Moreover, in a fashion that was analogous to the native toxin, the chimera was internalized into the cell by receptor-mediated endocytosis and, following acidification of the endocytic vesicle, the delivery of the catalytic domain to the cytosol of target cells resulted in the adenosine diphosphate ribosylation (ADPR) of elongation factor 2 [15].

It has been further shown that only those cells which display the high affinity form of the IL-2 receptor are sensitive to the inhibitory action of DAB_{486} IL-2 (16). Indeed, the early events following binding of DAB_{486} IL-2 to the high affinity IL-2 receptor have been shown to mimic those of recombinant IL-2. Within the first seven hours post-binding, there is an increase in messenger RNA's for IL-2, IL-2 receptor, c-*myc*, and interferon γ. This initial stimulation of mRNA synthesis is rapidly followed by a rapid inhibition of protein synthesis, such that by 12 hours post-exposure, the mRNA profile mimics that of cell cultures which have been exposed to cycloheximide [17]. Since the high affinity IL-2 receptor is only found on activated proliferating T cells, recently activated B cells, and activated monocytes, NK cells which constitutively express only the intermediate affinity form of the IL-2 receptor are resistant to the action of this fusion toxin.

Results

Clinical evaluation of DAB_{486} IL-2 hematologic malignancies

The high affinity form of the IL-2 receptor is an attractive target for cytotoxic therapy in both cancer and autoimmune diseases [18, 19]. While the high affinity receptor for IL-2 is

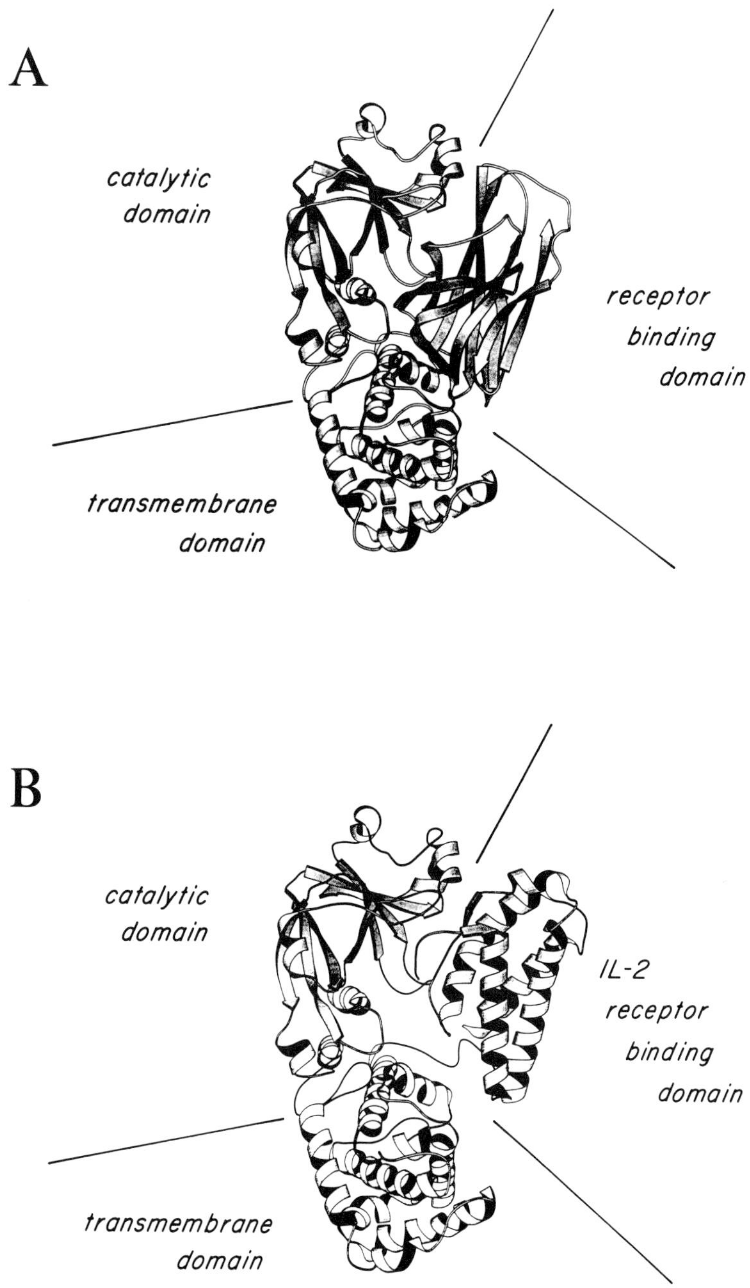

Fig. 1. (A) Ribbon diagram of the X-ray crystal structure of native diphtheria toxin (9); (B) Ribbon diagram of a molecular model of DAB_{389} IL-2. The relative positions of the catalytic, transmembrane, and receptor binding domains of the native toxin and the IL-2 receptor target fusion toxin are presented. The ribbon diagrams were generated using the software program MOLESCRIPT [36].

transiently expressed on T and B cells, it is not found on other normal tissue. More importantly, since the diphtheria toxin-based fusion proteins mediate their cytotoxic effect through the classic ADP-ribosylation of elongation factor 2, these chimeric proteins represented a new prototype biologic agent for cancer therapy.

A series of phase I/II clinical studies were developed to determine the safety, tolerability and pharmacokinetics of DAB_{486} IL-2 in patients with refractory hematologic malignancies [20–23]. This fusion was the first single chain chimeric toxin protein to be administered to man. In order to be eligible for study entry, patients had to present with a hematologic malignancy that could not be effectively treated with either standard or, in some cases, experimental therapy.

In addition, histologic confirmation of the malignancy and evidence of CD25 expression (i.e., p55, or α-chain of the IL-2 receptor) was also measured. Patients over the age of seventeen years of either sex could be treated if they had adequate hepatic and renal function and a Karnofsky performance score of 70% or greater.

The studies were designed as a three patient cohort dose escalation in which single and multiple doses were administered intravenously either as a bolus or a 90 minute infusion. The initial dose was 700 ng/kg/day, and, in the absence of adverse reactions, was followed by 3 daily doses, followed one week later by 7 daily doses. The dose was slowly escalated over eight cohort groups to 400 μg/kg/day. Importantly, patients who showed evidence of decreased tumor burden could be retreated following a minimum of four weeks after the first day of therapy. At each dose level patients were evaluated for safety and response. Toxicities and responses were scored according to the criteria established by the National Cancer Institute. Patients with no evidence of disease for at least four weeks duration were classified as complete responders (CR); patients whose tumor burden decreased by ≥ 50% for at least four weeks duration were classified as partial responders (PR), and patients whose tumor burden decreased between 25% to 50% were classified as minor responders (Table 1).

Table 1. Response to treatment with DAB_{389} IL-2 in patients with refractory leukemias/lymphomas.

Diagnosis	# Enrolled	IL-2R+	IL-2R?	CR	PR	MR	% Response
NHL, high grade	6	2	1	0	0	0	0
NHL, low & intermediate grade	43	20	12	1	4	4	25
Hodgkin's disease	14	8	5	1	0	1	15
CTCL	22	16	2	1	4	3	44
leukemias	19	16	1	0	1	1	16
miscellaneous	5	1	3	0	0	0	0

IL-2R+, biopsy specimen positive for CD25 antigen; IL-2R?, biopsy specimen either unavailable or presence of CD25 is questionable; NHL, non-Hodgkin's lymphoma; CTCL, cutaneous T cell lymphoma.

The intravenous administration of DAB_{486} IL-2 was found to be well tolerated at all dose levels in this group of patients with refractory malignancies. The adverse reactions were generally mild and included nausea/vomiting, hypersensitivity, fever/malaise/chills, and elevations in serum hepatic transaminases. Renal insufficiency defined the maximum tolerated dose at levels above 400 μg/kg/day. In all instances the adverse effects were transient, not cumulative, and did not preclude repeat administration of the fusion toxin in this patient population. Most importantly, no changes in lymphocyte function assays (e.g., blastogenesis), lymphocyte subsets, or opportunistic infections were observed in any of the patients that were treated.

The time course analysis of serum concentrations of biologically active DAB_{486} IL-2 following bolus administration, using a non-linear open mathematical model, showed that the clearance of the fusion toxin most closely followed a one-component model with a $t_{1/2}$ of approximately 11 minutes at dose levels of 200, 250, 300, and 400 ug/kg [23]. The pharmacokinetics of the fusion toxin did not change in a consistent fashion following multiple courses of administration.

In many patients, circulating proteins which could potentially bind DAB_{486} IL-2 and thereby decrease its potential effectiveness through enhanced clearance were detected. Increased serum levels of soluble IL-2 receptor (sIL-2R) were noted in a number of patients; however, there was no correlation between clearance rates of the fusion toxin in patients with high versus low levels of sIL-2R. Since the presence of up to 50,000 units of sIL-2R/ml failed to inhibit the cytotoxic action of DAB_{486} IL-2 *in vitro* [24], the appearance of elevated sIL-2R in serum was not expected to adversely affect the potential action of the fusion toxin *in vivo*. In addition, some patients had detectable levels of anti-diphtheria toxoid antibodies prior to entry into the clinical study. Following intravenous administration of DAB_{486} IL-2, approximately 60% of patients had an amnestic response to the diphtheria toxin component of the fusion protein and developed significant titers. In contrast, few patients had titers of anti-IL-2 antibodies in serum prior to study entry; however, after one or more courses of fusion toxin, approximately one-half of the patients developed a low level of anti-IL-2 antibody. As with sIL-2R, the presence of antibodies to either the diphtheria toxin-related or IL-2 components of the fusion toxin did not appear to prevent an antitumor response from being noted. Based upon the clinical experience with the monoclonal antibody OKT3 and the analysis of neutralizing and non-neutralizing monoclonal anti-toxin antibodies, it was not anticipated that pre-existing anti-diphtheria toxoid antibodies would interfere with the action of the fusion toxin *in vivo* [25–27].

The results from the phase I/II studies of the intravenous administration of DAB_{486} IL-2 to patients with refractory hematologic malignancies has now proven in principle the validity of fusion toxin therapy in man. Indeed, the cell surface receptor-specific intoxication of neoplastic cells through the catalytic ADP-ribosylation of elongation factor 2 is the prototype of a new class of biologic response modifiers that may be generally applicable. In those circumstances where either the *de novo* expression or up-regulation of a cell surface receptor can be associated with human disease (e.g., the up-regulation of the epidermal growth factor [EGF] receptor on breast cancer), it should be possible to genetically construct a diphtheria toxin-related / growth factor in order to produce an experimental biologic for the treatment of that malignancy. Indeed, the EGF receptor targeted fusion toxin, DAB_{389} EGF, has recently begun human phase I clinical trials.

Protein engineering of the diphtheria toxin-related cytokine fusion proteins

The genetic construction of this bacterial toxin cytokine fusion protein offers several advantages over the chemical assembly of the immunotoxins. Perhaps the most significant advantage is that the fusion toxins are expressed in recombinant *E. coli* as single chain proteins, rather than as a racemic mixture of monoclonal antibodies to which a protein toxophore has been chemically cross-linked. Since the fusion toxins are assembled at the level of the gene, the fusion junction between the toxophore and the ligand components are precisely defined by a peptide bond. As a result, the application of recombinant DNA and protein engineering methodologies to the fusion toxins has allowed for a detailed analysis of the structure function relationships that are essential for the intoxication of target cells.

Williams *et al.* [10] have demonstrated that the internal in-frame deletion of 97 amino acids from DAB_{486} IL-2 to form DAB_{389} IL-2 has resulted in a truncated form of the fusion toxin that is 10-fold more potent in the selective elimination of target cells *in vitro.* The increase in cytotoxic potency observed for DAB_{389} IL-2 is directly related to its increased binding affinity for the IL-2 receptor [28]. It is of interest to note that the increased activity of DAB_{389} IL-2 versus the 486 form of the IL-2 receptor targeted fusion toxin has also been observed with a number of other diphtheria toxin-based fusion toxins in which either IL-4, IL-6, EGF, α-MSH, or CD4 have been employed as the ligand component [29–33]. Additional in-frame deletion analysis of DAB_{389} IL-2 have demonstrated that amino acid 389 of recombinant diphtheria toxin is an optimal site for the genetic fusion of cell receptor binding ligands [34]. It is of interest to note that the recent solution of the X-ray crystal structure of native diphtheria toxin [9] has shown that amino acid 388 (i.e., amino acid 389 of the fusion toxins) is positioned at the end of a random coil separating the transmembrane domain from the receptor binding domain. Thus, the 389 family of diphtheria toxin-based fusion proteins represent chimeric molecules in which there is optimal receptor binding domain substitution (see Figs. 1a and 1b).

Bacha *et al.* [35] have also shown that DAB_{389} IL-2 is approximately 10-fold more effective in the treatment of an IL-2 receptor positive murine lymphoma *in vivo.* Recently, phase I/II clinical studies with DAB_{389} IL-2 have been initiated in patients presenting with refractory IL-2 receptor positive lymphoma. Preliminary results suggest that the intravenous administration of this second generation form of the IL-2 receptor targeted fusion toxins gives rise to fewer adverse reactions that the 486 prototype. While these studies are still underway, there are encouraging signs that the administration of DAB_{389} IL-2 is safe, better tolerated than the 486 form of the fusion toxin, and may give rise to remissions in IL-2R positive lymphomas.

Acknowledgements

The authors thank Drs. S. Choe and D. Eisenberg for providing the coordinants for the crystal structure of native diphtheria toxin at a resolution of 2.5 Å. J.R.M. is supported by Public Health Service grants AI-22882 and AI21628 from the National Institute of Allergy and Infectious Diseases, and CA-56889 and CA-60934 from the National Cancer Institute. J. vS. is supported by a fellowship from the Leukemia Society of America.

References

1. Chabner BA, Fine RL and Allegra CJ: Cancer chemotherapy: progress and expectations. Cancer, 54: 2599–2608, 1984
2. Goldie JH, Coldman AJ: Genetic instability in the development of drug resistance. Semin. Oncol., 12: 222–230, 1985
3. Pastan I, Gottesman M: Multiple-drug resistance in human cancer. N Engl J Med, 316: 1388–1393, 1987
4. Erlich P: The relationship existing between chemical constitution, distribution, and pharmacological action. In: Himmelweite IF, Dale Sir H. (eds). The Collected Papers of Paul Erlich, Vol. I Pergammon Press, London, 1956, pp. 596–618
5. Vitetta ES, Fulton RJ, May RD, Till M, and Uhr JW: Redesigning nature's poisons to create anti-tumor reagents. Science, 238: 1098–1104, 1987
6. Vitetta ES: Immunotoxins: new therapeutic reagents for autoimmunity, cancer, and AIDS. J Clin Immunol, 10: 15S–18S, 1990
7. Strom TB, Rubin-Kelley VE, Murphy JR, Nichols J, Woodworth TG: Interleukin-2 receptor directed therapies: Antibody or cytokine based targeting molecules. Ann Rev Med, 44: 343–353, 1993
8. Pastan I, Chaudhary V, FitzGerald DJ: Recombinant toxins as novel therapeutic agents. Annu Rev Biochem, 61: 331–354, 1992
9. Choe S, Bennett M, Fujii G, Curmi PM, Kantardjieff KA, Collier RJ, Eisenberg, D: The crystal structure of diphtheria toxin. Nature, 357: 216–222, 1992
10. Williams DP, Snider CE, Strom TB, Murphy JR: Structure function analysis of IL-2-toxin (DAB_{486} IL-2): Fragment B sequences required for the delivery of fragment A to the cytosol of target cells. J Biol Chem, 265: 11885–11889, 1990
11. Brandhuber B, Boone T, Kemmey WC, McKay DB: Three-dimensional structure of interleukin-2. Science, 238: 1707–1709, 1987
12. McKay DB: personal communication, 1992
13. Murphy JR, Bishai W, Borowski M, Miyanohara A, Boyd J, Nagle S: Genetic construction, expression, and melanoma selective cytotoxicity of a diphtheria toxin α-melanocyte stimulating hormone fusion protein. Proc Natl Acad Sci USA, 83: 8258–8262, 1986
14. Williams DP, Parker K, Bishai W, Borowski M, Genbauffe F, Strom

TB, Murphy JR: Diphtheria toxin receptor binding domain substitution with interleukin-2: genetic construction and properties of a diphtheria toxin-related interleukin-2 fusion protein. Protein Engn, 1: 493–498, 1987

15. Bacha P, Waters C, Williams J, Murphy JR, Strom TB: Interleukin-2 targeted cytotoxicity: Selective action of a diphtheria toxin-related interleukin-2 fusion protein J Exp Med, 167: 612–622, 1988
16. Waters CA, Schimke P, Snider CE, Itoh K, Smith KA, Nichols JC, Strom TB, Murphy JR: Interleukin-2 receptor targeted cytotoxicity: Receptor binding requirements for entry of IL-2 toxin into cells. Eur J Immunol, 20: 785–791, 1990
17. Walz G, Zanker B, Brand K, Swanlund D, Genbauffe F, Zeldis JC, Murphy JR, Strom TB: Sequential effects of interleukin-2/diphtheria toxin fusion protein on T-cell activation. Proc Natl Acad Sci USA 86: 9485–9488, 1989
18. Waldmann TA: The structure, function, and expression of interleukin-2 receptors on normal and malignant T cells. Science, 232: 727–732, 1986
19. Waldmann TA: The multichain interleukin-2 receptor. A target for immunotherapy in lymphoma, autoimmune disorders, and organ allografts. J Am Med Assoc, 263: 272–274, 1990
20. LeMaistre CF, Meneghetti C, Rosenbloom M, Reuben J, Parker K, Shaw J, Woodworth T, Parkinson D: Phase I trial of an interleukin-2 receptor (IL-2) fusion toxin (DAB_{486} IL-2) in hematologic malignancies expressing the IL-2 receptor. Blood 79: 2547–2554, 1992
21. Schwartz G, Tepler I, Charette J, Kadin L, Parker K, Woodworth T, and Schnipper L: Complete response of a Hodgkin's lymphoma in a phase I trial of DAB_{486} IL-2. Blood, 79: 175a, 1992
22. Hesketh P, Caguioa P, Koh H, Dewey H, Facada A, McCaffrey R, Parker K, Nylen P and Woodworth T: Clinical activity of a cytotoxic fusion protein in the treatment of cutaneous T cell lymphoma. J Clin Oncol, 11: 1682–1690, 1993
23. LeMaistre CF, Craig FE, Meneghetti C, McMullin B, Parker K, Reuben J, Boldt DH, Rosenblum M, Woodworth T: Phase I trial of a 90-minute infusion of the fusion toxin DAB_{486} IL-2 in hematological cancers. Cancer Res., 53: 3930–3934, 1993
24. Bacha, P. (personal communication).
25. Chatenoud L, Jonker M, Villemain F, Goldstein G, Bach JF: The human immune response to the OKT3 monoclonal antibody is oligoclonal. Science, 232: 1406–1408, 1986
26. Bacha P, Forte SE, Perper SJ, Trentham DE, Nichols JC: Anti-arthritic effects demonstrated by an interleukin-2 receptor-targeted cytotoxin (DAB_{486} IL-2) in rat adjuvant arthritis. Eur J Immunol, 22: 1673–1679, 1992
27. Zucker DR, Murphy JR: Monoclonal antibody analysis of diphtheria toxin. I. Localization of epitopes and neutralization of cytotoxicity. Molec Immunol, 21: 785–793, 1984
28. Kiyokawa T, Williams DP, Snider CE, Strom TB, Murphy JR: Protein engineering of diphtheria toxin-related interleukin-2 fusion toxins to increase biologic potency for high affinity interleukin-2 receptor bearing target cells. Protein Engn., 4: 463–468, 1991
29. Wen Z, Tao X, Lakkis F, Kiyokawa T, Murphy JR: Expression, purification, and α-melanocyte stimulating hormone receptor specific toxicity of DAB-α-MSH fusion toxins. J Biol Chem, 266: 12289–12293, 1991
30. Lakkis F, Steele A, Pacheco-Silva A, Kelley VE, Strom TB, Murphy JR: Interleukin-4 receptor targeted cytotoxicity: genetic construction and properties of diphtheria toxin-related interleukin-4 fusion toxins. Eur J Immunol, 21: 2253–2258, 1991
31. Jean L-F, Murphy JR: Diphtheria toxin receptor binding domain substitution with interleukin-6: genetic construction and interleukin-6 receptor specific action of a diphtheria toxin-related interleukin-6 fusion protein. Protein Engn, 4: 989–994, 1992
32. Aullo P, Alcani J, Popoff MR, Klatzmann DR, Murphy JR, and Boquet P: In vitro effects of a recombinant diphtheria-human CD4 fusion toxin on acute and chronically HIV-1 infected cells. EMBO J, 11: 575–583, 1992
33. Shaw JP, Akiyoshi DE, Arrigo DA, Rhoad AE, Sullivan B, Thomas J, Genbauffe FS, Bacha P, Nichols JC: Cytotoxic properties of DAB_{486} EGF and DAB_{389} EGF, epidermal growth factor (EGF) receptor-targeted fusion toxins. J Biol Chem, 266: 21118–21124, 1991
34. vanderSpek J, Cassidy D, Genbauffe F, Hunyh PD, Murphy JR: An intact transmembrane helix 9 is essential for the efficient delivery of the diphtheria toxin catalytic domain to the cytosol of target cells, manuscript submitted.
35. Bacha PA, Forte SE, McCarthy DM, Estis L, Yamada G, and Nichols JC: Impact of interleukin-2-receptor-targeted cytotoxins on a unique model of immune interleukin-2 receptor expressing malignancy. Int J Cancer, 49: 96–101, 1991
36. Kraulis PJ: Molescript: a program to produce both detailed and schematic plots of protein structures. J Appl Crystallo, 24: 946–950, 1991

Molecular and Cellular Biochemistry **138**: 157–166, 1994.

ADP-ribosylation factors: a family of ~20-kDa guanine nucleotide-binding proteins that activate cholera toxin

Catherine F. Welsh, Joel Moss and Martha Vaughan

Laboratory of Cellular Metabolism, National Heart, Lung, and Blood Institute, National Institutes of Health, Bethesda, MD 20892, USA

Abstract

ADP-ribosylation factors (ARFs) comprise a family of ~20 kDa guanine nucleotide-binding proteins that were discovered as one of several cofactors required in cholera toxin-catalyzed ADP-ribosylation of $G_{s\alpha}$, the guanine nucleotide-binding protein responsible for stimulation of adenylyl cyclase, and was subsequently found to enhance all cholera toxin-catalyzed reactions and to directly interact with, and activate the toxin. ARF is dependent on GTP or its analogues for activity, binds GTP with high affinity in the presence of dimyristoylphosphatidylcholine/cholate and contains consensus sequences for GTP-binding and hydrolysis. Six mammalian family members have been identified which have been classified into three groups (Class I, II, and III) based on size, deduced amino acid sequence identity, phylogenetic analysis and gene structure. ARFs are ubiquitous among eukaryotes, with a deduced amino acid sequence that is highly conserved across diverse species. They have recently been shown to associate with phospholipid and Golgi membranes in a GTP-dependent manner and are involved in regulating vesicular transport. (Mol Cell Biochem **138**: 157–166, 1994)

Key words: cholera toxin, adenylyl cyclase, ADP-ribosylation factors, guanine nucleotide-binding (G) proteins, vesicular trafficking

Abbreviations: ARF – ADP-ribosylation factor; sARF I and sARF II – soluble ADP-ribosylation factors purified from bovine brain; mARF – purified membrane-associated ARF; hARF – human ARF; bARF – bovine ARF; yARF – yeast ARF; rARF – bacterially-expressed recombinant ARF; gARF – *Giardia* ARF; dARF – *Drosophila* ARF; G protein – guanine nucleotide-binding protein; G_s – G protein responsible for stimulation of adenylyl cyclase; GTPγS – guanosine-5′-O-(3-thio-triphosphate); CTA1 – cholera toxin A1 subunit; DMPC – dimyristoylphosphatidylcholine; SDS – sodium dodecyl sulfate

Introduction

ADP-ribosylation factors are ~20 kDa guanine nucleotide-binding proteins that were initially discovered because they are cofactors for cholera toxin's enzymatic activities. Cholera toxin is a hetero-oligomeric protein secreted by *Vibrio cholerae* and composed of one A and five B subunits, the latter forming the cell membrane binding domain [1, 2]. The active A1 peptide (CTA1) is generated after reduction of a disulfide bond that links it to the A2 polypeptide [3, 4]. CTA1 is responsible for cholera toxin's enzymatic activities [5]. These include (i) NAD hydrolysis to ADP-ribose and nicotinamide [6]; (ii) transfer of ADP-ribose to free arginine and simple guanidino compounds [7, 8]; (iii) ADP-ribosylation of nonspecific proteins, presumably due to available arginine residues which serve as toxin substrates [9]; (iv) auto-ADP-ribosylation of CTA1 [10, 11]; and (v) ADP-ribosylation of $G_{s\alpha}$, the guanine nucleotide-binding protein responsible for stimulation of adenylyl cyclase [12–15]. $G_{s\alpha}$ is the substrate for cholera toxin *in vivo* and its modification by toxin underlies the pathogenesis of cholera by causing stimulation of adenylyl cyclase in intestinal mucosal cells [16, 17].

Address for offprints: C. Welsh, Room 5N-307, Building 10, National Institute of Health, Bethesda, MD 20892, USA

Activation of adenylyl cyclase by cholera toxin was shown to require NAD [18], GTP or analogues [19–22], and certain cellular factors [19]. In membrane preparations, cholera toxin-catalyzed ADP-ribosylation of $G_{s\alpha}$ also required GTP or nonhydrolyzable analogs and a cytosolic factor and correlated well with activation of adenylyl cyclase [23]. Two guanine nucleotide-binding sites were involved in this system: one responsible for activation of adenylyl cyclase and a distinct site which supported ADP-ribosylation but not cyclase activity [24].

Various cellular factors that augmented cholera toxin catalyzed ADP-ribosylation of $G_{s\alpha}$ and/or activation of adenylyl cyclase were described [19, 23, 25–35]. These factors, which were isolated from numerous tissues and species, varied in size, biochemical properties, and subcellular localization.

Purification and biochemical characterization of ARF

A membrane-associated ADP-ribosylation factor (mARF) was purified 1800-fold from rabbit liver membranes [32]. Protein stability required detergent throughout the purification. ARF migrated as a doublet on SDS polyacrylamide gel electrophoresis and exhibited a M_r of 21,500 [32]. GTP was required for ARF to stimulate cholera toxin-catalyzed ADP-ribosylation of $G_{s\alpha}$ and ARF appeared to act catalytically in the assay [32].

ARF purified from bovine brain membranes bound guanine nucleotides (GTP, GDP, and GTPγS), but not adenine nucleotides [33]. High-affinity binding occurred in the presence of dimyristoylphosphatidylcholine (DMPC), $MgCl_2$, and high ionic strength (0.8 M NaCl). The apparent dissociation constants for GTP and GDP were 90 and 40 nM, respectively. Purified ARF contained approximately equimolar quantities of bound GDP, however no GTPase activity was detected *in vitro*. ARF-GTP was active in supporting cholera toxin-catalyzed ADP-ribosylation of $G_{s\alpha}$ whereas ARF-GDP was inactive [33].

ARF also enhanced $G_{s\alpha}$-independent reactions catalyzed by the toxin A1 peptide (CTA1) [34, 35]. These included the hydrolysis of NAD to nicotinamide and ADP-ribose, the auto-ADP-ribosylation of CTA1, and ADP-ribosylation of simple guanidino compounds such as agmatine. In these reactions, ARF required GTP or a nonhydrolyzable analogue, whereas GDP and ATP analogues were ineffective [34, 35]. A model of a guanine nucleotide-binding protein cascade for cholera toxin activation of adenylyl cyclase has been proposed (Fig. 1). ARF, activated by GTP, interacts with, and activates the toxin A1 peptide which, in turn, catalyzes the ADP-ribosylation of $G_{s\alpha}$. In the presence of GTP, modified $G_{s\alpha}$ persistently stimulates the adenylyl cyclase catalytic subunit. In this model, there is no direct interaction between ARF and the toxin substrate.

Kinetic analysis of ARF-stimulated NAD:agmatine ADP-ribosyltransferase activity of CTA1 is consistent with direct interaction between ARF and the toxin subunit [36]. In the absence of detergents, ARF plus GTP lowered the apparent K_m for NAD and agmatine. In the presence of 0.003% SDS, the concentration at which activity was maximal, the apparent K_m values for both substrates were lowered further and the V_{max} was increased [36]. It was concluded that ARF is a direct allosteric activator of the toxin A1 subunit.

Since ARF's ability to stimulate cholera toxin-catalyzed reactions is dependent on GTP or its analogues, conditions optimal for high-affinity binding were determined [37]. Maximal binding of GTP by sARFII was observed in the presence of 3 mM DMPC and 0.2% cholate and required Mg^{2+} but not NaCl [37]. The apparent K_D for GTP was 70 nM [37]. In the presence of low concentrations of SDS, conditions optimal for ARF stimulation of cholera toxin ADP-ribosyltransferase activity, minimal high affinity GTP binding by sARF II was observed [37]. Consistent with the binding data, the EC_{50} for GTP in stimulation of cholera toxin ADP-ribosyltransferase activity was 5 μM in the presence of 0.003% SDS and 50 nM with DMPC/cholate. Thus, ARF stimulates cholera toxin-catalyzed reactions under conditions that support either high-(DMPC/cholate) or low-(SDS) affinity interactions with GTP.

ARF stimulation of cholera toxin reactions, both $G_{s\alpha}$-dependent and $G_{s\alpha}$-independent, was influenced by detergents and phospholipids [35–37]. DMPC was required for sARF plus GTP to stimulate maximally ADP-ribosylation of $G_{s\alpha}$ but not auto-ADP-ribosylation of CTA1, which was somewhat inhibited [35, 37]. Maximal ARF stimulation of NAD: agmatine ADP-ribosyltransferase activity was observed in the presence of nondenaturing concentrations of SDS [36, 37] as was CTA1 auto-ADP-ribosylation [36, 38].

In the presence of GTPγS and SDS, a fraction of ARF formed a complex with CTA1 [38]. Monomeric ARF separated from the complex by gel filtration was incapable of further complex formation. ARF-CTA1 complex stimulated auto-ADP-ribosylation to a greater extent than did the monomeric species, which exhibited clearly different substrate specificities [38]. No ARF-CTA1 complex was formed during incubation with GTPγS and DMPC/cholate. ARF, however, became aggregated under these conditions and was more active in stimulating cholera toxin ADP-ribosylation of added substrates than auto-ADP-ribosylation of CTA1 [38]. These studies may provide a biochemical basis for the observed effects of detergents and phospholipids on substrate specificity.

Six mammalian ARF family members have been identified through molecular cloning (see section on Molecular Characterization of ARF for complete discussion) and each of these has been expressed in bacteria [39–41]. Purified recombinant ARFs (rARFs) each stimulated cholera toxin-

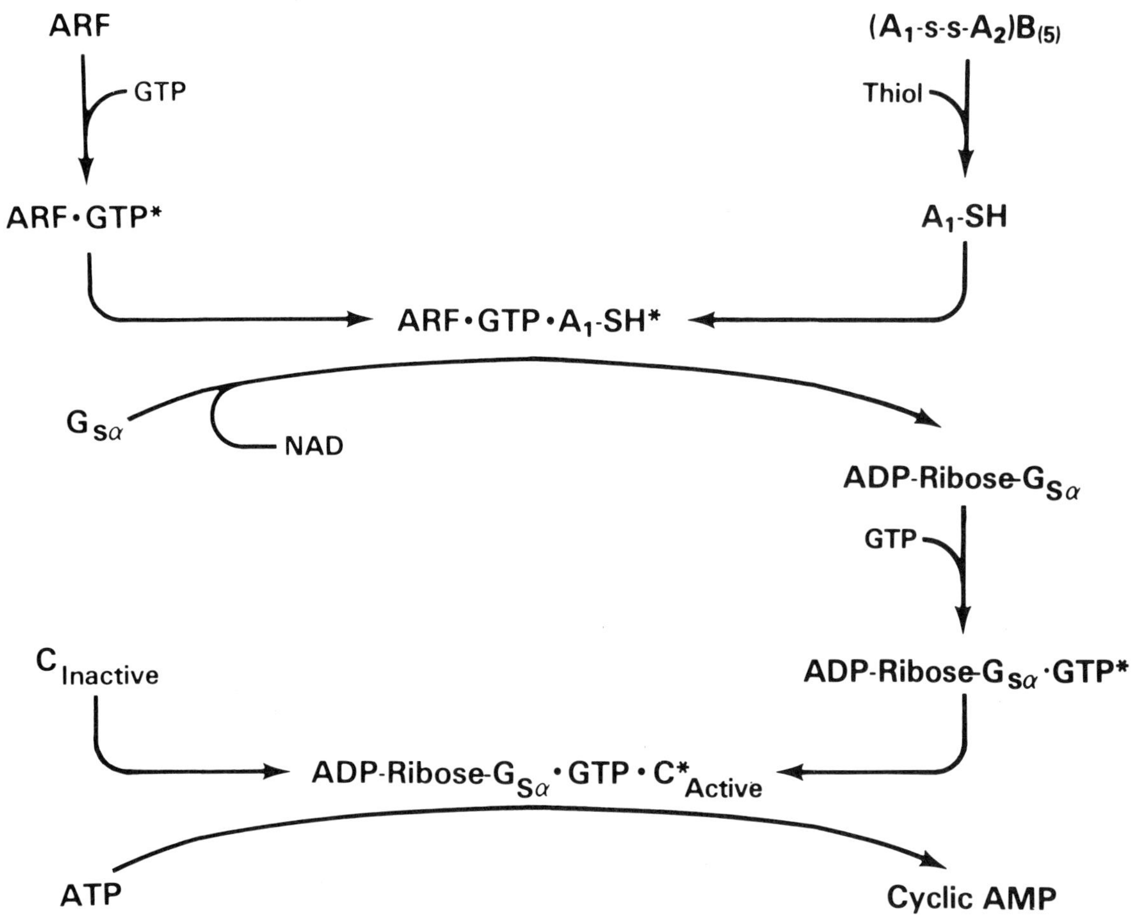

Fig. 1. ARF participation in a guanine nucleotide-binding protein cascade resulting in activation of adenylyl cyclase. Activated ARF (ARF-GTP*) interacts with toxin A1 protein (A_1-SH), generated from holotoxin (A_1-ss-A_2)$B_{(5)}$ by reduction with thiol. Activated toxin catalyzes the transfer of ADP-ribose from NAD to $G_{s\alpha}$. In the presence of GTP, modified $G_{s\alpha}$ persistently stimulates the adenylyl cyclase catalytic subunit (C), generating cyclic AMP from ATP. Data are from Ref. 35.

catalyzed ADP-ribosyltransferase activity, using either $G_{s\alpha}$ or agmatine as substrate. Individual differences were noted, however, in the conditions under which maximal activity was observed. Whereas rARFs 1–5 required DMPC/cholate for activity, rARF 6 was active in the presence of SDS, DMPC/cholate, cholate alone, and in the absence of detergents and lipids [41]. Other differences observed between individual ARFs included sensitivity to high ionic strength, the EC_{50} for GTP, and the requirement for Mg^{2+} in binding guanine nucleotides [39, 41]. Differences in biochemical properties noted during early characterization and purification of factors required for cholera toxin activation may be related to such differences among individual ARFs.

ARF developmental expression

Expression of ARF proteins differs with developmental stage in some tissues. sARF I and sARF II (two forms purified from bovine brain) expression was studied in post-natal rat brain cytosol [42]. The level of immunoreactive sARF II relative to sARF I increased between postnatal days 2 and 10. The level of ARF 3 mRNA increased with age while there were simultaneous decreases in levels of ARF 2 and 4 mRNAs and ARF 1, 5, and 6 mRNAs were unchanged [42].

Tissue-specific differences in ARF 4 mRNA during development have also been observed [43]. A short form of ARF 4 mRNA (~1.1 kb) was found only in the testis of several species, along with the ubiquitous 1.8-kb mRNA. The 1.1 kb-species resulted from use of the first of three putative polyadenylation signals in the 3′-untranslated region [43]. During development of rat testis, the 1.1-kb mRNA did not appear until the 35th postnatal day, suggesting that appearance of the shorter message may be associated with a late stage of spermatogenesis [43], a suggestion corroborated by *in situ* hybridization.

Molecular characterization of ARF

There are at least six mammalian ARFs based on isolation and sequencing of cDNA clones [40, 44–49]. ARF genes have also been cloned from yeast (yARFs 1 and 2) [44, 50], *Drosophila melanogaster* (dARF 1) [51], and the protozoan *Giardia lamblia* (gARF) [52]. With the exception of ARF 2, which has not been found in human or monkey [42, 48, 53], the individual ARFs appear to be highly conserved across species (Table 1). For example, the human and bovine deduced amino acid and coding region nucleotide sequences for ARF 1 are 100% and 91% identical, respectively [45]. The *Giardia* ARF gene encodes a protein of 191 amino acids 63–70% identical to mammalian ARFs, remarkable conservation considering that the divergence of this protozoan from other eukaryotes has been estimated to be ~1.5 billion years ago [52].

Table 1 highlights the conservation of nucleotide and deduced amino acid sequences among individual ARF family members. Based on these comparisons, the mammalian ARFs have been assigned to three classes. Class I ARFs (ARFs 1, 2, and 3) are 95–96% identical to each other in deduced amino acid sequence, 79–80% identical to Class II and 68–69% identical to Class III ARFs. Class I ARFs are 181 amino acids in length. Class II is composed of ARFs 4 and 5 which contain 180 amino acids and are 90% identical to each other. The sole known member of Class III is ARF 6, which has only 175 amino acids and is 64–69% identical to other mammalian ARFs. Sequence differences among Classes I, II, and III are located primarily in the amino terminal region and carboxyl half of the proteins (Fig. 2).

The classification of ARF proteins is corroborated by an evaluation of evolutionary relationships among ARF family members [48]. Phylogenetic analysis was performed using coding region nucleotide sequences of the individual ARFs and is shown in Fig. 3. Branch points, indicating evolutionary divergence, correspond to each class of mammalian ARF and yARF 1.

The ARFs represent distinct gene products. Each of the Class I ARF genes has been cloned and its structure determined (Fig. 4) [54–56]. Each of the genes contains five exons and four introns. The first exon is untranslated and translation initiation is located in exon 2. The consensus sequences for GTP binding and hydrolysis, which are identical in all mammalian ARFs, are encoded in separate exons. The sequence GLDAAGK, the putative phosphate binding loop, is encoded in exon 2 and DVGG, which is likely involved in Mg^{2+} coordination, in exon 3. The sequence encoding NKQD (important in guanine ring-binding specificity) is separated by intron 4 into exons 4 and 5 between the codons for glutamine and aspartate. Exon/intron boundaries occur at identical sites within the coding regions of all three genes (Fig. 4). Each of the 5′-untranslated regions contains multiple transcription initiation sites, multiple potential Sp1 binding sites, and no TATA box, similar to promoters of housekeeping genes [54–56].

A *Drosophila* ARF gene, dARF 1, was categorized as Class I based on a high degree of homology with mammalian ARFs 1–3 [51]. Two introns in the *Drosophila* gene contained splice sites very similar to the exon/intron boundaries in the mammalian Class I genes [51]. Gene structure in at least Class I ARFs, thus appears to be conserved across rather distantly related species.

The promoters of ARFs 2 and 3 both have regions that may be important for regulation [56, 57]. The ARF 2 promoter contains six inverted CCAAT motifs unlike those of ARFs 1 and 3 [54–56] and the functional regulatory region of the promoter includes a region containing one of these motifs. Whether this region is important in the species-specific and developmentally regulated expression patterns noted previously is an intriguing speculation, which requires further investigation. The ARF 3 promoter region contains a palindromic sequence important for transcription activity, which binds a protein from nuclear extracts [57]. A mutated palindromic sequence, which failed to compete for binding of the nuclear factor, had reduced transcription activity when linked

Table 1. Percentage identity of ARF deduced amino acid sequences and coding region nucleotide sequences

	hARF1	bARF1	bARF2	hARF3	hARF4	hARF5	hARF6	yARF1	gARF
hARF1	–	100	96	96	80	80	68	77	70
bARF1	91	–	96	96	80	80	68	77	70
bARF2	79	80	–	95	80	80	69	77	70
hARF3	84	84	80	–	79	79	68	76	69
hARF4	67	68	68	71	–	90	64	72	69
hARF5	75	73	71	73	77	–	64	69	69
hARF6	68	69	64	66	60	65	–	65	63
yARF1	64	66	66	65	67	64	60	–	62
gARF	65	67	62	64	61	66	62	67	–

Comparison of amino acid and nucleotide sequences of ARF family proteins. Percentage identity of deduced amino acid sequences is above the diagonal and of coding region nucleotide sequences below. See Fig. 2 for explanation of abbreviations and references. Data are from Refs. 48, 52.

```
                    10         20         30         40         50         60         70
hARF1      1    MGnifanLFk gLFGKKEMRI LMVGLDAAGK TTILYKLKLG EiVTTIPTIG FNVETVEYKN IsFTVWDVGG
bARF2      1    MGnvfekLFk sLFGKKEMRI LMVGLDAAGK TTILYKLKLG EiVTTIPTIG FNVETVEYKN IsFTVWDVGG
hARF3      1    MGnifgnLlk sLiGKKEMRI LMVGLDAAGK TTILYKLKLG EiVTTIPTIG FNVETVEYKN IsFTVWDVGG
hARF4      1    MGltiSsLFs rLFGKKqMRI LMVGLDAAGK TTILYKLKLG EiVTTIPTIG FNVETVEYKN IcFTVWDVGG
hARF5      1    MGltvSaLFs riFGKKqMRI LMVGLDAAGK TTILYKLKLG EiVTTIPTIG FNVETVEYKN IcFTVWDVGG
hARF6      1    MGkvlSk      iFGnKEMRI LMlGLDAAGK TTILYKLKLG qsVTTIPTvG FNVETVtYKN vkFnVWDVGG
yarf1      1    MGlfaSkLFs nLFGnKEMRI LMVGLDgAGK TTvLYKLKLG EviTTIPTIG FNVETVqYKN IsFTVWDVGG
yarf2      1    MGlyaSkLFs nLFGnKEMRI LMVGLDgAGK TTvLYKLKLG EviTTIPTIG FNVETVqYKN IsFTVWDVGG
gARF       1    MGqgaSkiFg kLFsKKEvRI LMVGLDAAGK TTILYKLmLG EvVTTvPTIG FNVETVEYKN InFTVWDVGG
CONSENSUS       MG                K   RI LM GLD AGK TT LYKL LG    TT PT G FNVETV YKN    F VWDVGG

                    80         90        100        110        120        130        140
hARF1     71    QDkIRPLWRH YfQNTQGLIF VVDSNDRE   Rv nEAReELmRM LaEDELRDAV LLVFANKQDL PnAMnaaEIT
bARF2     71    QDkIRPLWRH YfQNTQGLIF VVDSNDRE   Rv nEAReELtRM LaEDELRDAV LLVFvNKQDL PnAMnaaEIT
hARF3     71    QDkIRPLWRH YfQNTQGLIF VVDSNDRE   Rv nEAReELmRM LaEDELRDAV LLVFANKQDL PnAMnaaEIT
hARF4     71    QDrIRPLWkH YfQNTQGLIF VVDSNDRE   Ri qEvadELqkM LlvDELRDAV LLlFANKQDL PnAMaisEmT
hARF5     71    QDkIRPLWRH YfQNTQGLIF VVDSNDRE   Rv qEsadELqkM LqEDELRDAV LLVFANKQDm PnAMpvsElT
hARF6     67    QDkIRPLWRH YytgTQGLIF VVDcaDRd   Ri dEARqELhRi indrEmRDAi iLiFANKQDL PdAMkphEIq
yarf1     71    QDrIRsLWRH YyrNTeGvIF VVDSNDRs   Ri gEARevmqRM LnEDELRnAa wLVFANKQDL PeAMsaaEIT
yarf2     71    QDrIRsLWRH YyrNTeGvIF ViDSNDRs   Ri gEARevmqRM LnEDELRnAV wLVFANKQDL PeAMsaaEIT
gARF      71    QDsIRPLWRH YyQNTdaLIy ViDSaDlEpkRi edARnELhtl LgEDELRDAa LLVFANKQDL PkAMsttdlT
CONSENSUS       QD IR LW H Y    T   I V D   D     R                       E R A   L F NKQD  P AM

                   150        160        170             180
hARF1    141    dKLGLhSLRh RnWYIQATCA TSGdGLYEGL DWL  SNqL   rNqk           181
bARF2    141    dKLGLhSLRq RnWYIQATCA TSGdGLYEGL DWL  SNqL   KNqk           181
hARF3    141    dKLGLhSLRh RnWYIQATCA TSGdGLYEGL DWL  aNqL   KNkk           181
hARF4    141    dKLGLqSLRn RtWYvQATCA TqGtGLYEGL DWL  SNeLs  Kr             180
hARF5    141    dKLGLqhLRs RtWYvQATCA TqGtGLYdGL DWL  SheLs  Kr             180
hARF6    137    eKLGLtriRd RnWYvQpsCA TSGdGLYEGL tWL tSNy    Ks             175
yarf1    141    eKLGLhSiRn RpWfIQATCA TSGeGLYEGL eWL  SNsL   KNst           181
yarf2    141    eKLGLhSiRn RpWfIQsTCA TSGeGLYEGL eWL  SNnL   KNqs           181
gARF     143    erLGLqeLkk RdWYIQpTCA rSGdGLYqGL DWL  SdyifdkKNkkkgkkr      191
CONSENSUS         LGL        R W  Q  CA   G GLY GL  WL
```

Fig. 2. Deduced amino acid sequences of ARF proteins. Capital letters indicate amino acids identical in at least six ARFs. CONSENSUS indicates amino acids identical in all the ARFs. Sequences were aligned using Gene Works program and gaps were introduced where needed for optimal alignment. hARF1, 3–6, human ARFs [45, 47, 48]; bARF2, bovine ARF 2 [46]; dARF1, *Drosophila* ARF [51]; yARF1, 2, yeast ARFs 1 and 2 [44, 50]; and gARF, *Giardia* ARF [52]. Data are from Ref. 52.

to a reporter gene. The identity of this factor and its role in regulating ARF expression, and other TATA-less promoters, is unknown.

ARF protein structure

Through sequence comparisons with other GTPases, such as ras, certain functional domains in ARFs have been inferred. All of the ARF gene products contain the consensus sequences for GTP binding and hydrolysis [58]. These are 100% identical among the mammalian ARFs with minimal and usually conservative differences in yeast and *Giardia* ARFs (Fig. 2). The first GTP-binding sequence GLDAAGK, which spans amino acids 24–29 in hARF 1, corresponds to the phosphate-binding loop in the *ras* gene product. Aspartate 26 in ARF corresponds to Glycine 12 in c-Ha-ras [58], which is essential for normal GTP hydrolysis. In ras, replacement of Glycine 12 by almost any other amino acid markedly reduced GTPase activity and promotes oncogenic transformation [59–61]. The presence of aspartate in this position might help to explain the absence of detectable GTPase activity in the ARF proteins. However, arl, an ARF-like *Drosophila* protein, which also has aspartate in this position, has notably higher basal GTPase activity [62].

The second consensus site for GTP-binding is involved in Mg^{2+} ion coordination. This sequence in all of the known ARFs is DVGGQ, spanning amino acids 67–71 in hARF1. It is identical to the cognate sequence in the alpha subunits of the heterotrimeric G proteins but differs from those of other small GTPases [58]. Similarly, the TCAT sequence (amino acids 158–161 in hARF 1) is identical to that of the corresponding region in most of the heterotrimeric G proteins. In ras and other small GTPases, this sequence is SAK (amino acids 145–147 in c-Ha-ras) [58]. The consensus sequence of the region thought to participate in guanine ring binding is

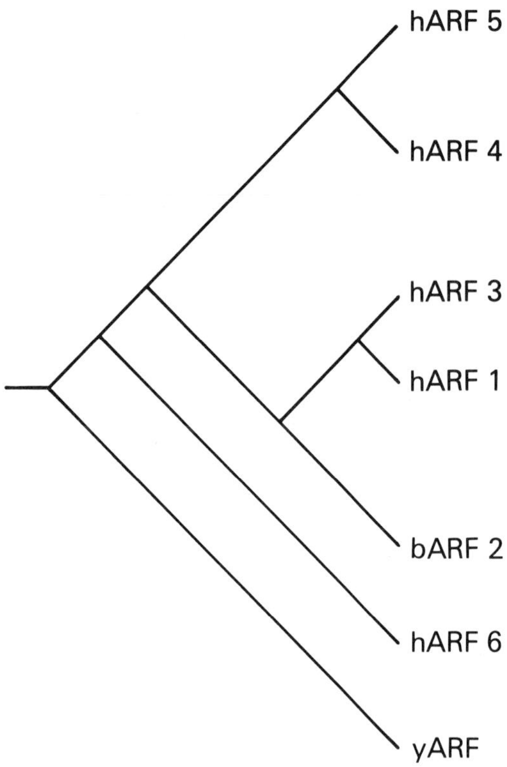

Fig. 3. Phylogenetic analysis of ARF cDNA sequences. Coding region nucleotide sequences of hARFs 1, 3 to 6, bARF 2, and yARF (yARF1) were analyzed using DNADIST and KITSCH programs to determine evolutionary relationships. Branches indicate points of evolutionary divergence. Data are from Ref. 48.

NKQD (amino acids 126–129 in hARF 1).

Like G_o and G_i, many of the ARFs undergo N-terminal modification by myristoyl CoA:protein N-myristoyltransferase (NMT) [63–65]. Each of the ARFs contains a glycine in the first position, following the initiating methionine, which is the required substrate for NMT, although neighboring amino acids also influence substrate suitability [66]. N-terminal myristoylation is necessary for ARF to associate with Golgi membranes [63] where it functions in protein trafficking via vesicular transport. In contrast, ras proteins undergo carboxyl terminal processing including farnesylation, carboxy methylation, and palmitoylation which is similarly required for membrane association. Thus, ARF appears to have more in common with G protein alpha subunits than with other small GTPases when comparing GTP-binding consensus sequences and post-translational modification.

ARF structure and function analysis has focused on the amino terminus since it is the site of covalent lipid modification. CD spectral studies revealed that a peptide composed of the first 16 amino acids of ARF 1 (excluding the initial methionine) would likely form an alpha helix in a hydrophobic but not an aqueous environment [67]. Myristoylation of the amino-terminal glycine of the peptide stabilized the α helical structure even in an aqueous solution. Furthermore, an ARF deletion mutant, lacking the amino-terminal 17 residues ([Δ1–17]mARF1p) retained the ability to bind GTPγS but no longer required DMPC/cholate to do so [67]. Amino-terminal α helix formation may, therefore, influence the binding or release of guanine nucleotides by ARF. To what extent the myristoylation requirement for membrane association [63] reflects the need for α helix formation is unknown.

ARD 1 is a 64-kDa protein, recently identified by cDNA cloning whose carboxyl terminal ~18 kDa is highly homologous to known mammalian ARFs but lacks sequence corresponding to the amino-terminal 15 amino acids [68]. Although previous studies had suggested that the amino terminus is critical for ARF activity [67], ARD 1 stimulated

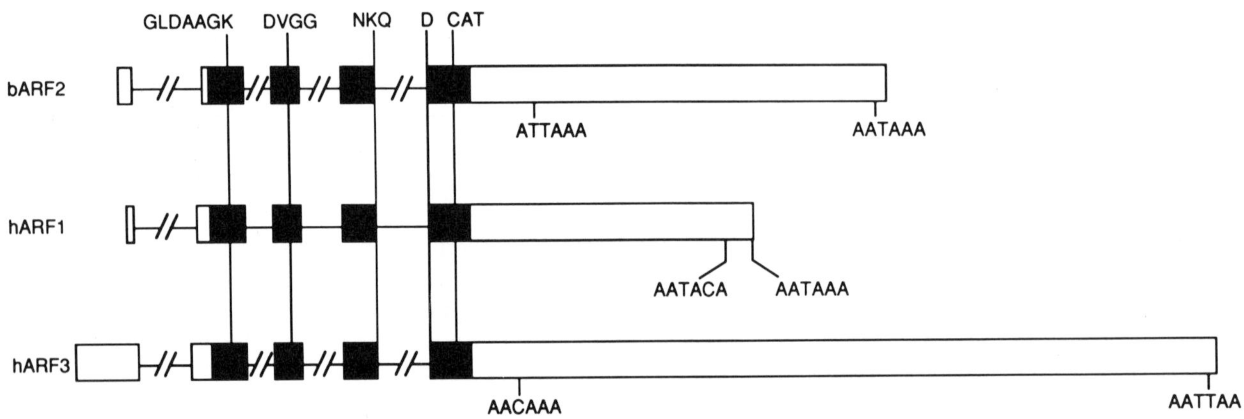

Fig. 4. Mammalian Class I ARF gene structures. Bovine ARF 2 (bARF2) [56], and human ARFs 1 and 3 (hARF1, 3) [54, 55] gene structures are compared. Exons are represented by boxes with coding regions shaded and untranslated regions open, introns by horizontal lines. Consensus sequences for GTP binding and hydrolysis are shown with locations indicated by vertical lines. Potential polyadenylation signal sequences and locations are indicated below each gene. Data are from Ref. 56.

cholera toxin ADP-ribosyltransferase activity in a GTP-dependent fashion in the presence of Tween 20 but not other lipids and detergents including DMPC/cholate. These studies clearly indicate that the amino terminal sequence is not responsible for ARF activity, although it may be involved with phospholipid interaction.

ARF's role in vesicular transport

Accumulating evidence has implicated ARF as one of a growing number of GTPases for which a role in intracellular protein transport has been recognized. In *Saccharomyces cerevisiae*, disruption of the yARF 1 gene, which accounts for 90% of ARF protein, caused a defect in invertase secretion with intracellular accumulation of incompletely glycosylated invertase [69].

In *in vitro* studies ARF shifts between cytosolic and membrane compartments in a guanine nucleotide-dependent cycle. In the presence of nonhydrolyzable GTP analogues, cytosolic ARF associates with membranes or anionic phospholipid vesicles [70, 71]. In NRK cells, ARF bound to Golgi membranes in the presence of GTPγS, an action inhibited by prior treatment with brefeldin A [72]. Brefeldin A is a fungal metabolite that disrupts Golgi architecture and blocks secretion [73, 74].

Serafini *et al.* [75] have identified ARF as a coat protein, along with α-, β-, γ-, and δ-COPs, major constituents of the non-clathrin coat of Golgi-derived vesicles. ARF was found associated with coated vesicles, which originate from the budding surface of Golgi membranes, and not with uncoated vesicles, which had bound to, but not fused with, the target Golgi membranes. ARF was also required for the GTPγS-dependent binding of β-COP, and therefore, presumably, the coat protein complex, the 'coatomer' [76].

Golgi-localized proteins that regulate the activation state of ARF and thus, its subcellular localization have been detected. A protein that enhanced the exchange of GDP bound to ARF with cytosolic GTP [77, 78] permitted ARF to associate with Golgi membranes. Rapid hydrolysis of the γ-phosphate of ARF-bound GTP occurred after the exchange reaction followed by release of ARF to the cytosolic fraction [77]. Subsequent studies revealed that individual ARFs differ in their subcellular localization and the conditions that promote membrane association [79, 80]. In addition, differences in the sensitivity to brefeldin A among ARF proteins have been reported [80].

The evidence supports a model for ARF's role in vesicular transport as proposed by Serafini *et al.* [75] in which ARF binds to the surface of Golgi membranes through interaction with a Golgi-localized exchange protein. ARF-GTP is generated which is competent to associate with membrane lipids. Coat assembly and budding ensues until destabilization of the vesicle coat is triggered by the formation of ARF-GDP by an ARF-specific GTPase-activating protein (GAP) at the target membrane. Components of the vesicle coat and ARF-GDP are released and recycled. ARF has also been implicated in other events involving vesicular fusion including endoplasmic reticulum to cis-Golgi transport [81], endosome fusion during endocytosis [82], and nuclear vesicle fusion following mitosis [83]. Specificity in targeting in each of these pathways may rely on individual ARFs interacting with specifically localized exchange proteins and GAPs.

Summary

ARF was discovered because of its role as a cofactor for cholera toxin-catalyzed ADP-ribosylation of $G_{s\alpha}$. ARF itself was found to be a GTP-binding protein and a direct allosteric activator of all of the toxin's catalytic activities. The ARFs are a family of highly homologous proteins that are conserved both structurally and functionally across distantly related eukaryotes and are distinct from other GTP-binding proteins. Biochemical features which distinguish ARFs include a relative dependence on phospholipids for GTP binding and activity, little or no intrinsic GTPase activity, and amino-terminal modification by N-myristoyltransferase. *In vivo*, ARF has been implicated in the regulation of vesicular transport through a guanine nucleotide-dependent cycle of association with and release from Golgi membranes. ARF has been identified as a coat protein and ARF-specific regulatory proteins which catalyze guanine nucleotide exchange and accelerate GTPase activity have been detected on Golgi membranes. Understanding ARF function and regulation will be important in elucidating mechanisms of vesicular transport.

Acknowledgements

We thank Mrs. Carol Kosh for expert secretarial assistance.

References

1. Gill DM: The arrangement of subunits in cholera toxin. Biochemistry 15: 1242–1248, 1976
2. Gill DM: Mechanism of action of cholera toxin. Adv Cyclic Nucleotide Res 8: 85–118, 1977
3. Moss J, Vaughan M: Participation of guanine nucleotide-binding protein cascade in activation of adenylyl cyclase by cholera toxin (choleragen). In: R. Iyengar and L. Birnbaumer (eds). G Proteins, Academic Press, San Diego, CA, 1990, pp 179–200
4. Mekalanos JJ, Collier RJ, Romig WR: Enzymic activity of cholera toxin. II. Relationships to proteolytic processing, disulfide bond reduction, and subunit composition. J Biol Chem 254: 5855–5861, 1979

5. Moss J, Stanley SJ, Lin MC: NAD glycohydrolase and ADP-ribosyltransferase activities are intrinsic to the A_1 peptide of choleragen. J Biol Chem 254: 11993–11996, 1979
6. Moss J, Manganiello VC, Vaughan M: Hydrolysis of nicotinamide adenine dinucleotide by choleragen and its A protomer: possible role in the activation of adenylate cyclase. Proc Natl Acad Sci USA 73: 4424–4427, 1976
7. Moss J, Vaughan M: Mechanism of action of choleragen. Evidence for ADP-ribosyltransferase activity with arginine as an acceptor. J Biol Chem 252: 2455–2457, 1977
8. Mekalanos JJ, Collier RJ, Romig WR: Enzyme activity of cholera toxin. I. New methods of assay and the mechanism of ADP-ribosyl transfer. J Biol Chem 254: 5849–5854, 1979
9. Moss J, Vaughan M: Isolation of an avian erythrocyte protein possessing ADP-ribosyltransferase activity and capable of activating adenylate cyclase. Proc Natl Acad Sci USA 75: 3621–3624, 1978
10. Moss J, Stanley SJ, Watkins PA, Vaughan M: ADP-ribosyltransferase activity of mono- and multi-(ADP-ribosylated) choleragen. J Biol Chem 255: 7835–7837, 1980
11. Trepel JB, Chuang D-M, Neff NH: Transfer of ADP-ribose from NAD to choleragen: A subunit acts as a catalyst and acceptor protein. Proc Natl Acad Sci USA 74: 5440–5442, 1977
12. Cassel D, Pfeuffer T: Mechanism of cholera toxin action: covalent modification of the guanyl nucleotide-binding protein of the adenylate cyclase system. Proc Natl Acad Sci USA 75: 2669–2673, 1978
13. Gill DM, Meren R: ADP-ribosylation of membrane proteins catalyzed by cholera toxin: basis of the activation of adenylate cyclase. Proc Natl Acad Sci USA 75: 3050–3054, 1978
14. Johnson GL, Kaslow HR, Bourne HR: Genetic evidence that cholera toxin substrates are regulatory components of adenylate cyclase. J Biol Chem 253: 7120–7123, 1978
15. Northup JK, Sternweis PC, Smigel MD, Schleifer LS, Ross EM, Gilman AG: Purification of the regulatory component of adenylate cyclase. Proc Natl Acad Sci USA 77: 6516–6520, 1980
16. Finkelstein RA: Cholera. CRC Crit Rev Microbiol 2: 553–623, 1973
17. Kelley MT: Cholera: a worldwide perspective. Pediat Infect Dis 5: S101–S105, 1986
18. Gill DM: Involvement of nicotinamide adenine dinucleotide in the action of cholera toxin *in vitro*. Proc Natl Acad Sci USA 72: 2064–2068, 1975
19. Enomoto K, Gill DM: Requirement for guanosine triphosphate in the activation of adenylate cyclase by cholera toxin. J Supramol Struct 10: 51–60, 1979
20. Moss J, Vaughan M: Choleragen activation of solubilized adenylate cyclase: requirement for GTP and protein activator for demonstration of enzymatic activity. Proc Natl Acad Sci USA 74: 4396–4400, 1977
21. Lin MC, Welton AF, Berman MF: Essential role of GTP in the expression of adenylate cyclase activity after cholera toxin treatment. J Cyclic Nucleotide Res 4: 159–168, 1978
22. Nakaya S, Moss J, Vaughan M: Effects of nucleoside triphosphates on choleragen-activated brain adenylate cyclase. Biochemistry 19: 4871–4874, 1980
23. Enomoto K, Gill DM: Cholera toxin activation of adenylate cyclase. Roles of nucleoside triphosphates and a macromolecular factor in the ADP-ribosylation of the GTP-dependent regulatory component. J Biol Chem 255: 1252–1258, 1980
24. Gill DM, Meren R: A secondary guanyl nucleotide-binding site associated with adenylate cyclase. Distinct nucleotides activate adenylate cyclase and permit ADP-ribosylation by cholera toxin. J Biol Chem 258: 11908–11914, 1983
25. LeVine H III, Cuatrecasas P: Activation of pigeon erythrocyte adenylate cyclase by cholera toxin. Partial purification of an essential macromolecular factor from horse erythrocyte cytosol. Biochim Biophys Acta 672: 248–261, 1981
26. Schleifer LS, Kahn RA, Hanski E, Northup JK, Sternweis PC, Gilman AG: Requirements for cholera toxin-dependent ADP-ribosylation of the purified regulatory component of adenylate cyclase. J Biol Chem 257: 20–23, 1982
27. Pinkett MO, Anderson WB: Plasma membrane-associated component(s) that confer(s) cholera toxin sensitivity to adenylate cyclase. Biochim Biophys Acta 714: 337–343, 1982
28. Gill DM, Woolkalis MJ: [^{32}P]ADP-ribosylation of proteins catalyzed by cholera toxin and related heat-labile enterotoxins. Methods Enzymol 165: 235–245, 1988
29. Schmidt GJ, Huber LJ, Weiter JJ: A-protein catalyzes the ADP-ribosylation of G-protein from cow rod outer segments. J Biol Chem 262: 14333–14336, 1987
30. Gill DM, Coburn J: ADP-ribosylation by cholera toxin: functional analysis of a cellular system that stimulates the enzymatic activity of cholera toxin fragment A_1. Biochemistry 26: 6364–6371, 1987
31. Woolkalis M, Gill DM, Coburn J: Assay and purification of cytosolic factor required for cholera toxin activity. Methods Enzymol 165: 246–249, 1988
32. Kahn RA, Gilman AG: Purification of a protein cofactor required for ADP-ribosylation of the stimulatory regulatory component of adenylate cyclase by cholera toxin. J Biol Chem 259: 6228–6234, 1984
33. Kahn RA, Gilman AG: The protein cofactor necessary for ADP-ribosylation of G_s by cholera toxin is itself a GTP binding protein. J Biol Chem 261: 7906–7911, 1986
34. Tsai S-C, Noda M, Adamik R, Moss J, Vaughan M: Enhancement of choleragen ADP-ribosyltransferase activities by guanyl nucleotides and a 19-kDa membrane protein. Proc Natl Acad Sci USA 84: 5139–5142, 1987
35. Tsai S-C, Noda M, Adamik R, Chang PP, Chen H-C, Moss J, Vaughan M: Stimulation of choleragen enzymatic activities by GTP and two soluble proteins purified from bovine brain. J Biol Chem 263: 1768–1772, 1988
36. Noda M, Tsai S-C, Adamik R, Moss J, Vaughan M: Mechanism of cholera toxin activation by a guanine nucleotide-dependent 19 kDa protein. Biochim Biophys Acta 1034: 195–199, 1990
37. Bobak DA, Blizotes MM, Noda M, Tsai S-C, Adamik R, Moss J: Mechanism of activation of cholera toxin by ADP-ribosylation factor (ARF): Both low- and high-affinity interactions of ARF with guanine nucleotides promote toxin activation. Biochemistry 29: 855–861, 1990
38. Tsai S-C, Adamik R, Moss J, Vaughan M: Guanine nucleotide dependent formation of a complex between choleragen (cholera toxin) A subunit and bovine brain ADP-ribosylation factor. Biochemistry 30: 3697–3703, 1991
39. Weiss O, Holden J, Rulka C, Kahn RA: Nucleotide binding and cofactor activities of purified bovine brain and bacterially expressed ADP-ribosylation factor. J Biol Chem 264: 21066–21072, 1989
40. Kahn RA, Kern FG, Clark J, Gelmann EP, Rulka C: Human ADP-ribosylation factors. A functionally conserved family of GTP-binding proteins. J Biol Chem 266: 2606–2614, 1991
41. Price SR, Welsh CF, Haun RS, Stanley SJ, Moss J, Vaughan M: Effects of phospholipid and GTP on recombinant ADP-ribosylation factors (ARFs). Molecular basis for differences in requirements for activity of mammalian ARFs. J Biol Chem 267: 17766–17772, 1992
42. Tsai S-C, Adamik R, Tsuchiya M, Chang PP, Moss J, Vaughan M: Differential expression during development of ADP-ribosylation factors, 20-kDa guanine nucleotide-binding protein activators of cholera toxin. J Biol Chem 266: 8213–8219, 1991
43. Mishima K, Price SR, Nightingale MS, Kousvelari E, Moss J, Vaughan M: Regulation of ADP-ribosylation factor (ARF) expression. Cross-species conservation of the developmental and tissue-specific alterna-

tive polyadenylation of ARF 4 mRNA. J Biol Chem 267: 24109–24116, 1992

44. Sewell JL, Kahn RA: Sequences of the bovine and yeast ADP-ribosylation factor and comparison to other GTP-binding proteins. Proc Natl Acad Sci USA 85: 4620–4624, 1988
45. Bobak DA, Nightingale MS, Murtagh JJ, Price SR, Moss J, Vaughan M: Molecular cloning, characterization, and expression of human ADP-ribosylation factors: two guanine nucleotide-dependent activators of cholera toxin. Proc Natl Acad Sci USA 86: 6101–6105, 1989
46. Price SR, Nightingale MS, Tsai S-C, Williamson KC, Adamik R, Chen H-C, Moss J, Vaughan M: Guanine nucleotide-binding proteins that enhance choleragen ADP-ribosyltransferase activity: nucleotide and deduced amino acid sequence of an ADP-ribosylation factor cDNA. Proc Natl Acad Sci USA 85: 5488–5491, 1988
47. Monaco L, Murtagh JJ, Newman KB, Tsai S-C, Moss J, Vaughan M: Selective amplification of an mRNA and related pseudogene for a human ADP-ribosylation factor, a guanine nucleotide-dependent protein activator of cholera toxin. Proc Natl Acad Sci USA 87: 2206–2210, 1990
48. Tsuchiya M, Price SR, Tsai S-C, Moss J, Vaughan M: Molecular identification of ADP-ribosylation factor mRNAs and their expression in mammalian cells. J Biol Chem 266: 2772–2777, 1991
49. Peng Z, Calvert I, Clark J, Helman L, Kahn R, Kung H-F: Molecular cloning, sequence analysis and mRNA expression of human ADP-ribosylation factor. Biofactors 2: 45–49, 1989
50. Stearns T, Kahn RA, Botstein D, Hoyt MA: ADP-ribosylation factor is an essential protein in *Saccharomyces cerevisiae* and is encoded by two genes. Mol Cell Biol 10: 6690–6699, 1990
51. Murtagh JJ Jr, Lee F-JS, Deak P, Hall LM, Monaco L, Lee C-M, Stevens LA, Moss J, Vaughan M: Molecular characterization of a conserved, guanine nucleotide-dependent ADP-ribosylation factor in *Drosophila melanogaster*. Biochemistry, 31: 6011–6018, 1993
52. Murtagh JJ Jr, Mowatt MR, Lee C-M, Lee F-JS, Mishima K, Nash TE, Moss J, Vaughan M: Guanine nucleotide-binding proteins in the intestinal parasite *Giardia lamblia*. Isolation of a gene encoding an ~20 kDa ADP-ribosylation factor. J Biol Chem 267: 9654–9662, 1992
53. Tsuchiya M, Price SR, Nightingale MS, Moss J, Vaughan M: Tissues and species distribution of mRNA encoding two ADP-ribosylation factors, 20-kDa guanine nucleotide binding proteins. Biochemistry 28: 9668–9673, 1989
54. Tsai S-C, Haun RS, Tsuchiya M, Moss J, Vaughan M: Isolation and characterization of the human gene for ADP-ribosylation factor 3, a 20-kDa guanine nucleotide-binding protein activator of cholera toxin. J Biol Chem 266: 23053–23059, 1991
55. Lee C-M, Haun RS, Tsai S-C, Moss J, Vaughan M: Characterization of the human gene encoding ADP-ribosylation factor 1, a guanine nucleotide-binding activator of cholera toxin. J Biol Chem 267: 9028–9034, 1992
56. Serventi IM, Cavanaugh E, Moss J, Vaughan M: Characterization of the gene for ADP-ribosylation factor (ARF) 2, developmentally regulated, selectively expressed member of the ARF family of ~20-kDa guanine nucleotide-binding proteins. J Biol Chem 268: 4863–4864, 1993
57. Haun RS, Moss J, Vaughan M: Characterization of the human ADP-ribosylation factor 3 promoter. Transcriptional regulation of a TATA-less promoter. J Biol Chem 268: 8793–8800, 1993
58. Price SR, Barber A, Moss J: Structure-function relationships of guanine nucleotide-binding proteins. In: J. Moss and M. Vaughan (eds). ADP-ribosylating Toxins and G Proteins: Insights Into Signal Transduction, American Society for Microbiology, Washington, DC, 1990, pp 397–424
59. Gibbs JB, Sigal IS, Poe M, Scolnick EM: Intrinsic GTPase activity distinguishes normal and oncogenic *ras* p21 molecules. Proc Natl Acad Sci USA 81: 5704–5708, 1984
60. McGrath JP, Capon DJ, Goeddel DV, Levinson AD: Comparative biochemical properties of normal and activated human *ras* p21 protein. Nature 310: 644–649, 1984
61. Sweet RW, Yokoyama S, Kamata T, Feramisco JR, Rosenberg M, Gross M: The product of *ras* is a GTPase and the T24 oncogenic mutant is deficient in this activity. Nature 311: 273–275, 1984
62. Tamkun JW, Kahn RA, Kissinger M, Brizuela BJ, Rulka C, Scott MP, Kennison JA: The arflike gene encodes an essential GTP-binding protein in *Drosophila*. Proc Natl Acad Sci USA 88: 3120–3124, 1991
63. Haun RS, Tsai S-C, Adamik R, Moss J, Vaughan M: Effect of myristoylation on GTP-dependent binding of ADP-ribosylation factor to Golgi. J Biol Chem 268: 7064–7068, 1993
64. Kahn RA, Goddard C, Newkirk M: Chemical and immunological characterization of the 21-kDa ADP-ribosylation factor of adenylate cyclase. J Biol Chem 263: 8282–8287, 1988
65. Kunz BC, Muczynski KA, Welsh CF, Stanley SJ, Tsai S-C, Adamik R, Chang PP, Moss J, Vaughan M: Characterization of recombinant and endogenous ADP-ribosylation factors synthesized in Sf9 insect cells. Biochemistry 32: 6643–6648, 1993
66. Towler DA, Adams SP, Eubanks SR, Towery DS, Jackson-Machelski E, Glaser L, Gordon JI: Purification and characterization of yeast myristoyl CoA:protein N-myristoyltransferase. Proc Natl Acad Sci USA 84: 2708–2712, 1987
67. Kahn RA, Randazzo P, Serafini T, Weiss O, Rulka C, Clark J, Amherdt M, Roller P, Orci L, Rothman JE: The amino terminus of ADP-ribosylation factor (ARF) is a critical determinant of ARF activities and is a potent and specific inhibitor of protein transport. J Biol Chem 267: 13039–13046, 1992
68. Mishima K, Tsuchiya M, Nightingale MS, Moss J, Vaughan M: ARD 1, a 64-kDa guanine nucleotide-binding protein with a carboxyl-terminal ADP-ribosylation factor domain. J Biol Chem 268: 8801–8807, 1993
69. Stearns T, Willingham MC, Botstein D, Kahn RA: ADP-ribosylation factor is functionally and physically associated with the Golgi complex. Proc Natl Acad Sci USA 87: 1238–1242, 1990
70. Regazzi R, Ullrich S, Kahn RA, Wollheim CB: Redistribution of ADP-ribosylation factor during stimulation of permeabilized cells with GTP analogues. Biochem J 275: 639–644, 1991
71. Walker MW, Bobak DA, Tsai S-C, Moss J, Vaughan M: GTP but not GDP analogues promote association of ADP-ribosylation factors, 20-kDa protein activators of cholera toxin, with phospholipids and PC-12 cell membranes. J Biol Chem 267: 3230–3235, 1992
72. Donaldson JG, Kahn RA, Lippincott-Schwartz J, Klausner RD: Binding of ARF and β-COP to Golgi membranes: Possible regulation by a trimeric G protein. Science 254: 1197–1199, 1991
73. Lippincott-Schwartz J, Yuan LC, Bonifacino JS, Klausner RD: Rapid redistribution of Golgi proteins into the ER in cells treated with Brefeldin A: evidence for membrane cycling from Golgi to ER. Cell 56: 801–813, 1989
74. Lippincott-Schwartz J, Donaldson JG, Schweizer A, Berger EG, Hauri H-P, Yuan LC, Klausner RD: Microtubule-dependent retrograde transport of proteins into the ER in the presence of Brefeldin A suggests an ER recycling pathway. Cell 60: 821–836, 1990
75. Serafini T, Orci L, Amherdt M, Brunner M, Kahn RA, Rothman JE: ADP-ribosylation factor is a subunit of the coat of Golgi-derived COP-coated vesicles: a novel role for a GTP-binding protein. Cell 67: 239–253, 1991
76. Donaldson JG, Cassel D, Kahn RA, Klausner RD: ADP-ribosylation factor, a small GTP-binding protein, is required for binding of the coatomer protein β-COP to Golgi membranes. Proc Natl Acad Sci USA 89: 6408–6412, 1992
77. Donaldson JG, Finazzi D, Klausner RD: Brefeldin A inhibits Golgi membrane-catalysed exchange of guanine nucleotide onto ARF pro-

tein. Nature 360: 350–352, 1992
78. Helms JB, Rothman JE: Inhibition by brefeldin A of a Golgi membrane enzyme that catalyses exchange of guanine nucleotide bound to ARF. Nature 360: 352–354, 1992
79. Tsai S-C, Adamik R, Haun RS, Moss J, Vaughan M: Differential interaction of ADP-ribosylation factors 1, 3, and 5 with rat brain Golgi membranes. Proc Natl Acad Sci USA 89: 9272–9276, 1992
80. Tsai S-C, Adamik R, Haun RS, Moss J, Vaughan M: Effects of brefeldin A and accessory proteins on association of ADP-ribosylation factors 1, 3, and 5 with Golgi. J Biol Chem 268: 10820–10825, 1993
81. Balch WE, Kahn RA, Schwaninger R: ADP-ribosylation factor is required for vesicular trafficking between the endoplasmic reticulum and the *cis*-Golgi compartments. J Biol Chem 267: 13053–13061, 1992
82. Lenhard JM, Kahn RA, Stahl PD: Evidence for ADP-ribosylation factor (ARF) as a regulator of *in vitro* endosome-endosome fusion. J Biol Chem 267: 13047–13052, 1992
83. Boman AL, Taylor TC, Melançon P, Wilson KL: A role for ADP-ribosylation factor in nuclear vesicle dynamics. Nature 358: 512–514, 1992

Molecular and Cellular Biochemistry **138**: 167–176, 1994.

Clostridial ADP-ribosylating toxins: effects on ATP and GTP-binding proteins

Klaus Aktories

Institut für Pharmakologie und Toxikologie der Universität des Saarlandes, D-66421 Homburg-Saar, Germany

Abstract

The actin cytoskeleton appears to be as the cellular target of various clostridial ADP-ribosyltransferases which have been described during recent years. *Clostridium botulinum* C2 toxin, *Clostridium perfringens* iota toxin and *Clostridium spiroforme* toxin ADP-ribosylate actin monomers and inhibit actin polymerization. *Clostridium botulinum* exoenzyme C3 and *Clostridium limosum* exoenzyme ADP-ribosylate the low-molecular-mass GTP-binding proteins of the Rho family, which participate in the regulation of the actin cytoskeleton. ADP-ribosylation inactivates the regulatory Rho proteins and disturbs the organization of the actin cytoskeleton. (Mol Cell Biochem **138**: 167–176, 1994)

Key words: ADP-ribosylation, actin, *Clostridium botulinum* C2 toxin, *Clostridium perfringens* toxin, *Clostridium botulinum* C3 exoenzyme, *Clostridium limosum* exoenzyme, gelsolin, low-molecular-mass GTP-binding proteins, Rho

Introduction

Mono-ADP-ribosylation of eukaryotic regulatory proteins is the underlying pathogenetic mechanism of a heterogenous family of bacterial protein toxins. Members of this toxin family include cholera toxin [1, 2], *E. coli* heat-labile toxins [3] and pertussis toxin [4], which ADP-ribosylate G-proteins, and diphtheria toxin [5] and *Pseudomonas* exotoxin A [6], which both mediate the ADP-ribosylation of elogantion factor 2. During the last years several clostridial ADP-ribosylating toxins have been described which can be divided into two groups with respect to their protein substrates. One group of toxins belongs to the family of actin-ADP-ribosylating toxins. The other group of transferases modify small GTP-binding proteins of the Rho family. More recent studies have shown that the latter group is not restricted to clostridia but is also produced by certain strains of *Bacillus cereus* and *Staphylococcus aureus*. In this mini-review, mainly the two groups of clostridial ADP-ribosyltransferases and their substrates are described. More comprehensive reviews focusing on Rho-ADP-ribosylating toxins [7, 8], on actin-ADP-ribosylating toxins [9–13] or dealing with both toxin groups [14–16] have been published.

Actin-ADP-ribosylating toxins

Several clostridial toxins were shown to ADP-ribosylate actin. These are *Clostridium botulinum* C2 toxin [17], *Clostridium perfringens* iota toxin [18, 19], *Clostridium spiroforme* toxin [20–22] and an ADP-ribosyltransferase produced by *Clostridium difficile*. The latter exoenzyme is clearly distinct from *Clostridium difficile* toxin A and B [23]. All these toxins are constructed according to the A-B model. The toxins consist of an enzyme component (A) which possesses ADP-ribosyltransferase activity and a binding component (B) that is responsible for the binding to the cell surface and the transport into the cell. However, in contrast to other bacterial transferases like cholera toxin [2] or diphtheria toxin [5], which have A-B structure, this family of clostridial ADP-ribosyltransferases is binary in structure [24, 25]. The toxin components are separate proteins and are linked by neither covalent nor non-covalent bonds. The interaction of the separate components occurs at the cell membrane. The binding component attaches to a specific cell surface acceptor of unknown nature, thereby inducing a binding site for the enzyme component. Then the biologically active enzyme component and the binding component

Address for offprints: K. Aktories, Institut für Pharmakologie und Toxikologie, D-66421 Homburg-Saar, Germany

are transferred into the cell, an event which most likely occurs by endocytosis [26, 27].

Depending on the structure of the binding components, the group of actin-modifying toxins can be divided into a subgroup of iota-like toxins (*Clostridium perfringens* iota toxin, *Clostridium spiroforme* toxin and *Clostridium difficile* ADP-ribosyltransferase) and *Clostridium botulinum* C2-toxin [18]. The binding components of the iota-like toxins are interchangeable, which is not true for C2 toxin [18].

Actin, substrate for ADP-ribosylation

Actin is the unique substrate for ADP-ribosylation by the toxins. It is one of the most abundant proteins in eukaryotic cells. Actin is a single chain protein of 374 or 375 amino acids with a molecular mass of 42 kDa. As deduced from crystallographic data, actin consists of four subdomains [28, 29]. Subdomains I–II and III–IV form a cleft in which one nucleotide (ATP or ADP) and one divalent cation Mg^{2+} or Ca^{2+} (*in vivo* most likely Mg^{2++}) are bound with high affinity. Actin possesses ATPase activity. During the polymerization process, the noncovalently bound ATP is hydrolyzed to ADP and phosphate [30]. But it appears that also globular monomeric actin is able to hydrolyze ATP (see below). Actin is involved in a large number of cellular functions (for review see [31, 32]). Besides its role in muscle contraction, it is the molecular basis for various cellular motile functions such as migration, intracellular transport, phagocytosis and secretion. Moreover, it is basically involved in the organization of the cellular architecture. All these functions depend on the ability of actin to polymerize and to form F-actin filaments. The filaments are polar structures having two non-equivalent ends. Actin polymerizes faster at the so-called 'barbed' end than at the 'pointed' end of filaments and the reverse is true for the depolymerization of F-actin. Thus, in non-muscle cells a dynamic equilibrium between F- and G-actin exists which is regulated by an increasing number of actin-binding proteins like profilin, β-thymosin, gelsolin and others [32–35].

In mammalian cells at least 6 actin isoforms exist: skeletal muscle α-actin, cardiac muscle α-actin, smooth muscle α- and γ-actin, and cytoplasmic β- and γ-actin, which differ in their isoelectric points and a limited number of amino acid exchanges. All vertebrate actin isoforms are about 95% homologous amongst each other [31, 36, 37].

Substrate for ADP-ribosylation by toxins is monomeric G-actin. Polymerized F-actin is not ADP-ribosylated by the toxins [17, 19]. Kinetic studies show that the ADP-bound form of actin is the preferred substrate for ADP-ribosylation by toxins [38]. By protein chemistry and site-directed mutagenesis it was shown that the acceptor amino acid for ADP-ribosylation is arginine-177 [39–41] and it appears that all toxins modify actin at the same site. However, differences were observed in the isoform substrate specificity of the toxins. Whereas *Clostridium perfringens* iota toxin is able to modify all actin isoforms studied so far (including mammalian α-actin isoforms), C2 toxin modifies only β-/γ-cytosplasmic actin and γ-smooth muscle actin in mammalian cells [17, 19, 42]. So far, the reason for this substrate specificity is not clear. However, it has been speculated that the N-terminal region of actin defines the substrate specificity [42], because γ-smooth muscle actin but not α-smooth muscle actin is modified by C2 toxin and both isoforms differ only in the N-terminus [36].

ADP-ribosylation of actin causes gross consequences in protein function (Fig. 1). First, ADP-ribosylated actin loses its ability to polymerize [17, 19]. This effect can be explained by the structure of F-actin. The ADP-ribosylation site of actin (arginine-177) is located near the axis of the helix which is formed by polymerized F-actin. The bulky ADP-ribose group attached to arginine-177 inhibits the actin polymerization most likely by steric hindrance [43]. On the

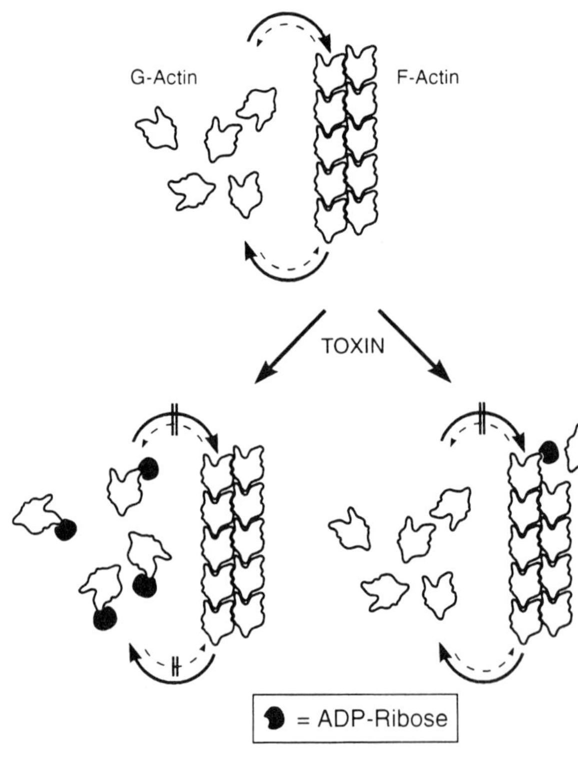

Fig. 1. Functional consequences of the ADP-ribosylation of actin. Monomeric G-actin and polymerized F-actin are in dynamic equilibrium which is regulated by various actin binding proteins. Polymerization of actin occurs at the so-called 'barbed' end of filaments and depolymerization at the 'pointed' end (the terms 'barbed' and 'pointed' are deduced from the arrow-head-like binding of myosin fragment S1 on F-actin). First, actin ADP-ribosylated by toxins is incapable of polymerization (left) and is trapped in the monomeric form. Second, ADP-ribosylated-G-actin binds to the 'barbed' ends of F-actin in a capping protein-like manner to inhibit the association of unmodified G-actin.

other hand, it is obvious that polymerized F-actin is a very poor substrate for the toxins, because the acceptor amino acid for ADP-ribose is buried in the F-actin filament molecule. Second, the ADP-ribosylated actin behaves like a capping protein, which binds with an affinity of 10^8 M^{-1} to the fast-growing 'barbed' end of actin filaments thereby inhibiting further actin polymerization [44]. In contrast, ADP-ribosylated actin does not interact with the slow-growing 'pointed' end of filaments. Again the atomic model of F-actin gives a clue as to the barbed-end-capping function of ADP-ribosylated actin, because the only place which accepts the ADP-ribose moiety is this end of the actin filament. Third, ADP-ribosylation of actin inhibits actin ATPase activity [45, 46]. This effect is most likely not secondary to inhibition of actin polymerization since it is also observed with actin in the quasi-monomeric state of the DNAse I-actin complex. This effect on actin ATPase may be explained by the fact that arginine-177, the acceptor amino acid of ADP-ribose, is located near to the γ-phosphate of ATP bound to actin [28]. Finally, ADP-ribosylation influences the functions of actin in complexes with actin-binding proteins. A recently studied example is the gelsolin-actin complex [47]. Gelsolin is a ubiquitous ~90 kDa actin-binding protein with F-actin-severing and capping activity. Moreover, it forms complexes with one and two actin monomers. These complexes possess polymerization nucleating activity [32, 48, 49]. Actins of the 1:1 and 2:1 actin-gelsolin complexes are non-equivalent. One actin is released with EGTA and the other with phosphatidylinositol phosphates. Recent studies showed that ADP-ribosylation of actin at the phosphatidylinositol phosphate-sensitive site had no effect on nucleation activity. In contrast, modification of actin at the EGTA-sensitive site inhibited nucleation activity of the gelsolin-actin complex (Fig. 2) [47]. These findings suggest that the pathophysiological effects of C2 toxin are not only based on modification of actin *per se* but also on modification of actin in complexes with actin-binding proteins. In fact, it was demonstrated recently, that the gelsolin-actin complex is a substrate in intact cells [50].

The effects of C2 toxin were studied in various tissues and cell types. The most striking effect observed in cultured cells is the breakdown of the microfilament network [51–53]. It was shown that C2 toxin increases the G-actin content and depolymerizes F-actin [54]. Finally, a complete depolymerization of F-actin occurs. Intermediate filaments are also redistributed by toxin [52]. However, this effect appears to be secondary to the destruction of the microfilament network. In contrast, the microtubule system is much less effected.

Thus, the F-actin cytoskeleton of non-muscle cells is destroyed by the toxins although F-actin is not a direct substrate for ADP-ribosylation. The toxins disturb the cellular equilibrium between F- and G-actin. ADP-ribosylated actin

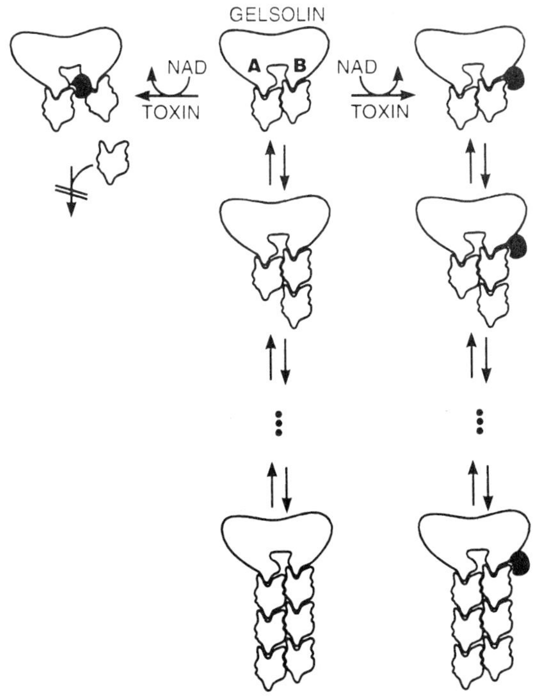

Fig. 2. Functional consequences of the ADP-ribosylation of actin-gelsolin complexes. Actin binds to gelsolin to form 1:1 and 1:2 (shown) complexes. The actin binding sites are non-equivalent. One site is EGTA-sensitive (A, low affinity site); the other site is sensitive to PIP_2 not EGTA (B, high affinity site). The gelsolin-actin complexes nucleate actin polymerization. ADP-ribosylation of actin at the high affinity site of gelsolin has no effect on nucleation. In contrast, ADP-ribosylation of actin at the EGTA sensitive site of gelsolin inhibits nucleation.

behaves like a capping protein to inhibit the association of G-actin at the barbed end of actin filaments. In contrast, the pointed end is not affected and, at this end, depolymerization occurs. The released G-actin is ADP-ribosylated by the toxin and, because, ADP-ribosylation blocks polymerization, the modified G-actin is trapped in the monomeric form. Furthermore, as discussed above, it is feasible that ADP-ribosylation of actin in complexes with actin-binding proteins participate in the cytopathic actions of the toxins.

Consequences of actin ADP-ribosylation on cell functions

Like cholera- and pertussis toxin for G-proteins, the actin-ADP-ribosylating toxins have been shown to be valuable tools to study the functions of actin. One advantage of the actin-ADP-ribosylating toxins is the high specificity of the reaction. No other substrates for ADP-ribosylation have been observed under *in vivo* or *in vitro* conditions. Neither G-proteins, which are ADP-ribosylated by cholera and pertussis toxins nor cytoskeletal proteins other than actin, like tubulin or intermediate filaments (vimentin) are modified by the toxin (unpublished observations). Moreover, the efficiency of ADP-ribosylation by toxins is very high. Using C2 toxin, the cellular F-actin can be completely depolymerized [55]. In contrast, depolymerization of F-actin by other available toxins like the cytochalasins, which have capping protein function and depolymerize actin in a large variety of cell systems, only cause partial F-actin depolmerization.

The influence of C2 toxin was studied on the barrier function of endothelial cells in monolayers [56]. These studies showed that the toxin increased the hydraulic conductivity of the cell monolayers in a concentration- and dose-dependent manner. The effects were paralleled by intracellular ADP-ribosylation of actin, depolymerization of F-actin and formation of interendothelial gaps. These findings indicate the involvement of actin in the vascular barrier function of the endothelium. Furthermore, the results are in agreement with earlier studies that the *in vivo* toxic effects of *Clostridium botulinum* C2 toxin are accompanied by increase in vascular permeability [57–59], massive decrease in blood pressure, and exudation in the lungs [60]. Thus, destruction of the barrier function of the endothelium by ADP-ribosylation of actin may be the pathophysiological mechanism underlying the toxicology of this group of toxins. Mauss *et al.* [61] showed that C2 toxin imparied the contraction of the isolated longitudinal muscle of the guinea-pig ileum. This effect was apparently caused by a direct action of the toxin on the smooth muscle cells. Because F-actin is not a substrate for C2 toxin, these findings give evidence for a role of G/F-actin transition in smooth muscle contraction.

In human polymorphonucleated leukocytes, it was shown that C2 toxin inhibited actin polymerization induced by chemoattractants like N-formyl-Met-Leu-Phe, leukotrien B_4 or complement split factor C5a [55, 62]. The toxin caused inhibition of leukocyte migration and increased superoxide anion production, and leukocyte enzyme secretion [55, 62–65]. Furthermore, in a series of experiments it was demonstrated that the destruction of actin by C2 toxin largely increased the stimulated production of lipid mediators [64, 65]. These data were interpreted as indicating that the intact actin cytoskeleton or the ability of actin to polymerize is involved in temporal and spatial limitation of leukocyte activation. In PC12 cells, noradrenaline release was affected by C2 toxin in a biphasic manner. A short term (<1h) treatment of cells with the toxin increased neurotransmitter release, whereas a decrease in noradrenaline release was detected after cell incubation with the toxin for more than 1h [66]. Moreover, in rat mast cells, C2 toxin inhibited histamine release stimulated by histamine-releasing factors like mastoparan or MCD peptide [67]. Thus, in this cell system, actin appears to be necessary for the maximal secretory response. Similarly, it was shown that in Y-1 cells steroid release was largely increased by C2 toxin and ADP-ribosylation of actin [68]. In this system the steroids are not released from vesicles and it was suggested that the destruction of the cytoskeleton may increase the substrate availability for production of steroids. Taken together, all these studies show that ADP-ribosylation causes drastic changes in actin function. On the other hand, the data give evidence that multiple actin compartments or pools exist, which are differentially regulated and involved in various motile cell functions like migration, secretion, phagocytosis, endocytosis or intracellular transport.

Eukaryotic ADP-ribosylation of actin

More recent investigations have focused on possible endogenous ADP-ribosylation of actin [69]. To meet this question, turkey erythrocyte ADP-ribosyltransferase type A, which specifically modifies arginine residues of various proteins, was studied with actin as substrate. Skeletal muscle actin and cytoplasmic actin were also ADP-ribosylated by the turkey erythrocyte transferase. Surprisingly, the eukaryotic enzyme modified two arginine groups of actin (Arg-95 and Arg-372). However, arginine-177 was not substrate for ADP-ribosylation by the erythrocyte transferase [70]. In contrast to the clostridial ADP-ribosyltransferases, modification of actin by the turkey enzyme did not inhibit polymerization and occurred with F-actin as substrate. More recently, the endogenous ADP-ribosylation of actin was studied in human platelets and neutrophils [71, 72]. However, the endogenous modification of actin turned out to be a non-enzymatic ADP-ribosylation of actin with ADP-ribose. In contrast to the toxin-induced ADP-ribosylation, the non-enzymatic ADP-ribosylation with ADP-ribose occurred at a cysteine residue [73]. Thus, these data indicate that the eukaryotic modification of actin by ADP-ribosylation is completely different from the toxin-induced enzyme reaction.

ADP-ribosylation of Rho proteins by C3-like clostridial toxins

Various clostridial exoenzymes have been described which ADP-ribosylate the small GTP-binding proteins of the Rho family. Among these transferases are *Clostridium botulinum* exoenzyme C3 [74, 75], and *Clostridium limosum* transferase [76]. Furthermore, two ~60–70% identical C3 isoenzymes have been described which are produced by *Clostridium botulinum* type C and D [77, 78]. Moreover, it was shown recently, that the ADP-ribosylation of Rho is not restricted to clostridial transferases. The epidermal differentiation inhibitor (EDIN), which is produced by *Staphylococcus aureus* [79, 80] and an exoenzyme from *Bacillus cereus* [81], are other transferases with properties similar to C3. All these transferases have molecular masses of about 25–30 kDa, are very basic proteins (IP>9) and appear to modify Rho proteins at the identical site. The transferase from *Staphylococcus aureus* shows only about 25% identity with C3. Based on partial amino acid sequencing, the *Bacillus cereus* transferase is even more distantly related to C3. However, the recent determination of the NAD-binding site of *Clostridium limosum* exoenzyme as glutamic acid-174 shows that this amino acid and the surrounding region is conserved in all C3-like exoenzymes including EDIN and *Bacillus cereus* transferase [82]. Furthermore, it appears that Glu-174 is equivalent to Glu-553 of *Pseudomonas aeruginosa* exotoxin A [83], Glu-148 of diphtheria toxin [84, 85] and Glu-129 of pertussis toxin [86] indicating that the C3-like exoenzymes possess a NAD-binding site which is common to all known bacterial ADP-ribosyltransferases.

Rho proteins as substrates for ADP-ribosylation by C3-like exoenzymes

Rho proteins belong to the family of small GTP-binding proteins which are about 30 to 40% homologous with Ras [87] and 50–90% identical between each other [88]. This family consists of the three Rho proteins, RhoA, B and C. Other members of the protein family are rac (rac 1,2), the Rac-related RhoG CDC42Hs and TC10 [88–91]. Closely related GTP-binding proteins have been described in yeast (RHO 1–4) [92, 93], Aplysia californica [94] and several other species, indicating the high conservation of the protein. Like other GTP-binding proteins, Rho proteins are regulated by a GTPase cycle [95]. They are active in the GTP-bound form and inactivated after hydrolysis of GTP to GDP by an intrinsic GTPase activity. Recently, several regulatory proteins have been described which appear to be involved in the control of the active state of Rho or Rho-related proteins. One group of proteins, which stimulates the intrinsic Rho GTPase (GTPase activating proteins (GAPs)), are comprised of RhoGAP [96, 97], and p190 [98, 99]. However, the specificity of these proteins is not clear. It appears that the GTPase of other GTP-binding proteins of the Rho family are also activated by these 'GAPs'. The same is true for the guanine nucleotide dissociation inhibitor and stimulator called GDI and GDS, respectively [100–103]. These proteins interact with Rho in its prenylated form and appear to regulate the GDP-GTP exchange and, furthermore, the cellular localization of the Rho proteins. For example, RhoGDI appears to inhibit the nucleotide exchange and induce the dissociation of Rho, Rac [104] and CDC42Hs [105] from the cell membrane. The substrate specificity of RhoGDS is even broader. It interacts with K-Ras, Rac, CDC42Hs and other low molecular mass GTP-binding proteins [104].

All mammalian Rho proteins (RhoA, B, C) are substrates for ADP-ribosylation by *Clostridium botulinum* C3 transferase [106–111]. Modification of Rho occurs in asparagine-41 [112]. There is evidence that the other C3-like exoenzymes also modify Rho at the identical position [76, 80]. Asparagine is unique as an ADP-ribose acceptor for mono-ADP-ribosylation by the C3-like transferases. The ADP-ribose-asparagine bond is particularly stable towards hydroxylamine treatment and mercury ions, which cleave the ADP-ribose-bonds formed with arginine or cysteine by cholera toxin (or actin-ADP-ribosylating toxins) and pertussis toxin, respectively [113].

Deduced from the structure of the related Ras protein, the acceptor amino acid for ADP-ribose (asparagine-41) is in or near the so-called effector region of the GTP-binding protein, roughly located between amino acid 30 (32) to 40 (42). Therefore, it is suggested that the inactivation of Rho function by ADP-ribosylation may be explained by inhibition of the interaction of ADP-ribosylated Rho with a putative effector protein (Fig. 3).

ADP-ribosylation seemed to be without effects on basal and GAP-stimulated single cycle GTPase of Rho. However, it has been reported that the ADP-ribosylation increased the steady state GTPase activity of the recombinant RhoA and RhoB proteins by about 1.8-fold [114], an effect which is rather small compared with several-fold GAP-induced stimulation of GTP-hydrolysis. ADP-ribosylation of Rho affects the interaction of the GTP-binding protein with GDI and, similarly, the GDI-Rho complex is a poor substrate for C3-like transferases.

All GTP-binding proteins of the Rho family possess asparagine at the equivalent position. However, it appears that Rho proteins are the only physiological substrate for ADP-ribosylation by C3-like exoenzymes [76]. This also holds true for the Rac proteins which were first cloned from the human promyelocytic leukemia cell line HL60 [89] and are believed to be involved in the control of the superoxide-generating NADPH oxidase complex in neutrophils [90, 115] and in membrane ruffling in fibroblasts [116]. 'Rac' means ras-related C3 botulinum toxin substrate [89], however, it was demonstrated recently, that Rac proteins are very poor substrates for ADP-ribosylation by C3 and other C3-like transferases [76].

Physiological role of Rho and functional consequences of its ADP-ribosylation

Several findings indicate that the Rho proteins are involved in the regulation of the actin cytoskeleton. It has been shown that the constitutively active Val14-Rho mutant protein induced formation of stress fibers in Swiss 3T3 cells [117–119]. Additional evidence for a regulatory role in actin

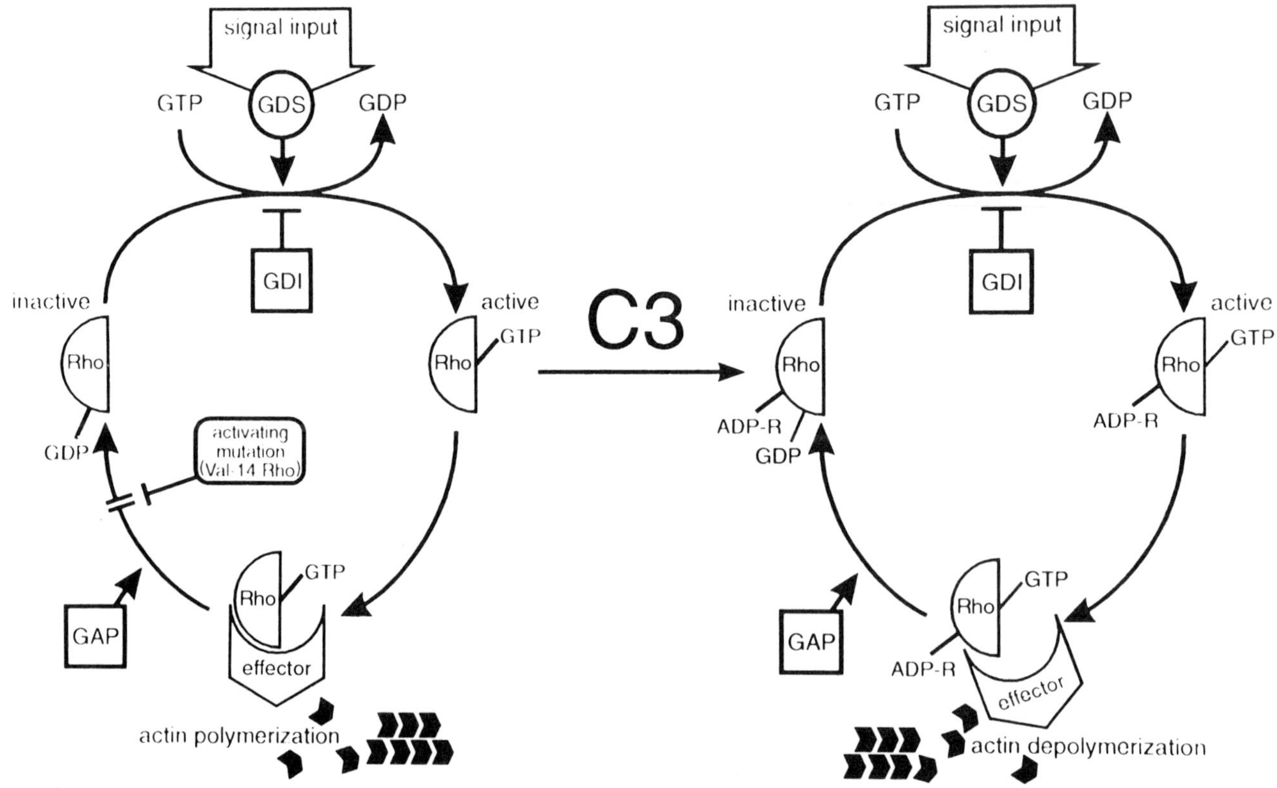

Fig. 3. Influence of C3-like ADP-ribosyltransferases on Rho function. The GTP-binding protein Rho is regulated by a GTPase cycle. It is inactive in the GDP-bound form and active with GTP bound. The GDP/GTP exchange is enhanced by guanine nucleotide dissociation stimulators (GDS) and blocked by the guanine nucleotide dissociation inhibitor (GDI). In the active form, Rho interacts with a putative effector, thereby inducing a cascade of events which finally results in polymerization of actin. The active state of Rho is terminated by GTPase activating proteins (GAP). Rho-mutants (Val-14-Rho), with blocked intrinsic GTPase, are constitutively active. ADP-ribosylation of Rho on asparagine-41 occurs in the so-called effector region of Rho, thereby blocking interaction with the putative effector. The functional consequence is the depolymerization of actin.

polymerization is based on the ADP-ribosylation of Rho by C3-like transferases. Treatment of cells with C3 causes depolymerization of the microfilament network in various cell types like PC12 cells [120], Vero cells [110] and FAO-cells [52]. These studies are hampered by the fact that C3-like enzymes hardly enter the intact cell. Therefore high toxin concentrations (> 10 μg/ml) are usually necessary to observe changes in the actin cytoskeleton, the morphological equivalent of which is rounding up of cells. Therefore, microinjection studies were performed showing that C3 causes depolymerization of actin within 15 min after toxin application. This indicates that changes of the cytoskeleton are most likely not secondary to the action of Rho [117, 118]. Studies with isolated proteins showed that Rho protein did not directly interact with actin to depolymerize F-actin or to inhibit the polymerization of G-actin (unpublished observations). Thus, it is believed that Rho participates in the concerted action of regulatory proteins which eventually induce actin polymerization. Excellent studies performed in the laboratory of A. Hall let suggest that Rho proteins are involved in stress fiber formation and focal adhesion assembly [118]. Addition of fetal calf serum or lyso-phosphatidic acid (LPA) to serum-starved swiss 3T3 cells caused immediate formation of actin fibers and induced assembly of focal adhesions. Microinjection of C3 inhibited both processes. C3 was shown to affect lymphocyte function. For example, ADP-ribosylation of RhoA by C3 inhibited lymphocyte-mediated cytotoxicity [121] and blocked the LFA-1-dependent aggregation of JY cells [122]. So far it is not clear whether both effects are related phenomena and depend on Rho effects on actin cytoskeleton. Recently, it was demonstrated that C3 reduced the growth rate of Swiss 3T3 cells. Flow cytometric analysis of the cells indicated that C3 caused the accumulation of cells in the G_1 phase of the cell cycle suggesting that RhoA plays a significant role in G_1-S transition in Swiss 3T3 cells [123]. Rather different aspects of Rho function were deduced from studies with EDIN, the C3-like exoenzyme from *Staphylococcus aureus*. EDIN inhibited the terminal differentiation of cultured keratinocytes and caused even stimulation of cell growth

[80]. Furthermore, EDIN induced changes in the Golgi structure of monkey kidney Vero cells that were similar to the changes induced by brefeldin A, a finding which was interpreted as indicating the involvement of Rho proteins in membrane trafficking between Golgi apparatus and the endoplasmic reticulum [124]. A different regulatory role for Rho protein has been suggested in smooth muscle function. In skinned smooth muscle from rabbit mesenteric arteries, Rho protein appeared to be involved in the GTP-enhanced Ca^{2+} sensitivity of smooth muscle contraction, an effect which was blocked by ADP-ribosylation of the GTP-binding protein and was overcome by addition of GTPγS-bound Rho [125]. Thus it appears that the ADP-ribosylations induced by C3-like exoenzymes have different effects in the various tissues and cells studied. This diversity may be based on the fact that the transferases modify all three Rho proteins, (Rho A, B, C) which are most likely involved in different cellular functions.

Concluding remarks

Recent studies greatly extented our knowledge of the pathobiochemical mechanisms underlying the cytopathic actions of clostridial ADP-ribosylating toxins. Furthermore, the toxins turned out to be very useful tools to study the physiological functions of their eukaryotic target proteins. This is true for actin-ADP-ribosylating toxins as well as the Rho-modifying exoenzymes. Several details about the toxin actions, their cellular entry mechanism and their role as virulence factors have still to be clarified. Further studies on the functions of eukaryotic toxin substrates are necessary. This is especially true for the elucidation of the precise regulatory function of the Rho proteins. Morever, one of the most fascinating questions remains whether the targets of toxins are also substrates for ADP-ribosylation by endogenous ADP-ribosyltransferases.

Acknowledgements

The critical reading of the manuscript by Dr. P. Wollenberg is thankfully acknowledged. Studies from the author's laboratory described herein were supported by the Deutsche Forschungsgemeinschaft (SFB 246, 249) and the Fonds der Chemischen Industrie).

References

1. Fishman PH: Mechanism of action of cholera toxin. In: J. Moss and M. Vaughan (eds). ADP-ribosylating toxins and G proteins. American Society for Microbiology, Washington DC, 1990, pp 127–140
2. Gill DM: Mechanism of action of cholera toxin. Adv Cyclic Nucleotide Res 8: 85–118, 1977
3. Spangler BD: Structure and function of cholera toxin and the related *Escherichia coli* heat-labile enterotoxin. Microbiol Rev 56: 622–647, 1992
4. Ui M, Pertussis toxin as a valuable probe for G-protein involvement in signal transduction. In: J. Moss and M. Vaughan (eds). ADP-ribosylating toxins and G proteins. American Society for Microbiology, Washington DC, 1990, pp 45–77
5. Collier RJ: Diphtheria Toxin: Structure and function of a cytocidal protein. In: J. Moss and M. Vaughan (eds). ADP-ribosylating toxins and G proteins. American Society for Microbiology, Washington DC, 1990, pp 3–19
6. Wick MJ, Iglewski BH: *Pseudomonas aeruginosa* Exotoxin A. In: J. Moss and M. Vaughan (eds). ADP-ribosylating toxins and G proteins. American Society for Microbiology, Washington DC, 1990, pp 31–43
7. Aktories K, Hall A: Botulinum ADP-ribosyltransferase C3: a new tool to study low molecular weight GTP-binding proteins. Trends Pharmacol Sci 10: 415–418, 1989
8. Aktories K, Mohr C, Koch G: *Clostridium botulinum* C3 ADP-ribosyltransferase. Curr Top Microbiol Immunol 175: 115–131, 1992
9. Aktories K, Just I: Botulinum C2 Toxin. In: J. Moss and M. Vaughan (eds). ADP-ribosylating toxins and G proteins. American Society for Microbiology, Washington DC, 1990, pp 79–95
10. Aktories K, Wille M, Just I: Clostridial actin-ADP-ribosylating toxins. Curr Top Microbiol Immunol 175: 97–113, 1992
11. Aktories K, Wegner A: ADP-ribosylation of actin by clostridial toxins. J Cell Biol 109: 1385–1387, 1989
12. Aktories K, Wegner A: Mechanisms of the cytopathic action of actin-ADP-ribosylating toxins. Mol Microbiol 6: 2905–2908, 1992
13. Considine RV, Simpson LL: Cellular and molecular actions of binary toxins possessing ADP-ribosyltransferase activity. Toxicon 29: 913–936, 1991
14. Aktories K: *Clostridium botulinum* C2 toxin and *C. botulinum* C3 ADP-ribosyltransferase. In: H. Herken and F. Hucho (eds). Handbook of Experimental Pharmacology, Vol 102, Selective Neurotoxicity, Springer-Verlag, Berlin Heidelberg, 1992, pp 841–854
15. Aktories K, Koch G, Just I: Molecular biology of clostridial ADP-ribosyltransferases and their substrates. In: M. Sebald (ed). Genetics and Molecular Biology of Anaerobic Bacteria, Brock/Springer, New York Heidelberg, 1993, pp 195–210
16. Aktories K: Clostridial ADP-ribosyltransferases – modification of low molecular weight GTP-binding proteins and of actin by clostridial toxins. Med Microbiol Immunol 179: 123–136, 1990
17. Aktories K, Bärmann M, Ohishi I, Tsuyama S, Jakobs KH, Habermann E: Botulinum C2 toxin ADP-ribosylates actin. Nature 322: 390–392, 1986
18. Simpson LL, Stiles BG, Zapeda HH, Wilkins TD: Molecular basis for the pathological actions of *Clostridium perfringens* Iota toxin. Infect Immun 55: 118–122, 1987
19. Schering B, Bärmann M, Chhatwal GS, Geipel U, Aktories K: ATP-ribosylation of skeletal muscle and non-muscle actin by *Clostridium perfringens* iota toxin. Eur J Biochem 171: 225–229, 1988
20. Popoff MR, Milward FW, Bancillon B, Boquet P: Purification of the *Clostridium spiroforme* binary toxin and activity of the toxin on HEp2 cells. Infect Immun 57: 2462–2469, 1989
21. Popoff MR, Boquet P: *Clostridium spiroforme* toxin is a binary toxin which ADP-ribosylates cellulr actin. Biochem Biophys Res Commun 152: 1361, 1988
22. Simpson LL, Stiles BG, Zepeda H, Wilkins TD: Production by *Clostridium spiroforme* of an iotalike toxin that possesses mono(ADP-ribosyl)transferae activity: Identification of a novel class of ADP-ribosyltransferases. Infect Immun 57: 255–261, 1989
23. Popoff MR, Rubin EJ, Gill DM, Boquet P: Actin-specific ADP-ribosyltransferase produced by a *Clostridium difficile* strain. Infect Immun 56: 2299–2306, 1988

24. Ohishi I, Iwasaki M, Sakaguchi G: Purification and characterization of two components of botulinum C2 toxin. Infect Immun 30: 668–673, 1980
25. Ohishi I: Response of mouse intestinal loop to botulinum C2 toxin: enterotoxic activity induced by cooperation of nonlinked protein components. Infect Immun 40: 691–695, 1983
26. Simpson LL: The binary toxin produced by *Clostridium botulinum* enters cells by receptor-mediated endocytosis to exert its pharmacologic effects. J Pharmacol Exp Ther 251: 1223–1228, 1989
27. Ohishi I, Yanagimoto A: Visualizations of binding and internalization of two nonlinked protein components of botulinum C2 toxin in tissue culture cells. Infect Immun 60: 4648–4655, 1992
28. Kabsch W, Mannherz HG, Suck D, Pai EF, Holmes KC: Atomic structure of the actin:DNase I complex. Nature 347: 37–44, 1990
29. Kabsch W, Vandekerckhove J: Structure and function of actin. Annu Rev Biophys Chem 21: 49–76, 1992
30. Carlier M-F: Actin: Protein Structure and filament dynamics. J Biol Chem 266: No. 1: 1–4, 1991
31. Bershadsky AD, Vasiliev JM: Cytoskeleton, New York, Plenum Press, 1988
32. Pollard TD, Cooper JA: Actin and actin-binding proteins. A critical evaluation of mechanisms and functions. Ann Rev Biochem 55: 987–1035, 1986
33. Selve N, Wegner A: Rate of treadmilling of actin filaments *in vitro*. J Mol Biol 187: 627–631, 1986
34. Hartwig JH, Kwiatkowski DJ: Actin-binding proteins. Curr Opin Cell Biol 3: 87–97, 1991
35. Vandekerckhove J: Actin-binding proteins. Curr Opin Cell Biol 2: 41–50, 1990
36. Vandekerckhove J, Weber K: The complete amino acid sequence of actins from bovine aorta, bovine heart, bovine fast skeletal muscle and rabbit slow skeletal muscle. Differentiation 14: 123–133, 1979
37. Vandekerckhove J, Weber K: At Least Six Different Actins are Expressed in a Higher Mammal: An Analysis Based on the Amino Acid Sequence of the Amino-terminal Tryptic Peptide. J Mol Biol 126: 783–802, 1978
38. Aktories K, Geipel U, Wille M, Just I: Characterization of the ADP-ribosylation of actin by *Clostridium botulinum* C2 toxin and *Clostridium perfringens* iota toxin. J Physiol, Paris 84: 262–266, 1990
39. Vandekerckhove J, Schering B, Bärmann M, Aktories K: *Clostridium perfringens* iota toxin ADP-ribosylates skeletal muscle actin in Arg-177. FEBS Lett 225: 48–52, 1987
40. Vandekerckhove J, Schering B, Bärmann M, Aktories K: Botulinum C2 toxin ADP-ribosylates cytoplasmic β/v-actin in aginine 177. J Biol Chem 263: 696–700, 1988
41. Just I, Hennessey ES, Drummond DR, Aktories K, Sparrow JC: ADP-ribosylation of Drosophila indirect flight muscle actin and arthrin by *Clostridium botulinum* C2 toxin and *Clostridium perfringens* iota toxin. Biochem J 291: 409–412, 1993
42. Mauss S, Chaponnier C, Just I, Aktories K, Gabbiani G: ADP-ribosylation of actin isoforms by *Clostridium botulinum* C2 toxin and *Clostridium perfringens* iota toxin. Eur J Biochem 194: 237–241, 1990
43. Holmes KC, Popp D, Gebhard W, Kabsch W: Atomic model of the actin filament. Nature 347: 44–49, 1990
44. Wegner A, Aktories K: ADP-ribosylated actin caps the barbed ends of actin filaments. J Biol Chem 263: 13739–13742, 1988
45. Geipel U, Just I, Aktories K: Inhibition of cytochalasin D-stimulated G-actin ATPase by ADP-ribosylation with *Clostridium perfringens* iota toxin. Biochem J 266: 335–339, 1990
46. Geipel U, Just I, Schering B, Haas D, Aktories K: ADP-ribosylation of actin causes increase in the rate of ATP exchange and inhibition of ATP hydrolysis. Eur J Biochem 179: 229–232, 1989
47. Wille M, Just I, Wegner A, Aktories K: ADP-ribosylation of the gelsolin-actin complex by clostridial toxins. J Biol Chem 267: 50–55, 1992
48. Harris HE, Weeds AG: Plasma gelsolin caps and severs actin filaments. FEBS Lett 177: 184–188, 1984
49. Janmey PA, Chaponnier C, Lind SE, Zaner KS, Stossel TP, Yin HL: Interactions of gelsolin and gelsolin-actin complexes with actin. Effects of calcium on actin nucleation, filament severing, and end blocking. Biochemistry 24: 3714–3723, 1985
50. Just I, Wille M, Chaponnier C, Aktories K: Gelsolin-actin complex is target for ADP-ribosylation by *Clostridium botulinum* C2 toxin in intact human neutrophils. Eur J Pharmacol 246: 293–297, 1993
51. Reuner KH, Presek P, Boschek CB, Aktories K: Botulinum C2 toxin ADP-ribosylates actin and disorganizes the microfilament network in intact cells. Eur J Cell Biol 43: 134–140, 1987
52. Wiegers W, Just I, Müller H, Hellwig A, Traub P, Aktories K: Alteration of the cytoskeleton of mammalian cells cultured *in vitro* by *Clostridium botulinum* C2 toxin and C3 ADP-ribosyltransferase. Eur J Cell Biol 54: 237–245, 1991
53. Ohishi I, Miyake M, Ogura H, Nakamura S: Cytopathic effect of botulinum C2 toxin on tissue-culture cells. FEMS Microbiol Lett 23: 281–284, 1984
54. Aktories K, Reuner K-H, Presek P, Bärmann M: Botulinum C2 toxin treatment increases the G-actin pool in intact chicken cells: a model for the cytopathic action of actin-ADP-ribosylating toxins. Toxicon 27: 989–993, 1989
55. Norgauer J, Just I, Aktories K, Sklar LA: Influence of botulinum C2 toxin on F-actin and N-formyl peptide receptor dynamics in human neutrophils. J Cell Biol 109: 1133–1140, 1989
56. Suttorp N, Polley M, Seybold J, Schnittler H, Seeger W, Grimminger F, Aktories K: Adenosine diphosphate-ribosylation of G-actin by botulinum C2 toxin increases endothelial permeability *in vitro*. J Clin Invest 87: 1575–1584, 1991
57. Ohishi I, Iwasaki M, Sakaguchi G: Vascular permeability activity of Botulinum C2 toxin elicited by cooperation of two dissimilar protein components. Infect Immun 31: 890–895, 1980
58. Ohishi I: Lethal and vascular permeability activities of botulinum C2 toxin induced by separate injections of the two toxin components. Infect Immun 40: 336–339, 1983
59. Ohishi I, DasGupta BR: Molecular structure and biological activities of *Clostridium botulinum* C2 toxin. In: *Avian Botulism*, edited by Eklund, M.W. and Dowell, V.R. Springfield: Thomas, 1987, pp 223–247
60. Simpson LL: A comparison of the pharmacological properties of *Clostridium botulinum* type C1 and C2 toxins. J Pharmacol Exp Ther 223: 695–701, 1982
61. Mauss S, Koch G, Kreye VAW, Aktories K: Inhibition of the contraction of the isolated longitudinal muscle of the guinea-pig ileum by botulinum C2 toxin: Evidence for a role of G/F-actin transition in smooth muscle contraction. Naunyn-Schmiedebergs Arch Pharmcol 340: 345–351, 1989
62. Norgauer J, Kownatzki E, Seifert R, Aktories K: Botulinum C2 toxin ADP-ribosylates actin and enhances O2 production and secretion but inhibits migration of activated human neutrophils. J Clin Invest 82: 1376–1382, 1988
63. Al-Mohanna FA, Hallett MB: Actin polymerization in neutrophils is triggered without a requirement for a rise in cytoplasmic Ca^{2+}. Biochem J 266: 669–674, 1990
64. Grimminger F, Sibelius U, Aktories K, Suttorp N, Seeger W: Inhibition of cytoskeletal rearrangement by botulinum C2 toxin amplifies ligand-evoked lipid mediator generation in human neutrophils. Mol Pharmacol 40: 563–571, 1991
65. Grimminger F, Sibelius U, Aktories K, Just I, Seeger W: Suppression of cytoskeletal rearrangement in activated human neutrophils by botulinum C toxin. Impact on cellular signal transduction. J Biol Chem 266: 19276–19282, 1991
66. Matter K, Dreyer F, Aktories K: Actin involvement in exocytosis from PC12 cells: studies on the influence of botulinum C2 toxin on stimulated noradrenaline release. J Neurochem 52: 370–376, 1989
67. Böttinger H, Reuner KH, Aktories K: Inhibition of histamine release from

rat mast cells by botulinum C2 toxin. Int Archs Allergy Appl Immun 84: 380–384, 1987
68. Considine RV, Simpson LL, Sherwin JR: Botulinum C toxin and steroid production in adrenal Y-1 cells: The role of microfilaments in the toxin-induced increase in steroid release. J Pharmacol Exp Ther 260: 859–864, 1992
69. Matsuyama S, Tsuyama S: Mono-ADP-ribosylation in brain: Purification and characterization of ADP-ribosyltransferases affecting actin from rat brain. J Neurochem 57: 1380–1387, 1991
70. Just I, Sehr P, Moss J: Actin is ADP-ribosylated by ADP-ribosyltransferase A from turkey erythrocytes. Naunyn-Schmiedebergs Arch Pharmcol Suppl 347: R47, 1993
71. Terashima M, Mishima K, Yamada K, Tsuchiya M, Wakutani T, Shimoyama M: ADP-ribosylation of actins by arginine-specific ADP-ribosyltransferase purified from chicken heterophils. Eur J Biochem 204: 305–311, 1992
72. Clancy RM, Leszczynska-Piziak J, Abramson SB: Nitric oxide stimulates the ADP-ribosylation of actin in human neutrophils. Biochem Biophys Res Commun 191: 847–852, 1993
73. Just I, Wollenberg P, Moss I, Aktories K: Cysteine-specific ADP-ribosylation of actin. Eur J Biochem 221: 1047–1054, 1994
74. Aktories K, Weller U, Chhatwal GS: *Clostridium botulinum* type C produces a novel ADP-ribosyltransferase distinct from botulinum C2 toxin. FEBS Lett 212: 109–113, 1987
75. Aktories K, Rösener S, Blaschke U, Chhatwal GS: Botulinum ADP-ribosyltransferase C3. Purification of the enzyme and characterization of the ADP-ribosylation reaction in platelet membranes. Eur J Biochem 172: 445–450, 1988
76. Just I, Mohr C, Schallehn G, Menard L, Didsbury JR, Vandekerckhove J, van Damme J, Aktories K: Purification and characterization of an ADP-ribosyltransferase produced by *Clostridium limosum*. J Biol Chem 267: 10274–10280, 1992
77. Nemoto Y, Namba T, Kozaki S, Narumiya S: *Clostridium botulinum* C3 ADP-ribosyltransferase gene. J Biol Chem 266: 19312–19319, 1991
78. Popoff MR, Hauser D, Boquet P, Eklund MW, Gill DM: Characterization of the C3 gene of *Clostridium Botulinum* types C and D and its expresion in *Escherichia coli*. Infect Immun 59: 3673–3679, 1991
79. Inoue S, Sugai M, Murooka Y, Paik S-Y, Hong Y-M, Ohgai H, Suginaka H: Molecular cloning and sequencing of the epidermal cell differentiation inhibitor gene from *Staphylococcus aureus*. Biochem Biophys Res Commun 174: 459–464, 1991
80. Sugai M, Hashimoto K, Kikuchi A, Inoue S, Okumura H, Matsumota K, Goto Y, Ohgai H, Moriishi K, Syuto B, Yoshikawa K, Suginaka H, Takai Y: Epidermal cell differentiation inhibitor ADP-ribosylates small GTP-binding proteins and induces hyperplasia of epidermis. J Biol Chem 267: 2600–2604, 1992
81. Just I, Schallehn G, Aktories K: ADP-ribosylation of small GTP-binding proteins by *Bacillus cereus*. Biochem Biophys Res Commun 183: 931–936, 1992
82. Jung M, Just I, van Damme J, Vandekerckhove J, Aktories K: NAD-binding site of the C3-like ADP-ribosyltransferase from *Clostridium limosum*. J Biol Chem 268: 23215–23218, (1993)
83. Carroll SF, Collier RJ: Active site of Pseudomonas aeruginosa exotoxin A. Glutamic acid 553 is photolabeled by NAD and shows functional homology with glutamic acid 148 of diphtheria toxin. J Biol Chem 262: 8707–8711, 1987
84. Carroll SF, McCloskey JA, Crain PF, Oppenheimer NJ, Marschner TM, Collier RJ: Photoaffinity labeling of diphtheria toxin fragment A with NAD: structure of the photoproduct at position 148. Proc Natl Acad Sci USA 82: 7237–7241, 1985
85. Carroll SF, Collier RJ: NAD binding site of diphtheria toxin: identification of a residue within the nicotinamide subsite by photochemical modification with NAD. Proc Natl Acad Sci USA 81: 3307–3311, 1984
86. Barbieri JT, Mende-Mueller M, Rappuoli R, Collier RJ: Photolabeling of glu-129 of the S-1 subunit of pertussis toxin with NAD. Infect Immun 57: 3549–3554, 1989
87. Drivas GT, Shih A, Coutavas E, Rush MG, D'Eustachio P: Characterization of four novel *ras*-like genes expressed in a human teratocarcinoma cell line. Mol Cell Biol 10: 1793–1798, 1990
88. Vincent S, Jeanteur P, Fort P: Growth-regulated expression of *rhoG*, a new member of the *ras* homolog gene family. Mol Cell Biol 12: 3138–3148, 1992
89. Didsbury J, Weber RF, Bokoch GM, Evans T, Snyderman R: rac, a novel ras-related family of proteins that are botulinum toxin substrates. J Biol Chem 264: 16378–16382, 1989
90. Knaus UG, Heyworth PG, Kinsella BT, Curnutte JT, Bokoch GM: Purification and characterization of Rac 2. A cytosolic GTP-binding protein that regulates human neutrophil NADPH oxidase. J Biol Chem 267: 23575–23582, 1992
91. Ménard L, Tomhave E, Casey PJ, Uhing RJ, Snyderman R, Didsbury JR: Rac1, a low-molecular-mass GTP-binding-protein with high intrinsic GTPase activity and distinct biochemical properties. Eur J Biochem 206: 537–546, 1992
92. Madaule P, Axel R, Myers AM: Characterization of two members of the rho gene family from the yeast Saccharomyces cerevisiae. Proc Natl Acad Sci USA 84: 779–783, 1987
93. Matsui Y, Toh-e A: Isolation and characterization of two novel *ras* superfamily genes in *Saccharomyces cerevisiae*. Gene 114: 43–49, 1992
94. Madaule P, Axel R: A novel ras-related gene family. Cell 41: 31–40, 1985
95. Bourne HR, Sanders DA, McCormick F: The GTPase superfamily: conserved structure and molecular mechanism. Nature 349: 117–127, 1991
96. Garrett MD, Self AJ, v.Oers C, Hall A: Identification of distinct cytoplasmic targets for ras, R-ras and rho regulatory proteins. J Biol Chem 264: 10–13, 1989
97. Morii N, Kawano K, Sekine A, Yamada T, Narumiya S: Purification of GTPase-activating protein specific for the *rho* gene. J Biol Chem 266: 7646–7650, 1991
98. Settleman J, Albright CF, Foster LC, Weinberg RA: Association between GTPase activators for Rho and Ras families. Nature 359: 153–154, 1992
99. Nobes C, Hall A: Regulation and function of the Rho subfamily of small GTPases. Curr Opin Genet Dev 4: 77–81, 1994
100. Isomura M, Kaibuchi K, Yamamoto T, Kawamura S, Katayama M, Takai Y: Partial purification and characterization of GDP dissociation stimulator (GDS) for the rho proteins from bovine brain cytosol. Biochem Biophys Res Commun 169: 652–659, 1990
101. Lelias J-M, Adra CN, Wulf GM, Guillemot J-C, Khagad M, Caput D, Lim B: cDNA cloning of a human mRNA preferentially expressed in hematopoietic cells and with homology to a GDP-dissociation inhibitor for the rho GTP-binding proteins. Proc Natl Acad Sci USA 90: 1479–1483, 1993
102. Fukumoto Y, Kaibuchi K, Hori Y, Fujioka H, Araki S, Ueda T, Kikuchi A, Takai Y: Molecular cloning and characterization of a novel type of regulatory protein (GDI) for the *rho* proteins, *ras* p21-like small GTP-binding proteins. Oncogene 5: 1321–1328, 1990
103. Ueda T, Kikuchi A, Ohga N, Yamamoto J, Takai Y: Purification and characterization from bovine brain cytosol of a novel regulatory protein inhibiting the dissociation of GDP from and the subsequent binding of GTP to *rho*B p20, a *ras* p21-like GTP-binding protein. J Biol Chem 265: 9373–9380, 1990
104. Hiraoka K, Kaibuchi K, Ando S, Mucha T, Takaishi K, Mizuno T, Asada M, Ménard L, Tomhave E, Didsbury J, Snyderman R, Takai Y: Both stimulatory and inhibitory GDP/GTP exchange proteins, *smg* GDS and *rho* GDI, are active on multiple small GTP-binding proteins. Biochem Biophys Res Commun 182: 921–930, 1992
105. Leonard D, Hart MJ, Platko JV, Eva A, Henzel W, Evans T, Cerione RA:

The identification and characterization of a GDP-dissociation inhibitor (GDI) for the CDC42Hs protein. J Biol Chem 267: 22860–22868, 1992
106. Kikuchi A, Yamamoto K, Fujita T, Takai Y: ADP-ribosylation of the bovine brain rho protein by botulinum toxin type C1. J Biol Chem 263: 16303–16308, 1988
107. Quilliam LA, Lacal J-C, Bokoch GM: Identification of rho as a substrate for botulinum toxin C3-catalyzed ADP-ribosylation. FEBS Lett 247: 221–226, 1989
108. Narumiya S, Sekine A, Fujiwara M: Substrate for botulinum ADP-ribosyltransferase, Gb, has an amino acid sequence homologous to a putative rho gene product. J Biol Chem 263: 17255–17257, 1988
109. Aktories K, Braun U, Rösener S, Just I, Hall A: The rho gene product expressed in *E. coli* is a substrate of botulinum ADP-ribosyltransferase C3. Biochem Biophys Res Commun 158: 209–213, 1989
110. Chardin P, Boquet P, Madaule P, Popoff MR, Rubin EJ, Gill DM: The mammalian G protein rho C is ADP-ribosylated by *Clostridium botulinum* exoenzyme C3 and affects actin microfilament in Vero cells. EMBO J 8: 1087–1092, 1989
111. Braun U, Habermann B, Just I, Aktories K, Vandekerckhove J: Purification of the 22 kDa protein substrate of botulinum ADP-ribosyltransferase C3 from porcine brain cytosol and its characterization as a GTP-binding protein highly homologous to the rho gene product. FEBS Lett 243: 70–76, 1989
112. Sekine A, Fujiwara M, Narumiya S: Asparagine residue in the rho gene product is the modification site for botulinum ADP-ribosyltransferase. J Biol Chem 264: 8602–86-5, 1989
113. Aktories K, Just I, Rosenthal W: Different types of ADP-ribose protein bonds formed by botulinum C2 toxin, botulinum ADP-ribosyltransferase C3 and pertussis toxin. Biochem Biophys Res Commun 156: 361–367, 1988
114. Mohr C, Koch G, Just I, Aktories K: ADP-ribosylation by *Clostridium botulinum* C3 exoenzyme increases steady state GTPase activities of recombinant rhoA and rhoB proteins. FEBS Lett 297: 95–99, 1992
115. Abo A, Pick E, Hall A, Totty N, Teahan CG, Segal AW: Activation of the NADPH oxidase involves the small GTP-binding protein p21rac. Nature 353: 668–670, 1991
116. Ridley AJ, Paterson HF, Johnston CL, Diekmann D, Hall A: The small GTP-binding protein rac regulates growth factor-induced membrane ruffling. Cell 70: 401–410, 1992
117. Paterson HF, Self AJ, Garrett MD, Just I, Aktories K, Hall A: Microinjection of recombinant p21 induces rapid changes in cell morphology. J Cell Biol 111: 1001–1007, 1990
118. Ridley AJ, Hall A: The small GTP-binding protein rho regulates the assembly of focal adhesions and actin stress fibers in response to growth factors. Cell 70: 389–399, 1992
119. Miura Y, Kikuchi A, Musha T, Kuroda S, Yaku H, Sasaki T, Takai Y: Regulation of morphology by *rho* p21 and its inhibitory GDP/GTP exchange protein (*rho* GDI) in Swiss 3T3 cells. J Biol Chem 268: 510–515, 1993
120. Nishiki T, Narumiya S, Morii N, Yamamoto M, Fujiwara M, Kamata Y, Sakaguchi G, Kozaki S: ADP-ribosylation of the rho/rac proteins induces growth inhibition, neurite. Biochem Biophys Res Commun 167: 265–272, 1990
121. Lang P, Guizani L, Vitté-Mony I, Stancou R, Dorseuil O, Gacon G, Bertoglio J: ADP-ribosylation of the *ras*-related, GTP-binding protein RhoA inhibits lymphocyte-mediated cytotoxicity. J Biol Chem 267: 11677–11680, 1992
122. Tominaga T, Sugie K, Hirata M, Morii N, Fukata J, Uchida A, Imura H, Narumiya S: Inhibition of PMA-induced, LFA-1-dependent lymphocyte aggregation by ADP-ribosylation of the small molecular weight GTP binding protein, *rho*. J Cell Biol 120: No. 6: 1529–1537, 1993
123. Yamamoto M, Marui N, Sakai T, Morii N, Kozaki S, Ikai K, Imamura S, Narumiya S: ADP-ribosylation of the rhoA gene product by botulinum C3 exoenzyme causes Swiss 3T3 cells to accumulate in the G phase of the cell cycle. Oncogene 8: 1449–1455, 1993
124. Sugai M, Chen C-H, Wu HC: Bacterial ADP-ribosyltransferase with a substrate specificity of the rho protein disassembles the Golgi apparatus in Vero cells and mimics the action of brefeldin A. Proc Natl Acad Sci USA 89: 8903–8907, 1992
125. Hirata K-i, Kikuchi A, Sasaki T, Kuroda S, Kaibuchi K, Matsuura Y, Seki H, Saida K, Takai Y: Involvement of *rho* p21 in the GTP-enhanced calcium ion sensitivity of smooth muscle contraction. J Biol Chem 267: 8719–8722, 1992

Molecular and Cellular Biochemistry **138**: 177–181, 1994.

Common structure of the catalytic sites of mammalian and bacterial toxin ADP-ribosyltransferases

Ian J. Okazaki and Joel Moss

Laboratory of Cellular Metabolism, National Heart, Lung and Blood Institute, National Institutes of Health, Bethesda, MD 20892, USA

Abstract

The amino acid sequences of several bacterial toxin ADP-ribosyltransferases, rabbit skeletal muscle transferases, and RT6.2, a rat T-cell NAD glycohydrolase, contain three separate regions of similarity, which can be aligned. Region I contains a critical histidine or arginine residue, region II, a group of closely spaced aromatic amino acids, and region III, an active-site glutamate which is at times seen as part of an acidic amino acid-rich sequence. In some of the bacterial ADP-ribosyltransferases, the nicotinamide moiety of NAD has been photo-crosslinked to this glutamate, consistent with its position in the active site. The similarities within these three regions, despite an absence of overall sequence similarity among the several transferases, are consistent with a common structure involved in NAD binding and ADP-ribose transfer. (Mol Cell Biochem **138**: 177–181, 1994)

Key words: ADP-ribosyltransferase, diphtheria toxin, cholera toxin, pertussis toxin, C3 exoenzyme

Introduction

ADP-ribosyltransferases catalyze the transfer of the ADP-ribose moiety of NAD to an acceptor protein. ADP-ribosylation of cellular proteins by bacterial toxins alters the activity of proteins in critical metabolic or regulatory pathways [1–5]. Diptheria toxin (DT) and *Pseudomonas* exotoxin A (ETA) ADP-ribosylate a modified histidine residue in elongation factor 2 resulting in inhibition of protein synthesis and cell death [2, 3]. Cholera toxin (CT) and the related heat-labile enterotoxin of *E. coli* (LT) ADP-ribosylate an arginine in Gsα, the stimulatory guanine nucleotide-binding (G) protein of the adenylyl cyclase system, resulting in its activation and leading to increased intracellular cAMP [1, 4]. ADP-ribosylation of a cysteine in several members of the family of G proteins (e.g., G_i, G_t, and G_0) by pertussis toxin (PT) alters adenylyl cyclase, phospholipid turnover and ion flux, by uncoupling the G protein from its receptor [5]. The ADP-ribosyltransferase of the nitrogen-fixing bacterium *Rhodospirillum rubrum* regulates the activitiy of the dinitrogenase reductase complex through ADP-ribosylation of an arginine [6]. Bacterial exoenzymes such as *Clostridium botulinum* C3 and *Clostridium limosum* exoenzyme ADP-ribosylate an asparagine in the Rho family of ~20 kDa GTP-binding proteins, leading to their inactivation [7–10].

Enzymes involved in ADP-ribose transfer reactions

Mammalian ADP-ribosyltransferases have been isolated from skeletal muscle [11] and the rabbit and human skeletal muscle ADP-ribosyltransferase cDNAs have been cloned [12, 13]. The latter is a glycosylphosphatidylinositol (GPI) anchored protein with an estimated MW of 36–40 kDa [13]. There is ~80% amino acid identity between the rabbit and human skeletal muscle transferases. Like CT, LT, and the *Rhodospirillum rubrum* enzyme, the muscle transferases specifically use the guanidino group of arginine as an ADP-

Address for offprints: I.J. Okazaki, National Institutes of Health, 9000 Rockville Pike, Building 10, Room 5N307, Bethesda, MD 20892, USA

ribose acceptor.

The mammalian ADP-ribosyltransferases are significantly similar in amino acid sequence to the rat RT6 family of proteins, RT6.1 and RT6.2 [14]. Since the 39-kDa muscle ADP-ribosyltransferases differ considerably in size from the ~25-kDa RT6.2 gene products, sequence alignment required the insertion of a significant number of gaps. The nonglycosylated RT6.2 and the nonglycosylated and variably glycosylated RT6.1, are significantly similar to each other in deduced amino acid sequences and are in turn similar to the mouse RT6 homologue [15].

The RT6.1 and RT6.2 proteins have apparent molecular weights of 24–26 kDa (RT6.2) and 25–35 kDa (RT6.1) by SDS-PAGE [16], with a calculated size of ~26 kDa from the deduced amino acid sequences [17]. Although RT6.1 and RT6.2 can be distinguished by allotype-specific antibodies, immunoprecipitation of both RT6.1 and RT6.2 with polyclonal rabbit antiserum demonstrated the presence of a common determinant on both antigens [16, 18]. Based on hydropathy profiles and physical properties, the proteins appear to be GPI-anchored [16, 19]. The sequence similarities between RT6.2 and the NAD:arginine ADP-ribosyltransferases were reflected in functional homologies as well. RT6.2 gene product, expressed in rat mammary adenocarcinoma (NMU) cells, exhibited an NAD glycohydrolase activity. In contrast to the much larger transferase, it did not synthesize ADP-ribosylagmatine. Further, in contrast to the lymphocyte antigen CD38, it neither synthesized nor degraded cyclic ADP-ribose. Consistent with it being GPI-linked, the expressed RT6.2 was found on the cell surface and released by phosphatidylinositol-specific phospholipase C [14].

Previous studies have shown that the bacterial toxin ADP-ribosyltransferases share three separate regions of similarity in their amino acid sequences [20]. These regions are also present in the muscle transferase and RT6.2 NADase. Alignment of these regions highlights conserved residues within the catalytic sites of the mammalian and bacterial toxin ADP-ribosyltransferases.

Region I

Region I is characterized by an essential nucleophilic amino acid; a histidine in DT and ETA, and an arginine in CT, LT and PT. As noted in Table 1, there is considerable sequence similarity among bacterial toxin and mammalian ADP-ribosyltransferases surrounding the critical residue. Modification of His21 of DT with diethylpyrocarbonate inhibited NAD binding, ADP-ribosyltransferase activity, and NAD glycohydrolase activity [21]. Replacement of histidine by tyrosine in position 426 of ETA resulted in a loss of ADP-ribosyltransferase activity, but did not affect NAD binding [22]. X-ray crystallographic data showed His426 of ETA to be positioned on the external surface of the active site cleft [22, 23]. Further study of the catalytic site of ETA revealed however, that His440 is analogous to His 21 of DT [24], suggesting that His440 of ETA is the residue that participates in both NAD binding and enzyme activity. Based on crystallographic data, Arg9 of PT and Arg7 of CT and LT are analogous to the critical histidine [25]. In PT, substituting an asparagine for His35 markedly decreased ADP-ribosyltransferase activity although not as completely as that resulting from the Arg9 to lysine substitution [26]. In the crystallographic structure of LT, His 44 is within an elongated crevice of the catalytic site [27]. However, sustituting Arg7 with lysine of CT and LT by site-directed mutagenesis resulted in complete loss of transferase activity consistent with the hypothesis that Arg9 of PT and Arg7 of CT and LT share a functional homology [25]. His83 of DRAT may be the corresponding region I histidine based on similarity of its surrounding residues to those in the other transferases (Table 1).

Table 1. Region I: critical nucleophilic amino acid-containing region

						*					
Trans	110	F	R	D	E	H	G	V	A	L	L
RT6.2	87	·	N	·	F	·	·	T	·	·	v
DRAT	79	E	q	q	d	·	R	P	g	Q	G
ETA	436	·	V	G	Y	·	·	T	F	·	E
DT	17	·	S	S	Y	·	·	T	K	P	G
PT	5	A	T	V	Y	R	Y	D	s	R	P
CT	3	D	k	L	Y	R	a	D	s	R	P
LT	3	D	·	L	Y	R	a	D	s	R	P

Alignment of ADP-ribosyltransferases centered around the critical histidine or arginine residue (asterisk). Trans, rabbit skeletal muscle ADP-ribosyltransferase; RT6.2, rat T-cell alloantigen RT6.2; DRAT, dinitrogenase reductase ADP-ribosyltransferase of *Rhodospirillum rubrum*; ETA, *Pseudomonas aeruginosa* exotoxin A; DT, diphtheria toxin; PT, pertussis toxin; CT, cholera toxin; LT, *Escherichia coli* heat-labile enterotoxin. · indicates amino acid identity with the rabbit skeletal muscle ADP-ribosyltransferase. Lower case letters indicate conservative replacements.

Region II

Region II is an aromatic amino acid-rich sequence. Characteristic regions are present in the muscle ADP-ribosyltransferase, RT6.2 NAD glycohydrolase, the bacterial toxin ADP-ribosyltransferases, and DRAT protein (Table 2). The crystallographic structure of the catalytic cleft of ETA showed that the amino acids near Glu553 (the active-site residue), which includes His440, Trp466, Phe469, Tyr470 and Tyr481, correspond to identical residues in the DT sequence [23, 24]. The iodination of tyrosine residues in ETA crystals by incubating ETA with I_3 revealed that the side chain of Tyr481 extends into the enzymatic cleft [28]. Iodination of Tyr481 also resulted in loss of ETA activity. The

Table 2. Region II: aromtic amino acid-rich segment

Trans	141	S R A H Y L Q H F S F K T L H F
rt6.2	116	N P g Q F – – · – – Y · a F · Y Y
DRAT	124	R F g L F P T F H R E P I T k F
ETA	464	a I W r G F – Y I a G D P A L A Y G Y
DT	49	D D W k G F – Y S t D N K Y D A A G Y
PT	82	g H F I G Y I Y E V R A D – – N N F Y
CT	79	g H s T Y Y I Y – V I A · A P N H F N
LT	79	g Y s T Y Y I Y – V I A · A P N H F N
C3	114	V F E Q V K A k F L K · D R T E Y G Y

Region of closely spaced aromatic amino acids (boxed) among ADP-ribosyltransferases. C3, Clostridium botulinum C3 exotoxin; abbreviation of the other enzymes as in Table 1. · indicates amino acid identity with the rabbit muscle transferase. Lower case letters indicate conservation of residues with those of the rabbit transferase. – indicates gaps to facilitate alignment.

model proposed for NAD binding by ETA, stacks the nicotinamide ring of NAD on the indole ring of Trp466, places the phenolic hydroxyl of Tyr470 adjacent to the nicotinamide-ribose bond, and has the phenolic ring of Tyr481 interacting with the purine ring of the adenosine moiety [28]. Likewise, PT, LT and CT each contain a stretch of closely spaced aromatic amino acids, which can be aligned with those in the other transferases. The crystal structure of the active site of LT was superpositioned with significant similarity on the enzymatic domain of ETA [26], despite virtually no similarity identity. The closely spaced aromatic residues may provide the active site with a conformation that facilitates NAD binding.

Region III

This region contains a critical glutamic acid initially detected in the bacterial toxins (Table 3). Photoaffinity labeling with

Table 3. Region III: acidic amino acid-rich segment containing the active-site glutamate

Trans	225	L G V P I Q G Y S F F P G E E E V L I P G
rt6.2	194	· · · Y · K E f · · R · D q · · · · · · ·
DRAT	246	F F N T l L t S H P L K · · G · Y · v I ·
ETA	539	· D A – · T · P E E E G · R L *E* T i L – ·
DT	136	· s l · F A E G · S S V – – – *E* Y i N N W
PT	119	· a G A l – a – t y – – – q S *E* Y · A H R
CT	101	· · A Y – – – – · P H · D · q · · S A L ·
LT	101	· · · Y – – – – · P H · Y · q · · S A L ·
C3	159	K · G Y · d P I · y · · · q L *E* · · l · R

The region containing the active-site glutamate of the bacterial toxins are aligned with a similar stretch of residues in the rabbit transferase, RT6.2 NADase and *Rhodospirillum rubrum* ADP-ribosyltransferase. Abbreviation of enzymes as in Table 2. · indicates amino acid identity with the rabbit muscle transferase. Lower case letters indicate conservation of residues with those of the rabbit transferase. – indicates gaps to facilitate alignment. Acidic region in ETA is underlined. The glutamates covalently linked to NAD in photocrosslinking experiments are shown in italics for ETA, DT, PT and C3 exoenzyme. Actual acidic residue in transferases remains to be determined.

NAD identified the essential glutamate in the active sites of DT, ETA and PT. Photolabeling results in the decarboxylation of glutamate and formation of a new bond between the γ-methylene carbon of glutamate and carbon 6 of the nicotinamide ring, with release of the ADP-ribose moiety of NAD [29]. The photolabeled DT exhibited negligible ADP-ribosyltransferase and NAD glycohydrolase activities [30]. In PT, Glu 129 was identified as the active-site residue by covalent crosslinking with NAD [31]. Similarly, photoaffinity-labeling of the *C. limosum* exoenzyme modified solely a glutamate at the catalytic site within a conserved sequence corresponding to Glu174 of the C3 exoenzyme [32]. Site-directed mutagenesis also provided evidence consistent with glutamate as the active site in many of the transferases. Replacement of Glu148 in DT by aspartate resulted in a >99% loss of ADP-ribosyltransferase activity, with little or no change in affinity for NAD and no change in immunoreactivity [33]. The substitution of aspartate for Glu553 of ETA decreased enzymatic activity more than 1000-fold [34]. Replacing Glu129 of PT with glycine or aspartate abolished transferase activity [35] and greatly diminished NAD glycohydrolase activity [31].

A single amino acid substitution in LT, substituting lysine for Glu112, resulted in the loss of biological activity without affecting immunoreactivity [36] or the ability of the mutant to interact with its allosteric activator, ADP-ribosylation factor (ARF), which, nevertheless, failed to enhance transferase activity [37]. The crystallographic structure of LT revealed that the active-site Glu112 is hydrogen-bonded to the hydroxyl group of Ser61 [26]; loss of activity was also observed with a Ser61 to phenylalanine mutation [38]. Thus mutagenesis at position 61 could significantly alter the catalytic site of LT. The amino acid sequences of LT and CT are >80% identical, with 100% identity in the active site region. They are structurally similar as well as immunologically cross-reactive [39, 40], consistent with the conclusion that Glu112 is equivalent in CT and LT. The cluster of acidic amino acids in the rabbit transferase (Glu236, Glu237 and Glu238) and RT6.2 (Asp207, Glu208, Glu209) may correspond to a similar sequence in ETA (Glu546, Glu547, Glu548). There is however, a stronger similarity among region III residues of CT and LT, C3 exoenzyme, the muscle transferase, RT6.2 NADase, and the region immediately surrounding Glu261 of the *Rhodospirillum rubrum* ADP-ribosyltransferase.

Summary

The regions of similarity among transferases are not as evident in other enzymes that utilize NAD as substrate, such as the poly(ADP-ribose)polymerase, and the lymphocyte differentiation antigen, CD38. CD38, a transmembrane glycoprotein, is a bifunctional enzyme exhibiting ADP-ribosyl cyclase activity which converts NAD to cyclic ADP-ribose (cADPR), an IP_3-independent calcium-mobilizing agent, and cyclic ADP-ribose hydrolase activity which hydrolyzes cADPR to ADP-ribose [41, 42]. The poly(ADP-ribose) polymerase catalyzes the ADP-ribosylation of proteins, the polymerization of the ADP-ribose units, and hydrolysis of NAD. The polymerase has been implicated in various cellular functions such as modulating chromatin structure, and DNA repair and replication [43]. The carboxy-terminal catalytic domain of the polymerase shares sequence homology with the NAD(P)-dependent leucine and glutamate dehydrogenases [44]. Site-directed mutagenesis of the polymerase demonstrated that Lys893 and Asp993 are critical for activity and are directly implicated in poly(ADP-ribosyl)ation initiation [44]. Similarly, Glu988 of the polymerase was found to be important for elongation of the ADP-ribose polymer but not for initiation of polymer synthesis [45]. Although CD38 may have regions with similar function and containing similar or identical amino acids to the mono-ADP-ribosyltransferases, the deduced amino acid sequence of CD38 [46] appears to lack the three readily identifiable conserved regions present in the transferases. Nonetheless, studies of the mammalian and bacterial toxin ADP-ribosyltransferases thus far indicate a common structure at the active site based on identical or similar amino acids, which together are involved in NAD binding and catalysis.

References

1. Moss J, Vaughan M: ADP-ribosylation of guanyl nucleotide-binding proteins by bacterial toxins. Adv Enzymol 61: 303–379, 1988
2. Collier JR: Diphtheria toxin: Structure and function of a cytocidal protein. In: J. Moss and M. Vaughan (eds). ADP-ribosylating toxins and G Proteins. Insights into signal transduction. American Society for Microbiology, Washington, DC, 1990, pp 3–19
3. Iglewski BH: Pseudomonas toxins. In: C. Hardegree, A.T. Tu (eds). Handbook of natural toxins. Vol 4. Bacterial toxins. Marcel Dekker, Inc, New York, NY, 1988, pp 249–265
4. Moss J, Vaughan M: Cholera toxin and *E. coli* enterotoxins and their mechanisms of action. In: C. Hardegree, A.T. Tu (eds). Handbook of natural toxins. Vol 4. Bacterial toxins. Marcel Dekker, Inc, New York, NY, 1988, pp 39–87
5. Ui M: Pertussis toxin as a valuable probe for G-protein involvement signal transduction. In: J. Moss and M. Vaughan (eds). ADP-ribosylating toxins and G proteins. Insights into signal transduction. American Society for Microbiology, Washington, DC, 1990, pp 45–77
6. Lowery RG, Ludden PW: Endogenous ADP-ribosylation in procaryotes. In: J. Moss and M. Vaughan (eds). ADP-ribosylating toxins and G proteins. Insights into signal transduction. American Society for Microbiology, Washington, DC, 1990, pp 459–477
7. Aktories K, Braun U, Habermann B, Rosener S: Botulinum ADP-ribosyltransferase C3. In: J. Moss and M. Vaughan (eds). American Society for Microbiology, Washington, DC, 1990, pp 97–115
8. Just I, Mohr C, Schallehn G, Menard L, Didsbury JR, Vandekerckhove J, van Damme J, Aktories K: Purification and characterization of an ADP-ribosyltransferase produced by *Clostridium limosum*. J Biol Chem 267: 10274–10280, 1992
9. Narumiya S, Sekine A, Fujiwara M: Substrate for botulinum ADP-ribosyltransferase, Gb, has an amino acid sequence homologous to a putative *rho* gene product. J Biol Chem 263: 17255–17257, 1988
10. Sekine A, Fujiwara M, Narumiya S: Asparagine residue in the *rho* gene product is the modification site for botulinum ADP-ribosyltransferase. J Biol Chem 266: 19312–19319, 1991
11. Peterson JE, Larew JS-A, Graves DJ: Purification and characterization of arginine-specific ADP-ribosyltransferase from skeletal muscle microsomal membranes. J Biol Chem 265: 17062–17069, 1990
12. Zolkiewska A, Nightingale MS, Moss J: Molecular characterization of NAD:arginine ADP-ribosyltransferase from rabbit skeletal muscle. Proc Natl Acad Sci USA 89: 11352–11356, 1992
13. Okazaki IJ, Zolkiewska A, Nightingale MS, Moss J: Immunological and molecular conservation of mammalian glycosylphosphatidylinositol-linked ADP-ribosyltransferase from skeletal muscle. FASEB J 7: A1042, 1993
14. Takada T, Iida K, Moss J: Expression of NAD glycohydrolase activity by rat mammary adenocarcinoma cells transformed with rat T-cell alloantigen RT6.2. J Biol Chem 269: 9420–9426, 1994
15. Koch F, Haag F, Thiele H-G: Nucleotide and deduced amino acid sequence for the mouse homologue of the rat T-cell differentiation marker RT6. Nucl Acids Res 18: 3636, 1990
16. Koch F, Kashan A, Thiele H-G: The rat T-cell differentiation marker RT6.1 is more polymorphic than its alloantigenic counterpart RT6.2. Immunology 65: 259–265, 1988
17. Koch F, Haag F, Kashan A, Thiele H-G: Primary structure of rat RT6.2, a nonglycosylated phosphatidylinositol-linked surface marker of postthymic T cells. Proc Natl Acad Sci USA 87: 964–967, 1990
18. Thiele H-G, Koch F, Hamann A, Arndt R: Biochemical characterization of the T-cell alloantigen RT6.2. Immunology 59: 195–201, 1986
19. Koch F, Thiele H-G, Low M: Release of the rat T cell alloantigen RT6.2 from cell membranes by phosphatidylinositol-specific phospholipase C. J Exp Med 164: 1338, 1986
20. Rappuoli R, Pizza M: Structure and evolutionary aspects of ADP-ribosylating toxins. In: J.E. Alouf and J.H. Freer (eds). Sourcebook of bacterial protein toxins. Academic Press Inc, San Diego, CA, 1991, pp 1–21
21. Papini E, Schiavo G, Sandona D, Rappuoli R, Montecucco C: Histidine 21 is at the NAD^+ binding site of diphtheria toxin. J Biol Chem 264: 12385–12388, 1989
22. Wozniak DJ, Hsu L-Y, Galloway DR: His-426 of the *Pseudomonas aeruginosa* exotoxin A is required for ADP-ribosylation of elongation factor II. Proc Natl Acad Sci USA 85: 8880–8884, 1988
23. Allured VS, Collier JR, Carroll SF, McKay DB: Structure of exotoxin A of *Pseudomonas aeruginosa* at 3.0 Angstrom resolution. Proc Natl Acad Sci USA 83: 1320–1324, 1986
24. Carroll SF, Collier RJ: Amino acid sequence homology between the enzymic domain of diphtheria toxin and *Pseudomonas aeruginosa* exotoxin A. Molec Microbiol 2: 293–296, 1988
25. Domenighini M, Magagnoli C, Pizza M, Rappuoli R: Common features of the NAD-binding and catalytic site of ADP-ribosylating toxins. Molec Microbiol, in press
26. Kaslow HR, Schlotterbeck JD, Mar VL, Burnette WN: Alkylation of cysteine 41, but not cysteine 200, decreases the ADP-ribosyltransferase activity of the S1 subunit of pertussis toxin. J Biol Chem 264: 6386–6390, 1989
27. Sixma TK, Pronk SE, Kalk KH, Wartna ES, van Zanten BAM, Witholt B, Hol WGJ: Crystal structure of a cholera toxin-related heat-labile enterotoxin from *E. coli*. Nature 351: 371–377, 1991
28. Brandhuber BJ, Allured VS, Falbel TG, McKay DB: Mapping the enzymatic active site of *Pseudomonas aeruginosa* exotoxin A. Proteins 3: 146–154, 1988
29. Carroll SF, McCloskey JA, Crain PF, Oppenheimer NJ, Marschner TM,

Collier JR: Photoaffinity labeling of diphtheria toxin fragment A with NAD: Structure of the photoproduct at position 148. Proc Natl Acad Sci USA 82: 7237–7241, 1985

30. Carroll SF, Collier JR: NAD binding site of diphtheria toxin: Identification of a residue within the nicotinamide subsite by photochemical modification with NAD. Proc Natl Acad Sci USA 81: 3307–3311, 1984
31. Barbieri JT, Mende-Meuller LM, Rappuoli R, Collier JR: Photolabeling of Glu-129 of the S-1 subunit of pertussis toxin with NAD. Infect Immun 57: 3549–3554, 1989
32. Jung M, Just I, van Damme J, Vandekerckhove J, Aktories K: NAD-binding site of the C3-like ADP-ribosyltransferase from *Clostridium limosum*. J Biol Chem 268: 23215–23218, 1993
33. Tweten RK, Barbieri JT, Collier RJ: Diphtheria toxin: effect of substituting aspartic acid for glutamic acid-148 on ADP-ribosyltransferase activity. J Biol Chem 260: 10392–10394, 1985
34. Douglas CM, Collier RJ: Exotoxin A of *Pseudomonas aeruginosa*: substitution of glutamic acid-553 with aspartic acid drastically reduces toxicity and enzymic activity. Infect Immun 169: 4967–4971, 1987
35. Pizza M, Bartoloni A, Prugnola A, Silvestri S, Rappuoli R: Subunit S1 of pertussis toxin: Mapping of the regions essential for ADP-ribosyltransferase activity. Proc Natl Acad Sci USA 85: 7521–7525, 1988
36. Tsuji T, Inoue T, Miyama A, Okamoto K, Honda T, Miwatani T: A single amino acid substitution in the A subunit of *Escherichia coli* enterotoxin results in a loss of its toxic activity. J Biol Chem 265: 22520–22525, 1990
37. Moss J, Stanley SJ, Vaughan M, Tsuji T: Interaction of ADP-ribosylation factor with *Escherichia coli* enterotoxin that contains an inactivating lysine 112 substitution. J Biol Chem 268: 6383–6387, 1993
38. Harford S, Dykes CW, Hobden AN, Read MJ, Halliday IJ: Inactivation of the *Escherichia coli* heat-labile enterotoxin by *in vitro* mutagenesis of the A subunit gene. Eur J Biochem 183: 311–316, 1989
39. Dallas WS, Falkow S: Amino acid sequence homology between cholera toxin and *Eschericia coli* heat-labile toxin. Nature 288: 499–501, 1980
40. Yamamoto T, Gojobori T, Yokota T: Evolutionary origin of pathogenic determinants in enterotoxigenic *Escherichia coli* and *Vibrio cholerae* O1. J Bacteriol 169: 1352–1357, 1987
41. Howard M, Grimaldi JC, Bazan JF, Lund FE, Santos-Argumedo L, Parkhouse RME, Walseth TF, Lee HC: Formation and hydrolysis of cyclic ADP-ribose catalyzed by lymphocyte antigen CD38. Science 262: 1056–1059, 1993
42. Zocchi E, Franco L, Guida L, Benatti U, Bargellesi A, Malavasi F, Lee HC, De Flora A: A single protein immunologically identified as CD38 displays NAD^+ glycohydrolase, ADP-ribosyl cyclase and cyclic ADP-ribose hydrolase activities at the outer surface of human erythrocytes. Biochem Biophys Res Commun 196: 1459–1465, 1993
43. Ueda K: Poly(ADP-ribose) Synthetase. In: J. Moss and M. Vaughan (eds). ADP-ribosylating toxins and G proteins. Insights into signal transduction. American Society for Microbiology, Washington, DC, 1990, pp 525–542
44. Simonin F, Poch O, Delarue M, de Murcia G: Identification of potential active-site residues in the human poly(ADP-ribose) polymerase. J Biol Chem 268: 8529–8535, 1993
45. Marsischky GT, Ikejima M, Suzuki H, Sugimura T, Esumi H, Miwa M, Collier RJ: Directed mutagenesis of glutamic acid 988 of poly(ADP-ribose) polymerase. In: G.G. Poirier, P. Moreau (eds). ADP-ribosylation reactions. Springer-Verlag, New York, NY, 1992, pp 47–52
46. Jackson DG, Bell JI: Isolation of a cDNA encoding the human CD38 (T10) molecule, a cell surface glycoprotein with an unusual discontinuous pattern of expression during lymphocyte differentiation. J Immunol 144: 2811–2815, 1990

PART V

INHIBITORS AND ACTIVATORS OF ADP-RIBOSYLATION

Molecular and Cellular Biochemistry **138**: 185–197, 1994.

Inhibitors and activators of ADP-ribosylation reactions

Marek Banasik and Kunihiro Ueda
Department of Clinical Science and Laboratory Medicine, Kyoto University Faculty of Medicine, Shogoin, Sakyo-ku, Kyoto 606-01, Japan

Abstract

ADP-ribosylation reaction, that is the transfer of the ADP-ribose moiety of NAD^+ to acceptor protein, is catalyzed by two classes of ADP-ribosyltransferases, *i.e.*, poly(ADP-ribose) synthetase and mono(ADP-ribosyl)transferases. These two types differ not only in the number of transferring ADP-ribose units but also in the acceptor amino acid(s) and protein. Their inhibitors, particularly those of poly(ADP-ribose) synthetase, have been successfully employed in studies on biological functions of the enzymes and other related fields of research. Recently, we found many potent and specific inhibitors of poly(ADP-ribose) synthetase, and broadened their chemical as well as biochemical variety. More recently, we found several potent inhibitors of arginine-specific mono(ADP-ribosyl)transferases and activators of poly(ADP-ribose) synthetase. (Mol Cell Biochem **138**: 185–197, 1994)

Key words: inhibitor, activator, poly(ADP-ribose) synthetase, mono(ADP-ribosyl)transferase

Introduction

Inhibitors of poly(ADP-ribose) synthetase

Poly(ADP-ribose) synthetase (or polymerase) (EC 2.4.2.30) is a nuclear enzyme that catalyzes the transfer of the ADP-ribose moiety of NAD^+ to various nuclear acceptor proteins with a concomitant release of nicotinamide, and then further additions to this protein-bound ADP-ribose, forming an $(ADP\text{-ribose})_n$ chain with branches. The enzymatic activity is unique in its absolute dependency on the presence of DNA with strand termini. Biological roles suggested for this modification of proteins include DNA repair, cell differentiation, cell cycle control, transformation, transcription, and alteration of chromatin architecture. Most of these findings were obtained from experiments employing inhibitors of the synthetase. Because nicotinamide is a product of ADP-ribosylation reaction and proved to be an inhibitor, it has been used as such from the very beginning of research in this field [1–18]. Other inhibitors used previously are thymidine [6, 8, 10–13, 15, 17, 19], theophylline [10, 13, 17], and 5-methylnicotinamide [5, 6, 12]. Most of these compounds soon proved to affect also other biological processes. For example, nicotinamide is an inhibitor of microsomal NADase, toxin ADP-ribosyltransferases, and adenosine 3′,5′-cyclic monophosphate phosphodiesterase (EC 3.1.4,17) [20], and is a substrate for nicotinamide *N*-methyltransferase, deaminase, and phosphoribosyltransferase. It is ultimately reutilized for NAD^+ biosynthesis, and leads to elevation of the concentrations of pyridine nucleotides. Furthermore, Berger and Sikorski [21] reported that nicotinamide stimulates repair of DNA damage in human lymphocytes. Nicotinamide and some other inhibitors inhibit deoxyglucose uptake in differentiated 3T3-L1 fat cells [22]. Thymidine, after conversion to a phosphorylated derivative, inhibits the biosynthesis of 2′-deoxycytidine phosphate, resulting in the arrest of cell division [23]. Theophylline is an inhibitor of cyclic nucleotide phosphodiesterase [24], and, interestingly, along with another methylxanthine, caffeine, it is an inhibitor of DNA repair [25–27]. In 1980, Purnell and Whish [28] reported that the substituted benzamides, 3-aminobenzamide and 3-methoxybenzamide, are potent and specific inhibitors

Address for offprints: K. Ueda, Department of Clinical Science and Laboratory Medicine, Kyoto University Faculty of Medicine, Shogoin, Sakyo-ku, Kyoto 606-01, Japan

of poly(ADP-ribose) synthetase. Around this time, Durkacz *et al.* [29] and Oikawa *et al.* [30] applied benzamides successfully to cellular studies, and showed a possible involvement of poly(ADP-ribosyl)ation reactions in DNA excision repair and SCEs (sister chromatid exchanges), respectively. Since that time, 3-aminobenzamide has been the most favored inhibitor and served as an indispensable tool in studies on biological roles of poly(ADP-ribosyl)ation. There is, however, growing evidence that, at least at high concentrations, 3-aminobenzamide and other benzamides have considerable side actions *in vivo*. For example, 3-aminobenzamide inhibits *de novo* synthesis of DNA purines in HeLa cells [31] or Chinese hamster ovary cells [32] and the metabolism of glucose and methionine into precursors of DNA in WIL-2 lymphoid cells [33] or C3H 10T½ mouse embryo fibroblasts [34]. On the other hand, Hunting *et al.* [35] using L1210 cells did not confirm these results, and Snyder [36] did not find such effects in human fibroblasts. Also in Chinese hamster fibroblasts, 3-aminobenzamide did not affect nucleotide synthesis [37]. Benzamide, its amino derivatives, and 5-methylnicotinamide inhibit nicotinamide *N*-methyltransferase (EC 2.1.1.1) in liver homogenates and in intact normal rat kidney cells [38]. It should be noted that benzamide at high concentrations inhibits *in vitro* activities of adenosine 3′,5′-cyclic monophosphate phosphodiesterase [20], chymotrypsin (EC 3.4.4.5) [39], and the esterase activity of carboxypeptidase A (EC 3.4.2.1) but activates its peptidase activity [40]. In view of these reports, it seems very important to take into account these possible side actions when we evaluate experimental results obtained with inhibitors.

Results

Many compounds have been shown to inhibit the poly(ADP-ribose) synthetase activity *in vitro* or *in vivo*. Table 1 is a list of inhibitors of poly(ADP-ribose) synthetase for which IC_{50} (the concentration that causes 50% inhibition) and/or K_i value(s) have been determined. The considerable variation in IC_{50} or K_i values reported for individual compounds indicates that the experimental conditions and enzyme preparations may profoundly effect these values.

The first systematic analysis of structure-activity relationship of inhibitors of poly(ADP-ribose) synthetase was performed by Sims *et al.* [66]. They concluded that an aromatic ring system and an intact carboxamide group are important for effective inhibition and that a ring nitrogen is not always necessary. Sestili *et al.* [63] reported that, among benzamides, the most powerful inhibitors were characterized by acylation of the amino function in the 3 position or by substitution in this position with hydroxy or methoxy groups. Rankin *et al.* [45] determined IC_{50} values of nineteen selected compounds for poly(ADP-ribose) synthetase and those of nine compounds for monomer/arginine-specific ADP-ribosyltransferase [mono(ADP-ribosyl)transferase] A (EC 2.4.2.31), showing that benzamide, 3-methoxybenzamide, and 3-aminobenzamide are highly specific for the synthetase. Recently, we made an extensive survey of inhibitors of poly-(ADP-ribose) synthetase, and discovered many strong inhibitors [41]. The four strongest were 4-amino-1,8-napthalimide, 6(5*H*)- and 2-nitro-6(5*H*)-phenanthridinones, and 1,5-dihydroxyisoquinoline. Their IC_{50} values, 0.18–0.39 μM, were about two orders of magnitude lower than that of 3-aminobenzamide. A common structural feature of these inhibitors was the presence of a carboxamide group built in a polyaromatic heterocyclic skeleton. The oxygen atom in the carbonyl group seems to serve as an electron donor, and the carbon atom as an electron acceptor in the interaction with the enzyme molecule [67]. The carbonyl group, however, is not indispensable for the inhibitory action; weaker inhibitors, such as phthalazine, quinazoline, norharman, and isoquinoline, do not have a carbonyl group, but have a C = N double bond in analogous positions [41]. Similarly, thiobenzamide [63] and thionicotinamide have a thiocarbamoyl group in place of a carbamoyl group. The aromatic ring also contributes to the inhibition, but is not absolutely required as exemplified by inhibitory action of cyclohexanecarboxamide [41, 63]. A big difference between structures of newly discovered inhibitors and benzamide or nicotinamide may suggest that they do not share the same side effects.

Comparative studies of the effects on poly(ADP-ribose) synthetase and arginine-specific mono(ADP-ribosyl)transferase from hen heterophils [68] revealed that most of the potent inhibitors are highly specific for the former, poly-(ADP-ribose) synthetase [41]. A comparison of IC_{50} values of selected inhibitors for poly(ADP-ribose) synthetase and another arginine-specific mono(ADP-ribosyl)transferase, that is, transferase A from turkey erythrocytes [69], is shown in Table 2. Most of very strong inhibitors of poly(ADP-ribose) synthetase, including 2-nitro-6(5*H*)-phenanthridinone, 3-hydroxybenzamide, and 1,5-dihydroxyisoquinoline, are highly specific for poly(ADP-ribose) synthetase, whereas coumermycin A_1 and novobiocin are specific for mono(ADP-ribosyl)transferase A. Unsaturated long-chain fatty acids are inhibitory to both enzymes. Apparently, this profile of inhibitors for turkey erythrocyte enzyme resembles, in large part, that for the hen heterophil enzyme reported previously [41]. These inhibitors with distinct specificity would be useful for selective inhibition of poly- or mono(ADP-ribosyl)-ation reaction *in vitro* as well as *in vivo*, although their effects on enzymes other than arginine-specific mono(ADP-ribosyl)transferases remain to be determined.

In vivo use of new inhibitors has already given seemingly very promising results. For example, al-1 cells, a transformed cell line obtained by transfection of NIH 3T3 cells with

Table 1. Effects of various compounds on poly(ADP-ribose) synthetase activity

Compound	K_i (μM)	IC_{50} (μM)	Inhibition (%) at[a] 0.1 mM[b]	Inhibition (%) at[a] 1 mM	Ref.
m-Acetamidoacetophenone	–[+]	930	20	65	41
2-Acetamidobenzamide	–	1000	–	50	41
3-Acetamidobenzamide	0.4	–	–	–	42
3-Acetamidobenzamide	–	12	80	97	41
8-Acetamidocarsalam*	–	1400	–	43	41
3-Acetamidosalicylamide*	–	2000	–	34	41
5-Acetamidosalicylamide	–	45	67	84	41
Acetophenone*	–	2300	–	35	41
3-Acryloylaminobenzamide	3.14	–	(5 μM) 47	–	43
m-Aminoacetophenone	–	1900	–	38	41
2-Aminobenzamide	71	–	(12 nM) 18	–	44
2-Aminobenzamide	–	100	–	–	45
2-Aminobenzamide	–	650	–	56	41
3-Aminobenzamide	4.4	–	–	–	29
3-Aminobenzamide	1.8	–	(50 μM) 90	–	28
3-Aminobenzamide	–	≈ 250	–	73	34
3-Aminobenzamide	2.6	–	–	–	42
3-Aminobenzamide	4.3	–	–	–	46
3-Aminobenzamide	12	–	(12 nM) – 125[$]	–	44
3-Aminobenzamide	–	5.4	–	–	45
3-Aminobenzamide	–	33	68	88	41
3-Aminobenzamide	11.3	–		–	43
4-Aminobenzamide	65	–	(12 nM) – 20	–	44
4-Aminobenzamide	75	–	–	–	47
4-Aminobenzamide	–	400	–	–	45
4-Aminobenzamide	–	1800	–	37	41
6-Amino-1,2-benzopyrone	≈ 28	–	–	–	48
6-Amino-1,2-benzopyrone	47	–	–	90	49
6-Amino-1,2-benzopyrone*	–	850	–	53	41
2-Amino-3-chloro-1,4-naphthoquinone*	–	820	–	> 53 #	41
4-Amino-1,8-naphthalimide*	–	0.18	99	> 99	41
6-Aminonicotinamide	–	1100	–	48	41
5-Aminosalicylamide	–	100	50	79	41
A2′pA2′pA	50	–	(50 μM) 38	–	50
Apigenin*	–	< 1500 ##	–	> 42	51
Arachidonic acid*	–	44	67	> 67	52
Benzamide	< 1.0	–	–	–	53
Benzamide	1.0	–	–	–	42
Benzamide	2.6	–	–	–	46
Benzamide	39	–	(30 nM) – 163	–	44
Benzamide	2	–	–	–	47
Benzamide	–	3.3	–	–	45
Benzamide	1.5	–	92	–	49
Benzamide	–	22	73	92	41
1,2-Benzopyrone	–	47	≈ 85	–	54
1,2-Benzopyrone*	–	2800	–	33	41
1,4-Benzoquinone	–	400	–3	78	41
Benzoyleneurea*	–	8.1	81	98	41
2-Bromobenzamide	–	2900	–	22	41
3-Bromobenzamide*	–	55	59	87	41
4-Bromobenzamide*	–	2200	–	36	41
5-Bromo-2′-deoxyuridine	–	15	–	–	45
3-(3-Bromopropionyl)aminobenzamide	1.73	–	(5 μM) 58	–	43
5-Bromouracil	–	160	42	74	41
5-Bromouridine	–	210	33	75	41
Caffeine	244	–	–	–	53
Caffeine	–	1400	–	–	45
Carsalam*	–	460	–	66	41

Table 1. Effects of various compounds on poly(ADP-ribose) synthetase activity (continued)

Compound	K_i (μM)	IC_{50} (μM)	Inhibition (%) at[a] 0.1 mM[b]	1 mM	Ref.
2-Chlorobenzamide	–	1000	–	50	41
3-Chlorobenzamide	–	22	73	92	41
4-Chlorobenzamide	–	300	27	67	41
N-(2-Chloroethyl)-1,8-naphthalimide*	–	< 1800	–	> 37	41
3-(2-Chloropropionyl)aminobenzamide	2.01	–	(5 μM) 53	–	43
3-(3-Chloropropionyl)aminobenzamide	1.88	–	(5 μM) 58	–	43
5-Chlorosalicylamide*	–	190	36	77	41
5-Chlorouracil	–	270	33	71	41
Chlorthenoxazin*	–	8.5	80	98	41
4-Chromanone*	–	720	–	55	41
Chromone-2-carboxylic acid	–	560	–	62	41
Coenzyme Q_0	–	3900	–	−2	52
Cyclohexanecarboxamide	–	620	–	59	41
trans-1-Decalone*	–	4300	–	22	41
Diadenosine 5′,5′′′-p[1],p[2]-diphosphate	–	–	50	–	55
Diadenosine 5′,5′′′-p[1],p[4](-1,N^6-ethenyl-)tetraphosphate	13.9	≈ 20	–	–	56
Diadenosine 5′,5′′′-p[1],p[4](-p[1],p[2]-methylene-)tetraphosphate	22.6	≈ 20	–	–	56
Diadenosine 5′,5′′′-p[1],p[4]-tetraphosphate	5.1	–	88	–	55
Diadenosine 5′,5′′′-p[1],p[4]-tetraphosphate	7.7	–	–	–	56
4-Diazoniobenzamide-dGMP	150	–	–	–	47
3,5-Dibromosalicylamide*	–	560	–	62	41
2,3-Dichloro-1,4-naphthoquinone*	–	260	36	> 62	41
2,6-Difluorobenzamide	–	180	33	73	41
10,11-Dihydrodibenz[*b,f*][1,4]-oxazepin-11-one*	–	< 2300	–	34	51
1,5-Dihydroxyisoquinoline	–	0.39	99	100	41
1,3-Dihydroxynaphthalene*	–	1300	–	43	41
3,5-Dimethoxybenzamide*	–	1200	–	46	41
3-(*N,N*-Dimethylamino)benzamide	–	120	47	81	41
3,5-Dinitrobenzamide	–	2500	–	36	41
Ethidium bromide	143.8	–	–	–	57
Flavone*	–	22	72	88	41
2-Fluorobenzamide*	–	120	47	85	41
3-Fluorobenzamide	–	20	73	92	41
4-Fluorobenzamide*	–	200	–	71	41
Formycin B	68.9	–	–	–	57
Formycin B	75	–	–	–	58
Harmine hydrochloride	–	< 3500	–	> 18	51
m-Hydroxyacetophenone	–	600	21	65	41
2-Hydroxybenzamide	–	82	53	78	41
3-Hydroxybenzamide	1.0	–	–	–	42
3-Hydroxybenzamide	–	9.1	81	95	41
4-Hydroxybenzamide	–	280	35	74	41
4-Hydroxy-1,2-benzopyrone*	–	570	17	62	41
1-Hydroxyisoquinoline	–	7.0	82	98	41
4-Hydroxy-2-methylquinoline*	–	74	55	87	41
N-Hydroxynaphthalimide sodium salt	–	450	8	> 96	41
4-Hydroxypyridine	–	2300	–	34	41
4-Hydroxyquinazoline	–	9.5	80	92	41
4-Hydroxyquinoline*	–	80	59	86	41
Hypoxanthine	–	1700	–	–	45
1-Indanone*	–	810	–	53	41
5-Iodouracil	–	71	55	84	41
5-Iodouridine	–	43	66	89	41
Isatoic anhydride*	–	< 3900	–	4	51
3-Isobutyl-1-methylxanthine	–	3100	–	–	45
Isoluminol*	–	290	22	70	41
Isonicotinamide	–	990	–	51	41
Isonicotinate hydrazide	–	4800	–	–	45

Table 1. Effects of various compounds on poly(ADP-ribose) synthetase activity (continued)

Compound	K_i (µM)	IC_{50} (µM)	Inhibition (%) at[a] 0.1 mM[b]	1 mM	Ref.
Juglone*	–	250	12	> 93	41
Kynurenic acid*	–	670	–	61	41
Lawsone*	–	330	26	74	41
Linoleic acid*	–	48	57	85	52
Linolenic acid*	–	110	44	> 70	52
γ-Linolenic acid*	–	120	21	> 80	41
Luminol*	–	23	74	96	41
Menadione*	–	420	8	> 70	52
Menadione sodium bisulfite	–	720	–	58	41
2-Mercapto-4(3*H*)-quinazolinone*	–	44	69	94	41
2-Methoxybenzamide	–	20	74	89	41
3-Methoxybenzamide	1.5	–	(50 µM) 98	–	28
3-Methoxybenzamide	0.6	–	–	–	42
3-Methoxybenzamide	2.9	–	–	–	46
3-Methoxybenzamide	–	3.4	–	–	45
3-Methoxybenzamide	–	17	75	92	41
4-Methoxybenzamide	–	1100	–	47	41
1-Methyladenine	226.6	–	–	–	57
2-Methylbenzamide	–	1500	–	43	41
3-Methylbenzamide	–	19	77	95	41
4-Methylbenzamide	–	1800	–	37	41
2-Methylchromone	–	45	60	83	41
1-Methylnicotinamide	–	1700	–	–	45
5-Methylnicotinamide	200	–	–	–	5
5-Methylnicotinamide	30.0	–	–	–	53
5-Methylnicotinamide	–	70	–	–	45
5-Methylnicotinamide	–	350	–	66	41
8-Methylnicotinamide	–	7800	–	–	45
1-Methylnicotinamide chloride	–	3800	–	28	41
2-Methyl-4(3*H*)-quinazolinone*	–	5.6	86	99	41
NADH	55	–	–	–	14
NADP	460	–	(200 µM) 37	–	15
1,8-Naphthalimide*	–	1.4	98	98	41
1,4-Naphthoquinone*	–	250	24	87	41
Nicotinamide	52	–	–	48	2
Nicotinamide	20	–	–	–	5
Nicotinamide	≈ 14	–	–	–	7
Nicotinamide	14.3	–	–	–	8
Nicotinamide	5.7	–	–	–	59
Nicotinamide	50	–	–	–	14
Nicotinamide	5.6	–	–	–	60
Nicotinamide	23	–	(200 µM) 56	–	15
Nicotinamide	–	≈ 100	–	–	18
Nicotinamide	15	–	–	–	61
Nicotinamide	13.0	–	–	–	53
Nicotinamide	–	31	–	–	45
Nicotinamide	–	210	41	72	52
3-Nitrobenzamide	9.8	–	–	–	42
3-Nitrobenzamide	–	160	43	74	41
2-Nitro-6(5*H*)-phenanthridinone*	–	0.35	92	> 93	41
3-Nitrophthalhydrazide*	–	72	56	90	41
4-Nitrophthalhydrazide*	–	510	–	63	41
3-Nitrosalicylamide*	–	1600	–	41	41
6-Nitroso-1,2-benzopyrone	40	–	–	–	62
5-Nitrouracil*	–	430	20	65	41
Norharman	–	4700	–	21	41
Novobiocin	–	2200	– 3	16	41
Oleic acid*	–	82	53	86	41

Table 1. Effects of various compounds on poly(ADP-ribose) synthetase activity (continued)

Compound	K_i (μM)	IC_{50} (μM)	Inhibition (%) at[a] 0.1 mM[b]	Inhibition (%) at[a] 1 mM	Ref.
Palmitoleic acid*	–	95	51	86	41
6(5*H*)-Phenanthridinone*	–	0.30	99	> 99	41
Phthalamide	–	1000	–	50	41
m-Phthalamide*	–	50	64	89	41
Phthalazine	–	150	44	78	41
1(2*H*)-Phthalazinone*	–	12	77	98	41
Phthalhydrazide*	–	30	70	95	41
α-Picolinamide	–	≈ 100	–	–	18
α-Picolinamide	–	250	–	68	41
Plumbagin*	–	700	21	> 62	41
pppA2′pA2′pA	5	≈ 10	(50 μM) 85	–	50
Pyrazinamide	–	130	–	–	45
Pyridoxal 5-phosphate*	–	4250	–	13	52
Quinazoline	–	2000	–	38	41
Reserpine**	–	790	–	> 66	41
trans-Retinal*	–	450	10	> 70	52
Showdomycin	107.8	–	–	–	57
α-Tetralone*	–	310	–	67	41
Theobromine	15.2	–	–	–	53
Theobromine	–	110	–	–	45
Theophylline	29.8	–	–	–	53
Theophylline	–	46	–	–	45
Thiobenzamide	–	620	–	57	41
Thionicotinamide	–	1800	–	43	41
Thiophene-3-carboxamide	–	≈ 50	–	–	63
Thymidine	32.5	–	–	–	8
Thymidine	140	–	(200 μM) 30	–	15
Thymidine	25	–	–	–	61
Thymidine	13.3	–	–	–	53
Thymidine	–	43	–	–	45
Thymidine	–	180	36	77	41
Thymine	–	290	30	73	41
2-Trichloromethyl-4(3*H*)-quinazolinone*	–	2200	6	32	41
Trp-P-1 (3-Amino-1,4-dimethyl-5*H*-pyrido[4,3-*b*]indole)	–	220	–	91	64
Trp-P-2 (3-Amino-1-methyl-5*H*-pyrido[4,3-*b*]indole)*	–	2200	–	– 134	64
Vitamin K_1**	–	520	–	> 72	52
Vitamin K_5*	–	1300	–	> 31	41
Xanthurenic acid*	–	190	33	77	41
$ZnCl_2$	–	10	≈ 90	–	65
$ZnCl_2$	–	77	–	–	45

[a] Test conditions may vary.
[b] Unless otherwise specified.
[+] Not determined.
* 2% (final) Me_2SO.
** 10% (final) Me_2SO.
≈ Approximately.
[$] Stimulation.
\# Minimum value estimated under the condition of limited solubility.
\#\# Maximum value estimated under the condition of limited solubility.

Table 2. Comparison of inhibitory potencies of various compounds on mono(ADP-ribosyl)transferase A from turkey erythrocytes and poly(ADP-ribose) synthetase

Compound	IC_{50} (μM) Mono(ADP-ribosyl)transferase (a)	Poly(ADP-ribose) synthetase (b)	(a/b)
2-Nitro-6(5*H*)-phenanthridinone*	230	0.35	657
3-Hydroxybenzamide	4600	9.1	505
Benzamide	9600	22	436
1,5-Dihydroxyisoquinoline	130	0.39	333
2-Methylchromone	9500	45	211
1-Hydroxyisoquinoline	1400	7.0	200
6(5*H*)-Phenanthridinone*	47	0.30	157
4-Hydroxyquinazoline	1000	9.5	105
2-Methyl-4(3*H*)-quinazolinone*	260	5.6	46
4-Amino-1,8-naphthalimide*	7.1	0.18	39
5-Iodouridine	1300	43	30
Flavone*	360	22	16
1(2*H*)-Phthalazinone*	210	12	18
Benzoyleneurea*	39	8.1	5
1,8-Naphthalimide*	3.1	1.4	2.2
Linoleic acid (C18:2, *cis*-9,12)*	67	48	1.4
Arachidonic acid (C20:4, *cis*-5,8,11,14)*	52	44	1.2
Linolenic acid (C18:3, *cis*-9,12,15)*	102	110	0.93
Palmitoleic acid (C16:1, *cis*-9)*	67	95	0.71
Oleic acid (C18:1, *cis*-9)*	56	82	0.68
γ-Linolenic acid (C18:3, *cis*-6,8,12)*	66	120	0.55
Novobiocin	370	2200	0.17
Coumermycin A_1	27	≈ 250	≈ 0.11

All marks are the same as in Table 1. The assay conditions for poly(ADP-ribose) synthetase were as described previously [43]. The reaction mixture for mono(ADP-ribosyl)transferase A contained 50 mM potassium phosphate (pH 7.0), 100 μM [*Ade*-U-^{14}C]NAD$^+$ (255,000 cpm), 250 μg/ml histone H2B, and enzyme in 200 μl. Incubation was for 30 min at 30°C.

human c-Ha-*ras* T^{24} (hc-Ha-*ras*T^{24}), were converted to morphologically normal flat cells following a 2-week culture in the presence of the synthetase inhibitor, luminol or benzamide. Concomitant with these morphological changes was the loss of the exogenous and amplified hc-Ha-*ras* T^{24} oncogene [70]. In another study, inhibitors of the synthetase, including 4-hydroxyquinazoline and benzamide, caused an efficient loss of endogenous and amplified human c-*myc* genes from human leukemic HL-60 cells and concomitantly induced granulocytic differentiation [71]. It is also of interest that luminol induced SCEs (sister chromatid exchanges) in cultured Chinese hamster cells about ten times more effectively than 3-aminobenzamide and that this induction of SCEs was restricted to the S-phase of the cell cycle [72]. *C*-nitroso-substituted ligands of poly(ADP-ribose) synthetase, 6-nitroso-1,2-benzopyrone and 3-nitrosobenzamide, that are known to oxidize one of the zinc fingers of the enzyme [62], were shown to be cytostatic and apoptotic toward malignant human cells [73]. They also inhibit infection by human immunodeficiency virus (HIV) of human lymphocytes by ejecting zinc ions from zinc fingers of nucleocapsid or intact HIV-1 virions [74].

Mode of action on poly(ADP-ribose) synthetase is not clear for most of the inhibitors. The compounds documented as competitive inhibitors with respect to NAD$^+$ include 3-acryloylaminobenzamide [43], 3-aminobenzamide [28, 29, 46], 4-aminobenzamide [47], 6-amino-1,2-benzopyrone [48, 49], benzamide [44, 46, 47, 49, 53], 3-bromoacetylaminobenzamide [43], 3-(3-bromopropionyl)aminobenzamide [43], caffeine [53], 3-chloroacetylaminobenzamide [43], 3-(4-chlorobutanyl)aminobenzamide [43], 3-(2- or 3-chloropropionyl)aminobenzamide [43], 3-crotonylaminobenzamide [43], 2′-deoxythymidine 5′-monophosphate [19], ethidium bromide [57], formycin B [57, 58], 3-methoxybenzamide [28, 46], 1-methyladenine [57], 5-methylnicotinamide [53], 3-(*N*-methyl-*N*-nitroso)ureidobenzamide [43], NADH [14, 75], NADP [15], nicotinamide [2, 8, 14, 15, 17, 53, 75], poly(dT) [75], showdomycin [57], theobromine [53], theophylline [53], and thymidine [8, 15, 19, 53, 75]. However, we showed that most inhibitors, including several listed above, such as nicotinamide and 3-aminobenzamide, exhibit mixed-type inhibition with respect to the substrate, NAD$^+$, at its micromolar concentrations [41]. The discrepancy may be due to the difference in the reaction conditions or the enzyme preparation. On the other hand, xanthurenic acid (at 50 and 100 μM) and 5-nitrouracil proved to exhibit competitive inhibition with respect to NAD^+ [41]. Xanthurenic acid, however, exhibits mixed-type inhibition at 200 μM, resembling dual inhibition exerted by arachidonic acid [52]. Mixed-type inhibition was previously reported for a natural nucleotide, diadenosine 5′, 5′′′-p^1, p^4-tetraphosphate, a ligand of a subunit of DNA polymerase α [55, 56], and two of its analogues, diadenosine 5′, 5′′′-p^1, p^4(–1, N^6-ethenyl)-tetraphosphate and diadenosine 5′, 5′′′-p^1, p^4(–p^1, p^2-methylene-)tetraphosphate [56]. Two 2′,5′-oligoadenylates, A2′pA2′pA and pppA2′pA2′pA [50], 1,2-benzopyrone [54], and 1,10-phenanthroline [76] have been documented as noncompetitive inhibitors.

Recently, it was reported that the exposure of monolayers of human kidney T1-cells to a strong static magnetic field reduced poly(ADP-ribosyl)ation of total cellular proteins to about 60% of its normal metabolic level. It was suggested that the magnetic field interacts with some cellular membrane components and affects the transfer of signal(s) controlling the intranuclear poly(ADP-ribosyl)ation of proteins [77]. It is of interest that several other interventions on the T1-cell monolayer, such as mechanical scraping, trypsinization, or agitation, increased poly(ADP-ribosyl)ation of cellular proteins to varying degrees [77].

Activators of poly(ADP-ribose) synthetase

Poly(ADP-ribose) synthetase is a DNA-dependent enzyme. It was shown that short double-stranded DNA fragments, such as an octamer d(GGAATTCC), but not a hexamer d(GTTAAC), can potently stimulate the enzymatic activity, and that the latter, d(GTTAAC), becomes highly stimulatory in the presence of histone H1 [78]. It seems, therefore, that the minimum length of a double-stranded DNA (dsDNA) necessary for enzyme activation is between 6 and 8 base pairs. Single-stranded DNA and RNA are not stimulatory [61], but DNA-RNA hybrids are partially effective [79]. Free DNA strand ends, either blunt, 5′ or 3′ extruding, are necessary for activation of the enzyme [61, 78]. Phosphorylated ends are more effective than dephosphorylated ones. Covalently closed dsDNA has no activating effect unless it is opened by endonuclease treatment [17, 61].

Mg^{2+} stimulates the activity of poly(ADP-ribose) synthetase in the presence of both DNA and histones. However, in the presence of a large excess of histones over DNA, Mg^{2+} exhibits an inhibitory effect [80, 81]. It is of interest that the automodification reaction is dependent on the presence of Mg^{2+}; in the absence of Mg^{2+}, exogenously added acceptor proteins are primarily modified, while auto(ADP-ribosyl)-ation of the synthetase is minimal [80, 82]. With some purified enzyme preparations, Mn^{2+} [15, 83] or Ca^{2+} [15] has been shown to replace Mg^{2+} in the stimulation of the synthetase activity. Kun *et al.* [84] noticed that, in 9L-gliosarcoma cells, Ca^{2+} is a more active cation than Mg^{2+}. Using a purified enzyme preparation and no exogenous acceptor, we found that Ca^{2+} is significantly more effective than Mg^{2+} with respect to the optimal concentration and the extent of activation [85]. Other metal ions such as Cu^+, Cu^{2+}, Zn^{2+}, Hg^{2+}, and Cd^{2+} are highly inhibitory [14]. Several possible mechanisms have been proposed for the action of Mg^{2+}. Shall *et al.* [7, 86] postulated formation of an active substrate complex, NAD^+-Mg^{2+}, based on the finding that the optimal Mg^{2+} concentration depended on the NAD^+ concentration in mouse fibroblast nuclei. On the other hand, stabilization of the DNA-enzyme complex was proposed by others [87, 88]. Byrne *et al.* [89] noticed that Mg^{2+} enhanced the formation of histone H1-poly(ADP-ribose) complexes. Competition of Mg^{2+} with polycations such as histones and polyamines for interaction with DNA was also proposed [90].

Polyamines (spermine, spermidine, putrescine) were reported to stimulate poly(ADP-ribose) synthesis in isolated nuclei [57, 89, 91–94] or purified enzyme systems [90]. The effect of polyamines appears to be cooperative partly with Mg^{2+} and partly with histones. With respect to both the extent of stimulation and the optimal concentration, spermine is most effective, followed by spermidine; putrescine often exhibits no appreciable effect [93, 95].

Organic solvents such as ethanol and methanol also stimulate the enzymatic activity. Kristensen and Holtlund [96] added 9.5% ethanol in routine assays of the enzyme purified from Ehrlich ascites tumor cells; in its absence, the enzymatic activity decreased to 53%, and this decrease was reversed by the addition of 10% methanol or ethylene glycol. Berger *et al.* [12] observed potent (about 3.4-fold) stimulation of poly(ADP-ribose) synthesis in nucleotide-permeable cells by 7% ethanol, but inhibition at ethanol concentrations above 15%. Rickwood and Osman [60] observed 33% stimulation of the enzymatic activity by 0.5 mM ethanol in *Dictyostelium discoideum* nuclei. Another study showed that poly(ADP-ribose) synthesis was almost unaffected by ethanol up to 5% in undamaged cells, but stimulated 3.5-fold by 1% ethanol in lightly damaged cells [97]. Higher concentrations of ethanol than 1% also stimulated polymer synthesis, but less effectively. The authors postulated that ethanol directly stimulates poly(ADP-ribose) synthetase by increasing its affinity for damaged DNA [97]. Juarez-Salinas *et al.* [98] observed that the intracellular level of poly(ADP-ribose) in cultured mouse cells could be elevated by hyperthermic treatment and this increase was dramatically potentiated by the presence of 1% ethanol.

Metal chelators such as 1,10-phenanthroline or 2,2′-dipyridyl were found to stimulate purified enzyme in the presence of Mg^{2+} but inhibit the enzyme in its absence [41]. These effects may be related to the fact that poly(ADP-ribose) synthetase is a zinc-metalloprotein [76]. The enzyme molecule contains two zinc ions localized in two zinc fingers, FI and FII [99]. It was found that disruption of the metal binding ability of FII dramatically reduces the binding to target DNA, while mutation of zinc ligands in FI only slightly affects the DNA binding activity [100]. In another study, destabilization of Zn^{2+} coordination in poly(ADP-ribose) synthetase by 6-nitro-1,2-benzopyrone was shown to cause a loss of 99% of enzymatic activity but not the DNA-binding function [62]. Evidently, the stimulation by 1,10-phenanthroline in the presence of Mg^{2+} is not due to destabilization of Zn^{2+} in the enzyme molecule; the mechanism of activation has yet to be established.

In the presence of 25 nM NAD^+, potent activation of the initiation reaction of poly(ADP-ribose) synthesis was reported for nicotinic acid-NAD, acetylpyridine-NAD, aminopyridine-NAD, benzamide, and some other compounds at nanomolar concentrations [44] (Table 1). Potent activation was also observed at 50 nM 3-acetamidobenzamide, some other benzamides and nicotinamides in permeabilized L1210 cells or intact cells [101]. The same compounds at 50 μM concentration were inhibitory. Noteworthy in this regard is the observation that 3-aminobenzamide reduces strand-break frequencies and increases repair replication of DNA in human lymphoid cells damaged by methyl methanesulfonate [102]. This finding is consistent with an earlier report that 3-aminobenzamide stimulates DNA repair in human lymphocytes following treatment with ultraviolet light or dimethyl sulfate [103].

Recently, we found that some carboline compounds affect the activity of poly(ADP-ribose) synthetase in a complex way [64]. Trp-P-1 is a fairly strong inhibitor, whereas Trp-P-2 is a strong activator at, and only at, around 1 mM, but it is inhibitory at higher concentrations [see Fig. 1 in Ref. 64]. These two γ-carbolines are pyrolytic products of tryptophan and known to be very mutagenic and carcinogenic [104]. The only structural difference between Trp-P-1 and Trp-P-2 is the absence or presence of a methyl group at the 4-position.

Activators and inhibitors of mono(ADP-ribosyl)transferases

Four arginine-specific mono(ADP-ribosyl)transferases were purified from turkey erythrocytes; A [69] and B [105] from cytosol, C from membrane fraction [106], and D (or A′) from nucleus [106]. Transferase A is strongly stimulated by inorganic salts [107], histones [108], certain phospholipids [109], and detergents [109]. The order of effectiveness for anions is $SCN^- > Br^- > Cl^- > F^- > PO_4^{3-}$ [102]. The effect (>10-fold stimulation) of salt on transferase A activity depends on the ADP-ribose acceptor [107]. Increasing concentrations of NaCl increase the rate of transfer of ADP-ribose to water, low-molecular-weight guanidino compounds, and histones, but not to ovalbumin [107, 110]. In the absence of salt, the enzymatic activity is stimulated by histone in micromolar concentrations; other proteins are ineffective. ADP-ribosylation of histones is reduced by the addition of agmatine due to competitive formation of ADP-ribosyl-agmatine [110]. Arginine methyl ester, but not lysine, inhibits the ADP-ribosylation of other proteins, probably, by a similar mechanism [111]. Lysolecithins containing long-chain fatty acids such as palmitoyl (C16:0) and stearoyl (C18:0) are more effective (4–6-fold stimulation) than those with shorter chains; *viz*., C16:0 > C18:0 > C14.0 > C12:0 [109]. On the other hand, we found that unsaturated fatty acids inhibit transferase A activity more strongly than saturated ones; *vis*., C20:4 > C18:1 > γ C18:3 > C18:2 = C16:1 > C14:0 > C18:3 > C16:0 > C18:0 > C20:0 > C12:0 (M. Banasik and K. Ueda, unpublished data). It was postulated that the activators mentioned above convert the transferase from an inactive oligomeric form to an active monomeric form. Nucleoside triphosphates stimulate the ADP-ribosylation of several proteins up to about twofold. With lysozyme as ADP-ribose acceptor, the order of effectiveness at 10 mM was ATP > ITP = GTP > CTP = UTP [112]. Nicotinamide, theophylline, and thymidine, at 1 mM, inhibit the activity of transferase A by 52%, 52%, and 43%, respectively [69]. Nicotinamide was shown to be a competitive inhibitor with K_i value of 14 mM [113]. Rankin *et al.* [45] determined IC_{50} values of nine compounds for transferase A, showing that the inhibitors of poly(ADP-ribose) synthetase inhibit also the transferase activity but to much lesser extent. They examined inhibition curves of benzamide, 3-aminobenzamide, and 3-methoxybenzamide for several other endogenous transferases and found that the inhibition patterns are similar to those for transferase A. Table 2 presents IC_{50} values of various compounds for transferase A compared with those for poly(ADP-ribose) synthetase. It is evident that novobiocin and coumermycin A_1 are rather specific for the transferase A.

Transferase B is inhibited by salt and unaffected by histone over a concentration range that fully activates transferase A [105]. Transferase C is moderately (up to about twofold) stimulated by salt or histone [106], whereas transferase D is stimulated more than tenfold by histone or salt [106].

Arginine-specific mono(ADP-ribosyl)transferase purified from chicken erythrocytes is stimulated by low (1–10 mM) but inhibited by higher (100–500 mM) concentrations of salts [114]. 1,4-Naphthoquinone, 5,8-dihydroxy-1,4-naphthoquinone, 4-amino-1-naphthol hydrochloride, and 1,2-naphthoquinone are competitive inhibitors with K_i values of 4.6 μM, 29.0 μM, 32.0 μM, and 49.0 μM, respectively [114]. Formylluminol, 3-formylaminobenzamide, and 3-ureidobenzamide were found to be weaker inhibitors, whereas

3-methoxybenzamide and 3-aminobenzamide were not inhibitory [114].

3-Aminobenzamide was shown to be a weak competitive inhibitor with respect to NAD^+ (K_i = 14 mM) and a noncompetitive inhibitor with respect to L-arginine methyl ester for arginine-specific mono(ADP-ribosyl)transferase from rabbit skeletal muscle [115].

References

1. Fujimura S, Hasegawa S, Shimizu Y, Sugimura T: Polymerization of the adenosine 5′-diphosphate-ribose moiety of nicotinamide-adenine dinucleotide by nuclear enzyme. I. Enzymatic reactions. Biochim Biophys Acta 145: 247–259, 1967
2. Römer V, Lambrecht J, Kittler M, Hilz H: Identity of nuclear NAD nucleosidase with a polyADP-ribose forming enzyme in Ehrlich ascites tumor cells. Hoppe-Seyler's Z Physiol Chem 349: 109–112, 1968
3. Sugimura T, Fujimura S, Hasegawa S, Shimizu Y, Okuyama H: Polymerization of ADPR moiety of NAD by nuclear enzyme preparation. J Vitaminol 14: 135–142, 1968
4. Nishizuka Y, Ueda K, Nakazawa K, Reeder RH, Honjo T, Hayaishi O: Poly adenosine diphosphate ribose synthesis and nicotinamide adenine dinucleotide transglycosidases. J Vitaminol 14: 143–152, 1968
5. Clark JB, Ferris GM, Pinder S: Inhibition of nuclear NAD nucleosidase and poly ADP-ribose polymerase activity from rat liver by nicotinamide and 5′-methyl nicotinamide. Biochim Biophys Acta 238: 82–85, 1971
6. Preiss J, Schlaeger R, Hilz H: Specific inhibition of poly ADPribose polymerase by thymidine and nicotinamide in HeLa cells. FEBS Lett 19: 244–246, 1971
7. Shall S, Brightwell M, O'Farrell MK, Stone P, Whish WJD: Properties of poly(ADP-ribose) polymerase in *Physarum polycephalum* and mouse fibroblasts. Hoppe-Seyler's Z Physiol Chem 353: 846–847, 1972
8. Stone PR, Shall S: Poly(adenosine diphosphoribose) polymerase in mammalian nuclei. Characterization of the activity in mouse fibroblasts (LS cells). Eur J Biochem 38: 146–152, 1973
9. Ueda K, Fukushima M, Okayama H, Hayaishi O: Nicotinamide adenine dinucleotide glycohydrolase from rat liver nuclei. Isolation and characterization of a new enzyme. J Biol Chem 250: 7541–7546, 1975
10. Claycomb WC: Poly(adenosine diphosphate ribose) polymerase activity and nicotinamide adenine dinucleotide in differentiating cardiac muscle. Biochem J 154: 387–393, 1976
11. Müller WEG, Zahn RK: Poly ADP-ribosylation of DNA-dependent RNA polymerase I from quail oviduct. Dependence on progesterone stimulation. Mol Cell Biochem 12: 147–159, 1976
12. Berger NA, Weber G, Kaichi AS: Characterization and comparison of poly(adenosine diphosphoribose) synthesis and DNA synthesis in nucleotide-permeable cells. Biochim Biophys Acta 519: 87–104, 1978
13. Levi V, Jacobson EL, Jacobson MK: Inhibition of poly(ADP-ribose) polymerase by methylated xanthines and cytokinins. FEBS Lett 88: 144–146, 1978
14. Ito S, Shizuta Y, Hayaishi O: Purification and characterization of poly(ADP-ribose) synthetase from calf thymus. J Biol Chem 254: 3647–3651, 1979
15. Niedergang C, Okazaki H, Mandel P: Properties of purified calf thymus poly(adenosine diphosphate ribose) polymerase. Comparison of the DNA-independent and the DNA-dependant enzyme. Eur J Biochem 102: 43–57, 1979
16. Terada M, Fujiki H, Marks PA, Sugimura T: Induction of erythroid differentiation of murine erythroleukemia cells by nicotinamide and related compounds. Proc Natl Acad Sci USA 76: 6411–6414, 1979
17. Ohgushi H, Yoshihara K, Kamiya Y: Bovine thymus poly(adenosine diphosphate ribose) polymerase. Physical properties and binding to DNA. J Biol Chem 255: 6205–6211, 1980
18. Yamamoto H, Okamoto H: Protection by picolinamide, a novel inhibitor of poly(ADP-ribose) synthetase, against both streptozotocin-induced depression of proinsulin synthesis and reduction of NAD content in pancreatic islets. Biochem Biophys Res Commun 95: 474–481, 1980
19. Hayaishi O, Ueda K: On the roles of DNA and DNA fragments in the enzymic synthesis and degradation of poly(ADP-ribose). In: M. Harris (ed). Poly(ADP-Ribose). An International Symposium. Fogarty International Center Proceedings No. 26, Bethesda, Maryland, 1974, pp 69–76
20. Shimoyama M, Kawai M, Nasu S, Shioji K, Hoshi Y, Ueda I: Inhibition of adenosine 3′,5′-monophosphate phosphodiesterase by nicotinamide and its homologues *in vitro*. Physiol Chem & Physics 7: 125–132, 1975
21. Berger NA, Sikorski GW: Nicotinamide stimulates repair of DNA damage in human lymphocytes. Biochem Biophys Res Commun 95: 67–72, 1980
22. Grunfeld C, Shigenaga JK: Nicotinamide and other inhibitors of ADP-ribosylation block deoxyglucose uptake in cultured cells. Biochem Biophys Res Commun 123: 785–791, 1984
23. Morris NR, Reichard P, Fischer GA: Studies concerning the inhibition of cellular reproduction by deoxyribonucleotides. II. Inhibition of the synthesis of deoxycytidine by thymidine, deoxyadenosine and deoxyguanosine. Biochim Biophys Acta 68: 93–99, 1963
24. Johnson GS, D'Armiento M, Carchman RA: N^6-Substituted adenines induce cell elongation irrespective of the intracellular cyclic AMP levels. Exp Cell Res 85: 47–56, 1974
25. Cleaver JE: Repair replication of mammalian cell DNA: effects of compounds that inhibit DNA synthesis or dark repair. Radiation Res 37: 334–348, 1969
26. Cleaver JE, Thomas GH: Single strand interruptions in DNA and the effects of caffeine in Chinese hamster cells irradiated with ultraviolet light. Biochem Biophys Res Commun 36: 203–208, 1969
27. Lehmann AR, Kirk-Bell S: Post-replication repair of DNA in ultraviolet-irradiated mammalian cells. No gaps in DNA synthesized late after ultraviolet irradiation. Eur J Biochem 31: 438–445, 1972
28. Purnell MR, Whish WJD: Novel inhibitors of poly(ADP-ribose) synthetase. Biochem J 185: 775–777, 1980
29. Durkacz BW, Omidiji O, Gray DA, Shall S: $(ADP\text{-ribose})_n$ participates in DNA excision repair. Nature (London) 283: 593–596, 1980
30. Oikawa A, Tohda H, Kanai M, Miwa M, Sugimura T: Inhibitors of poly-(adenosine diphosphate ribose) polymerase induce sister chromatid exchanges. Biochem Biophys Res Commun 97: 1311–1316, 1980
31. Cleaver JE, Bodell WJ, Morgan WF, Zelle B: Differences in regulation by poly(ADP-ribose) of repair of DNA damage from alkylating agents and ultraviolet light according to cell type. J Biol Chem 258: 9059–9068, 1983
32. Cleaver JE: Differential toxicity of 3-aminobenzamide to wild-type and 6-thioguanine-resistant Chinese hamster cells by interference with pathways of purine biosynthesis. Mutat Res 131: 123–127, 1984
33. Milam KM, Cleaver JE: Inhibitors of poly(adenosine diphosphate-ribose) synthesis: effect on other metabolic processes. Science 223: 589–591, 1984
34. Borek C, Morgan WF, Ong A, Cleaver JE: Inhibition of malignant transformation *in vitro* by inhibitors of poly(ADP-ribose) synthesis. Proc Natl Acad Sci USA 81: 243–247, 1984
35. Hunting DJ, Gowans BJ, Henderson JF: Specificity of inhibitors of poly-(ADP-ribose) synthesis. Effects on nucleotide metabolism in cultured cells. Mol Pharmacol 28: 200–206, 1985
36. Snyder RD: 3-Aminobenzamide does not alter DNA repair in human fibroblasts through modulation of deoxynucleoside triphosphate pools. Biochem Biophys Res Commun 124: 457–461, 1984
37. Ben-Hur E, Chen C-C, Elkind MM: Inhibitors of poly(adenosine diphosphoribose) synthetase, examination of metabolic perturbations, and enhancement of radiation response in Chinese hamster cells. Cancer Res 45: 2123–2127, 1985

38. Johnson GS: Benzamide and its derivatives inhibit nicotinamide methylation as well ADP-ribosylation. Biochem Int 2: 611–617, 1981
39. Bender ML, Gibian MJ, Whelan DJ: The alkaline pH dependence of chymotrypsin reactions: postulation of a pH-dependent intramolecular competitive inhibition. Proc Natl Acad Sci USA 56: 833–839, 1966
40. Davies RC, Auld DS, Vallee BL: The effect of modifiers on the hydrolysis of esters and peptides by carboxypeptidase A. Biochem Biophys Res Commun 31: 628–633, 1968
41. Banasik M, Komura H, Shimoyama M, Ueda K: Specific inhibitors of poly(ADP-ribose) synthetase and mono(ADP-ribosyl)transferase. J Biol Chem 267: 1569–1575, 1992
42. Purnell MR, Kidwell WR, Minshall L, Whish WJD: Specificity of poly-(ADP-ribose) synthetase inhibitors. In: F.R. Althaus, H. Hilz, and S. Shall (eds). ADP-Ribosylation of Proteins. Springer-Verlag, Berlin, 1985, pp 98–105
43. Tavassoli M, Shall S: Covalent modification of poly(ADP-ribose) polymerase by reactive benzamides. In: G.G. Poirier and P. Moreau (eds). ADP-Ribosylation Reactions. Springer-Verlag, New York, 1992, pp 290–296
44. Bauer PI, Hakam A, Kun E: Mechanisms of poly(ADP-ribose) polymerase catalysis; mono-ADP-ribosylation of poly(ADP-ribose) polymerase at nanomolar concentrations of NAD. FEBS Lett 195: 331–338, 1986
45. Rankin PW, Jacobson EL, Benjamin RC, Moss J, Jacobson MK: Quantitative studies of inhibitors of ADP-ribosylation *in vitro* and *in vivo*. J Biol Chem 264: 4312–4317, 1989
46. Farzaneh F, Shall S, Michels P, Borst P: ADP-ribosyltransferase activity in Trypanosoma brucei. In: F.R. Althaus, H. Hilz, and S. Shall (eds). ADP-Ribosylation of Proteins. Springer-Verlag, Berlin, 1985, pp 367–371
47. McLick J, Bauer PI, Hakam A, Kun E: Covalent binding of 4-carbamoylbenzenediazonium chloride to deoxyguanine bases of DNA resulting in apparent irreversible inhibition of poly(adenosine diphosphoribose) polymerase at the nicotinamide binding site. Biochemistry 26: 2226–2231, 1987
48. Hakam A, McLick J, Buki K, Kun E: Catalytic activities of synthetic octadeoxyribonucleotides as coenzymes of poly(ADP-ribose) polymerase and the identification of a new enzyme inhibitory site. FEBS Lett 212: 73–78, 1987
49. Bauer PI, Kline K, Kun E: Apparent role of adenosine diphosphoribosyl transferase in the development of Mytilus edulis and the inhibition of differentiation by ligands of the enzyme protein. Proc Soc Exp Biol Med 196: 396–400, 1991
50. Leone E, Suzuki H, Farina B, Pivazian AD, Karpeisky MYA: Inhibition of ADP-ribosylation reaction by 2′,5′-oligoadenylates. In: F.R. Althaus, H. Hiltz, and S. Shall (eds). ADP-Ribosylation of Proteins. Springer-Verlag, Berlin, 1985, pp 106–110
51. Banasik M, Komura H, Ueda K: Specific inhibitors of poly(ADP-ribose) synthetase. In: G.G. Poirier and P. Moreau (eds). ADP-Ribosylation Reactions. Springer-Verlag, New York, 1992, pp 343–350
52. Banasik M, Komura H, Ueda K: Inhibition of poly(ADP-ribose) synthetase by fatty acids, vitamins and vitamin-like substances. FEBS Lett 263: 222–224, 1990
53. Shall S: ADP-ribosylation, DNA repair, cell differentiation and cancer. In: M. Miwa, O. Hayaishi, S. Shall, M. Smulson, and T. Sugimura (eds). ADP-Ribosylation, DNA Repair and Cancer. Japan Sci Soc Press, Tokyo/VNU Science Press BV, Utrecht, The Netherlands, 1983, pp 3–25
54. Tseng A Jr, Lee WMF, Kirsten E, Hakam A, McLick J, Buki K, Kun E: Prevention of tumorigenesis of oncogene-transformed rat fibroblasts with DNA site inhibitors of poly(ADP-ribose) polymerase. Proc Natl Acad Sci USA 84: 1107–1111, 1987
55. Tanaka Y, Matsunami N, Yoshihara K: Inhibition of ADP-ribosylation of histone by diadenosine 5′, 5′′′-p^1, p^4-tetraphosphate. Biochem Biophys Res Commun 99: 837–843, 1981
56. Suzuki H, Tanaka Y, Buonamassa DT, Farina B, Leone E: Inhibition of ADP-ribosylation of histone H1 by analogs of diadenosine 5′, 5′′′-p^1, p^4-tetraphosphate. Mol Cell Biochem 74: 17–20, 1987
57. Müller WEG, Zahn RK: Influence of agents that act on DNA and RNA synthesis on the activity of poly(ADP-rib) polymerase. Experientia 31: 1014–1015, 1975
58. Müller WEG, Rohde HJ, Steffen R, Maidhof A, Lachmann M, Zahn RK, Umezawa H: Influence of formycin B on polyadenosine diphosphoribose synthesis *in vitro* and *in vivo*. Cancer Res 35: 3673–3681, 1975
59. Brightwell MD, Leech CE, O'Farrell MK, Whish WJD, Shall S: Poly-(adenosine diphosphate ribose) polymerase in *Physarum polycephalum*. Biochem J 147: 119–129, 1975
60. Rickwood D, Osman MS: Characterisation of poly(ADP-rib) polymerase activity in nuclei from the slime mould Dictyostelium discoideum. Mol Cell Biochem 27: 79–84, 1979
61. Benjamin RC, Gill DM: Poly(ADP-ribose) synthesis *in vitro* programmed by damaged DNA. A comparison of DNA molecules containing different types of strand breaks. J Biol Chem 255: 10502–10508, 1980
62. Buki KG, Bauer PI, Mendeleyev J, Hakam A, Kun E: Destabilization of Zn^{2+} coordination in ADP-ribose transferase (polymerizing) by 6-nitroso-1,2-benzopyrone coincidental with inactivation of the polymerase but not the DNA binding function. FEBS Lett 290: 181–185, 1991
63. Sestili P, Spadoni G, Balsamini C, Scovassi I, Cattabeni F, Duranti E, Cantoni O, Higgins D, Thomson C: Structural requirements for inhibitors of poly(ADP-ribose) polymerase. J Cancer Res Clin Oncol 116: 615–622, 1990
64. Ueda K, Banasik M: Inhibition of poly(ADP-ribose) synthetase activity by tryptophan metabolites. In: I. Ishiguro, R. Kido, T. Nagatsu, Y. Nagamura, and Y. Ohta (eds). Advances in Tryptophan Research 1992. Fujita Health University Press, Japan, 1992, pp 141–144
65. Larsen AG, Østvold AC, Holtlund J, Kristensen T, Laland SG: The inhibitory effect of Zn^{2+} on poly(ADP-ribose) polymerase activity and its reversal. Biochem J 203: 511–513, 1982
66. Sims JL, Sikorski GW, Catino DM, Berger SJ, Berger NA: Poly(adenosinediphosphoribose) polymerase inhibitors stimulate unscheduled deoxyribonucleic acid synthesis in normal human lymphocytes. Biochemistry 21: 1813–1821, 1982
67. Banasik M, Komura H, Saito I, Abed NAN, Ueda K: New inhibitors of poly(ADP-ribose) synthetase. In: M.K. Jacobson and E.L. Jacobson (eds). ADP-Ribose Transfer Reactions: Mechanisms and Biological Significance, Springer-Verlag, New York, 1989, pp 130–133
68. Tanigawa Y, Tsuchiya M, Imai Y, Shimoyama M: ADP-ribosyltransferase from hen liver nuclei. Purification and characterization. J Biol Chem 259: 2022–2029, 1984
69. Moss J, Stanley SJ, Watkins PA: Isolation and properties of an NAD- and guanidine-dependent ADP-ribosyltransferase from turkey erythrocytes. J Biol Chem 255: 5838–5840, 1980
70. Nakayasu M, Shima H, Aonuma S, Nakagama H, Nagao M, Sugimura T: Deletion of transfected oncogenes from NIH 3T3 transformants by inhibitors of poly(ADP-ribose) polymerase. Proc Natl Acad Sci USA 85: 9066–9070, 1988
71. Shima H, Nakayasu M, Aonuma A, Sugimura T, Nagao M: Loss of the *MYC* gene amplified in human HL-60 cells after treatment with inhibitors of poly(ADP-ribose) polymerase or with dimethyl sulfoxide. Proc Natl Acad Sci USA 86: 7442–7445, 1989
72. Ikushima T: Bimodal induction of sister-chromatid exchanges by luminol, an inhibitor of poly(ADP-ribose) synthetase, during the S-phase of the cell cycle. Chromosoma 99: 360–364, 1990
73. Rice WG, Hillyer CD, Harten B, Schaeffer CA, Dorminy M, Lackey III DA, Kirsten E, Mendeleyev J, Buki KG, Hakam A, Kun E: Induction of endonuclease-mediated apoptosis in tumor cells by C-nitroso-substituted ligands of poly(ADP-ribose) polymerase. Proc Natl Acad Sci USA 89: 7703–7707, 1992
74. Rice WG, Schaeffer CA, Harten B, Villinger F, South TL, Summers MF, Henderson LE, Bess JW Jr, Arthur LO, McDougal JS, Orloff SL,

Mendeleyev J, Kun E: Inhibition of HIV-1 infectivity by zinc-ejecting aromatic *C*-nitroso compounds. Nature (London) 361: 473–475, 1993
75. Benjamin RC, Cook PF, Jacobson MK: Kinetic mechanism of poly(ADP-ribose) polymerase. In: F.R. Althaus, H. Hilz, and S. Shall (eds). ADP-Ribosylation of Proteins. Springer-Verlag, Berlin, 1985, pp 93–97
76. Zahradka P, Ebisuzaki K: Poly(ADP-ribose) polymerase is a zinc metalloenzyme. Eur J Biochem 142: 503–509, 1984
77. Schneeweiss FHA, Xia F, Sharan RN, Feinendegen LE: A strong static magnetic field inhibits the poly-ADP-ribosylation of proteins in human kidney T1-cells. Bioelectrochem Bioenergetics 30: 111–117, 1993
78. Berger NA, Petzold SJ: Identification of minimal size requirements of DNA for activation of poly(ADP-ribose) polymerase. Biochemistry 24: 4352–4355, 1985
79. Yoshihara K, Koide SS: Influence of polynucleotides on poly(adenosine diphosphate ribose) synthetase activity of rat liver. FEBS Lett 30: 261–264, 1973
80. Tanaka Y, Hashida T, Yoshihara H, Yoshihara K: Bovine thymus poly-(ADP-ribose) polymerase histone-dependent and Mg^{2+}-dependent reaction. J Biol Chem 254: 12433–12438, 1979
81. Yoshihara K, Tanaka Y, Yoshihara H, Hashida T, Ohgushi H, Arai R, Kamiya T: Effect of histone, DNA, and Mg^{2+} on purified poly(ADP-ribose) polymerase reaction: the histone-dependent and the Mg^{2+}-dependent reaction. In: M.E. Smulson and T. Sugimura (eds). Novel ADP-Ribosylations of Regulatory Enzymes and Proteins. Elsevier, North-Holland, 1980, pp 33–44
82. Tanaka Y, Yoshihara K, Ohashi Y, Itaya A, Nakano T, Ito K, Kamiya T: A method for determining oligo- and poly(ADP-ribosyl)ated enzymes and proteins *in vitro*. Anal Biochem 145: 137–143, 1985
83. Caplan AI, Niedergang C, Okazaki H, Mandel P: Poly ADP-ribose polymerase: self-ADP-ribosylation, the stimulation by DNA, and the effects on nucleosome formation and stability. Arch Biochem Biophys 198: 60–69, 1979
84. Kun E, Minaga T, Kirsten E, Jackowski G, McLick J, Peller L, Oredsson SM, Martin L, Pattabiraman N, Milo G: Biochemical basis of the regulatory role of polyadenosine diphosphoribose. Adv Enzyme Regul 21: 177–199, 1983
85. Ueda K, Banasik M: Activators of poly(ADP-ribose) synthetase. Oji International Seminar on ADP-ribosylation Reaction: Mechanism and Biological Function. 1992, Abstracts, p 19
86. Althaus FR: Poly(ADP-ribose) biosynthesis. In: F.R. Althaus and C. Richter (eds). ADP-Ribosylation of Proteins: Enzymology and Biological Signficance. Springer-Verlag, Berlin, 1987, pp 12–37
87. Yoshihara K, Hashida T, Tanaka Y, Matsunami N, Yamaguchi A, Kamiya T: Mode of enzyme-bound poly(ADP-ribose) synthesis and histone modification by reconstituted poly(ADP-ribose) polymerase-DNA-cellulose complex. J Biol Chem 256: 3471–3478, 1981
88. Zahradka P, Ebisuzaki K: A shuttle mechanism for DNA-protein interactions. The regulation of poly(ADP-ribose) polymerase. Eur J Biochem 127: 579–585, 1982
89. Byrne RH, Stone PR, Kidwell WR: Effect of polyamines and divalent cations on histone H1-poly(adenosine diphosphate ribose) complex formation. Exp Cell Res 115: 277–283, 1978
90. Kawamura M, Tanigawa Y, Kitamura A, Miyake Y, Shimoyama M: Effect of polyamines on purified poly(ADP-ribose) synthetase from rat liver nuclei. Biochim Biophys Acta 652: 121–128, 1981
91. Tanigawa Y, Kawamura M, Shimoyama M: Effect of polyamines on ADP-ribosylation of nuclear proteins from rat liver nuclei. Biochem Biophys Res Commun 76: 406–412, 1977
92. Tanigawa Y, Kawakami K, Imai Y, Shimoyama M: Effect of polyamines on ADP-ribosylation by chick-embryo-liver nuclei. Biochim Biophys Acta 608: 82–95, 1980
93. Perrella FW, Lea MA: Spermine-induced variations in the adenosine 5′-diphosphate ribosylation patterns of nuclear proteins from rat liver and hepatoma. Cancer Res 39: 1382–1389, 1979
94. Whitby AJ, Stone PR, Whish WJD: Effect of polyamines and Mg^{++} on poly(ADP-ribose) synthesis and ADP-ribosylation of histones in wheat. Biochem Biophys Res Commun 90: 1295–1304, 1979
95. Perrella FW, Lea MA: Polyamine induced changes in the ADP-ribosylation of nuclear proteins from rat liver. Biochem Biophys Res Commun 82: 575–581, 1978
96. Kristensen T, Holtlund J: Poly(ADP-ribose) polymerase from Ehrlich ascites tumor cells. Properties of the purified polymerase. Eur J Biochem 88: 495–501, 1978
97. Sims JL, Benjamin RC: Mechanism of ethanol stimulation of poly(ADP-ribose) synthetase. In: F.R. Althaus, H. Hilz, and S. Shall (eds). ADP-Ribosylation of Proteins. Springer-Verlag, Berlin, 1985, pp 124–128
98. Juarez-Salinas H, Duran-Torres G, Jacobson MK: Alteration of poly(ADP-ribose) metabolism by hyperthermia. Biochem Biophys Res Commun 122: 1381–1388, 1984
99. Mazen A, de Murcia JM, Molinete M, Simonin F, Gradwohl G, Poirier G, de Murcia G: Poly(ADP-ribose) polymerase: a novel finger protein. Nucleic Acids Res 17: 4689–4698, 1989
100. Gradwohl G, de Murcia JM, Molinete M, Simonin F, Koken M, Hoeijmakers JHJ, de Murcia G: The second zinc-finger domain of poly-(ADP-ribose) polymerase determines specificity for single-stranded breaks in DNA. Proc Natl Acad Sci USA 87: 2990–2994, 1990
101. Jones J, Patel BN, Skidmore CJ: Benzamides can stimulate as well as inhibit the activity of nuclear ADP-ribosyltransferase. Carcinogenesis 9: 2023–2026, 1988
102. Cleaver JE, Morgan WF: 3-Aminobenzamide, an inhibitor of poly(ADP-ribose) polymerase, is a stimulator, not an inhibitor, of DNA repair. Exp Cell Res 172: 258–264, 1987
103. Bohr V, Klenow H: 3-Aminobenzamide stimulates unscheduled DNA synthesis and rejoining of strand breaks in human lymphocytes. Biochem Biophys Res Commun 102: 1254–1261, 1981
104. Wakabayashi K, Nagao M, Esumi H, Sugimura T: Food-derived mutagens and carcinogens. Cancer Res (Suppl) 52: 2092s–2098s, 1992
105. Yost DA, Moss J: Amino acid-specific ADP-ribosylation. Evidence for two distinct NAD:arginine ADP-ribosyltransferases in turkey erythrocytes. J Biol Chem 258: 4926–4929, 1983
106. West RE, Moss J: Amino acid specific ADP-ribosylation: specific NAD:arginine mono-ADP-ribosyltransferases associated with turkey erythrocyte nuclei and plasma membranes. Biochemistry 25: 8057–8062, 1986
107. Moss J, Stanley SJ, Osborne JC Jr: Effect of self-association on activity of an ADP-ribosyltransferase from turkey erythrocytes. Conversion of inactive oligomers to active protomers by chaotropic salts. J Biol Chem 256: 11452–11456, 1981
108. Moss J, Stanley SJ, Osborne JC Jr: Activation of an NAD:arginine ADP-ribosyltransferase by histone. J Biol Chem 257: 1660–1663, 1982
109. Moss J, Osborne JC Jr, Stanley SJ: Activation of an erythrocyte NAD:arginine ADP-ribosyltransferase by lysolecithin and nonionic and zwitterionic detergents. Biochemistry 23: 1353–1357, 1984
110. Moss J, Stanely SJ: Histone-dependent and histone-independent forms of an ADP-ribosyltransferase from human and turkey erythrocytes. Proc Natl Acad Sci USA 78: 4809–4812, 1981
111. Moss J, Stanley SJ: Isolation of an avian erythrocyte protein possessing ADP-ribosyltransferase activity and capable of activating adenylate cyclase. Proc Natl Acad Sci USA 75: 3621–3624, 1978
112. Watkins PA, Moss J: Effects of nucleotides on activity of a purified ADP-ribosyltransferase from turkey erythrocytes. Arch Biochem Biophys 216: 74–80, 1982
113. Osborne JC Jr, Stanley SJ, Moss J: Kinetic mechanisms of two NAD:arginine ADP-ribosyltransferases: the soluble, salt-stimulated transferase from turkey erythrocytes and choleragen, a toxin from *Vibrio cholerae*. Biochemistry 24: 5235–5240, 1985

114. Sabir J, Tavassoli M, Shall S: Purification and characterization of NAD:arginine mono ADP-ribosyl transferase from chicken erythrocytes; identification of some enzyme inhibitors. In: G.G. Poirier and P. Moreau (eds). ADP-Ribosylation Reactions. Springer-Verlag, New York, 1992, pp 397–401
115. Larew JS-A, Peterson JE, Graves DJ: Determination of kinetic mechanism of arginine-specific ADP-ribosyltransferases using a high performance liquid chromatographic assay. J Biol Chem 266: 52–57, 1991

PART VI

DERIVITIZATION OF PROTEINS WITH ADP-RIBOSE, NAD AND THEIR ANALOGUES

Molecular and Cellular Biochemistry **138**: 201–206, 1994.

Nitric oxide and NAD-dependent protein modification

Lee J. McDonald and Joel Moss

Laboratory of Cellular Metabolism, National Heart, Lung, and Blood Institute, National Institutes of Health, Building 10/ 5N-307, Bethesda, MD 20892, USA

Abstract

Nitric oxide (NO) has been suggested to act as a regulator of endogenous intracellular ADP-ribosylation, based on radiolabelling of proteins in tissue homogenates incubated with [^{32}P]NAD and NO. After the NO-stimulated modification was replicated in a defined system containing only the purified acceptor protein, glyceraldehyde-3-phosphate dehydrogenase (GAPDH), the hypothesis of NO-stimulation of an endogenous ADP-ribosyltransferase became moot. The NO-stimulated, NAD-dependent modification of GAPDH was recently characterized as covalent binding of the whole NAD molecule to the enzyme, not ADP-ribosylation. With this result, along with the knowledge that GAPDH is stoichiometrically S-nitrosylated, the role of NO in protein modification with NAD may be viewed as the conferring of an unexpected chemical reactivity upon GAPDH, possibly due to nitrosylation of a cysteine in the enzyme active site. (Mol Cell Biochem **138**: 201–206, 1994)

Key words: nitric oxide, NAD, ADP-ribosylation, cysteine

Introduction

Nitric oxide (NO) is a novel biological messenger, unique in the sense that it is a gas and a free radical. Nitric oxide is produced in animal cells by the enzyme NO synthase, in response to a variety of signals [1]. As research has intensified in this area, an increasing number of biological functions have been ascribed to NO. Contraction of blood vessel smooth muscle is regulated by endothelium-derived relaxation factor, which is equivalent to NO itself, or more likely, NO in the form of a nitrosothiol [1, 2]. In neural tissue, NO plays a role in several processes, including long term potentiation and NMDA-mediated neurotoxicity [1, 3]. Other functions for NO include platelet inhibition, and as a component of the cytotoxic response of macrophages [1, 2].

The intracellular targets of nitric oxide action include heme proteins, proteins with iron-sulfur centers, and protein or free thiols [1, 4]. Intracellular soluble guanylate cyclase is activated by coordination of NO with the enzyme heme [1]. Other heme proteins also bind NO, including hemoglobin, cytochrome P450, and both the constitutive and inducible isozymes of nitric oxide synthase, which are inhibited by NO [1, 4–7]. Thiols react readily with NO to form nitrosothiols [4, 8]. Proteins that are S-nitrosylated include serum albumin, tissue-type plasminogen activator (t-PA), glyceraldehyde-3-phosphate and other dehydrogenases, cathepsin B, and the NMDA receptor-channel complex [1, 9–11]. Nitrosothiols are more stable than NO in solution; the half-life is in the order of seconds for free NO and up to minutes or hours for nitrosothiols [4, 8, 12]. S-Nitrosylation is reversible [4, 8], so that nitrosothiols, for example the nitrosylated forms of serum albumin, t-PA, and N-acetylcysteine, can dissociate to release free NO with all its biological functions [13–15].

ADP-ribosylation is a potentially regulatory protein modification catalyzed by a family of amino acid-specific ADP-ribosyltransferases [16, 17]. During the enzyme-catalyzed transfer of ADP-ribose to the protein acceptor, the nicotinamide-ribose bond of NAD is hydrolyzed and free nicotinamide is released [16, 17]. The study of endogenous ADP-ribosylation processes in complex systems such as cell or tissue homogenates is often monitored by the transfer of

Address for offprints: L.J. McDonald, Laboratory of Cellular Metabolism; National Heart, Lung, and Blood Institute; National Institutes of Health; Building 10/ 5N-307; Bethesda, MD 20892, USA

[adenylate-^{32}P]ADP-ribose from [adenylate-^{32}P]NAD to acceptor proteins in the sample. A new, possibly regulatory, role for nitric oxide has been suggested by recent reports: involvement in ADP-ribosyltion in eukaryotic cells. This review summarizes research in this area, and describes current understanding of the role of NO in NAD-dependent protein modification.

NO-stimulated, NAD-dependent protein modification in cells and cell-free systems

NO-stimulated modification of protein by NAD was initially found by Brüne and Lapetina in human platelet cytosol, in which radiolabeling of a 39-kDa protein was stimulated by the addition of a NO-donating agent [18, 31, 32]. NAD-dependent, NO-stimulated modification of proteins from 36- to 41-kDa was subsequently described in broken cells or subcellular fractions from HL-60 cells [19], rat brain [20–22], L929 fibroblasts [23], human erythrocytes [24], *Dictyostelium discoideum* [25], human neutrophils [26], RINm5F cells [27], and bovine retinal rod outer segments [28]. Modification of these proteins with [^{32}P]NAD was stimulated by NO derived from sodium nitroprusside (SNP) [18, 20–24, 27, 28], 3-morpholinosydnonimine (SIN-1) [18, 25, 27], or nitric oxide in solution [26]. NAD-dependent modification of the ~39-kDa protein was also induced by NO produced endogenously by NO synthase stimulated by the addition of substrates or cofactors [19, 28], by the addition of tumor necrosis factor α to L929 cells [23], by the addition of interleukin-1β to RINm5F cells [27], or by the addition of excess purified NO synthase to rat brain cytosol [22].

Evidence that NO stimulated the modification of the same ~39-kDa protein in intact cells was recently reported [27]. Data supporting this hypothesis were obtained by the 'back ADP-ribosylation' method, in which immediately following treatment of whole cells with NO, the cells were broken and a cell-free labeling reaction was performed with added NO and [^{32}P]NAD. A decrease in the labeling with radioactive NAD in the cell-free reaction is taken to indicate that some of the sites had been modified in the intact cell. In RINm5F cells treated with the NO-donor nitroprusside or stimulated to produce NO with interleukin-1β, decreased protein labeling with [^{32}P]NAD was observed in the subsequent cell-free reaction in the presence of NO, suggesting that these agents stimulated the protein modification in the whole cell [27].

The NAD-dependent linkage to the ~39 kDa protein acceptor was suggested to be mono-ADP-ribosylation rather than poly-ADP-ribosylation, based on identification of 5′-AMP as the major product of treatment of the modified protein with snake venom phosphodiesterase [18, 20, 24, 25]. The chemical linkage was characterized by its reactivity; it was resistant to hydroxylamine and sensitive to Hg^{2+} [24, 25], indicating it likely involved a cysteine [29, 30].

Modification of the ~39-kDa protein was postulated to be due to the stimulation of an endogenous ADP-ribosyltransferase by NO [18, 19, 31, 32]. Several properties of the NO-stimulated protein modification have been reported. One of the major known effects of NO is the production of cyclic GMP in cells [1], but modification of the ~39-kDa protein was independent of cyclic GMP [18, 21, 25, 32]. NADPH also stimulated modification of the protein in the presence of SNP, and although the mechanism was unclear, it was not due to stimulation of endogenous NO synthase [32].

Modification of proteins other than the ~39-kDa protein with [^{32}P]NAD in the presence of NO has also been reported. In bovine retinal rod outer segments, NO stimulated modification of the α- and β-subunits of transducin with NAD [28, 33]. In the plasma membrane fraction of human neutrophils, actin was modified in reactions with [^{32}P]NAD and NO [26]. In each of these reports, the existence of an endogenous, NO-stimulated ADP-ribosyltransferase was proposed to be responsible for the modifications observed.

NO-stimulated, NAD-dependent modification of purified proteins

The protein in subcellular fractions of rat brain [22] and human platelets [34] that was modified when incubated with NAD and NO, was purified and sequenced and identified as GAPDH. The modified protein from human erythrocytes was also identified as GAPDH based on reactivity and immunoprecipitation with monoclonal antibodies to GAPDH [24].

The action of an endogenous ADP-ribosyltransferase to catalyze the NO-stimulated modification of GAPDH by NAD was ruled out by finding that purified GAPDH incubated with NAD and NO was modified similarly to that in cell homogenates [22, 24, 34]. Subsequent work demonstrated that pure GAPDH from several different sources could be modified in this manner, including that from human erythrocytes [24, 34], chicken muscle [22], *Bacillus stearothermophilus* [22], human platelets [34], and rabbit muscle [10, 34–36]. GAPDH is a homotetrameric enzyme of 36,300-Da subunits, that binds four moles of NAD with negative cooperativity, and reduces NAD in the oxidation of glyceraldehyde-3-phosphate to 1,3-bisphosphoglycerate [37, 38]. The affinity for NAD of the first two binding sites is below 50 nM [37], and as isolated, GAPDH contains NAD or ADP-ribose in two or three of the four NAD sites [38].

NO derived from several sources supported modification of GAPDH during incubation with NAD. These included SNP [22, 24, 34–36], NO gas in solution [10, 34], SIN-1 [34, 38], S-nitroso-dithiothreitol [36], and NO generated by NO synthase [22].

The NO-stimulated modification of pure GAPDH was

suggested to be auto-ADP-ribosylation, implying that GAPDH possesses a latent ADP-ribosyltransferase activity that is activated by NO. This hypothesis was rejected when the NO-stimulated, NAD-dependent modification of GAPDH was characterized as the binding of the whole NAD molecule to the protein [36]. Both [adenine-^{14}C]NAD and [nicotinamide-^{14}C]NAD were incorporated equally into the protein in the presence of NO [36]. Incorporation of nicotinamide is inconsistent with ADP-ribosylation, in which case the nicotinamide is released as the nicotinamide-ADP-ribose bond is hydrolyzed [16, 17]. In two other studies, radiolabel from [nicotinamide-^{14}C]NAD was found not to be incorporated into GAPDH while radioactivity from [adenine-^{14}C]NAD was incorporated [24, 35], although in one case the data were not shown and details sufficient to evaluate the experiment were not provided [24]. In the other report, there may have been a significant miscalculation of the specific activities of the two types of radiolabeled NAD, so that during the labeling reaction, the specific activity of [nicotinamide-^{14}C]NAD was much lower than that of [adenine-^{14}C]NAD [35]. Also supporting the conclusion that the entire NAD molecule was bound to GAPDH, treatment of modified GAPDH with Hg^{2+} released a product that also contained both adenine and nicotinamide, showing that these components of NAD were together in the linkage to the protein [36].

The sensitivity of the NAD-modification in pure GAPDH to treatment with Hg^{2+} strongly suggested a thiol was involved. Sensitivity to Hg^{2+} was found in all cases in which it was tested [10, 22, 24, 34–36].

Although the labeling of GAPDH in reactions with [adenylate-^{32}P]NAD was readily detected by autoradiography after SDS-PAGE, even in whole cell homogenates, the extent of modification was actually very limited, ranging from 0.1 to 5% [10, 22, 24, 36]. There was one report of ~60% modification [35], but many of the other studies in which the extent of modification was much lower were performed under conditions that were similar to this one. Since GAPDH binds NAD so avidly, care must be taken to denature the enzyme sufficiently to release non-covalently bound NAD. Boiling for 5 min in 1–2% SDS or 6 M urea is sufficient to allow full separation of free NAD from that covalently bound to GAPDH by column chromatography or SDS-PAGE [36].

Under conditions used in the labeling with NAD, the NO present leads to S-nitrosylation of GAPDH, probably at the active site cysteine [10]. GAPDH was stoichiometrically S-nitrosylated to 4 mol NO per mol GAPDH tetramer [10]. Inhibition of GAPDH activity was also observed under the conditions used for modification with NAD [10, 30, 34, 36, 38]. Inhibition was >60% in most of these reports, suggesting that the stoichiometric level of nitrosylation was more likely to explain the inhibition than the much lower extent of covalent modification (~1%).

The NO-stimulated modification by NAD may be specific for GAPDH, in that a number of other dehydrogenases were not labeled to nearly the same extent as GAPDH. Modification of alcohol dehydrogenase was stimulated by NO, but at a much lower level than GAPDH [22, 39]. Other dehydrogenases, including lactate-, glutamate-, malate-, aldehyde-, and glycerol-3-phosphate-dehydrogenases, and several other enzymes with reactive cysteine groups, were not discernably modified with NAD in the presence of NO [22, 30, 39]. An unrelated protein, bovine serum albumin (BSA), was modified in incubations with NAD and NO [35]. BSA is S-nitrosylated [9, 12, 13], and it was suggested that the modification of BSA with NAD could be a model for the modification of S-nitroso-GAPDH [35]. It seems unlikely that BSA could provide a model for the reaction of NAD with nitrosylated GAPDH as closely related nitrosylated proteins with NAD-binding sites do not react with NAD. In fact, the modification of BSA did include the whole NAD molecule, but the linkage was not sensitive to Hg^{2+}, and so probably was not the same type of thiol-associated linkage as found in modified GAPDH.

NO and poly-ADP-ribosylation

In addition to mono-ADP-ribosylation, poly-ADP-ribosylation is another major NAD-dependent protein modification. This modification is synthesized by poly(ADP-ribose) polymerase (PARP) in the nucleus, where proteins are modified during repair of DNA damage, during replication, and possibly in other nuclear processes [40]. Several reports describing a connection between NO and poly-ADP-ribosylation have appeared. Two C-nitroso compounds, 6-nitroso-1,2-benzopyrone and 3-nitrobenzamide, inhibited PARP by oxidation of one of the zinc fingers in the enzyme [41]. A pair of recent reports demonstrated that inhibitors of PARP (nicotinamide, 3-aminobenzamide, and 3-methoxybenzamide), prevented induction of nitric oxide synthase in TNFα-stimulated L929 cells [23] and in lipopolysaccharide-stimulated mouse macrophages [42]. The mechanism was suggested to be the blockage of transcription of inducible NO synthase mRNA by inhibiting PARP's proposed role in transcription. In another report, toxicity and death of rat pancreatic islet cells caused by NO was prevented by inhibition of PARP (by nicotinamide or 3-aminobenzamide) [43]. This effect was probably due to the blocking of PARP-mediated NAD depletion in the cells, which occurs in cases of extreme DNA damage [40]. In these cells, the DNA damage was suggested to be caused by the NO treatment [43]. Together, these reports reveal several points of interaction between the effects of NO on cells and the function of PARP, but of a different nature than the effect of NO on covalent binding of NAD to GAPDH.

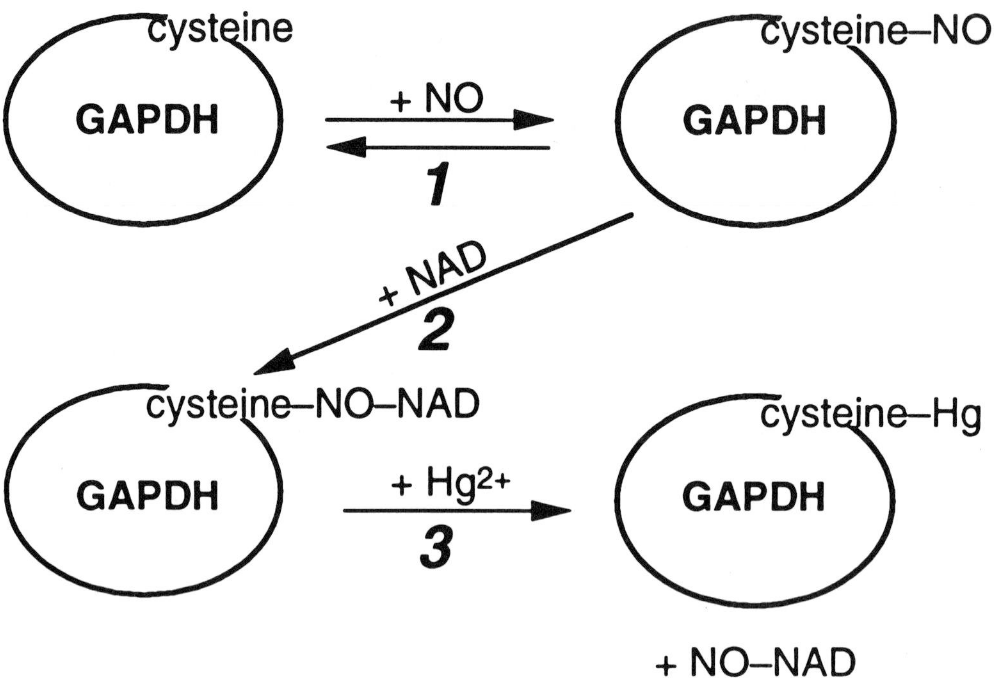

Fig. 1. Proposed scheme for the reaction of GAPDH with NO and NAD. Details are in the text.

Current understanding

The current understanding of the NO- and NAD-dependent modification of GAPDH is summarized in the scheme shown in Fig. 1. The first step in the process involved nitrosylation of a cysteine in GAPDH, a reversible process [4, 8]. In the presence of NO, GAPDH is S-nitrosylated to a level of four moles nitrosothiol per mole of GAPDH homotetramer [10]. This modification likely involves the catalytic, reactive cysteine in each of the subunits [10]. Thus nitrosylation itself could be sufficient to inhibit the activity of GAPDH, without invoking covalent modification with NAD as a means of inhibition. Since GAPDH avidly binds NAD at the active site and NAD is present during assay of GAPDH activity, some effort is required to distinguish between the effects of NO alone and those of NO and NAD together. Several arguments support the conclusion that nitrosylation is itself sufficient for inhibition of GAPDH activity.

1. NO inhibited GAPDH containing only bound NADH, not NAD, which then was assayed in the reverse reaction with NADH as the reactant; inhibition was ~60% [10].
2. The extent of modification, in the range of 0.1 to 5% in most reports, is much too low to account for the much higher level of inhibition, which is usually >60%. Covalent modification with NAD at the active site would be expected to inhibit enzyme activity, but only by an amount comparable to the extent of modification.
3. A related enzyme, ALDH, was inhibited by NO to the same extent as GAPDH, but was not modified at all by NAD [30]. Yet another dehydrogenase, alcohol dehydrogenase, was modified with NAD in the presence of NO to a much lower extent than GAPDH [22, 38, 44], but was inhibited by NO to nearly same extent as GAPDH [38].

Taken together, these points conclusively argue that covalent modification of GAPDH with NAD does not account for the majority of the inhibition observed in the presence of NO, and that nitrosylation alone is sufficient to inhibit the activity of these dehydrogenases.

The second step in the reaction scheme (Fig. 1) is the covalent addition of NAD to a small fraction of the nitrosylated GAPDH. As described above, the modification includes the entire NAD molecule rather than just ADP-ribose addition [36]. Most studies concluded that the extent of modification of GAPDH was less than 5% [10, 22, 24, 36], although it was noted that this could be an underestimation because covalent addition of the non-radiolabelled NAD already present in the GAPDH active sites would not be measured [22]. There is evidence against the problem of underestimation, however. Covalent binding of NAD to GAPDH was measured with apo-GAPDH (NAD-free GAPDH) that was loaded completely with [^{32}P]NAD and then treated with NO, and the extent of covalent modifiction of this sample was still <1% (10, and unpublished data).

The nature of the linkage of NAD with GAPDH is still not defined. NAD could be linked directly to the nitrosothiol group in GAPDH, or alternatively, nitrosylation of NAD could allosterically alter the chemical reactivity of a different cysteine in the enzyme so that it reacted covalently with NAD. The first alternative was favored by data from the analysis of the products released by Hg^{2+} from NAD-modified GAPDH [36]. The released compound contained both adenine and nicotinamide, but was not NAD (or NADH) as would be expected from Hg^{2+} cleavage of a direct cysteine-NAD linkage. The release of a modified form of NAD could be explained by the binding of NAD to GAPDH through the nitrosothiol group. Then the compound released by Hg^{2+} treatment would be an NO-NAD adduct, as shown in step 3 (Fig. 1).

Prospects

A major question is whether the NAD-modification of GAPDH is a minor epi-phenomenom secondary to enzyme S-nitrosylation or a physiologically important protein modification. The sole report of enzyme modification stimulated by NAD in whole cells indicated that the enzyme was modified in a similar manner as *in vitro* [27]. If the extent of modification of GAPDH with NAD in whole cells were at a stoichiometric level, this would argue for its significance as an NO-mediated, irreversible mechanism of GAPDH inactivation in cells. Another question is whether the modification is reversible in cells, for example by the action of a hydrolytic enzyme that would regenerate native GAPDH.

It will be important to define chemically the linkage of NAD to the thiol in GAPDH, including determining whether NO is included in the linkage. Although the inability of other related enzymes to serve as effective acceptors for NO and NAD modification might indicate an absolute specificity for GAPDH, the covalent binding of NAD to S-nitrosylated GAPDH could represent an example of a new class of chemical reaction of nitrosothiols.

Acknowledgements

The authors thank Dr. Martha Vaughan for critical reading of this manuscript.

References

1. Nathan C: Nitric oxide as a secretory product of mammalian cells. FASEB J 6: 3051–3064, 1992
2. Lowenstein CJ, Snyder SH: Nitric oxide, a novel biologic messenger. Cell 70: 705–707, 1992
3. Bredt DS, Snyder SH: Nitric oxide, a novel neuronal messenger. Neuron 8: 3–11, 1992
4. Stamler JS, Singel DJ, Loscalzo J: Biochemistry of nitric oxide and its redox-activated forms. Science 258: 1898–1902, 1992
5. Wink DA, Osawa Y, Darbyshire JF, Jones CR, Eshenaur SC, Nims RW: Inhibition of cytochromes P450 by nitric oxide and a nitric oxide-releasing agent. Arch Biochem Biophys 300: 115–123, 1993
6. Rogers NE, Ignarro LJ: Constitutive nitric oxide synthase from cerebellum is reversibly inhibited by nitric oxide formed from L-arginine. Biochem Biophys Res Commun 189: 242–249, 1992
7. Assreuy J, Cunha FQ, Liew FY, Moncada S: Feedback inhibition of nitric oxide synthase activity by nitric oxide. Br J Pharmacol 108: 833–837, 1993
8. Girard P, Potier P: NO, thiols and disulfides. FEBS Lett 320: 7–8, 1993
9. Stamler JS, Simon DI, Osborne JA, Mullins ME, Jaraki O, Michel T, Singel DJ, Loscalzo J: S-Nitrosylation of proteins with nitric oxide: synthesis and characterization of biologically active compounds. Proc Natl Acad Sci USA 89: 444–448, 1992
10. Molina y Vedia L, McDonald B, Reep B, Brüne B, Di Silvio M, Billiar TR, Lapetina EG: Nitric oxide-induced S-nitrosylation of glyceraldehyde-3-phosphate dehydrogenase inhibits enzymatic activity and increases endogenous ADP-ribosylation. J Biol Chem 267: 24929–24932, 1992
11. Lei SZ, Pan Z-H, Aggarwal SK, Chen H-SV, Hartman J, Sucher NJ, Lipton SA: Effect of nitric oxide production on the redox modulatory site of the NMDA receptor-channel complex. Neuron 8: 1087–1099, 1992
12. Stamler JS, Jaraki O, Osborne J, Simon DI, Keaney J, Vita J, Singel D, Valeri CR, Loscalzo J: Nitric oxide circulates in mammalian plasma primarily as an S-nitroso adduct of serum albumin. Proc Natl Acad Sci USA 89: 7674–7677, 1992
13. Keaney JF Jr, Simon DI, Stamler JS, Jaraki O, Scharfstein J, Vita JA, Loscalzo J: NO forms an adduct with serum albumin that has endothelium-derived relaxing factor-like properties. J Clin Invest 91: 1582–1589, 1993
14. Stamler JS, Simon DI, Jaraki O, Osborne JA, Francis S, Mullins M, Singel D, Loscalzo J: S-nitrosylation of tissue-type plasminogen activator confers vasodilatory and antiplatelet properties on the enzyme. Proc Natl Acad Sci USA 89: 8087–8091, 1992
15. Myers PR, Minor RL Jr., Guerra R Jr., Bates JN, Harrison DG: Vasorelaxant properties of the endothelium-derived relaxing factor more closely resemble S-nitrosocysteine than nitric oxide. Nature 345: 161–163, 1990
16. Moss J, Vaughan M: ADP-ribosylation of guanyl nucleotide-binding regulatory proteins by bacterial toxins. Adv Enzymol 61: 303–379, 1988
17. Williamson K, Moss J: Mono-ADP-ribosyltransferases and ADP-ribosylarginine hydrolases: a mono-ADP-ribosylation cycle in animal cells. In: J. Moss and M. Vaughan (eds). ADP-ribosylating Toxins and G Proteins: Insights into Signal Transduction. American Society for Microbiology, Washington DC, 1990, pp 493–510
18. Brüne B, Lapetina EG: Activation of a cytosolic ADP-ribosyltransferase by nitric oxide-generating agents. J Biol Chem 264: 8455–8458, 1989
19. Dimmeler S, Brüne B: L-Arginine stimulates an endogenous ADP-ribosyltransferase. Biochem Biophys Res Commun 178: 848–855, 1991
20. Duman RS, Terwilliger RZ, Nestler EJ: Endogenous ADP-ribosyltion in brain: initial characterization of substrate proteins. J Neurochem 57: 2124–2132, 1991
21. Williams MB, Li X, Gu X, Jope RS: Modulation of endogenous ADP-ribosylation in rat brain. Brain Res 592: 49–56, 1992
22. Zhang J, Snyder SH: Nitric oxide stimulates auto-ADP-ribosylation of glyceraldehyde-3-phosphate dehydrogenase. Proc Natl Acad Sci USA 89: 9382–9385, 1992
23. Hauschildt S, Scheipers P, Bessler WG, Mülsch A: Induction of nitric oxide synthase in L929 cells by tumour-necrosis factor α is prevented by inhibitors of poly(ADP-ribose) polymerase. Biochem J 288: 255–260, 1992
24. Kots AYa, Skurat AV, Sergienko EA, Bulargina TV, Severin ES: Nitroprusside stimulates the cysteine-specific mono (ADP-ribosylation) of glyceraldehyde-3-phosphate dehydrogenase from human erythrocytes. FEBS Lett 300: 9–12, 1992
25. Tao Y, Howlett A, Klein C: Nitric oxide stimulates the ADP-ribosylation

of a 41-kDa cytosolic protein in *Dictyostelium discoideum*. Proc Natl Acad Sci USA 89: 5902–5906, 1992

26. Clancy RM, Leszczynska-Piziak J, Abramson SB: Nitric oxide stimulates ADP-ribosylation of actin in human neutrophils. Biochem Biophys Res Commun 191: 847–852, 1993
27. Dimmeler S, Ankarcrona M, Nicotera P, Brüne B: Exogenous Nitric Oxide (NO) generation or IL-1β-induced intracellular NO production stimulates inhibitory auto-ADP-ribosylation of glyceraldehyde-3-phosphate dehydrogenase in RINm5F cells. J Immunol 150: 2964–2971, 1993
28. Pozdnyakov N, Lloyd A, Reddy VN, Sitaramayya A: Nitric oxide-regulated endogenous ADP-ribosylation of rod outer segment proteins. Biochem Biophys Res Commun 192: 610–615, 1993
29. Meyer T, Koch R, Fanick W, Hilz H: ADP-ribosyl proteins formed by pertussis toxin are specifically cleaved by mercury ions. Biol Chem Hoppe-Seyler 369: 579–583, 1988
30. McDonald LJ, Moss J: Nitric oxide-independent, thiol-associated ADP-ribosylation inactivates aldehyde dehydrogenase. J Biol Chem 268: 17878–17882, 1993
31. Brüne B, Lapetina EG: Properties of a novel nitric oxide-stimulated ADP-ribosyltransferase. Arch Biochem Biophys 279: 286–290, 1990
32. Brüne B, Dimmeler S, Lapetina EG: NADPH: A stimulatory cofactor for nitric oxide-induced ADP-ribosylation reaction. Biochem Biophys Res Commun 182: 1166–1171, 1992
33. Ehret-Hilberer S, Nullans G, Aunis D, Virmaux N: Mono ADP-ribosylation of transducin catalyzed by rod outer segment extract. FEBS Lett 309: 394–398, 1992
34. Dimmeler S, Lottspeich F, Brüne B: Nitric oxide causes ADP-ribosylation and inhibition of glyceraldehyde-3-phosphate dehydrogenase. J Biol Chem 267: 16771–16774, 1992
35. Dimmeler S, Brüne B: Characterization of a nitric-oxide-catalysed ADP-ribosylation of glyceraldehyde-3-phosphate dehydrogenase. Eur J Biochem 210: 305–310, 1992
36. McDonald LJ, Moss J: Stimulation by nitric oxide of a novel linkage of NAD to glyceraldehyde-3-phosphate dehydrogenase. Proc Natl Acad Sci USA 90: 6238–6241, 1993
37. De Vijlder JJM, Slater EC: The reaction between NAD^+ and rabbit-muscle glyceraldephosphate dehydrogenase. Biochim Biophys Acta 167: 23–34, 1968
38. Bloch W, MacQuarrie RA, Bernhard SA: The nucleotide and acyl group content of native rabbit muscle glyceralehyde 3-phosphate dehydrogenase. J Biol Chem 246: 780–790, 1971
39. Dimmeler S, Brüne B: Nitric oxide preferentially stimulates auto-ADP-ribosylation of glyceraldehyde-3-phosphate dehydrogenase compared to alcohol or lactate dehydrogenase. FEBS Lett 315: 21–24, 1993
40. Cleaver JE, Morgan WF: Poly(ADP-ribose)polymerase: a perplexing participant in cellular responses to DNA breakage. Mutation Res 257: 1–8, 1991
41. Rice WG, Hillyer CD, Harten B, Schaeffer CA, Dorminy M, Lackey DA III, Kirsten E, Mendeleyev J, Buki KG, Hakam A, Kun E: Induction of endonuclease-mediated apoptosis in tumor cells by C-nitroso-substituted ligands of poly(ADP-ribose) polymerase. Proc Natl Acad Sci USA 89: 7703–7707, 1992
42. Hauschildt S, Scheipers P, Bessler WG: Inhibitors of poly(ADP-ribose) polymerase suppress lipopolysaccharide-induced nitrite formation in macrophages. Biochem Biophys Res Commun 179: 865–871, 1991
43. Kallmann B, Vurkat V, Kröncke K-D, Kolb-Bachofen V, Kolb H: Toxicity of chemically generated nitric oxide towards pancreatic islet cells can be prevented by nicotinamide. Life Sci 51: 671–678, 1992
44. McDonald LJ, Moss J: Pleiotropic effects of nitric oxide on ADP-ribosylation, covalent binding of NAD, and catalytic activity of GAPDH and aldehyde dehydrogenase. Trans Assoc Amer Phys 106: 155–161, 1993

Molecular and Cellular Biochemistry **138**: 207–212, 1994.

Glycation of proteins by ADP-ribose

Elaine L. Jacobson[1,3], Daniel Cervantes-Laurean[2], and Myron K. Jacobson[2,3]

[1]*Department of Clinical Sciences, College of Allied Health Professions,* [2]*Division of Medicinal Chemistry and Pharmaceutics, College of Pharmacy, and* [3]*Markey Cancer Center, University of Kentucky, Lexington, KY. 40536, USA*

Abstract

Numerous metabolic pathways generate free ADP-ribose at many locations within cells. The metabolic fates of this nucleotide are poorly understood and measurement of it *in situ* is technically difficult at present. Yet considerable evidence has accumulated implicating that protein glycation by ADP-ribose can occur. This evidence is reviewed here along with recent developments in characterizing the chemistry of this reaction and the application of this information to the identification of this posttranslational modification in protein *in situ*. (Mol Cell Biochem **138**: 207–212, 1994)

Key words: free ADP-ribose, protein glycation, model conjugates of ADP-ribosyl lysine, histone glycation, non-enzymatic ADP-ribosylation, stability of ADP-ribosyl amino acids

Introduction

The oxidized form of nicotinamide adenine dinucleotide (NAD) is the substrate for enzymes that catalyze cleavage of the linkage between nicotinamide and ribose and transfer of adenosine diphosphoribose (ADP-ribose) to specific nucleophilic acceptors [1]. Eucaryotic cells contain several distinct classes of enzymes that catalyze ADP-ribose transfer to many different acceptors. The nuclear enzyme, poly-(ADP-ribose) polymerase, catalyzes the modification of proteins with polymers of ADP-ribose [2]. A second class is comprised of a family of protein: mono-ADP-ribosyltransferases that catalyze protein modification by single ADP-ribose residues [3]. A third class catalyzes an intramolecular ADP-ribose transfer to form cyclic ADP-ribose [4]. The longest known ADP-ribosyltransferases are the NAD glycohydrolases, which convert NAD to nicotinamide and free ADP-ribose [5]. The NAD glycohydrolases have been considered a distinct class of ADP-ribose transfer enzyme, but recent studies suggest that at least some NAD glycohydrolases are members of the third class since they are bifunctional enzymes that catalyze both the synthesis and hydrolysis of cyclic ADP-ribose [6]. Recently, the possibility of another class of ADP-ribosyltransferase has been suggested by the description of an enzyme involved in transfer RNA splicing that generates ADP-ribose 1′′–2′′ cyclic phosphate [7].

Metabolism related to ADP-ribose transfer reactions leads to free ADP-ribose as a metabolic intermediate as shown in Fig. 1. Polymers of ADP-ribose generated in response to deoxyribonucleic acid (DNA) strand breaks are rapidly turned over by the action of poly(ADP-ribose) glycohydrolase [8, 9]. Following exposure of cells to agents that cause DNA damage, this nuclear metabolism can cause rapid turnover of the NAD pool [10, 11], providing the potential for the formation of high nuclear concentrations of free ADP-ribose. Free ADP-ribose is formed from products of monomeric ADP-ribose transfer reactions by the action of enzymes that catalyze the removal of ADP-ribose from proteins [3] and by the hydrolysis of cyclic ADP-ribose [6]. The rate of formation of ADP-ribose related to monomeric ADP-ribose transfer reactions has not been well characterized, but these reactions result in the turnover of the NAD pool in cultured cells with a half life of approximately 8 to 10 hours [12]. Thus, these reactions also result in the formation of considerable amounts of free ADP-ribose.

The levels of free ADP-ribose present *in vivo* are poorly understood. The direct measurement of intracellular free

Address for offprints: E.L. Jacobson, Department of Clinical Sciences, University of Kentucky, Lexington, KY 40536-0080, USA

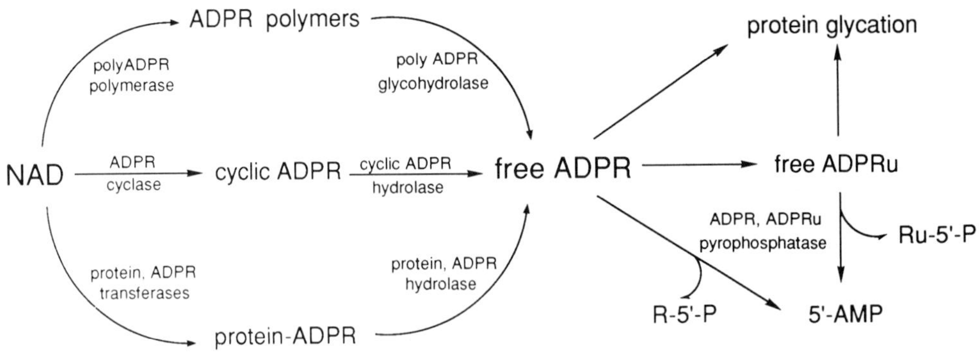

Fig. 1. Metabolic pathways for the formation and utilization of free ADP-ribose.

ADP-ribose is complicated by serious technical problems. Acid and alkaline conditions usually used for the extraction of nucleotides from cells result in the degradation of NADH and NAD, respectively, to ADP-ribose [13, 14]. Extractions at neutral pH are usually ineffective in inactivating NADases that can rapidly convert NAD to ADP-ribose. Recently, Guida *et al.* [15] reported the presence of free ADP-ribose in erythrocytes. In these cells, the presence of hemoglobin results in the rapid oxidation of NADH under conditions of acid extraction, thereby reducing the formation of ADP-ribose from NADH. They showed that 99.7% of the NADH was recovered as NAD following TCA extraction of the erythrocytes. Even with this efficient conversion of NADH to NAD, a potential problem in analysis of free ADP-ribose still remains due to the large amount of NADH relative to ADP-ribose present in cells. For example, conversion of the remaining 0.3% of the NADH to free ADP-ribose during acid extraction would result in an amount of ADP-ribose that was approximately the amount reported for free ADP-ribose in erythrocytes [15]. The utilization of acid conditions to obtain estimates of free ADP-ribose in cells not containing hemoglobin does not appear feasible. Thus, methodology to distinguish between free ADP-ribose present *in vivo* and that resulting from extraction conditions remains a pressing need.

Results and discussion

The metabolic fate of free ADP-ribose is not known, but recent studies have described enzymes that putatively are involved in its metabolism (Fig. 1). An ADP-ribose pyrophosphatase activity with a high affinity for free ADP-ribose has been described [16] and enzymatic activities that catalyze conversion of ADP-ribose to ADP-ribulose and ADP-ribulose to adenosine monophosphate (AMP) have been reported [17].

The steady state level of free ADP-ribose *in vivo* under different physiological conditions is determined by the relative rates of formation and breakdown of this nucleotide. The intracellular concentration of free ADP-ribose is an important question since ADP-ribose contains a reactive aldehydic group with the potential for protein glycation. Protein glycation by hexoses has been implicated in the pathophysiology of many diseases as well as the aging process [18]. The chemistry of glycation involving the reaction of hexoses with the ε-amino groups of protein lysine residues has been studied widely [19–26]. As shown in Fig. 2, this interaction results in the initial formation of a Schiff base that can be stabilized by the formation of either a furanose or pyranose form of an aminoglycoside or be converted via an Amadori rearrangement to a ketoamine. The ketoamine also can be stabilized by formation of furanose or pyranose ring structures. The ketoamine has been reported to be more stable than the aminoglycosides but can react intra- or intermolecularly with other amino groups leading to adducts called advanced glycosylation products [27, 28] that can result in additional protein structural changes including cross linking.

The potential for ADP-ribose (and ADP-ribulose) to participate in protein glycation has stimulated studies of protein glycation by these nucleotides *in vitro*. Kun *et al.* reported the glycation by ADP-ribose of histones and polylysine [29]. Ueda *et al.* showed that histone H1 modified by ADP-ribose glycation could serve as a site of elongation for poly(ADP-ribose) polymerase [30]. Ikejima and Gill [31] demonstrated that poly(ADP-ribose) polymerase was glycated by ADP-ribose when incubated with NAD at alkaline pH. Richter *et al.* [32, 33] and Hilz *et al.* [34] have demonstrated that specific mitochondrial proteins were glycated by ADP-ribose. Tanaka *et al.* have reported the selective glycation of cytosolic proteins in skeletal muscle extracts [35]. Recent studies by Franco *et al.* have also provided evidence for the glycation of erythrocyte membrane proteins by ADP-ribulose [17].

Despite reports of protein glycation by ADP-ribose over the last 20 years, the chemistry of this modification has not been well characterized. The lack of information concern-

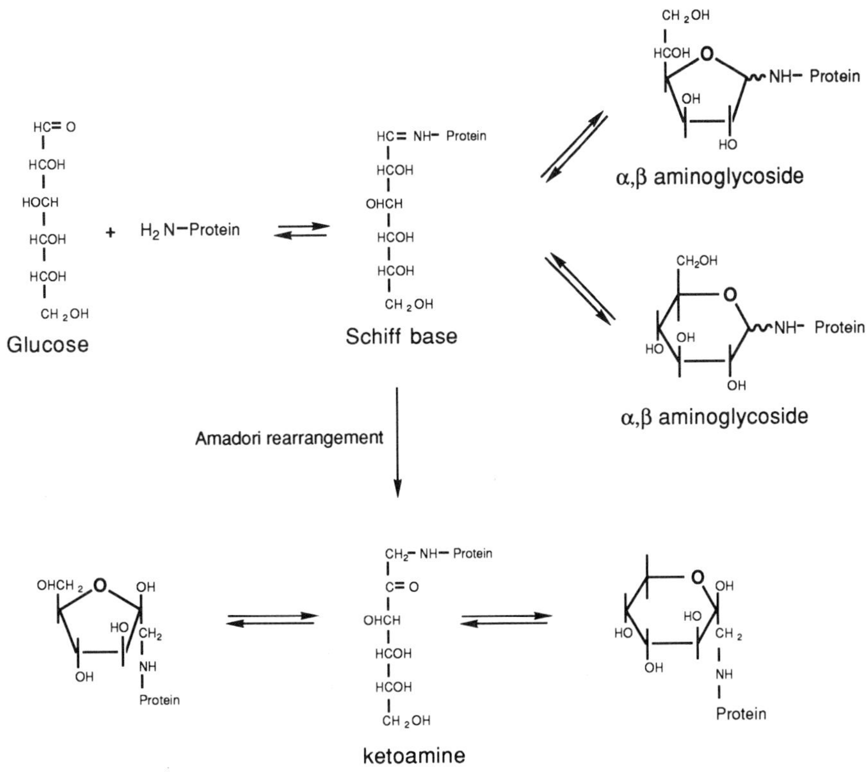

Fig. 2. Products formed between glucose and ε-amino groups of lysine residues in protein.

ing ADP-ribose glycation is due in part to the fact that when proteins or polypeptides are used as acceptors, structural characterization of the products formed by high resolution methods such as nuclear magnetic resonance (NMR) is difficult. Our laboratory has attempted to overcome these difficulties by synthesizing and characterizing low molecular weight model conjugates for protein glycation by ADP-ribose. To model the glycation of lysine residues, we studied the reaction of ADP-ribose with *n*-butylamine. These studies have been described in detail [36] and will be reviewed briefly here. Two stable conjugates were formed when ADP-ribose was reacted with *n*-butylamine. These were isolated, identified, and characterized [36]. Both chemical and ^{13}C NMR spectral characterizations demonstrated the presence of a carbonyl group in each conjugate, indicating that two different ketoamines were formed, which we termed ketoamines 1 and 2. Stable oximes were generated from both ketoamines by reaction with hydroxylamine and the oximes were characterized by ^{1}H NMR. The data showed that ketoamine 1 contained a keto group at position 2′′ and that ketoamine 2 contained a keto group at either position 3′′ or 4′′. Fig. 3 shows a postulated mechanism for the formation of multiple ketoamines derived from ADP-ribose and *n*-butylamine. This mechanism involves the initial formation of a Schiff base. Although the Schiff base could conceivably undergo ring closure to form an aminoriboside, we have not

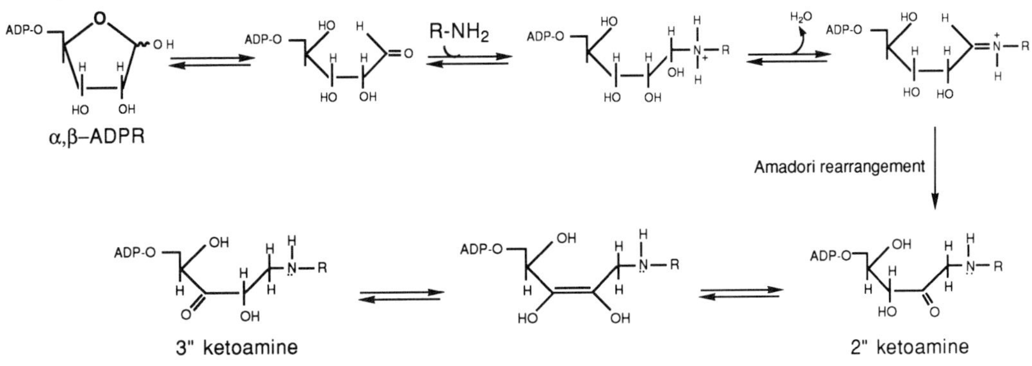

Fig. 3. Proposed mechanism for the formation of ketoamine products from ADP-ribose and *n*-butylamine.

yet detected aminoribosides derived from ADP-ribose. These compounds may be less stable than the hexose counterpart. Instead, an Amadori rearrangement yields a ketoamine with a carbonyl at position 2′′. Migration of the carbonyl group through an enol intermediate can result in ketoamines with a carbonyl group at position 3′′. A similar mechanism can result in migration of the carbonyl group to position 4′′ (not shown). The formation of ketoamines with carbonyl groups at multiple locations within the molecule is likely due to the fact that it is not possible to stabilize the ketoamines derived from ADP-ribose with ring structures, as is the case for hexoses (Fig. 2).

A major objective of our studies has been the development of technology to allow the differentiation of ADP-ribose lysine glycation products from the other ADP-ribosyl amino acids that can occur in proteins by the action of protein: mono-ADP-ribosyltransferases. Previous studies from our laboratory have characterized the chemical stability of linkages between ADP-ribose and amino acid side chains modified by protein: mono-ADP-ribosyltransferases [1, 37, 38]. Therefore we studied the chemical stability of the ketoamine model conjugates. Table 1 summarizes chemical stability information for ketoamines 1 and 2 and compares them to linkages between other amino acid side chains and ADP-ribose. Conjugates were considered stable if their $T_{½}$ values were at least 10 hours. Both ketoamines 1 and 2 were stable in 44% formic acid with $T_{½}$ values of 13 and 16 hours, respectively. They were stable in the presence of Hg^{+2} and hydroxylamine, but extremely unstable in 1 M NaOH. The key feature differentiating the ketoamines from ADP-ribosyl amino acid conjugates formed enzymatically was the lability of both ketoamines at pH 9.0. The products released when the ketoamines were incubated at pH 9.0 were analyzed by high pressure liquid chromatography (HPLC). The released material did not co-migrate with either ADP-ribose, AMP, or ADP, indicating that unique products are formed upon hydrolysis [36]. The structures of these hydrolysis products are currently being investigated, however, their unique migration on HPLC is diagnostic for differentiating protein glycation by ADP-ribose from enzymatic ADP-ribosylation of proteins.

Table 1. Stability of the linkage between amino acid side chains and ADP-ribose

ADP-Ribosyl linkage to:	Formic Acid (44%)	Hydroxylamine (1M, pH 7.0)	Hg^{2+} (10 mM)	CHES (pH 9.0)	NaOH (1M)
Arginine	Stable	Released	Stable	Stable	Released
Cysteine	Stable	Stable	Released	Stable	Released
Histidine (Diphthamide)	Stable	Stable	Stable	Stable	Stable
Asparagine	Stable	Stable	Stable	Stable	Stable
Serine, Threonine	Released	Stable	Stable	Stable	Stable
Lysine (ketoamines)	Stable	Stable	Stable	Released	Released

Having established chemical tools for identifying glycation of amino groups by ADP-ribose, we investigated the glycation of protein by ADP-ribose *in vitro*. We chose to initially study interaction with histones since these proteins have a high lysine content and have been reported to be mono ADP-ribosylated *in vivo* following DNA damage [39], when ADP-ribose polymer turnover is occurring at a rapid rate, making glycation of these proteins *in vivo* a likely possibility. For these experiments, ^{32}P-ADP-ribose was incubated with histone H1 and the histone H1 was subjected to electrophoresis on acid-urea gels followed by autoradiography. Fig. 4 shows that radiolabel co-migrating with protein was readily observed. Further, the chemical stability of the radiolabel attached to protein was studied. Table 2 shows that the protein conjugates displayed stability characteristics very similar to the ketoamine model conjugates. The products released at pH 9.0 also were shown by HPLC to co-migrate with products released at pH 9.0 from the ketoamines formed from ADP-ribose and *n*-butylamine. These results show that protein glycation by ADP-ribose readily occurs and that the stable glycation products are very likely ketoamine derivatives.

The chemical stability of the ketoamines formed from ADP-ribose and *n*-butylamine is very similar to the stability of the mitochondrial protein conjugates reported by Hilz *et al.* [40] and the cytosolic protein conjugates reported by Tanaka *et al.* [35]. This suggests that lysine was the site of modification in these proteins and that the stable products formed were one or more ketoamine(s). Recent evidence has shown that glycation by glucose occurs at specific lysine residues in albumin *in vivo* [26]. The preferential sites observed for glycation such as Lys-Lys and Lys-His-Lys led to the proposal that a basic amino acid adjacent to the site of

Table 2. Chemical stability of glycated histone H1 and *n*-butylamine model conjugates

ADP-ribosyl conjugate	Formic acid (44%)	Hydroxylamine (1 M, pH 7.0)	CHES buffer, (pH 9.0) ($t_{1/2}$, min)
Histone H1	Stable	Stable	37
Ketoamine 1	Stable	Stable	16
Ketoamine 2	Stable	Stable	32

glycation promotes the Amadori rearrangement via combined acid base catalysis [26]. This proposal is supported also by the observation that the site of glycation in hemoglobin A_{IC} contains an adjacent histidine residue [41]. The specific glycation of mitochondrial and cytosolic proteins by ADP-ribose [34, 35] may occur by a similar mechanism. The question of sequence specificity for protein glycation by ADP-ribose is an interesting area for further study.

Cellular recovery from DNA damage involves the activation by DNA strand breaks of poly(ADP-ribose) polymerase. This results in high rates of formation of ADP-ribose polymers in chromatin and their rapid turnover to free ADP-ribose [1]. The generation of high local concentrations of free ADP-ribose in close proximity to histones rich in lysine

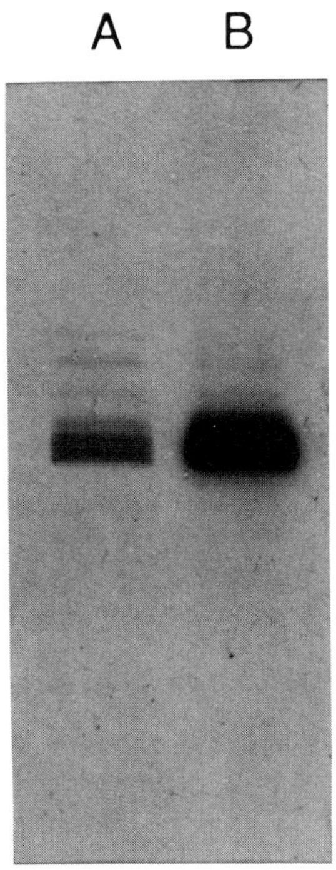

Fig. 4. Glycation of histone H1 by ADP-ribose. A mixture of 50 µM ^{32}P-ADP-ribose and 500 µM histone H1 in 50 mM sodium phosphate buffer, pH 7.0, 10 mM sodium pyrophosphate, was incubated for 8 hours at 37°C. The protein was subjected to electrophoresis on acid-urea polyacrylamide gels, stained with Coomasie blue, destained, dried and subjected to autoradiography. Panel A is the stained gel and panel B is an autoradiograph of the gel.

residues raises the likelihood of histone glycation. Kreimeyer *et al.* [39] have reported the occurrence of monomeric ADP-ribose conjugates of histone H1 in large excess over ADP-ribose polymer conjugates following DNA damage in hepatoma cells. While proteins carrying monomeric ADP-ribose residues may occur as a consequence of ADP-ribose polymer turnover, these residues are linked to the acceptor proteins by carboxylate ester linkages that are very sensitive to hydroxylamine [42]. In contrast, the histone H1 conjugates observed by Kreimeyer *et al.* [39] were stable in hydroxylamine. Comparing the stability of these conjugates with the stability of the ADP-ribose glycation model conjugates described here, suggests that these histone modifications represent *in vivo* glycation by ADP-ribose.

Hilz *et al.* [34] have reported glycation by ADP-ribose of polyhistidine and polyarginine in addition to polylysine. When we incubated ADP-ribose with low molecular weight imidazoyl or guanidino model compounds that did not contain primary amino groups, using conditions similar to Hilz *et al.*, no products were observed. This difference may be due to some chemistry unique to polymers of histidine or arginine. Alternatively it is possible that the glycation observed with polyhistidine and polyarginine occurred at α-amino groups of these homopolymers. These differences will require further investigation for clarification.

While this review has focused on glycation by ADP-ribose at lysine ε-amino groups, it should be noted that glycation by ADP-ribose has recently been reported to occur at cysteine thiol moieties of acetaldehyde dehyrogenase *in vitro* [43]. It is interesting that this product has the same chemical stability as that formed by enzyme catalyzed ADP-ribosylation at cysteine [36, 38, 43].

A variety of nonenzymic modifications of proteins occurs as a function of age including protein glycation [18]. A growing body of evidence suggests that oxidative damage to DNA and proteins is involved in the pathophysiology of cancer and the aging process [44]. Oxidative damage to DNA results in the formation of free ADP-ribose in the nucleus from ADP-ribose polymer metabolism. The occurrence and consequences of protein glycation by ADP-ribose in biological systems have not been demonstrated but can now by studied as this modification can be distinguished from other classes of protein modification by ADP-ribose. Based on studies of glycation *in vitro*, it is likely that glycation by ADP-ribose occurs *in vivo* as well and that cells possess mechanisms to minimize this modification of proteins.

Acknowledgements

This work was supported by the National Institutes of Health Grant CA43894 and by the National Science Foundation Grant DMB 8917274.

References

1. Jacobson MK, Aboul-Ela N, Cervantes-Laurean D, Loflin PT, Jacobson EL: ADP-ribose levels in animal cells. In: J. Moss and M. Vaughan (eds). ADP-Ribosylating Toxins and G Proteins: Insights Into Signal Transduction. American Society of Microbiology, Washington DC, 1990, pp 479–492
2. Althaus FR, Richter C: ADP-ribosylation of Proteins. Springer-Verlag, Heidelberg, 1987
3. Williamson KC, Moss J: Mono-ADP-ribosyltransferases and ADP-ribosylarginine hydrolases: a mono-ADP-ribosylation cycle in animal cells. In: J. Moss and M. Vaughan (eds). ADP-Ribosylating Toxins and G Proteins: Insights Into Signal Transduction. American Society of Microbiology, Washington DC, 1990, pp 493–506
4. Lee HC, Aarhus R: ADP-ribosyl cyclase: an enzyme that cyclizes NAD^+ into a calcium-mobilizing metabolite. Cell Reg 2: 203–209, 1991
5. Price SR, Pekala PH: Pyridine nucleotide-linked glycohydrolases. In: D. Dolphin, R. Poulson and O. Avramovic (eds). Pyridine Nucleotide Coenzymes: Chemical, Biological and Medical Aspects, Part B. Wiley-Interscience, New York, 1987, pp 513–548
6. Kim H, Jacobson EL, Jacobson MK: Synthesis and degradation of cyclic ADP-ribose by NAD glycohydrolases. Science 261: 1330–1331, 1993
7. Culver GM, McCraith SM, Zillman M, Kierzek R, Michaud N, LaReau RD, Turner DH, Phizicky EM: An NAD derivative produced during transfer RNA splicing: ADP-ribose 1′′–2′′ cyclic phosphate. Science 261: 206–208, 1993
8. Juarez-Salinas H, Sims JL, Jacobson MK: Poly(ADP-ribose) levels in carcinogen-treated cells. Nature (London) 282: 740–741, 1979
9. Miwa M, Tanaka M, Matsushima T, Sugimura T: Purification and properties of a glycohydrolase from calf thymus splitting ribose-ribose linkages of poly(adenosine diphosphate ribose). J Biol Chem 249: 3475–3482, 1974
10. Wielckens K, Schmidt A, George E, Bredehorst R, Hilz H: DNA fragmentation and NAD depletion. J Biol Chem 257: 12872–12877, 1982
11. Jacobson EL, Antol KM, Juarez-Salinas H, Jacobson MK: Poly(ADP-ribose) metabolism in UV irradiated human fibroblasts. J Biol Chem 258: 103–107, 1983
12. Jacobson MK, Jacobson EL, Hilz H: Involvement of ADP-ribosylation reactions in NAD turnover. Fed Proc 43: 315A, 1984
13. Burch HB, Bradley ME, Lowry OH: The measurement of triphosphopyridine nucleotide and reduced triphosphopyridine nucleotide and the role of hemoglobin in producing erroneous triphosphopyridine nucleotide values. J Biol Chem 242: 4546–4554, 1967
14. Jacobson MK, Payne DM, Alvarez-Gonzalez R, Juarez-Salinas H, Sims JL, Jacobson EL: Determination of *in vivo* levels of polymeric and monomeric ADP-ribose by fluorescence methods. Methods Enzymol 106: 483–494, 1984
15. Guida L, Zocchi E, Franco L, Benatti U, De Flora A: Presence and turnover of adenosine diphosphate ribose in human erythrocytes. Biochem Biophys Res Comm 188: 402–408, 1992
16. Miro A, Costas MJ, Garcia-Diaz M, Hernandez MT, Cameselle JC: A specific, low K_m ADP-ribose pyrophosphatase from rat liver. FEBS Lett 244: 123–126, 1989
17. Franco L, Guida L, Zocchi E, Silvestro L, Benatti U, De Flora A: Adenosine diphosphate ribulose in human erythrocytes: a new metabolite with membrane binding properties. Biochem Biophys Res Commun 190: 1143–1148
18. Stadtman E: Covalent modification reactions are making steps in protein turnover. Biochemistry 29: 6323–6331, 1990
19. Maillard LC: Action des acides aminés sur les sucres; formation des mélanoïdines par voie methodique. Comp Rend Acad Sci (Paris) 154: 66–68, 1912
20. Lea CH, Hannan RS: Biochemical and nutritional significance of the reaction between proteins and reducing sugars. Nature (London) 165: 438–439, 1950
21. Hodge JE, Rist CE: The Amadori rearrangement under new conditions and its significance for non-enzymatic Browning reactions. J Am Chem Soc 75: 316–322, 1953
22. Reynolds TM: The chemistry of nonenzymatic browning: I Adv Food Res 12: 1–52, 1963
23. Bunn HF, Haney DN, Gabbay KH, Gallop PM: Further identification of the nature and linkage of the carbohydrate in hemoglobin A_{Ic} Biochem Biophys Res Commun 67: 103–109, 1975
24. Monnier VM, Cerami A: Nonenzymatic glycosylation and browning in diabetes and aging. Diabetes 31 (Suppl. 3): 57–63, 1982
25. Watala C, Zawodniak M, Bryszewska M, Nowak S: Nonenzymatic protein glycosylation I. lowered erythrocyte membrane fluidity in juvenile diabetes. Ann Clin Res 17: 327–330, 1985
26. Iberg N, Fluckiger R: Non-enzymatic glycosylation of albumin *in vivo*: identification of multiple glycosylated sites. J Biol Chem 261: 13542–13545, 1986
27. Cerami A, Crabbe MJC: Recent advances in ocular cataract research. Trends in Pharmacol Sci 7: 271–274, 1986
28. Cerami A, Vlassara H, Brownlee MJ: Role of nonenzymatic glycosylation in atherogenesis. Cell Biochem 30: 111–120, 1986
29. Kun E, Chang ACY, Sharma ML, Ferro AM, Nitecki D: Covalent modification of proteins by metabolites of NAD^+. Proc Natl Acad Sci USA 73: 3131–3135, 1976
30. Ueda K, Kawaichi M, Okayama H, Hayaishi O: Poly(ADP-ribosyl)ation of nuclear proteins: enzymatic elongation of chemically synthesized ADP-ribose-histone adduct. J Biol Chem 254: 679–687, 1979
31. Ikejima M, Gill DM: Nonenzymatic adenosine 5′ diphosphate ribosylation of poly(adenosine diphosphate ribose). Biochemistry 24: 5039–5045, 1985
32. Richter C, Winterhalter KH, Baumhüter S, Lötscher H-R, Moser B: ADP-ribosylation in inner membrane of rat liver mitochondria. Proc Natl Sci USA 80: 3188–3192, 1983
33. Frei B, Richter C: Mono(ADP-ribosylation) in rat liver mitochondria. Biochemistry 27: 529–535, 1988
34. Hilz H, Koch R, Fanick W, Klapproth K, Adamietz P: Nonenzymic ADP-ribosylation of specific mitochondrial polypeptides. Proc Natl Acad Sci USA 81: 3929–3933, 1984
35. Tanaka Y, Yoshihara K, Kamiya T: Enzymic and nonenzymic mono ADP-ribosylation of proteins in skeletal muscle. Biochem Biophys Res Commun 163: 1063–1070, 1989
36. Cervantes-Laurean D, Minter DE, Jacobson EL, Jacobson MK: Protein glycation by ADP-ribose: studies of model conjugates. Biochemistry 32: 1528–1534, 1993
37. Payne DM, Jacobson EL, Moss J, Jacobson MK: Modification of proteins by ADP-ribose *in vivo*. Biochemistry 24: 7540–7549, 1985
38. Jacobson MK, Loflin PT, Aboul-Ela N, Mingmuang M, Moss J, Jacobson EL: Modification of plasma membrane protein cysteine residues by ADP-ribose *in vivo*. J Biol Chem 265: 10825–10828, 1990
39. Kreimeyer A, Wielckens K, Adamietz P, Hilz H: DNA repair-associated ADP-ribosylation *in vivo*. J Biol Chem 259: 890–896, 1984
40. Hilz H, Koch R, Kreimeyer A, Adamietz P, Jacobson MK: Mono(ADP-ribosyl)ation and phospho(ADP-ribosylation) reactions in eukaryotic cells. In: F.R. Althaus, H. Hilz and S. Shall (eds). ADP-Ribosylation of Proteins. Springer-Verlag, Berlin, Heidelberg, 1985, pp 518–525
41. Koenig RJ, Cerami A: Hemoglobin A_{Ic} and diabetes mellitus. Annu Rev Med 31: 29–34, 1980
42. Nishisuka Y, Ueda K, Yoshihara K, Yamamura H, Takeda M, Hayaishi O: Enzymic adenosine diphosphoribosylation of nuclear proteins. Cold Spring Harbor Symp Quant Biol 34: 781–786, 1969
43. McDonald LJ, Moss J: Nitric oxide-independent, thiol-associated ADP-ribosylation inactivates aldehyde dehydrogenase. J Biol Chem 268: 17878–17882, 1993
44. Stadtman ER: Protein oxidation and aging. Science 257: 1220–1224, 1992

Molecular and Cellular Biochemistry **138**: 213–219, 1994.

DeoxyNAD and deoxyADP-ribosylation of proteins

Rafael Alvarez-Gonzalez

The Department of Microbiology and Immunology, University of North Texas Health Science Center at Fort Worth, Fort Worth, TX 76107-2699, USA

Abstract

Recently, two deoxyribose analogs of βNAD^+ (2´-deoxy and 3´-deoxyNAD$^+$) have been synthesized and purified in this laboratory. Whereas 2´-deoxyNAD$^+$ was an efficient substrate for arg-specific mon(ADP-ribosyl) transferases, it was not a substrate for poly(ADP-ribose) polymerase (PARP). Instead, it was a non-competitive inhibitor of βNAD^+ in the ADP-ribose polymerization reaction catalyzed by PARP. Thus, 2´-deoxyNAD$^+$ has been utilized to distinguish between mono(ADP-ribose) and poly(ADP-ribose) acceptor proteins. 2´-deoxyNAD$^+$ has also been used to characterize the arg-specific mono(2´-deoxyADP-ribosyl)ation reaction of PARP with cholera toxin or avian mono(ADP-ribosyl)transferase. By contrast, 3´-deoxyNAD$^+$ can effectively be utilized as a substrate by PARP. However, while the estimated Km and Kcat of polymerization with 3´-deoxyNAD$^+$ were 20 μM and 0.11 moles/sec, the Km and Kcat with βNAD^+ as a substrate were 59 μM and 1.29 moles/sec, respectively. Determination of the average size of 3´-deoxyADP-ribose polymers indicated that chains no larger than four residues are synthesized with this substrate. Thus, the utilization of 3´-deoxyNAD$^+$ has facilitated the electrophoretic identification of poly(ADP-ribose) acceptor proteins in mammalian chromatin. (Mol Cell Biochem **138**: 213–219, 1994)

Key words: 2´-deoxyNAD$^+$, mono(ADP-ribosyl) transferase(s), mono(2´-deoxyADP-ribosyl)ation, 3´-deoxyNAD$^+$, poly(ADP-ribose) polymerase, poly(3´-deoxyADP-ribosyl)ation

Introduction

Classification of ADP-ribosyl transferases

At present, two major classes of βNAD:protein ADP-ribosyl transferases have been identified in living organisms [1]. Enzymes that transfer a single unit of ADP-ribose from βNAD to a protein acceptor are classified as mono(ADP-ribosyl)transferases. These enzyme(s) have been identified in both procaryotes and eucaryotes [1]. A second class of ADP-ribosyltransferase activity corresponds to the ADP-ribose polymerizing activity associated with the chromatin fraction of eucaryotes [2, 3]. This enzyme is known as poly-(ADP-ribose) polymerase [PARP] [EC 2.4.2.30], a highly versatile DNA-binding protein [2, 3]. Whereas the former class of ADP-ribosylating activities comprises a group of enzymes that modulate the biological activity of the acceptor protein(s) with high specificity [4], the latter is a DNA-dependent enzyme that assembles protein-bound ADP-ribose polymers in the nucleus of eucaryotic cells. In fact, PARP appears to modulate changes in chromatin architecture in nuclear functions that involve DNA-strand breaking and rejoining [5, 6].

Mono(ADP-ribosyl)ation of proteins

Three distinct classes of procaryotic mono(ADP-ribosyl)-transferase activities that modify specific amino acid acceptor sites have been identified to date. These include: (i) diphtheria toxin fragment A [7] and pseudomonas exotoxin [8] which mono(ADP-ribosyl)ate a hypermodified histidine residue of eucaryotic elongation factor 2; (ii) cholera toxin A (choleragen) [9] and *Escherichia coli* heat-labile enterotoxin

Address for offprints: R. Alvarez-Gonzalez, Ph.D., Department of Microbiology and Immunology, Texas College of Osteopathic Medicine, 3500 Camp Bowie Boulevard, Fort Worth, TX 76107-2699, USA

[9] which modify an arginine residue of a GTP-binding protein (Gsα); and (iii) pertussis toxin, a cysteine-specific mono(ADP-ribosyl)transferase [10]. More recently, mono-(ADP-ribosyl)transferases with amino acid specificities for histidine [11], arginine [12–14], and cysteine [15] have also been identified in animal tissues. Whereas histidine-specific [11] and cysteine-specific [15] mono(ADP-ribosyl)transferases appear to be confined to the cytosol and plasma membrane fractions, respectively, arginine-specific mono(ADP-ribosyl)transferases have been found in the cytosol [12], the plasma membrane [13], and the nucleus [13, 14].

Whereas cellular fractionation techniques are helpful to identify, isolate, and characterize some of these enzymes, the co-existence of arginine-specific mono(ADP-ribosyl)transferase(s) with PARP in the nuclear compartment of eucaryotic cells [2, 3, 14] complicates the identification, isolation, and characterization of critical protein targets for mono and poly(ADP-ribosyl)ation. As discussed below, we have recently developed specific ADP-ribosylation substrates to identify mono [16] and poly(ADP-ribose) [17] acceptor proteins.

Poly(ADP-ribosyl)ation of proteins

An additional complication in the characterization of (ADP-ribosyl)ated-proteins in chromatin is the well-established ability of PARP to catalyze at least four distinct chemical reactions. These are: (i) the auto-poly(ADP-ribosyl)ation reaction (automodification) [18, 19]; (ii) a glutamate-specific mono(ADP-ribosyl)ation reaction (ADP-ribose chain initiation) [20]; (iii) the protein-distal ADP-ribose polymerization reaction (elongation reaction) [21–23]; as well as (iv) the polymeric branching reaction [24]. More importantly, some of the multiple catalytic activities of this enzyme happen to complicate the enzymological characterization of PARP even further. For example, the automodification reaction of PARP at multiple amino acid acceptor sites rapidly induces its own catalytic inactivation as a result of electrostatic repulsion with co-enzymic DNA [25, 26].

As mentioned above, incubation of eucaryotic chromatin with [^{32}P]NAD may result in the formation of arginine-bound [14] and glutamate-bound [20] monomeric and polymeric ADP-ribose units, respectively. Thus, it is of paramount importance to clearly distinguish between the acceptor proteins for each reaction. In addition, the ability of PARP to synthesize enzyme-bound polymers of over 200 ADP-ribose residues in size containing 5 to 6 branching points per molecule [27] complicates the characterization of (ADP-ribosyl)ated-proteins in eucaryotic chromatin. The covalent attachment of large and highly branched polymers to an acceptor protein results in the unavoidable alteration of the electrophoretic and chromatographic properties of the target molecule. Thus, new poly(ADP-ribosyl)ation experimental protocols should be helpful in dissecting the multiple reactions catalyzed by PARP.

To distinguish between mono and poly(ADP-ribose) target proteins in eucaryotic chromatin as well as to limit ADP-ribose polymerization to oligomeric ADP-ribose chains with no branches, we have synthesized, purified, and biochemically characterized β2′-deoxyNAD$^+$ and β3′-deoxyNAD$^+$. These analogs of βNAD have been found to be specific ADP-ribosylation substrates for the monomeric [16] and oligomeric [17] ADP-ribosylation of chromatin proteins, respectively.

Materials and methods

Synthesis and purification of 2′-deoxyNAD$^+$ and 3′-deoxyNAD$^+$

Deoxyribose analogs of NAD$^+$, (Fig. 1) have been enzymatically synthesized with commercially available NMN$^+$:-adenyltransferase (NAD pyrophosphorylase) [EC 2.7.7.18], βNMN$^+$, and [^{32}P]2′-deoxyATP or [^{32}P]3′-deoxyATP (cordy-

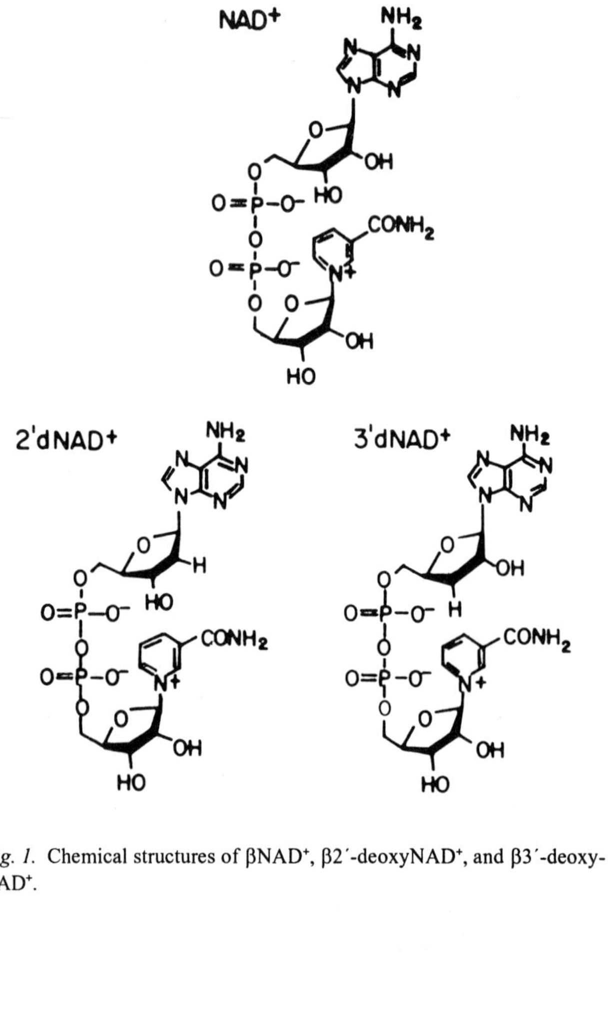

Fig. 1. Chemical structures of βNAD$^+$, β2′-deoxyNAD$^+$, and β3′-deoxyNAD$^+$.

cepin) as adenylate transfer substrates [29]. DeoxyNAD's are subsequently purified to a single nucleotide component by affinity chromatography on a boronate resin [30] followed by strong anion exchange-HPLC [29]. Fractions containing [^{32}P] radiolabeled deoxyNAD are pooled, lyophilized to dryness, dissolved in water, and stored at –20°C. Purity of the deoxyribose pyridine dinucleotide(s) is always checked by strong anion exchange-high pressure liquid chromatography (SAX-HPLC) [29] before use.

Results

Mono(2′deoxyADP-ribosyl)ation of proteins

Knowing that the 2′-hydroxyl group of either mono or oligo-(ADP-ribosyl)ated-proteins is the chemical target for the protein-distal ADP-ribose chain elongation [17, 21–23], it was hypothesized that 2′-deoxyNAD$^+$, a 2-deoxyribose analog of βNAD$^+$, should not be a substrate for enzymatic ADP-ribose polymerization. In contrast, amino acid-specific mono(ADP-ribosyl)transferases, whether procaryotic or eucaryotic should not distinguish between βNAD$^+$ and β2′-deoxyNAD$^+$ as mono(ADP-ribosyl)ation substrates.

As expected, it was recently observed that β2′-deoxyNAD$^+$ was not used as an ADP-ribosylation substrate by highly purified PARP, both in the absence (automodification) or presence of histone proteins (heteromodification) [16]. More interestingly, it was observed that this chemical analog of βNAD$^+$ was a potent inhibitor of ADP-ribose polymerization [16]. In fact, it was a non-competitive inhibitor versus βNAD$^+$, the ADP-ribose donor molecule, in the ADP-ribose transfer reaction catalyzed by PARP [16]. Therefore, 2′-deoxyNAD$^+$ may be useful inhibitor to characterize the multiple catalytic activities of PARP.

In marked contrast, pure 2′-deoxyNAD$^+$ was efficiently utilized as a mono(ADP-ribosyl)ation substrate by an arginine-specific mono(ADP-ribosyl)transferase purified from turkey erythrocytes [16, 31]. Thus, [^{32}P] radiolabeled 2′-deoxyNAD$^+$ should also prove useful in distinguishing between mono and poly(ADP-ribose) acceptor proteins in eucaryotic chromatin.

It is important to note that other investigators reported that 2′-deoxyNAD$^+$ was a good ADP-ribosylation substrate for PARP [32–34]. However, in the late 70's, these investigators did not have pure enzyme available. In their studies, they used either isolated nuclei [32, 33] or permeabilized HeLa cells [34] as a source of poly(ADP-ribose) polymerase activity. While they identified several ADP-ribosylated-proteins following polyacrylamide gel electrophoresis and autoradiography, it was possible that these conjugates were generated by nuclear mono(ADP-ribosyl)transferases [13, 14]. From our experimental observations [16], we conclude that the enzymatic ADP-ribosylation of nuclear proteins reported in earlier deoxyADP-ribosylation studies [32–34], must have been the result of mono(ADP-ribosyl)transferase activities. Support for this interpretation is the fact that arginine-specific mono(ADP-ribosyl)transferases have been identified in the nucleus of turkey erythrocytes [13] and liver cells [14].

Recent experimental data from this laboratory [35], as well as others [34], indicates that 2′-deoxyNAD$^+$ is also a good ADP-ribosylation substrate for cholera toxin and pertussis toxin. Therefore, it is clear that both procaryotic and eucaryotic mono(ADP-ribosyl)transferases may use either βNAD$^+$ or 2′-deoxyNAD$^+$. Table 1 below shows the current list of mono(ADP-ribosyl)transferase activities that efficiently catalyze the amino acid-specific mono(2′-deoxyADP-ribosyl)-ation of proteins.

Mono(2′deoxyADP-ribosyl)ation of poly(ADP-ribose) polymerase

As discussed above, the main protein target for poly(ADP-ribosyl)ation is PARP itself. This highly efficient automodification reaction occurs intermolecularly and two molecules of PARP are required for autopoly(ADP-ribosyl)ation [23]. Since protein-protein interactions between two units of PARP are also required for the auto-mono(ADP-ribosyl)ation reaction of this enzyme, it is also possible that PARP interacts with nuclear arginine-specific mono(ADP-ribosyl)transferases and thereby generates PARP-mono(ADP-ribose) adducts having arginine residues as the amino acid acceptor sites.

To examine this possibility, one may incubate highly purified PARP with an arginine-specific mono(ADP-ribosyl)transferase under protein(ADP-ribosyl)ation conditions. However, utilization of [^{32}P]-radiolabeled βNAD$^+$ as the ADP-ribosylation substrate is precluded since this dinucle-

Table 1. Utilization of 2′-deoxyNAD$^+$ as a substrate for protein-(ADP-ribosyl)ation

Enzyme	ADP-ribosylation Target	2′-deoxyNAD$^+$ as a Substrate	Reference
Cholera toxin	Gsα	yes	35
	PARP	yes	36
	rARF2	yes	36
	choleragen	yes	36
Pertussis toxin	Giα	yes	35
Diphtheria toxin	Elongation Factor 2	yes	34
Avian mono(ADP-ribosyl)transferase	PARP	yes	36
NADase	NAD hydrolysis	yes	34
PARP	PARP	No	16,29
	Histones	No	16

otide would function as an ADP-ribose donor for both enzymes and it would be quite difficult to distinguish between the homologous (automodification) poly(ADP-ribosyl)ation reaction and the heterologous mono(ADP-ribosyl)ation event, even in the absence of DNA. To avoid these technical complications, we utilized [^{32}P]-radiolabeled 2′-deoxyNAD$^+$ as a specific mono(ADP-ribosyl)ation substrate for either cholera toxin or an arginine-specific mono(ADP-ribosyl)-transferase isolated from avian tissues [36]. In these experiments, we observed that both procaryotic and eucaryotic mono(ADP-ribosyl)transferases were able to covalently attach 2′-deoxyADP-ribose monomers to surface-exposed arginine residues of PARP [36, 37]. It is also of interest to note that while cholera toxin modified the DNA-binding domain only [36], the avian enzyme mono(ADP-ribosyl)ated multiple arginine residues in the DNA-binding and the NAD-binding domains of PARP [37]. Thus, it seems possible that some of the multiple amino acid acceptor sites involved in the poly(ADP-ribosyl)ation of PARP [19, 38] are generated by heterologous poly(ADP-ribose) initiation events. The physiological relevance of this intriguing possibility is further emphasized by the fact that arginine-specific mono-(ADP-ribosyl)ation of PARP by the avian enzyme results in the catalytic inactivation of PARP [38]. Further experiments are in progress in this laboratory to characterize the functional role(s) of potential mono(ADP-ribosyl)ation sites of PARP involving arginine residues on the multiple catalytic activities of this highly versatile enzyme.

Oligo(3′-deoxyADP-ribosyl)ation of chromatin proteins

The second deoxyribose analog of βNAD$^+$, 3′-deoxyNAD$^+$, was synthesized to address specific questions about the biochemistry and enzymology of protein-poly(ADP-ribosyl)-ation. This poly(ADP-ribosyl)ation substrate turned out to be a powerful tool to specifically probe some of the reactions catalyzed by PARP [17, 28].

For example, we observed that the ADP-ribose polymerizing activity asociated with either crude chromatin fractions or highly purified enzyme [17], resulted in the synthesis of non-branched oligomeric ADP-ribose chains of no more than four residues in size. Considering the fact that a significant obstacle to the biochemical characterization of poly(ADP-ribosyl)ated-proteins in eucaryotic chromatin is the molecular size of the ADP-ribose polymers generated with βNAD$^+$ as the ADP-ribosylation substrate [27], it was very advantageous to have a way to limit the ADP-ribose polymerization to small oligomers that would not change the electrophoretic and/or chromatographic properties of the acceptor molecule. Furthermore, the automodification reaction of PARP with small oligomers of 3′-deoxyADP-ribose did not lead to self-inactivation since oligo(3′-deoxyADP-ribosyl)ated-PARP was still able to bind DNA [17].

Interestingly, [^{32}P]-radiolabeled 3′-deoxyNAD$^+$ was recently utilized to determine fluctuations in the ADP-ribosylation of specific chromatin proteins in liver cells following feeding of rats with a number of hepatocarcinogens that induce the formation of DNA-strand breaks following DNA-damage [28].

Another important application of 3′-deoxyNAD$^+$ was its utilization in a hot-cold pulse-chase protocol to determine the direction of ADP-ribose chain growth [17]. In this experiment highly purified PARP was incubated with 40 μM [^{32}P] radiolabeled 3′-deoxyNAD$^+$ for 5 min at 37°C and subsequently with 1 mM [^{3}H]βNAD$^+$ to elongate the initiated ADP-ribose chains. Subsequently, radiolabeled molecules were detached from the acceptor proteins with 0.1 NaOH, 20 mM EDTA [39] and the free polymers were enzymatically hydrolyzed with snake venom phosphodiesterase in the presence of magnesium ions [17]. Resolution of radiolabeled nucleotides by SAX-HPLC prior to and after the chase revealed that protein-bound ADP-ribose is elongated via a protein-distal addition mechanism [17]. Thus, oligomers of 3′-deoxyADP-ribose are good ADP-ribose elongation acceptors when using βNAD$^+$ as the ADP-ribose donor.

Comparison between βNAD and β3′-deoxyNAD as poly(ADP-ribose) polymerase substrates

Incubation of mammalian chromatin with increasing concentrations of βNAD$^+$ typically results in a marked reduction of the electrophoretic mobility of the poly(ADP-ribose) covalent acceptor protein. Fig. 2 shows the autoradiographic profile of rat liver chromatin proteins that become poly(ADP-ribosyl)ated following incubation with either 1, 10, 100, or 200 μM [^{32}P]βNAD$^+$ in the absence or presence of 1 mM benzamide. In this experiment, four major chromatin proteins corresponding to PARP itself (automodification reaction), histone H1, histone H2b, and a 42 kDa protein become poly(ADP-ribosyl)ated. Addition of 1 mM benzamide completely inhibits the modification of these proteins suggesting that they comprise poly and not mono(ADP-ribose) acceptors since higher concentrations of benzamide are required to inhibit protein-mono(ADP-ribosyl)ation [40]. With concentrations as low as 10 μM [^{32}P]βNAD$^+$, it is obvious that the mobility of the acceptor proteins is slowed down (Fig. 2). Higher concentrations of [^{32}P]βNAD$^+$ result in the formation of protein-(ADP-ribose)n conjugates that do not migrate into the polyacrylamide gel. Therefore, identification of poly(ADP-ribose) targets at physiological concentrations of NAD$^+$, i.e., 200–400 μM [41], becomes difficult. In marked contrast, the utilization of increasing concentrations of [^{32}P]β3′-deoxyNAD$^+$, as a poly(ADP-ribosyl)ation substrate, from 1 to 100 μM, resulted in the formation of the

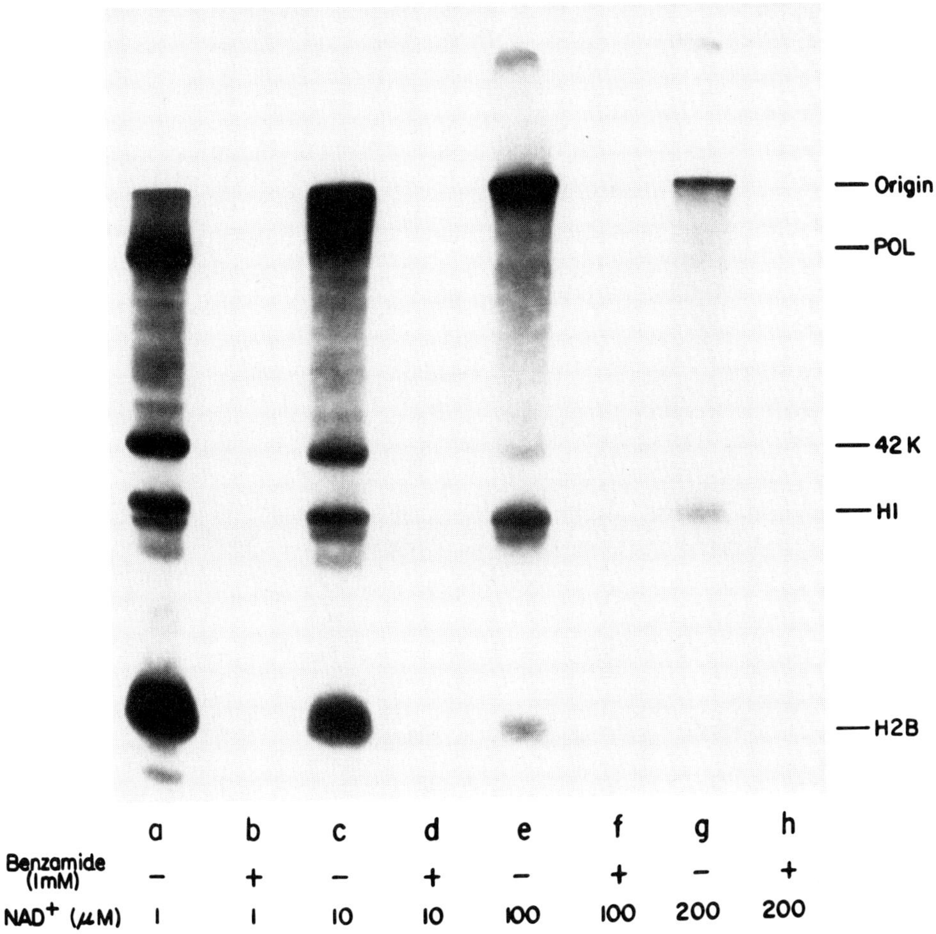

Fig. 2. Autoradiographic analysis of [^{32}P]poly(ADP-ribosyl)ated proteins from rat liver chromatin following incubation with increasing concentrations of [^{32}P]βNAD$^+$ with or without 1 mM benzamide.

same four chromatin protein-(3′-deoxyADP-ribose)n conjugates [17] generated with [^{32}P]βNAD$^+$ as a substrate. Once again, addition of 1 mM benzamide to the incubation reaction mixture resulted in complete inhibition of protein-poly-(ADP-ribosyl)ation. In the absence of benzamide, however, increasing concentrations of 3′-deoxyNAD$^+$ did not affect the electrophoretic mobility of the acceptor proteins (Fig. 3). Therefore, this NAD$^+$-substrate analog should prove useful in the future identification and characterization of poly(ADP-ribose) acceptor proteins in crude chromatin extracts.

References

1. Althaus FR, Richter CR: ADP-ribosylation of proteins. Mol Biol Biochem Biophys 37: 1–126, 1987
2. de Murcia G, Huletsky A, Poirier GG: Modulation of chromatin structure by poly(ADP-ribosyl)ation. Biochem Cell Biol 66: 626–636, 1988
3. Leduc Y, de Murcia G, Lamarre D, Poirier GG: Visualization of poly(ADP-ribose) synthetase associated with polynucleosomes by immunoelectron microscopy. Biochim Biophys Acta 885: 248–255, 1986
4. Moss J, Vaughan M: ADP-ribosylating Toxins and G Proteins: Insights Into Signal Transduction, American Society for Microbiology. Washington DC, 1990
5. Poirier GG, de Murcia G, Jongstra-Bile J, Niedergang C, Mandel P: Poly-(ADP-ribosyl)ation of polynucleosomes causes relaxation of chromatin structure. Proc Natl Acad Sci USA 79: 3423–3427, 1982
6. de Murcia G, Huletsky A, Lamarre D, Gaudreu A, Poyet J, Daune M, Poirier GG: Modulation of chromatin superstructure induced by poly(ADP-ribose) synthesis and degradation. J Biol Chem 261: 7011–7017, 1986
7. Honjo T, Nishizuka Y, Hayaishi O, Kato I: Diphtheria toxin-dependent adenosine diphosphate ribosylation of aminoacyl transferase II and inhibition of protein synthesis by diphtheria toxin. J Biol Chem 243: 3553–3555, 1968
8. Iglewski BH, Liu PV, Kabat D: Mechanism of action of *Pseudomonas aeroginosa* exotoxin A: adenosine diphosphate-ribosylation of mammalian elongation factor 2 *in vitro* and *in vivo*. Infect Immun 15: 138–144, 1977
9. Moss J, Vaughan M: Mechanism of action of choleragen and *E. coli* heat-labile enterotoxin: activation of adenylate cyclase by ADP-ribosylation. Mol Cell Biochem 37: 75–90, 1981

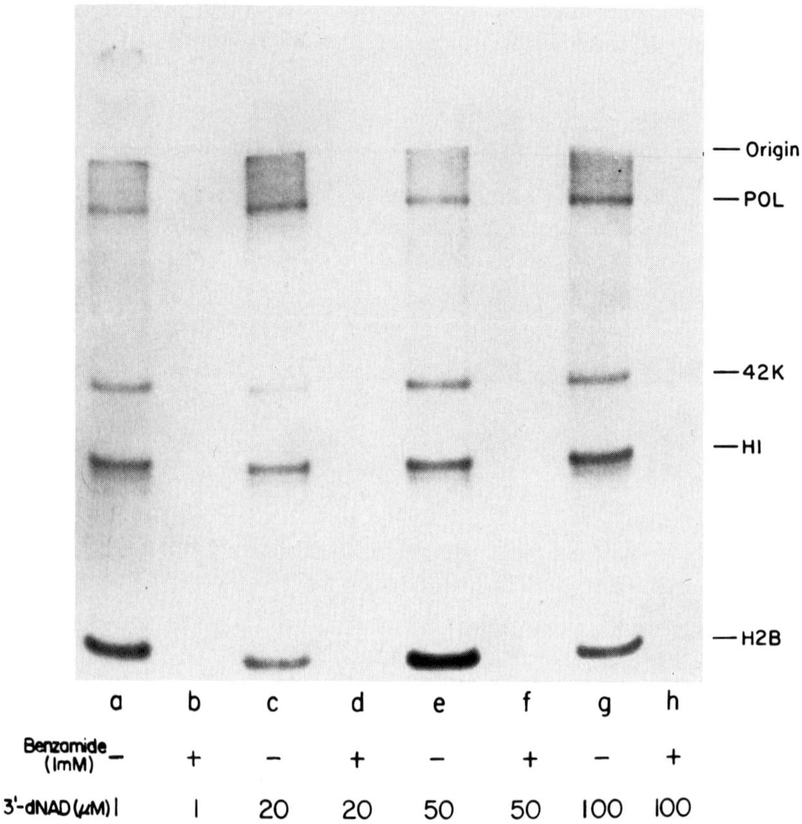

Fig. 3. Autoradiographic analysis of [^{32}P]poly(3′-deoxyADP-ribosyl)ated proteins from rat liver chromatin following incubation with increasing concentrations of [^{32}P]β3′-deoxyNAD$^+$ with or without 1 mM benzamide.

10. Katada T, Ui M: Direct modification of the membrane adenylate cyclase system by islet-activating protein due to ADP-ribosylation of a membrane protein. Proc Natl Acad Sci USA 79: 3129–3133, 1982
11. Iglewski WJ, Lee H, Muller P: ADP-ribosyltransferase from beef liver which ADP-ribosylates elongation factor 2. FEBS Letters 173: 113–118, 1984
12. Moss J, Vaughan M: Isolation of an avian erythrocyte protein possessing ADP-ribosyltransferase activity and capable of activating adenylate cyclase. Proc Natl Acad Sci USA 75: 3621–3624, 1978
13. West RE Jr, Moss J: Amino acid specific ADP-ribosylation: specific NAD:arginine monoADP-ribosyltransferases associated with turkey erythrocyte nuclei and plasma membranes. Biochemistry 25: 8057–8062, 1986
14. Tanigawa Y, Tsuchiya M, Imai Y, Shimoyama M: ADP-ribosyl transferase from hen liver nuclei. J Biol Chem 259: 2022–2029, 1984
15. Tanuma S-I, Kawashima K, Endo H: An NAD:cysteine ADP-ribosyltransferase is present in human erythrocytes. J Biochem 101: 821–824, 1987
16. Alvarez-Gonzalez R, Moss J, Niedergang C, Althaus FR: Selective probing of ADP-ribosylation reactions with oxidized 2′-deoxynicotinamide adenine dinucleotide. Biochemistry 27: 5378–5383, 1988
17. Alvarez-Gonzalez R: 3′-deoxy-NAD$^+$ as a substrate for poly(ADP-ribose) polymerase and the reaction mechanism of poly(ADP-ribose) elongation. J Biol Chem 263: 17690–17696, 1988
18. Yoshihara K, Hashida T, Yoshihara H, Tanaka Y, Ohgushi H: Enzyme-bound early product of purified poly(ADP-ribose) polymerase. Biochem Biophys Res Commun 78: 1281–1288, 1977
19. Kawaichi M, Ueda K, Hayaishi O: Multiple autopoly(ADP-ribosyl)ation of rat liver poly(ADP-ribose) synthetase: Mode of modification and properties of automodified synthetase. J Biol Chem 256: 9483–9489, 1981
20. Kawaichi M, Ueda K, Hayaishi O: Initiation of poly(ADP-ribosyl) histone synthesis by poly(ADP-ribose) synthetase. J Biol Chem 255: 816–819, 1980
21. Ueda K, Kawaichi M, Okayama H, Hayaishi O: Poly(ADP-ribosyl)ation of nuclear proteins: enzymatic elongation of chemically synthesized ADP-ribose histone adducts. J Biol Chem 254: 679–687, 1979
22. Taniguchi T: Reaction mechanism for automodification of poly(ADP-ribose) synthetase. Biochem Biophys Res Commun 147: 1008–1012, 1987
23. Mendoza-Alvarez H, Alvarez-Gonzalez R: Poly(ADP-ribose) polymerase is a catalytic dimer and the automodification reaction is intermolecular. J Biol Chem 268: 22575–22580, 1993
24. Miwa M, Saikawa N, Yamaizumi Z, Nishimura S, Sugimura T: Structure

of poly(adenosine diphosphate ribose): identification of 2′-(1′′-ribosyl-2′′-(or 3′′-) (1′′′-ribosyl)adenosine-5′,5′′,5′′′-tris-(phosphate) as a branch linkage. Proc Natl Acad Sci USA 76: 595–599, 1979

25. Ferro AM, Olivera BM: Poly(ADP-ribosylation) *in vitro*: Reaction parameters and enzyme mechanism. J Biol Chem 257: 7808–7813, 1982
26. Zahradka P, Ebisuzaki K: A shuttle mechanism for DNA-protein interactions. The regulation of poly(ADP-ribose) polymerase. Eur J Biochem 127: 579–585, 1982
27. Alvarez-Gonzalez R, Jacobson MK: Characterization of polymers of adenosine diphosphate ribose generated *in vitro* and *in vivo*. Biochemistry 26: 3218–3224, 1987
28. Alvarez-Gonzalez R, Panzeter P, Ringer DP, Mendoza-Alvarez HO: Poly-(3′-dADP-ribosyl)ation of Proteins in Liver Chromatin Isolated from Rats Fed with Hepatocarcinogens. In: G.G. Poirier, P. Moreau (eds). ADP-ribosylation Reactions. Springer-Verlag, New York, Berlin-Heidelberg, 1992, pp 149–152
29. Alvarez-Gonzalez R: Synthesis and purification of deoxyribose analogues of NAD^+ by affinity chromatography and strong anion exchange-high pressure liquid chromatography. J Chromatography 444: 89–95, 1988
30. Alvarez-Gonzalez R, Juarez-Salinas H, Jacobson E, Jacobson MK: Evaluation of immobilized boronates for studies of adenine and pyridine nucleotide metabolism Anal Biochem 135: 69–77, 1983
31. Moss J, Stanley SJ, Watkins PA: Isolation and properties of an NAD- and guanidine-dependent ADP-ribosyltransferase from turkey erythrocytes. J Biol Chem 255: 5838–5840, 1980
32. Suhadolnik RJ, Baur R, Lichtenwalner DM, Uematsu T, Roberts JH, Sudhakar S, Smulson M: ADP-ribosylation of Isolated Nuclei from He-La Cells, Rat Liver, and Novikoff Hepatoma: Effect of Nicotinamide Adenine Dinucleotide Analogs on Template Activity for DNA Synthesis, Incorporation Into Nuclear Proteins, and a New 1′′-3′ Osidic Linkage. J Biol Chem 252: 4134–4144, 1977
33. Lichtenwalner DM, Suhadolnik RJ: Adenosine Diphosphate Ribosylation of Histone and Non-histone Chromosomal Proteins with Oxidized Nicotinamide Adenine Dinucleotide and 2′-Deoxynicotinamide Adenine Dinucleotide Using Nuclei from Rat Liver and He-La Cells. Biochemistry 18: 3749–3755, 1979
34. Ikejima M, Marsishky G, Gill DM: Direction of elongation of poly(ADP-ribose) chains. J Biol Chem 262: 17641–17650, 1987
35. Alvarez-Gonzalez R, Martinez-Cadena Ma.G: Bacterial Toxin Catalyzed ADP-ribosylation of G-Proteins with 2′-deoxyNAD. ASM Abstracts *ISSN 0067-2777*: 219 (Abstract K30), 1991
36. Martinez M, Price SR, Moss J, Alvarez-Gonzalez R: Cholera Toxin-Catalyzed ADP-ribosylation of Poly(ADP-ribose)polymerase. Biophysical J 61: A35 (Abstract 197), 1992
37. Alvarez-Gonzalez R, Martinez-Cadena MaG, Stanley SJ, Moss J: Inhibition of Poly(ADP-ribose)polymerase by Arginine-Specific Mono(ADP-ribosyl)ation. Biophysical J 61: A282 (Abstract 1619), 1992
38. Desmarais Y, Menard L, Lagueux J, Poirier GG: Enzymological properties of poly(ADP-ribose)polymerase: characterization of automodifiction sites and NADase activity. Biochim Biophys Acta 1078: 179–186
39. Alvarez-Gonzalez R, Mendoza-Alvarez HO: Chemical Stability of Protein-(ADP-ribose)n Chromatin Conjugates at pH 9.0 and pH 13.0 in the Absence or Presence of Magnesium. FASEB J 7, A1094 (Abstract 242), 1993
40. Rankin PW, Jacobson EL, Benjamin RC, Moss J, Jacobson MK: Quantitative studies of inhibitors of ADP-ribosylation *in vitro* and *in vivo*. J Biol Chem 264: 4312–4317, 1989
41. Loetscher P, Alvarez-Gonzalez R, Althaus FR: Poly(ADP-ribose) may signal changing metabolic conditions to the chromatin of mammalian cells. Proc Natl Acad Sci USA 84: 1286–1289, 1987

Molecular and Cellular Biochemistry **138**: 221–226, 1994.

Enzymatic and nonenzymatic ADP-ribosylation of cysteine

Lee J. McDonald and Joel Moss

Laboratory of Cellular Metabolism, National Heart, Lung, and Blood Institute, National Institutes of Health, Building 10/ 5N-307, Bethesda, MD 20892, USA

Abstract

Mono-ADP-ribosylation is a protein modification that occurs at a number of different amino acids, dictated by the specificity of the individual ADP-ribosyltransferases. A specific cysteine in several guanine nucleotide-binding regulatory proteins is ADP-ribosylated by the bacterial protein pertussis toxin. Recent purification of an ADP-ribosylcysteine hydrolase and NAD:cysteine ADP-ribosyltransferase, and detection of ADP-ribose-cysteine linkages in tissue samples has raised hope that an endogenous regulatory cysteine-specific ADP-ribosylation pathway exists. A current goal is the identification of such a pathway for ADP-ribosylation of cysteine within animal cells. Interpretation of the data in this field has been complicated by recent reports that revealed several unforeseen chemical reactions of NAD and its metabolites with free cysteine and cysteine in proteins. This mini-review covers the latest understanding of the ADP-ribosylation reactions associated with cysteine, and provides a set of criteria for future research to establish positively the existence of an endogenous cysteine-specific mono-ADP-ribosyltransferase. (Mol Cell Biochem **138**: 221–226, 1994)

Key words: ADP-ribosylation, cysteine, pertussis toxin, mercury

Introduction

Mono-ADP-ribosylation is a post-translational regulatory modification in which ADP-ribose is transferred, by mono-ADP-ribosyltransferases, from the donor β-NAD^+ to specific amino acids in acceptor proteins. Mono-ADP-ribosyltransferases are a family of enzymes that include members from both bacteria and eukaryotes. The known enzymes are specific for the transfer of ADP-ribose to a single amino acid. For example cholera toxin, *E. coli* entertotoxins, dinitrogenase reductase ADP-ribosyltransferase from *Rhodospirillum rubrum*, and other bacterial and animal cell enzymes are specific for the transfer of ADP-ribose to protein arginine [1–5]. Diphtheria toxin, *Pseudomonas* exotoxin A, and an animal cell enzyme associated with EF-2 all ADP-ribosylate diphthamide, a modified histidine, in EF-2 [1, 2, 6]. Other mono-ADP-ribosyltransferases are specific for cysteine (pertussis toxin) or asparagine (botulinum C3-ADP-ribosyltransferase and a brain enzyme) [1–3, 7].

In recent years, research into endogenous ADP-ribosylation pathways in animal cells has lead to the identification of ADP-ribosyltransferases and enzymes that catalyze the reverse reaction, ADP-ribosyl-amino acid hydrolases. Particular progress, including molecular cloning of cDNAs that encode the enzymes involved, has been made in the pathways of arginine-specific ADP-ribosylation [4, 8, 9]. Research into endogenous ADP-ribosylation specific for cysteine has shown that ADP-ribosylcysteine linkages are formed *in vitro* and *in vivo* [10–13]. Enzymes possibly responsible for this reaction pathway, NAD:cysteine ADP-ribosyltransferase and ADP-ribosylcysteine hydrolase, were identified in animal cells [11, 14, 15]. Investigation of the ADP-ribosylation of cysteine in our laboratory has uncovered a number of complications due to the unexpected chemical reactivity of ADP-ribose and NAD with free cysteine and protein cysteine residues [16–18]. These possible reactions need to be taken into account in future investigations of endogenous cysteine-specific ADP-ribosylation pathways.

Address for offprints: L.J. McDonald, Laboratory of Cellular Metabolism, National Heart, Lung, and Blood Institute, National Institutes of Health, Building 10/ 5N-307, Bethesda, MD 20892, USA

Nonenzymatic reaction of ADP-ribose and cysteine

For many ADP-ribosyltransferases, especially those newly-defined in animal cells, the physiological substrate proteins are unknown or are not available in large amounts. In these cases, use of a model acceptor compound can be very advantageous for characterizing the ADP-ribosyltransferase. This strategy has been of particular use for the family of NAD: arginine ADP-ribosyltransferases, including both cholera toxin and animal cell enzymes, which readily ADP-ribosylate free arginine and related guanidino compounds [3, 4, 9]. Use of low molecular weight thiol compounds as substrates for cysteine-specific ADP-ribosyltransferases has been reported. A putative human erythrocyte NAD:cysteine ADP-ribosyltransferase was purified on the basis of its ability to utilize cysteine methyl ester as an acceptor [11]. It was suggested that pertussis toxin, the best characterized NAD:cysteine ADP-ribosyltransferase, can ADP-ribosylate dithiothreitol and cysteine [19].

Analysis of pertussis toxin-catalyzed ADP-ribosylation of the model acceptor cysteine revealed that cysteine reacted with ADP-ribose in the absence of any enzyme [16]. The product, which formed readily in mixtures containing 50 mM L-cysteine and 0.2 mM ADP-ribose, was separated from the precursors by anion-exchange HPLC. Of several potential acceptors examined for their ability to react with ADP-ribose, there was specificity for compounds with both amino and sulfhydryl groups. Thus, L-cysteine, D-cysteine, cysteamine and L-cysteine methyl ester reacted with ADP-ribose, but dithiothreitol, 2-mercaptoethanol, glutathione, N-acetyl-L-cysteine, alanine, serine, lysine, and arginine did not. These properties were inconsistent with a linkage via a thioglycoside bond as is formed by pertussis toxin-catalyzed ADP-ribosylation of cysteine in G-proteins in the presence of NAD. The reaction of ADP-ribose with cysteine was analogous to the reported reaction of cysteine with aldehydes or aldoses to form thiazolidines, in which the amino and sulfhydryl groups of cysteine are linked to the aldose anomeric carbon [20–23].

Sensitivity to neutral hydroxylamine distinguished ADP-ribosyl-thiazolidine from the thioglycoside-type ADP-ribosylcysteine produced by pertussis toxin in the presence of NAD and G-proteins. The known ADP-ribose-amino acid linkages differ greatly in their relative stabilities in neutral hydroxylamine [24–26]. ADP-ribosylthiazolidine was rapidly degraded in 0.5 M hydroxylamine at pH 7, with an exponential decay of half-life 13 min. Fig. 1 shows the proposed reaction scheme for the formation of ADP-ribosylthiazolidine from cysteine and ADP-ribose [16]. The first step involves a reversible attack by the free amino group of cysteine on the anomeric carbon of the ribose, which proceeds to a Schiff base. The thiolate then attacks to form the thiazolidine

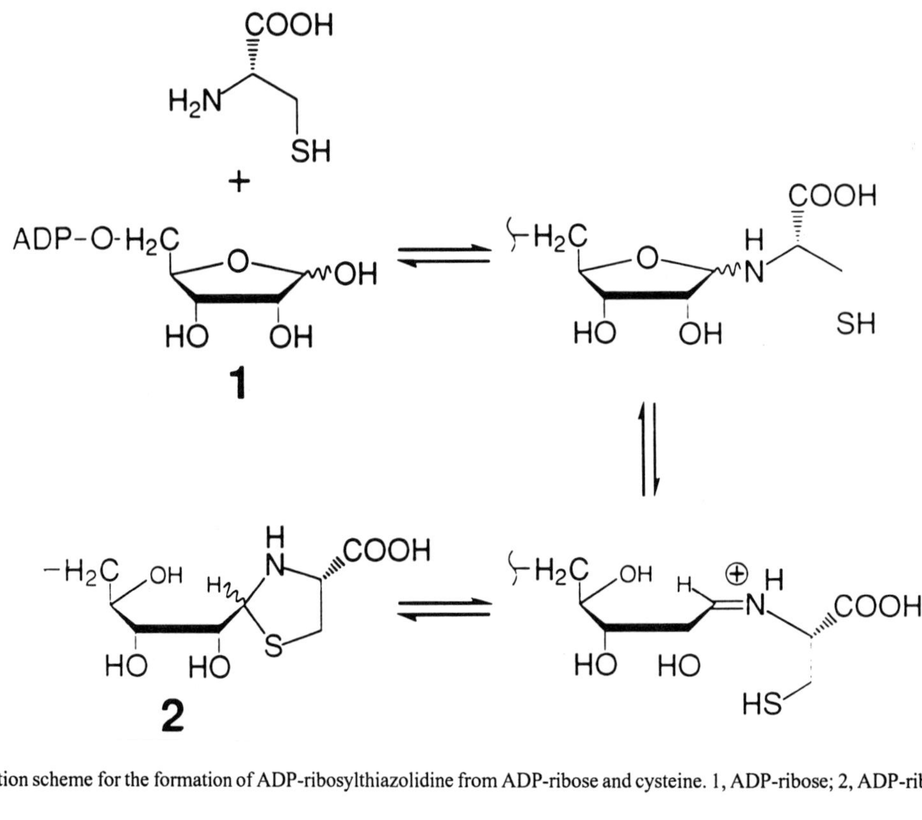

Fig. 1. Proposed reaction scheme for the formation of ADP-ribosylthiazolidine from ADP-ribose and cysteine. 1, ADP-ribose; 2, ADP-ribosylthiazolidine. [From ref. 16]

ring. In this scheme, thiol compounds lacking a primary amine should not be able to react with ADP-ribose, as was observed with the potential acceptors tested.

The identification of a nonenzymatic reaction product of ADP-ribose and cysteine suggests that this compound could form in the assay of a putative NAD:cysteine ADP-ribosyltransferase using cysteine as a substrate, especially in crude tissue samples that can contain substantial NAD glycohydrolase activity. Therefore in reactions using cysteine or related thiols as ADP-ribose acceptor, the product should be shown to be stable in hydroxylamine to exclude the thiazolidine product. Alternatively, a thiol acceptor lacking a free amine and therefore unable to react with ADP-ribose, could be used as a model acceptor.

Pertussis toxin ADP-ribosyltransferase

Pertussis toxin, also called islet activating protein, is a causative agent in the pathogenesis of whooping cough [27]. The first description of a cysteine-associated ADP-ribosylation was in the α-subunit of the retinal photo-transduction G-protein transducin, in which ADP-ribosylation of cys-347 was catalyzed by pertussis toxin [28]. Pertussis toxin ADP-ribosylates certain other G-protein α-subunits, notably those of the G_i family, and G_o [1–3]. The site of modification of the other G-protein acceptors is presumably the analogous carboxyl-terminal region cysteine that is present in the predicted amino acid sequences of these proteins [29–31].

Pertussis toxin has a very strict substrate specificity for the mentioned G-protein α-subunit cysteines over other cysteines in the same and different proteins. Alterations within the carboxyl-terminal region of pertussis toxin substrate proteins greatly diminished their ability to serve as substrates, further demonstrating the specificity of this enzyme [32, 33]. Recently, a model acceptor for the transferase activity of pertussis toxin was described, a synthetic peptide consisting of the 10 carboxyl-terminal amino acids of the α-subunit of transducin [34]. This is the only known substrate for the toxin other than the complete G-protein α-subunits mentioned.

Pertussis toxin was reported to catalyze the ADP-ribosylation of the free thiol compounds L-cysteine and dithiothreitol when incubated with NAD, but the reaction products were not identified [19]. Following the published procedure, we incubated radiolabeled NAD and thiols with pertussis toxin. The product detected from the reaction with L-cysteine was not ADP-ribosylcysteine, but was ADP-ribosylthiazolidine [16]. In the reaction containing dithiothreitol as the acceptor, the only product observed was ADP-ribose [16]. These results indicate that under the stated conditions, neither cysteine nor dithiothreitol was directly ADP-ribosylated by pertussis toxin; i.e., the toxin did not produce the expected hydroxylamine-stable thioglycoside linkage between the thiol and ADP-ribose. Instead, the NAD glycohydrolase activity of pertussis toxin produced ADP-ribose, which reacted nonenzymatically with cysteine, yielding ADP-ribosylthiazolidine with a hydroxylamine-sensitive linkage [16].

Cysteine-associated ADP-ribosylation endogenous to animal cells

Endogenous ADP-ribosylation pathways have been best described for the modification of arginine; both an NAD:arginine ADP-ribosyltransferase and ADP-ribosylarginine hydrolase have been purified, and their cDNAs cloned [8, 9]. Investigation of endogenous cysteine-associated ADP-ribosylation has not progressed to that extent, but there have been a number of reports supporting the existence of this type of protein modification in cells.

In several cell-free systems, proteins that were modified with added radiolabeled NAD contained linkages with chemical properties consistent with ADP-ribosylcysteine [10, 35, 36]. The ADP-ribosylcysteine linkage is sensitive to hydroxide and Hg^{2+} but is resistant to hydroxylamine [23, 24, 26, 37, 38]. In mitochondria, a hydroxylamine-stable, hydroxoxide-sensitive fraction of proteins that incorporated label from NAD was identified [10]. Proteins with ADP-ribosylcysteine linkages were present exclusively in the plasma membrane fraction, among liver proteins ADP-ribosylated *in vivo* without the addition of labeled NAD [13].

Together, these reports indicate that ADP-ribosylation of cysteine occurs endogenously, but does not demonstrate an enzyme-catalyzed modification to the exclusion of an nonenzymatic glycation reaction. Evidence for an enzymatic pathway for the formation of ADP-ribosylcysteine linkages was obtained in the purification of a putative NAD:cysteine ADP-ribosyltransferase from human erthrocytes [11]. This enzyme was purified on the basis of its ability to ADP-ribosylate cysteine methylester, and migrated at 27 kDa on SDS-PAGE [11, 14]. The enzyme was reported to ADP-ribosylate the α-subunit of G_i in erythrocyte and platelet membranes with a functional effect of blocking isoproteronol-mediated inhibition of adenylate cyclase activity in platelet membranes [11, 39].

The product of incubation of the erythrocyte enzyme with NAD and cysteine methylester was not conclusively identified as authentic ADP-ribosylcysteine methylester to the exclusion of the possible thiazolidine product. The assay for ADP-ribosylation of cysteine methylester involved binding to anion-exchange cellulose filters [11, 14]. In our hands, this assay was not quantitative, and produced anomalous results in reactions of increasing ionic strength (unpublished data). Binding of ADP-ribosylthiazolidine to the filters reached a plateau in reactions with increasing cysteine concentration, so that a saturation curve was obtained that provided an apparent K_m of approximately 60 mM cysteine, closely matching that reported from the kinetic analysis of the erythrocyte putative NAD:cysteine ADP-ribosyltransferase [11]. Thus the enzyme kinetics and other data for the erythrocyte enzyme with cysteine methyl ester as a substrate could be explained by an NAD glycohydrolase activity that generated free ADP-ribose followed by nonenzymatic condensation of ADP-ribose with the thiol compound to form a thiazolidine that was detected in the assay. This means the erythrocyte enzyme may not ADP-ribosylate free cysteine or cysteine methyl ester, but does not exclude the ability of the enzyme to ADP-ribosylate cysteine in protein. As discussed above, pertussis toxin does not ADP-ribosylate cysteine, but will ADP-ribosylate specific cysteine residues in proteins.

Evidence that ADP-ribosylation of cysteine in animal cells is a reversible modification was obtained with the finding of an ADP-ribosylcysteine hydrolase activity, which was partially purified from the cytosol of human erythrocytes [15]. The enzyme removed the radiolabel that was incorporated from NAD into the α-subunit of G_i by the erythrocyte ADP-ribosyltransferase, and was dependent on Ca^{2+} for activity. The product released from labeled G_i by the action of the hydrolase was identified as ADP-ribose [15].

Nonenzymatic ADP-ribosylation of protein cysteine

ADP-ribose, produced by hydrolysis of NAD, is reactive through the anomeric carbon of the terminal ribose. Nonenzymatic ADP-ribosylation occurs at several different amino acids, primarily at lysine amines through a Schiff base linkage [26, 40], but also at other amino acids [10]. Known examples of the nonenzymatic reaction of ADP-ribose with protein appear to be non-specific, as the modified proteins are among the most abundant in the cell, and the extent of ADP-ribose incorporation is low [40, 41]. Studies of nonenzymatic ADP-ribosylation, notably in the mitochondrial system, have identified a fraction of the total that cannot be attributed to modification of lysines, and has been suggested to be associated with cysteine, because of its stability in hydroxylamine [10].

A specific enzyme containing a cysteine reactive with ADP-ribose is aldehyde dehydrogenase [17]. The mitochondrial isozyme of ALDH, a tetrameric enzyme, reacted with free ADP-ribose to an extent of over 2 moles of ADP-ribose per mole of ALDH. Modification was associated with a 90% loss of enzyme activity [17]. The ADP-ribose linkages with ALDH were >95% sensitive to treatment with Hg^{2+}, indicating a very high specificity of cysteine in the reaction with this enzyme. Evidence from ADP-ribose inhibition kinetics, and from prevention of modification by the substrates NAD and acetaldehyde suggested that ADP-ribose reacted at the ALDH active site [17], which contains a cysteine involved in catalysis [43]. The chemical sensitivity of the ADP-ribosylcysteine linkage in ALDH resembled the thioglycoside linkage synthesized by pertussis toxin from NAD in the guanine nucleotide-binding regulatory proteins [24, 25], in that it was stable in hydroxylamine, but sensitive to Hg^{2+} [17]. A number of other dehydrogenases and other enzymes with reactive cysteines were modified to 10% or less of the extent of ALDH. The sensitivity to Hg^{2+} was also lower, suggesting that amino acids other than cysteine were modified. The modification of an essential cysteine in ALDH appears to be a special case of an enzyme with a reactive cysteine in a coenzyme site that has an affinity for ADP-ribose, resulting in a very high extent of modification. Recently, modification of actin with ADP-ribose was reported; the linkage was sensitive to Hg^{2+}, so probably occurred at a cysteine [42].

Nitric oxide- and NAD-dependent modification of protein cysteine

Nitric oxide (NO) is a biological messenger that has a role in vascular relaxation, neurotransmission, and other processes [44–46]. Recent reports suggested that NO stimulated the ADP-ribosylation, on a cysteine, of purified glyceraldehyde-3-phosphate dehydrogenase [47–50]. The mechanism was suggested to be auto-ADP-ribosylation of GAPDH, that was stimulated by NO. More recently, the automodification of glyceraldehyde-3-phosphate dehydrogenase was found to be not auto-ADP-ribosylation, but instead was covalent binding the entire NAD molecule to the enzyme, probably via a nitrosocysteine group in the protein [18]. Other reports of NO-stimulated ADP-ribosylation reactions, which for the most part are suggested to be cysteine-associated, include modification of actin [51], the α-subunit of transducin [52, 53], and BSA [54]. None of these modifications has yet been positively identified as ADP-ribosylation and not NAD binding.

Prospects

A suggested set of criteria for demonstrating the legitimacy of a putative NAD:cysteine mono-ADP-ribosyltransferase is shown in Fig. 2. Several control experiments are necessary to exclude the possible alternate reactions described above. An initial requirement is the demonstration of a thioglycoside linkage in the product formed, to the exclusion of a thiazolidine linkage, whether the substrate is a low molecular weight thiol, a peptide, or protein substrate. This requires affirming sensitivity to mercuric ion and resistance to neutral

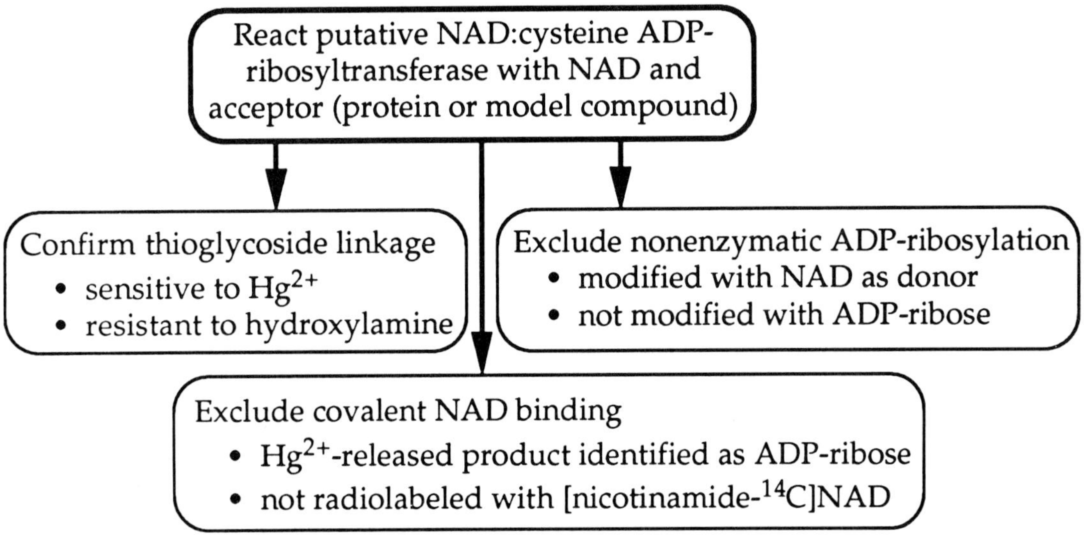

Fig. 2. Method for legitimizing a putative NAD:cysteine ADP-ribosyltransferase.

hydroxylamine. Other necessary controls include ruling out a nonenzymatic reaction of the thiol group with free ADP-ribose, by directly testing for the reaction with radiolabeled ADP-ribose, rather than NAD, and the thiol substrate. This control would be especially important in reactions that include tissue samples containing substantial NAD glycohydrolase activity, in which much of the radiolabeled NAD is broken down to radiolabeled ADP-ribose during the labeling reaction. A final test to confirm that the modification was ADP-ribosylation rather than covalent NAD binding, for example, is identification of the product released by treatment with Hg^{2+}; the product released should conform to standard ADP-ribose.

Conclusion

To date, mono-ADP-ribosyltransferases with specificity for four amino acids have been characterized, enzymes that use arginine, diphthamide, asparagine, or cysteine. The known mono-ADP-ribosyltransferases include bacterial enzymes for each amino acid [1] and animal enzymes positively-identified for arginine [4, 9], diphthamide [6], and asparagine [7], and a putative transferase specific for cysteine [11]. The reverse reaction, regeneration of free cysteine from ADP-ribosylcysteine, is catalyzed by an animal enzyme [15]. In this field, then, the current goals are positive identification of an animal cell NAD:cysteine ADP-ribosyltransferase, by including the appropriate controls as described above, followed by definition of the substrates and regulatory mechanism of this modification pathway.

Acknowledgements

The authors thank Dr. Martha Vaughan for critical reading of this manuscript.

References

1. Rappuoli R, Pizza M: Structure and evolutionary aspects of ADP-ribosylating toxins. In: J.E. Alouf and J.H. Freer (eds). Sourcebook of Bacterial Protein Toxins. Academic Press, London, 1991, pp 1–21
2. Boquet P, Gill DM: Modulation of cell functions by ADP-ribosylating bacterial toxins. In: J.E. Alouf and J.H. Freer (eds). Sourcebook of Bacterial Protein Toxins. Academic Press, London, 1991, pp 23–44
3. Moss J, Vaughan M: ADP-ribosylation of guanyl nucleotide-binding regulatory proteins by bacterial toxins. Adv Enztmol 61: 303–379, 1988
4. Williamson K, Moss J: Mono-ADP-ribosyltransferases and ADP-ribosylarginine hydrolases: a mono-ADP-ribosylation cycle in animal cells. In: J. Moss and M. Vaughan (eds). ADP-ribosylating Toxins and G Proteins: Insights into Signal Transduction. American Society for Microbiology, Washington D.C., 1990, pp 493–510
5. Lowery RG, Ludden PW: Purification and properties of dinitrogenase reductase ADP-ribosyltransferase from the photosynthetic bacterium *Rhodospirillum rubrum*. J Biol Chem 263: 16714–16719, 1988
6. Lee H, Iglewski WJ: Cellular ADP-ribosyltransferase with the same mechanism of action as diphtheria toxin and *Pseudomonas* toxin. A. Proc Natl Acad Sci USA 81: 2703–2707, 1984
7. Maehama T, Takahashi K, Ohoka Y *et al*: Identification of a botulinum C_3-like enzyme in bovine brain that catalyzes ADP-ribosylation of GTP-binding proteins. J Biol Chem 266: 10062–10065, 1991
8. Moss J, Stanley SJ, Nightingale MS *et al*: Molecular and immunological characterization of ADP-ribosylarginine hydrolases. J Biol Chem 267: 10481–10488, 1992
9. Zolkiewska A, Nightingale MS, Moss J: Molecular characterization of NAD:arginine ADP-ribosyltransferase from rabbit skeletal muscle. Proc

Natl Acad Sci USA 89: 11352–11356, 1992

10. Frei B, Richter C: Mono(ADP-ribosylation) in rat liver mitochondria. Biochem 27: 529–535, 1988
11. Tanuma S, Kawashima K, Endo H: Eukaryotic mono(ADP-ribosyl)transferase that ADP-ribosylates GTP-binding regulatory G_i protein. J Biol Chem 263: 5485–5489, 1988
12. Kots AY, Skurat AV, Bulargina TV: Study of substrates of ADP-ribosyltransferase from human erythrocytes. Biochem (New York) 55: 873–878, 1990
13. Jacobson MK, Loflin PT, Aboul-Ela N, Mingmuang M, Moss J, Jacobson EL: Modification of plasma membrane protein cysteine residues by ADP-ribose *in vivo*. J Biol Chem 265: 10825–10828, 1990
14. Tanuma S, Kawashima K, Endo H: An NAD:cysteine ADP-ribosyltransferase is present in human erythrocytes. J Biochem 101: 821–824, 1987
15. Tanuma S, Endo H: Identification in human erythrocytes of mono(ADP-ribosyl) protein hydrolase that cleaves a mono(ADP-ribosyl) G_i linkage. FEBS Lett 261: 381–384, 1990
16. McDonald LJ, Wainschel LA, Oppenheimer NJ, Moss J: Amino acid-specific ADP-ribosylation: structural characterization and chemical differentiation of ADP-ribose-cysteine adducts formed nonenzymatically and in a pertussis toxin-catalyzed reaction. Biochem 31: 11881–11887, 1992
17. McDonald LJ, Moss J: Nitric oxide-independent, thiol-associated ADP-ribosylation inactivates aldehyde dehydrogenase. J Biol Chem 268: 17878–17882, 1993
18. McDonald LJ, Moss J: Stimulation by Nitric Oxide of a Novel Linkage of NAD to Glyceraldehyde 3-Phosphate Dehydrogenase Proc. Natl. Acad. Sci. USA 90: 6238–6241, 1993
19. Lobban MD, van Heyningen S: Thiol reagents are substrates for the ADP-ribosyltransferase activity of pertussis toxin. FEBS Lett 233: 229–232, 1988
20. Schubert MP: Compounds of thiol acids with aldehydes. J Biol Chem 114: 341–350, 1936
21. Schubert MP: The combination of cysteine with sugars. J Biol Chem 130: 601–603, 1939
22. Ratner S, Clarke HT: The action of formaldehyde upon cysteine. J Am Chem Soc 59: 200–206, 1947
23. Horton D, Hutson D: Developments in the chemistry of thio sugars. Adv Carbohydrate Chem 18: 123–199, 1963
24. Hsia JA, Tsai S-C, Adamik R, Yost DA, Hewlett EL, Moss J: Amino acid-specific ADP-ribosylation: sensitivity to hydroxylamine of [cysteine(ADP-ribose)]protein and [arginine(ADP-ribose)] protein linkages. J Biol Chem 260: 16187–16191, 1985
25. Payne DM, Jacobson EL, Moss J, Jacobson MK: Modification of proteins by mono(ADP-ribosylation) *in vivo*. Biochem 24, 7540–7549, 1985
26. Cervantes-Laurean D, Minter DE, Jacobson EL, Jacobson MK: Protein glycation by ADP-ribose: studies of model conjugates. Biochemistry 32: 1528–1534, 1993
27. Olson LC: Pertussis. Medicine 54: 427–469, 1975
28. West RE Jr., Moss J, Vaughan M, Lui T, Lui T-Y: Pertussis toxin-catalyzed ADP-ribosylation of transducin: cysteine 347 is the ADP-ribose acceptor site. J Biol Chem 260: 14428–14430, 1985
29. Medynski DC, Sullivan K, Smith D *et al*: Amino acid sequence of the α subunit of transducin deduced from the cDNA sequence. Proc Natl Acad Sci USA 82: 4311–4315, 1985
30. Van Meurs KP, Angus CW, Lavu S *et al*: Deduced amino acid sequence of bovine retinal $G_{o\alpha}$: similarities to other guanine nucleotide-binding proteins. Proc Natl Acad Sci USA 84: 3107–3111, 1987
31. Itoh H, Toyama R, Kozasa T, Tsukamoto T, Matsuoka M, Kaziro Y: Presence of three distinct molecular species of G_i protein α subunit: structure of rat cDNAs and genomic DNAs. J Biol Chem 263: 6656–6664, 1988
32. Avigan J, Murtagh JJ Jr., Stevens LA, Angus CW, Moss J, Vaughan M: Pertussis toxin-catalyzed ADP-ribosylation of $G_{o\alpha}$ with mutations at the carboxyl terminus. Biochem 31: 7736–7740, 1992
33. Freismuth M, Gilman AG: Mutations of $G_{s\alpha}$ designed to alter the reactivity of the protein with bacterial toxins: substitutions at Arg ^{187} result in loss of GTPase activity. J Biol Chem 264: 21907–21914, 1989
34. Graf R, Codina J, Birnbaumer L: Peptide inhibitors of ADP-ribosylation by pertussis toxin are substrates with affinities comparable to those of the trimeric GTP-binding proteins. Mol Pharm 42: 760–764, 1992
35. Kun E, Zimber PH, Chang ACY, Puschendorf B, Grunicke H: Macromolecular enzymatic product of NAD^+ in liver mitochondria. Proc Natl Acad Sci USA 72: 1436–1440, 1975
36. Bredehorst R, Wielckens K, Gartemann A, Lengyel H, Klapproth K, Hilz H: Two different types of bonds linking single ADP-ribose residues covalently to proteins: quantification in eukaryotic cells. Eur J Biochem 92: 129–135, 1978
37. Krantz MJ, Lee YC: Quantitative hydrolysis of thioglycosides. Anal Biochem 71: 318–321, 1976
38. Meyer T, Koch R, Fanick W, Hilz H: ADP-ribosyl proteins formed by pertussis toxin are specifically cleaved by mercury ions. Biol Chem Hoppe-Seyler 369: 579–583, 1988
39. Tanuma S, Endo H: Mono(ADP-ribosyl)ation of Gi by eukaryotic cysteine-specific mono(ADP-ribose)transferase attenuates inhibition of adenylate cyclase by epinephrine. Biochem Biophys Acta 1010: 246–249, 1989
40. Kun E, Chang ACY, Sharma ML, Ferro AM, Nitecki D: Covalent modification of proteins by metabolites of NAD^+. Proc Natl Acad Sci USA 73: 3131–3135, 1976
41. Tanaka Y, Yoshihara K, Kamiya T: Enzymic and nonenzymic mono ADP-ribosylation of proteins in skeletal muscle. Biochem Biophys Res Commun 163: 1063–1070, 1989
42. Just I, Koch G, Aktories K: Endogenous ADP-ribosylation of actin. Naunyn-Schmiedeberg's Arch Pharmacol 345: R19, 1992 (abstract)
43. Saigal D, Cunningham SJ, Farrés J, Weiner H: Molecular cloning of the mitochondrial aldehyde dehydrogenase gene of *Saccharomyces cerevisiae* by genetic complementation. J Bacteriol 173: 3199–3208, 1991
44. Nathan C: Nitric oxide as a secretory product of mammalian cells. FASEB J 6: 3051–3064, 1992
45. Lowenstein CJ, Snyder SH: Nitric oxide, a novel biologic messenger. Cell 70: 705–707, 1992
46. Stamler JS, Singel DJ, Loscalzo J: Biochemistry of nitric oxide and its redox-activated forms. Science 258: 1898–1902, 1992
47. Kots, AY, Skurat AV, Sergienko EA, Bulargina TV, Severin ES: Nitroprusside stimulates the cysteine-specific mono(ADP-ribosylation) of glyceraldehyde-3-phosphate dehydrogenase from human erythrocytes. FEBS Lett 300: 9–12, 1992
48. Dimmeler S, Lottspeich F, Brüne B: Nitric oxide causes ADP-ribosylation and inhibition of glyceraldehyde-3-phosphate dehydrogenase. J Biol Chem 267: 16771–16774, 1992
49. Zhang J, Snyder SH: Nitric oxide stimulates auto-ADP-ribosylation of glyceraldehyde-3-phosphate dehydrogenase. Proc Natl Acad Sci USA 89: 9382–9385, 1992
50. Molina y Vedia L, McDonald B, Reep B, Brüne B, Di Silvio M, Billiar TR, Lapetina EG: Nitric oxide-induced S-nitrosylation of glyceraldehyde-3-phosphate dehydrogenase inhibits enzymatic activity and increases endogenous ADP-ribosylation. J Biol Chem 267: 24929–24932, 1992
51. Clancy RM, Leszczynska-Piziak J, Abramson SB: Nitric oxide stimulates ADP-ribosylation of actin in human neutrophils. Biochem Biophys Res Commun 191: 847–852, 1993
52. Ehret-Hilberer S, Nullans G, Aunis D, Virmaux N: Mono ADP-ribosylation of transducin catalyzed by rod outer segment extract. FEBS Lett 309: 394–398, 1992
53. Pozdnyakov N, Lloyd A, Reddy VN, Sitaramayya A: nitric oxide-regulated endogenous ADP-ribosylation of rod outer segment proteins. Biochem Biophys Res Commun 192: 610–615, 1993
54. Dimmeler S, Brüne B: Characterization of a nitric-oxide-catalyzed ADP-ribosylation of glyceraldehyde-3-phosphate dehydrogenase. Eur J Biochem 210: 305–310, 1992

PART VII

CYCLIC ADP-RIBOSE, NAD HYDROLYSIS AND ADP-RIBOSE SYNTHESIS

Molecular and Cellular Biochemistry **138**: 229–235, 1994.

Cyclic ADP-ribose: A calcium mobilizing metabolite of NAD$^+$

Hon Cheung Lee

Department of Physiology, 6-182 Lyon Laboratory, University of Minnesota, Minneapolis, MN 55455, USA

Abstract

Mobilization of Ca^{+2} from intracellular stores is a signalling mechanism that is of fundamental importance to many cellular processes. It is mediated by two major mechanisms, the inositol 1,4,5-trisphosphate pathway and the Ca^{+2}-induced Ca^{+2} release process. A naturally occurring metabolite of NAD$^+$ called cyclic ADP-ribose has been discovered recently and shown to be as effective as inositol 1,4,5-trisphosphate in mobilizing Ca^{+2} stores in sea urchin eggs, a marine invertebrate cell, as well as several mammalian cells. This article reviews the accumulating evidence that indicates cyclic ADP-ribose may function as a physiological regulator of the Ca^{+2}-induced Ca^{+2} release process and the current knowledge about its receptor as well as the enzymes involved in its metabolism. (Mol Cel Biochem **138**: 229–235, 1994)

Key words: cyclic ADP-ribose, inositol 1,4,5-trisphosphate, Ca^{+2}-induced Ca^{+2} release, Ca^{+2} mobilization, ADP-ribosyl cyclase, CD38, cyclic ADP-ribose hydrolase, cyclic ADP-ribose receptor, ryanodine receptor

Introduction

Mobilization of Ca^{+2} from intracellular stores is of fundamental importance to such diverse cellular processes as fertilization, cell division, muscle contraction, secretion and morphogenesis. It is mediated by two major mechanisms, the inositol trisphosphate (IP3) pathway and the Ca^{+2}-induced Ca^{+2} release (CICR) process. Since the first demonstration ten years ago that IP3 can release Ca^{+2} from internal stores of permeabilized pancreatic acinar cells [1], concerted efforts have yielded a wealth of information about the pathway, from the metabolism of IP3 to the details of how it interacts with its receptor (reviewed in 2). The CICR, on the other hand, has not been as intensely studied. It is known that the process is mediated by the ryanodine receptor (RyR), which shares a considerable sequence homology with the IP3-receptor [3]. So far, three different types of ryanodine-receptors have been identified [4]. The skeletal and cardiac form of the receptor have been purified and reconstituted into black lipid membranes and they exhibit Ca^{+2} channel activity that can be activated by micromolar concentrations of Ca^{+2} [5]. Although exogenous substances (e.g. ryanodine, caffeine, ruthenium red) can modulate the channel activity of the RyR, it is not known if the channel is regulated by endogenous molecules. In addition to mediating muscle contraction, the CICR mechanism has been proposed to play a major role in propagation of Ca^{+2} waves as well as generation of Ca^{+2} oscillations in cells. The dominant role of Ca^{+2} in regulating cellular functions underscores the likelihood of multiple mechanisms and messengers participating in the signalling process. This is further strengthened by the realization of the complexity and the dynamic nature of the intracellular Ca^{+2} stores which may express either, both or neither of the IP3-receptor and the RyR. Increasing evidence indicates that cyclic ADP-ribose (cADPR), a Ca^{+2} mobilizing metabolite of NAD$^+$, acts through the CICR mechanism and therefore may function as a physiological regulator of the RyR [6–8]. Cyclic ADP-ribose is naturally occurring and its endogenous level in various tissues has been determined [9]. The metabolite can mobilize Ca^{+2} from sea urchin eggs [10], a marine invertebrate cell, as well as several mammalian cells [11–14], indicating the generality of its action. Indeed, cADPR has recently been suggested to represent a new pathway for regulation of intracellular Ca^{+2} concentration [15].

Address for offprints: Hon Cheung Lee, Department of Physiology, 6-182 Lyon Laboratory, University of Minnesota, Minneapolis, MN 55455, USA

Enzymes involved in the metabolism of cyclic ADP-ribose

ADP-ribosyl cyclase

The enzyme that synthesizes cADPR, ADP-ribosyl cyclase (ADPR cyclase), was first identified in sea urchin eggs [10]. It was found that incubation of NAD^+ with egg extracts generated an active metabolite which could elicit Ca^{+2} release from sea urchin egg microsomes as effective as IP3 [16]. An elaborate HPLC procedure was devised to purify enough of the active substance for structural determination. Collaborating data from radioactive precursor labeling, phosphate determination, HPLC, NMR and mass spectrometry analysis, indicate that the active substance is a cyclic compound with the adenine ring of the NAD^+ molecule forming an N-glycosidic linkage with C1' of the terminal ribosyl unit and displacing the nicotinamide group [17]. This metabolite of NAD^+ is quite stable. Spontaneous hydrolysis of the linkage between the adenine group and the terminal ribose takes about 2–3 weeks at room temperatture even under acidic conditions. HPLC, NMR and mass spectrometry analyses show definitely that the hydrolysis product is ADP-ribose. Since the metabolite is a cyclic compound and can be hydrolyzed to ADP-ribose, the name cyclic ADP-ribose was proposed [17]. Fig. 1 shows the structure of cADPR. The finer details of the structure remain to be determined, such as the confirmation of the linkage between the adenine and the terminal ribose (α- or β-linkage) and that of the adenine ring (syn- or anti-conformation). In its pure form, cADPR can be readily crystallized into large plate-like crystals (Lee, unpublished observation, 1993). X-ray crystallography of cADPR would be of interest for confirming the structure and for determining the finer aspects of its structure.

ADP-ribosyl cyclase is an ubiquitous enzyme present not only in sea urchin eggs but also in a variety of mammalian tissues [18]. As shown in Table 1, it is also present in tissues from avian (chicken), amphibian (salamander), marine

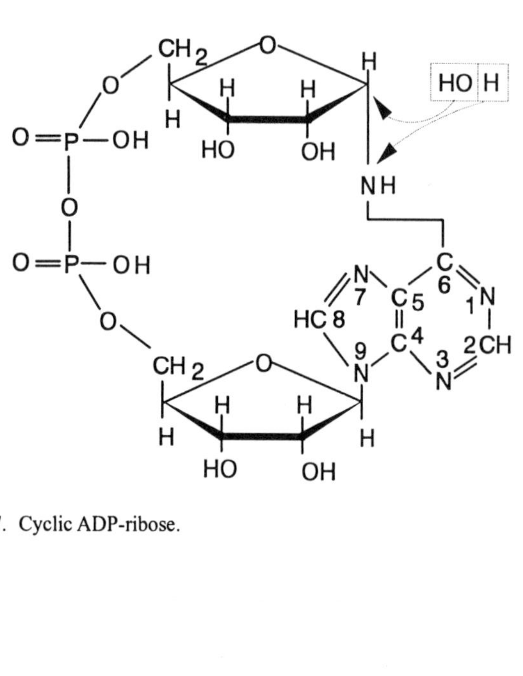

Fig. 1. Cyclic ADP-ribose.

Table 1. Widespread occurrence in animal tissues of ADP-ribosyl cyclase and cADPR hydrolase

Tissue	ADP-ribosyl cyclase (nmol/mg/hr)	cADPR hydrolase (nmol/mg/hr)
Human RBC ghosts[23]	8.4	5.4
Dog brain[22]	62.7 ± 6.2	165.4 ± 29.4
Chick embryonic brain[22]	20.6 ± 2.8	21.2 ± 5.8
Salamander brain[22]	4.5 ± 0.1	4.6 ± 0.5
Sea urchin egg[22]	0.3 ± 0.05	3.8 ± 1.6
Aplysia ovotestis[19]	1.3×10^6	–

Results summarized in this table are from references #19, 22 and 23. Experimental details are described in the references. Red blood cell is abbreviated as RBC. cADPR hydrolase activity in Aplysia ovotestis extracts was not determined.

invertebrates (sea urchin and *Aplysia*) and insect (*Drosophila* embryos) (Lee, unpublished observations, 1993). In most of these tissues, the ADPR cyclase is membrane bound. There is, however, a soluble form of the enzyme in the testes of dog, sea urchin and *Aplysia*. Both forms of the enzyme cyclize NAD^+ with high degree of substrate specificity. Several close analogs of NAD^+ including α-NAD^+ and ADP-ribose can not substitute for β-NAD^+ as substrate [18, 19]. β-NADH is the only nucleotide tested so far which can serve as a substrate for the cyclase but with lower efficiency. The fact that ADP-ribose is not a substrate for the reaction indicates the cyclase directly cyclizes NAD^+ to cADPR instead of going through ADP-ribose as an intermediate step. The steriospecificity of the cyclase reaction suggests that the adenyl-NH_2 group may attack the anomeric carbon of the terminal ribose from the α-direction to displace the nicotinamide group.

It is of interest to note that although the ADPR cyclase activity is lowest in sea urchin egg extracts, as shown in Table 1, it was the first to be identified and its identification had led to the discovery of cADPR [10, 17]. The highest cyclase activity is in *Aplysia* ovotestis extracts, and is a million times higher than that found in any other tissues tested. The *Aplysia* ADPR cyclase was previously purified based on its NAD^+-utilizing activity and was identified as a NAD glycohydralase or NADase [20, 21], which is classified as an enzyme that hydrolyzes NAD^+ to produce ADP-ribose and nicotinamide. However, analyses of the reaction products using HPLC, NMR and Ca^{+2} release activity definitely proved that the *Aplysia* ADPR cyclase does not catalyze the production of ADP-ribose [19]. Instead, it cyclizes each molecule of NAD^+ into a molecule of cADPR with the release of a molecule of nicotinamide. Although it uses NAD^+ as substrate, it is definitely not a NADase since no ADP-ribose is produced. Indeed, previous studies of NADase that rely on following either the disappearance of NAD^+ or the production of nicotinamide may require reexamination, since these two assays can not distinguish between NADase and ADPR cyclase. Furthermore, because cADPR is a recent discovery and because it is quite similar to ADP-ribose, some of the assays for ADP-ribose, may not be able to distinguish between the two. In any case, it should be clear that ADPR cyclase is enzymatically and functionally different from the degradative NADase that produces ADP-ribose, a metabolite with no Ca^{+2} mobilizing activity.

The *Aplysia* ADPR cyclase is a soluble protein and is highly basic. A very efficient method using cation exchange chromotography has been developed to routinely purify mg quantities of the cyclase from the ovotestis extracts [19]. ADPR cyclase has been crystallized and X-ray diffraction data to the resolution of about 3A° has been obtained (Levitt, personal communication). The purified enzyme appears as a single band of about 29 kDa on SDS-PAGE but can be further separated into multiple peaks by high-resolution cation-exchange chromatography. All of the protein peaks have enzymatic activity, indicating that the enzyme has multiple forms differing by charge [19, 20].

Cyclic ADP-ribose hydrolase

In addition to the synthetic pathway based on ADPR cyclase, the degradation pathway of cADPR through the action of cADPR hydrolase has also been identified [22, 23]. As shown in Table 1, the cADPR hydrolase is as widely distributed as the ADPR cyclase and is particularly active in brain extracts. The enzyme hydrolyzes the linkage between the adenine group and the terminal ribosyl unit in cADPR and produces ADP-ribose. That the product is ADP-ribose was verified by NMR and HPLC [22]. Because the enzyme hydrolyzes a rather unique linkage, it was named cADPR hydrolase [22]. Indeed, although a commercial NADase (Sigma, N-9629) effectively hydrolyzes NAD^+, it does not have any detectable cADPR hydrolase activity even when assayed at six times higher concentration (Lee, unpublished observation). The hydrolase characterized in dog brain extracts has a Km of 0.16 mM cADPR and a Vmax of 1.2 μmol/mg/hr. The brain enzyme is membrane bound and it uses cADPR as substrate without the requirement for other co-factors such as Mg^{+2}. The enzymatic activity has a broad pH optimum around neutrality. Although cADPR is a relatively stable molecule, it does spontaneous hydrolyze to ADP-ribose at a very slow rate. The pH dependence of the rate of spontaneous hydrolysis is highest in acidic range (pH 2 or lower) in contrast to that of the enzymatic reaction [22]. It thus appears that cADPR hydrolase catalyzes the same hydrolysis reaction in a greatly accelerated rate and can do so at neutral pH.

In human red blood cell (RBC) ghosts, both ADPR cyclase and cADPR hydrolase activities can be detected (Table 1 and ref. #23). The unique simplicity of the RBC provides a good model system for determining the locations of these enzymes. When the activities of both enzymes in right-side-out resealed ghosts and the detergent-treated ghosts were compared, they were found to be virtually identical [23]. This indicates that both enzymes are membrane bound and that the catalytic sites are exposed on the extracellular side of the RBC membrane. This was confirmed by the fact that both enzyme activities could be measured in intact RBC in the absence of detectable hemolysis [23]. That the ADPR cyclase is an ecto-enzyme in RBC is consistent with the previously noted sequence homology between *Aplysia* ADPR cyclase and a human lymphocyte surface antigen called CD38 [24]. Analyses of the cDNA clone of *Aplysia* ADPR cyclase indicates that the mature enzyme contains 285 amino acids [21]. Of these, 86 are identical to human CD38 and an additional 10 amino acid residues are conservative substitutions [24]. The overall homology between the two molecules is 69%. The putative transmembrane domain of CD38 consisting of 23 hydrophobic amino acids sequence is absent in the mature *Aplysia* ADP-ribosyl cyclase. This is consistent with the *Aplysia* cyclase being a soluble protein. The ten cysteine residues in the mature cyclase are perfectly aligned with those on the CD38 molecules. This suggests that the tertiary structures of the cyclase and CD38, with all the disulfide linkages formed, may be very similar. There are four potential glycosylation sites in the CD38 sequence and these are absent in the cyclase sequence [24]. The fact that ADPR cyclase in human RBC is an ecto-enzyme and that the cyclase has close homology with CD38 suggest that CD38 may have ADPR cyclase activity.

Calcium mobilization by cyclic ADP-ribose

Sea urchin egg homogenate as a biological assay for calcium release activators

The characteristics of Ca^{+2} release induced by cADPR have been extensively investigated in sea urchin eggs [6, 8, 10, 16]. In addition, several mammalian cell systems have since been shown to be responsive to cADPR [11–14]. Because of space limitation I will focus mainly in the sea urchin egg studies. This marine invertebrate cell has an extensive network of endoplasmic reticulum (ER) that forms a cell wide, interconnected compartment [25]. The ER components of the eggs can be easily isolated by gentle homogenization and are remarkably stable. Large quantities of the microsomes can be prepared and stored frozen. Loading of the microsomes with Ca^{+2} is achieved by simply incubating them with ATP, which results in activation of the endogenous Ca^{+2}-pump of the microsomes [26]. Movements of Ca^{+2} can be easily monitored by Ca^{+2} indicators. The ease of preparation and the stability of the microsomes make the egg homogenate system ideally suited for use as a biological assay for Ca^{+2} release activators. Indeed, the preparation is responsive not only to IP3 [26], an established Ca^{+2} messenger, but also to other active molecules such as cADPR and a derivative of NADP which has not yet been fully characterized [10]. It was the careful analyses of the Ca^{+2} release activity in the egg microsomes that had led to the discovery of cADPR [10]. Among cellular systems that have been shown to be sensitive to cADPR the egg microsomes is still, by far, the most responsive system. It is likely that the final unraveling of the details of the molecular machinery involved in the cADPR-sensitive Ca^{+2} release will come from analyses of the egg microsomal system.

Ca^{+2} release in sea urchin eggs can be studied at both the single cell level and biochemically, at the microsomal level. The eggs are large cells of about 100 μ in diameter. They are sturdy and amenable to microinjection. Intracellular Ca^{+2} changes in a single cell can be monitored by fluorescent Ca^{+2} indicators microinjected into the cell. Also, large quantities of eggs can be readily obtained for biochemical analyses. Ca^{+2} release can be studied in total egg homogenates, which provide a good representation of various intracellular stores and release mechanisms, or at the level of specific organelles by fractionation. The plasma membrane of the eggs with the associated cortical granules can be isolated as cell ghosts and removed from total homogenates by low speed centrifugation. The ER-type of microsomes can be cleanly separated from mitochondria by Percoll density gradient centrifugation [10, 26, 29]. The ER fraction also contains the Ca^{+2} sequestration system. In the presence of ATP, the microsomes accumulate Ca^{+2} and appear as a tight membrane band in the gradient. In the absence of ATP, the microsomes lose Ca^{+2} to varying extents and the band broadens [26]. Both the IP3- and the cADPR-sensitive Ca^{+2} release activity are found only in this fraction.

Characteristics of cADPR-sensitive calcium release

Cyclic ADP-ribose is as effective as IP3 in releasing Ca^{+2} in microsomes from *L. pictus* eggs. Both activators released similar amounts of Ca^{+2} but the half maximal effective concentration of cADPR was about 18 nM in the *L. pictus* microsomes, which was seven times lower than that of IP3 [16]. Analyses show that cADPR does not act as a nonspecific Ca^{+2}-ionophore or as an inhibitor of the microsomal Ca^{+2}-transport, instead, it is likely to be operating through a specific receptor system. In addition to being active in releasing Ca^{+2} in the cell-free system, microinjection of cADPR into live sea urchin eggs can induce a transient increase in intracellular Ca^{+2} and elicit a massive cortical granules exocytotic reaction, very much similar to those observed during normal fertilization. The injected eggs also undergo multiple cycles of nuclear membrane breakdown and reformation as well as DNA synthesis. The periods of the nuclear cycling correlate well with cell divisions in fertilized eggs [16]. Comparisons between the dose-response relationships of IP3 and cADPR show that both activators are equally effective in mobilizing Ca^{+2} in intact cells and neither of them requires external Ca^{+2} [16]. These results indicate the Ca^{+2} release mechanism activated by cADPR in egg microsomes is not an artifact of the cell-free system, but is fully operative in live cells.

Although the cADPR-sensitive Ca^{+2} release mechanism shares some similarity with that of the IP3-sensitive mechanism in terms of effectiveness and maximal amounts of Ca^{+2} released, the two systems are totally independent. Heparin, a competitive inhibitor of the IP3-receptor, has no effect on cADPR-induced Ca^{+2} release, although it can totally block the IP3-sensitive release in egg microsomes [16]. Conversely, 8-amino-cADPR, a specific antagonist of cADPR, inhibits the cADPR-dependent Ca^{+2} release without affecting the release activated by IP3 [27]. Finally, high concentrations of either cADPR and IP3 can induce desensitization in the microsomes to itself but not to the other agent [10, 16]. Similar desensitizing effects are seen in some mammalian systems [11, 13, 14]. Therefore, cells can possess two different Ca^{+2} release systems which are distinguishable pharmacologically and functionally.

Cyclic ADP-ribose as a modulator of the calcium-induced calcium release mechanism

CICR is another Ca^{+2} mobilization mechanism operating in cells. That a CICR-like mechanism is present in sea urchin eggs is shown by several lines of results. First, caffeine and ryanodine, modulators of CICR, can induce Ca^{+2} release from egg microsomes in a concentration-dependent manner [6]. Second, both modulators can also mobilize Ca^{+2} in intact eggs [6, 28]. Third, high concentrations of divalent cations such as strontium (>50 μM) can stimulate further Ca^{+2} release from egg homogenates [8]. Fourth, submaximal concentrations of caffeine can sensitize the egg homogenates such that even a relatively low concentration of Ca^{+2} induces further Ca^{+2} release from the egg homogenates [6]. Fifth, the Ca^{+2} releasing activity of caffeine, but not IP3, can be blocked by classical inhibitors of CICR such as procaine and ruthenium red [6].

Mounting evidence indicates cADPR may be operating through the CICR mechanism [6–8]. Thus, both procaine and ruthenium red can inhibit selectively the cADPR-sensitive Ca^{+2} release in egg homogenates without affecting the IP3-sensitive release [6]. High concentrations of caffeine or ryanodine activate Ca^{+2} release from egg microsomes and desensitize them so that subsequent addition of cADPR can no longer produce Ca^{+2} release. The IP3-induced release, on the other hand, is not affected. Conversely, high concentrations of cADPR also desensitize the microsomes to subsequent action of the two modulators of the CICR but not to IP3 [6]. The desensitization could be at the level of the Ca^{+2} release mechanism or it could be due to depletion of a common Ca^{+2} store. Further support for cADPR and caffeine activating the same Ca^{+2} release mechanism comes from the finding of cross potentiation between them [8]. Caffeine, at low concentrations which are not sufficient to elicit Ca^{+2} release, sensitizes the egg microsomes to cADPR by reducing its half maximal concentration about 10-fold. Similarly, low concentrations of cADPR also potentiate the Ca^{+2} release activity of caffeine. Therefore, inhibitors of CICR block the cADPR-sensitive Ca+2 release while an agonist of CICR potentiates it. Taken together, these results show that cADPR and caffeine not only release Ca^{+2} from the same store but also act on the same release mechanism.

Direct effects of cADPR on the divalent cation-induced Ca^{+2} release have been demonstrated [8]. In the presence of low concentrations of cADPR, addition of divalent cations to egg homogenates stimulate further release of Ca^{+2} in a concentration-dependent manner. The selectivity of this process is $Ca^{+2} > Sr^{+2} >> Ba^{+2} \approx 0$. In the absence of cADPR, only high concentrations of strontium (> 50 μM) can induce Ca^{+2} release. The effective concentrations of Sr^{+2}, however, are reduced 10–20 fold in the presence of low concentrations of cADPR. These results are consistent with cADPR being a modulator of CICR by sensitizing the mechanism to divalent cations. In this respect, the action of cADPR and caffeine are similar, except that cADPR is five orders of magnitude more effective.

Antagonists of cyclic ADP-ribose

That cADPR is active in the nanomolar range suggests that it is operating through a specific receptor system. The structure-function relationship between cADPR and its receptor has been investigated by synthesizing a series of analogs modified at the 8-position of the adenine group [27]. Substitutions of the 8-hydrogen of the adenine ring with a bromo-, amino- or azido-group all result in inactivity, indicating that the 8-position is critically important to the Ca^{+2} release activity of cADPR. All these analogs, however, are able to compete for specific ^{32}P-cADPR binding to egg microsomes. Indeed, 8-amino-cADPR is as effective as cADPR itself in competing for microsomal binding [27]. Therefore, occupation of the receptor is a necessary condition for Ca^{+2} release but not a sufficient one. In addition, the receptor has to interact with cADPR at the 8-position of the adenine ring in a specific manner before the Ca^{+2} release channel can be activated.

The fact that the 8-substituted analogs can bind to the receptor but not activate Ca^{+2} release suggests that they may function as reversible antagonists. This is the case since all of them inhibit cADPR from releasing Ca^{+2} with different potency, even though none of them release Ca^{+2} by themselves. Among them, 8-amino-cADPR, is the most effective. Pre-treatment of the egg microsomes with 9 nM of 8-amino-cADPR reduced in half the rate of Ca^{+2} release induced by 135 nM cADPR [27]. As expected, the inhibitory effect of 8-amino-cADPR is reversible and can be overcome by higher concentrations of cADPR. The overall effect of the antagonist is to shift the dose-response of cADPR toward higher cADPR concentrations, increasing the half-maximal effective concentration of cADPR by 6–7 fold. Thus, high concentrations of cADPR can displace 8-amino-cADPR from the receptor site and recover the Ca^{+2} release activity. The converse is also true. Addition of higher concentrations of the antagonist after the cADPR-induced Ca^{+2} release has commenced immediately stops the release process [27]. Therefore, continuous occupation of the receptor by cADPR is a necessary condition for maintaining the activation of the Ca^{+2} release mechanism.

Cyclic ADP-ribose receptors

Direct demonstration of the presence of cADPR-receptor in egg microsomes comes from ligand binding studies [29]. Radioactive cADPR with high specific activity can be synthesized from ^{32}P-NAD^+ using *Aplysia* ADPR cyclase. Specific binding of ^{32}P-cADPR to egg ER microsomes is much higher than to mitochondria fractions. As expected, IP3 is more than a 1,000 times less effective than cADPR in competing for the binding sites. Furthermore, heparin, at a concentration that blocks the IP3-dependent Ca^{+2} release, has very little effect on cADPR binding, indicating the cADPR-receptor is different from that of IP3 [29]. Scatchard analyses show that the receptor density is about 25 fmol/mg protein and the affinity is about 17 nM. The low receptor density is due to the presence of an inhibitor in the egg homogenates. This inhibitor is of low molecular weight and is heat labile. Removal of the inhibitor from the egg homogenates by dialysis increases the receptor density and makes the demonstration of specific binding to egg microsomes much easier [27, 30].

Photoaffinity labeling has been used to identify the receptor for cADPR in sea urchin egg microsomes [30]. The antagonist, 8-azido-cADPR, is photoactive. It does not release Ca^{+2} by itself but can effectively compete for cADPR binding sites. A ^{32}P-labeled antagonist with high specific activity has been synthesized and specific binding to egg microsomes has been demonstrated. Photoaffinity labeling shows that a 140 kDa and a 100 kDa protein are specifically labeled by ^{32}P-8-azido-cADPR [30]. The presence of nanomolar concentrations of cADPR totally blocks the labeling of these proteins, while other adenine nucleotides, such as ATP, ADP, AMP and cAMP have no effect even at micromolar concentrations. As expected, IP3 also has no effect on the labeling. On the other hand, caffeine, at millimolar concentrations, preferentially blocks the labeling of the 100 kDa protein without much effect on the labeling of the 140 kDa, suggesting that the 100 kDa protein may be the target site of caffeine.

The molecular weights of the two labeled proteins in egg microsomes are much smaller than the skeletal and cardiac RyR [4, 5], which is believed to be responsible for mediating the CICR. The possibility that the 140 kDa and the 100 kDa proteins represent variant forms of RyR can not be definitely ruled out at the present time. Immunoblot analysis has detected a protein of 380 kDa in sea urchin eggs that cross reacts with antibodies raised against skeletal RyR [31]. Whether this protein is a RyR remained to be established, since it has not been shown that the protein can, in fact, bind ryanodine or function as a Ca^{+2} channel.

It is possible that cADPR does not act on RyR directly but, instead, binds first to the 140 kDa and 100 kDa proteins forming a complex. The activated complex then interacts with RyR resulting in Ca^{+2} release. In addition to the fact that the cADPR binding proteins have molecular weights different from RyR, two other lines of evidence are also consistent with such a notion. First, modulators of the RyR such as, ryanodine, ruthenium red and procaine, have very little effects on the photoaffinity labeling of the 140 kDa and 100 kDa proteins [30]. Second, the antagonist, 8-amino-cADPR, blocks cADPR-dependent Ca^{+2} release effectively but it does not inhibit Ca^{+2} release induced by ryanodine or caffeine. The potentiation of the caffeine and divalent cation-induced Ca^{+2} release by cADPR, however, can be blocked by the antagonist [27]. In any case, the success of photoaffinity labeling the cADPR-binding proteins should set the stage for their eventual purification and identification. It may turn out that the Ca^{+2} release activated by cADPR is a multi-step process, much more complicated than the action of IP3.

Physiological roles of cyclic ADP-ribose

Fertilization

Mobilization of intracellular Ca^{+2} stores is associated with fertilization of sea urchin eggs. The resulting transient increase in internal Ca^{+2} triggers massive exocytosis, the cortical reaction, leading to formation of the fertilization envelope, a protein membrane responsible for prevention of polyspermy. The Ca^{+2} change together with alkalinization of internal pH are the two ionic signals that initiate development [32]. The similarity between sperm-egg interaction during fertilization and ligand binding to surface receptors has led to the proposal that IP3 is the second messenger mediating the Ca^{+2} changes. Indeed, phosphoinositide metabolism increases during fertilization [33, 34] and microinjection of IP3 into unfertilized eggs can activate Ca^{+2} changes similar to those occurring during fertilization [35]. However, the proposal is contradicted by the fact that eggs preloaded with heparin, an antagonist of the IP3-receptor, exhibit delayed, but otherwise normal Ca^{+2} changes during fertilization [36–38]. Similarly, eggs microinjected with 8-amino-cADPR, an antagonist of the cADPR pathway, also exhibit normal Ca^{+2} changes at fertilization [38]. These results indicate that either a third messenger, which has not yet been identified, is responsible for mediating the Ca^{+2} mobilization at fertilization or that both the cADPR- and the IP3-sensitive mechanisms are involved. The last possibility has proven to be correct. Eggs preloaded with both 8-amino-cADPR and heparin fail to show Ca^{+2} changes after fertilization [38]. The cortical reaction associated with fertilization is also blocked by the presence of both antagonists. Sperm incorporation into these eggs is normal and, in fact, many of them show polyspermy due to failure of the formation of the fertilization envelope [38]. Similar results were obtained using ruthenium red and heparin to block the cADPR- and IP3-sensitive mechanism respectively (Galione, personal communication, 1993). These studies provide the first direct evidence that cADPR is involved in mediating Ca^{+2} mobilization during fertilization.

Perspective

Since the discovery of cADPR in 1987 [10] there has been a steady accumulation of information concerning this novel metabolite of NAD^+. Its structure has been determined [17] and its natural occurrence demonstrated [9]. The metabolite can now be produced in large quantity and has been crystallized (Lee, unpublished results, 1993). Much progress has been made in identifying the metabolic enzymes of cADPR. Both its synthesizing enzyme, ADPR cyclase [18], and its degradation enzyme, cADPR hydrolase [22], have been shown to be ubiquitous in animal tissues. A soluble form of ADPR cyclase from *Aplysia* has been purified [19, 20], sequenced [21] and crystallized (Levitt, personal communication, 1993).

Since the original demonstration of the Ca^{+2} mobilizing activity of cADPR in sea urchin eggs [10], several mammalian systems have since been shown to also be responsive [11–14]. In all cases, cADPR is as effective as IP3 in releasing Ca^{+2} but through a totally independent mechanism [16]. Specific binding of cADPR to egg microsomes has been demonstrated [29] and photoaffinity labeling has identified a 140 kDa and a 100 kDa proteins as specific cADPR binding proteins [30]. Antagonists that can specifically block the cADPR-sensitive Ca^{+2} release have been developed [27] and used to provide direct evidence that cADPR is involved in mediating Ca^{+2} mobilization at fertilization [38]. Accumulating evidence indicates the Ca^{+2} release system activated by cADPR is pharmacologically and functionally very similar to CICR [6, 8]. It is proposed that cADPR may either act as a modulator of CICR or is itself a Ca^{+2} messenger. In the former case, cADPR would be constitutively present at low levels in cells, sensitizing the CICR system to Ca^{+2} in a manner similar to that seen in sea urchin egg microsomes [8]. The availability of antagonists that block cADPR from potentiating CICR [27] should be useful for investigating this aspect of the cADPR function. Alternatively, cADPR can be a Ca^{+2} messenger. In this case, cellular levels of cADPR would increase in response to stimuli. Indeed, in pancreatic islets, metabolic stimulation by glucose apparently increases cADPR levels and activates insulin secretion [13]. The best assay for cADPR currently available is a bio-assay based on its Ca^{+2} release activty [9]. Future development of an easier and more sensitive assay, such as a radio-immuno assay, would facilitate the measurements of the changes in the endogenous levels of cADPR in cells. Although cADPR was discovered because of its Ca^{+2} mobilizing activity [10] this may not be its only function. The fact that ADPR cyclase and cADPR hydrolase in RBC are ecto-enzymes, suggests that cADPR may have extracellular functions. Similar to other metabolites, such as cAMP and ATP, these functions may or may not be related to its intracellular function of mobilizing Ca^{+2}.

Acknowledgements

I thank Robert Aarhus, Richard Graeff and Timothy F. Walseth for critical reading of the manuscript. My research is partially supported by a grant from NIH (HD17484).

References

1. Streb H, Irvine RF, Berridge M, Schulz I: Release of Ca^{+2} from a non-mitochondrial intracellular store in pancreatic acinar cells by inositol 1,4,5-trisphosphate. Nature 306: 67–69, 1983
2. Berridge M: Inositol trisphosphate and calcium signalling. Nature 361: 315–325, 1993
3. Furuichi T, Yoshikawa S, Miyawaki A, Wada K, Maeda N, Mikoshiba K: Primary structure and functional expression of the inositol 1,4,5-trisphosphate-binding protein P_{400}. Nature 342: 32–38, 1989
4. Sorrentino V, Volpe P: Ryanodine receptors: how many, where and why? Trends Pharmacol Sci 14: 98–103, 1993
5. Smith JS, Imagawa T, Ma J, Fill M, Campbell KP, Coronado R: Purified ryanodine receptor from rabbit skeletal muscle is the calcium-release channel of sarcoplasmic reticulum. J Gen Physiol 92: 1–26, 1988
6. Galione A, Lee HC, Busa WB: Ca^{+2}-induced Ca^{+2} release in sea urchin egg homogenates and its modulation by cyclic ADP-ribose. Science 253: 1143–1146, 1991
7. Galione A: Ca^{+2}-induced Ca^{+2} release and its modulation by cyclic ADP-ribose. Trends Pharm Sci 13: 304–306, 1992
8. Lee HC: Potentiation of calcium- and caffeine-induced calcium release by cyclic ADP-ribose. J Biol Chem 268: 293–299, 1993
9. Walseth TF, Aarhus R, Zeleznikar RJ Jr, Lee HC: Determination of endogenous levels of cyclic ADP-ribose in rat tissues. Biochem Biophys Acta 1094: 113–120, 1991
10. Clapper DL, Walseth TF, Dargie PJ, Lee HC: Pyridine nucleotide metabolites stimulate calcium release from sea urchin egg microsomes desensitized to inositon trisphosphate. J Biol Chem 262: 9561–9568, 1987
11. Koshiyama H, Lee HC, Tashjian AH Jr: Novel mechanism of intracellular calcium release in pituitary cells. J Biol Chem 266: 16985–16988, 1991
12. Currie KPM, Swann K, Galione A, Scott RH: Activation of Ca^{+2} dependent currents in cultured rat dorsal ganglion neurones by a sperm factor and cyclic ADP-ribose. Mol Biol Cell 3: 1415–1425, 1992
13. Takasawa S, Nata K, Yonekura H, Okamoto H: Cyclic ADP-ribose in insulin secretion from pancreatic β cells. Science 259: 370–373, 1993
14. White MA, Watson SP, Galione A: Cyclic ADP-ribose-induced Ca^{+2} release from rat brain microsomes. FEBS letters 318: 259–263, 1993
15. Galione A: Cyclic ADP-ribose: A new way to control calcium. Science 259: 325–326, 1993
16. Dargie PJ, Agre MC, Lee HC: Comparison of Ca^{+2} mobilizing activities of cyclic ADP-ribose and inositol trisphosphate. Cell Regulation 1: 279–290, 1990
17. Lee HC, Walseth TF, Bratt GT, Hayes RN, Clapper DL: Structural determination of a cyclic metabolite of NAD^+ with intracellular Ca^{+2} mobilizing activity. J Biol Chem 264: 1608–1615, 1989
18. Rusinko N, Lee HC: Widespread occurrence in animal tissues of an enzyme catalyzing the conversion of NAD^+ into a cyclic metabolite with intracellular Ca^{+2} mobilixing activity. J Biol Chem 264: 11725–11731, 1989
19. Lee HC, Aarhus R: ADP-ribosyl cyclase: an enzyme that cyclazes NAD^+ into a calcium-mobilizing metabolite. Cell Regulation 2: 203–209, 1991
20. Hellmich MR, Strumwasser F: Purification and characterization of a molluscan egg-specific NADase, a second-messenger enzyme. Cell Regulation 2: 193–202, 1991
21. Glick DL, Hellmich MR, Beushausen S, Tempst P, Bayley H, Strumwasser F: Primary structure of a molluscan egg-specific NADase, a second-messenger enzyme. Cell Regulation 2: 211–218, 1991
22. Lee HC, Aarhus R: Wide distribution of an enzyme that catalyzes the hydrolysis of cyclic ADP-ribose. Biochem Biophys Acta 1164: 68–74, 1993
23. Lee HC, Zocchi E, Guida L, Franco L, Benatti U, De Flora A: Production and hydrolysis of cyclic ADP-ribose at the outer surface of human erythrocytes. Biochem Biophys Res Commun 191: 639–645, 1993
24. States DJ, Walseth TF, Lee HC: Similarities in amino acid sequences in *Aplysia* ADP-ribosyl cyclase and human lymphocyte antigen CD38. Trens Biochem Sci 17: 495, 1992
25. Terasaki M, Jaffe LA: Organization of the sea urchin egg endoplasmic reticulum and its reorganization at fertilization. J Cell Biol 114: 929–940, 1991
26. Clapper DL, Lee HC: Inositol trisphosphate induces Ca^{+2} release from non-mitochondrial stores in sea urchin egg homogenates. J Biol Chem 260: 13947–13954, 1985
27. Walseth TF, Lee HC: Synthesis and characterization of antagonists of cyclic ADP-ribose. Biochem Biophys Acta 1178: 235–242, 1993
28. Buck WR, Rakow TL, Shen SS: Synergistic release of calcium in sea urchin eggs by caffeine and ryanodine. Exptl Cell Res 202: 59–66, 1992
29. Lee HC: Specific binding of cyclic ADP-ribose to calcium-storing microsomes from sea urchin eggs. J Biol Chem 266: 2276–2281, 1991
30. Lee HC, Walseth TF, Aarhus R, Levitt D: Cyclic ADP-ribose as a messenger for mobilizing intracellular stores. J Reproduct Develop 39: 63–64, 1993
31. McPherson SM, McPherson PS, Mathews L, Campbell KP, Longo FJ: Cortical localization of a calcium release channel in sea urchin eggs. J Cell Biol 116: 1111–1121, 1992
32. Whitaker MJ, Steinhardt RA: Ionic regulation of egg activation. Quart Rev Biophys 15: 593–666, 1982
33. Turner PR, Sheetz MP, Jaffe LA: Fertilization increases the polyphosphoinositide content of sea urchin eggs. Nature 310: 464–475, 1984
34. Ciapa B, Whitaker M: Two phases of inositol polyphosphate and diacylglycerol production at fertilization. FEBS lett 195: 347–351, 1986
35. Whitaker M, Irvine RF: Microinjection of inositol trisphosphate activates sea urchin eggs. Nature 312: 636–638, 1984
36. Rakow T, Shen SS: Multiple stores of calcium are released in the sea urchin egg during fertilization. Proc Natl Acad Sci 87: 9285–9289, 1990
37. Crossley I, Whalley T, Whitaker MJ: Guanosine 5'-thiotriphosphate may stimulate phosphoinositide messenger production in sea urchin eggs by a different route than the fertilizing sperm. Cell Regul 2: 121–133, 1991
38. Lee HC, Aarhus R, Walseth TF: Calcium mobilization by dual receptors during fertilization of sea urchin eggs. Science 261: 352–355, 1993

Molecular and Cellular Biochemistry **138**: 237–243, 1994.

NAD Glycohydrolases: A possible function in calcium homeostasis

Hyuntae Kim[1], Elaine L. Jacobson[2,3] and Myron K. Jacobson[2,3]

[1] *Division of Medicinal Chemistry and Pharmaceutics, College of Pharmacy,* [2] *Department of Clinical Sciences, College of Allied Health Professions and* [3] *Lucille P. Markey Cancer Center, University of Kentucky, Lexington, KY 76017, USA*

Abstract

NAD glycohydrolases are the longest known enzymes that catalyze ADP-ribose transfer. The function of these ubiquitous, membrane-bound enzymes has been a long standing puzzle. The NAD glycohydrolases are briefly reviewed in light of the discovery by our laboratory that NAD glycohydrolases are bifunctional enzymes that can catalyze both the synthesis and hydrolysis of cyclic ADP-ribose, a putative second messenger of calcium homeostasis. (Mol Cell Biochem **138**: 237–243, 1994)

Key words: NAD glycohydrolases, cyclic ADP-ribose, ADP-ribose transfer, calcium homeostasis, nicotinamide adenine dinucleotide

Abbreviations: NADase – nicotinamide adenine dinucleotide glycohydrolase, NAD – nicotinamide adenine dinucleotide; ADP-ribose – adenosine diphosphoribose, cADPR – cyclic adenosine diphosphoribose

Introduction

Nicotinamide adenine dinucleotide glycohydrolases (NADase, EC 3.2.2.5) were the first ADP-ribose transferring enzymes discovered. Their description in animal tissues by Handler and Klein in 1942 [1] preceded the detection of other classes of ADP-ribosyltransferases by more than 20 years. Like other ADP-ribosyltransferases, NADases catalyze the cleavage of the nicotinamide-ribose bond of oxidized β-NAD and the transfer of the ADP-ribose moiety to an acceptor nucleophile. NADases are distinguished from the other classes of ADP-ribose transferring enzymes in that they efficiently use water as the ADP-ribose acceptor.

Although NADases have been known for more than 50 years, the physiological function of these enzymes in cellular metabolism remains unknown. Their ability to efficiently catalyze hydrolysis of NAD has been a long standing enigma considering the metabolic expense of the hydrolysis of the nicotinamide ribose bond, which has a high negative free energy of hydrolysis. Fig. 1 shows the metabolic conversions required for the resynthesis of NAD following its conversion to ADP-ribose. The group transfer potential of at least 6 phosphoanhydride linkages are required to resynthesize NAD. Energetic considerations alone suggest that the ability of NADases to utilize other acceptors is likely related to their physiological function. In this chapter, we briefly review the NADases in light of the recent discover that NADases can catalyze both the synthesis and degradation of a cyclic form of ADP-ribose (cADPR) [2], a putative second messenger involved in the regulation of intracellular Ca^{2+}, signalling [3–6]. For more comprehensive information about NADases, the reader is referred to the excellent review of Price and Pekala [7].

Results and discussion

NADases are widely distributed in both procaryotes and eucaryotes. The microbial NADases are soluble enzymes;

Address for offprints: M. K. Jacobson, College of Pharmacy, University of Kentucky, Lexington, KY 40536-0082, USA

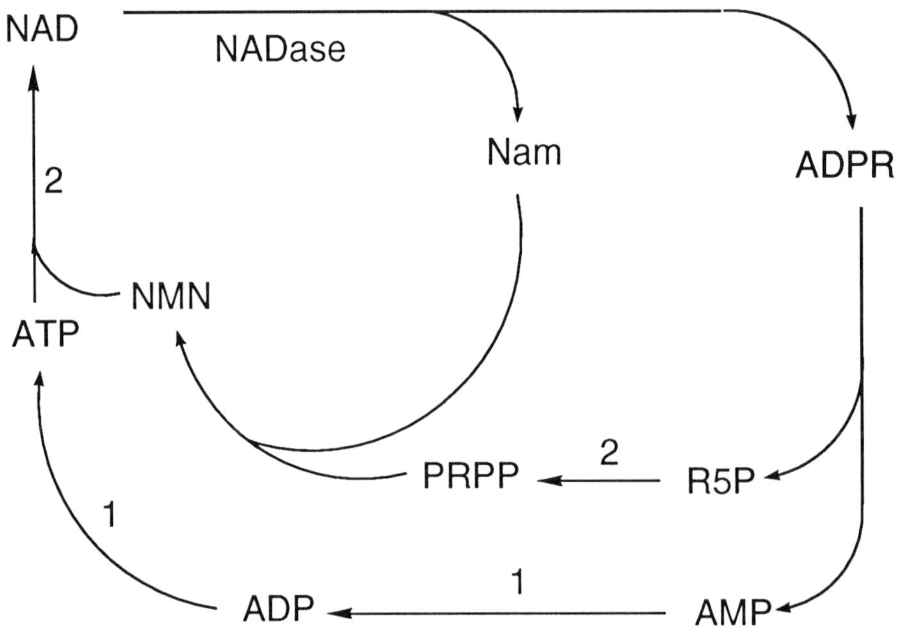

Fig. 1. Metabolic conversions for the resynthesis of NAD following its conversion to nicotinamide and ADP-ribose. The numbers beside the arrows indicate the number of moles of phosphanhydride linkage necessary to effect the conversion of one mole of substrate. Abbreviations used are: Nam, nicotinamide; ADPR, ADP-ribose; R5P, ribose 5'-phosphate; AMP, adenosine 5' monophosphate; PRPP, 5-phosphoribosyl 1 pyrophosphate; ADP, adenosine 5'-diphosphate; ATP, adenosine 5'-triphosphate; NMN, nicotinamide 5'-mononucleotide.

but all mammalian NADases described are membrane associated, except those present in seminal fluid [7]. Much of the activity of the mammalian enzymes is associated with the plasma membrane fraction although activity has been shown also in microsomal and mitochondrial membranes [7, 8]. Most erythrocyte NADase activity is located on the outer plasma membrane surface and the enzyme has been shown to be anchored to the membrane via a glycosylphosphatidylinositol linkage [9]. Membrane-bound NADase activity generally can be solubilized with detergents such as Triton X-100, Emulphogene BC-720, or deoxycholate, or by treatment with pancreatic lipase. The erythrocyte NADase can be specifically solubilized with phosphatidylinositol-specific phospholipase C [9]. The possibility that membrane-bound NADases from other tissues also may be phosphatidylinositol-anchored has yet to be investigated.

NADases have been isolated from many different sources. The enzyme was first purified to apparent homogeneity from acetone powders of porcine brain [10]. NADases have also been purified to electrophoretic homogeneity from other mammalian tissues including rat liver, calf spleen and canine spleen [2, 11, 12]. The general approach to purification of the membrane bound enzymes involves initial solubilization of membrane fractions with detergents or pancreatic lipase followed by a series of chromatographic fractionations. Dye-ligand column chromatography has proven to be particularly useful for purification of NADases [2, 13]. It is noteworthy that isonicotinate hydrazide has been used to elute the canine spleen enzyme from Red A Sepharose resulting in a high degree of purification [2]. Soluble NADases have been purified from *N. crassa*, a number of bacterial strains, *B. facsiatus* venom, and bovine seminal fluid [13–16].

A wide range of molecular weight values have been observed among NADases [17]. Within each species, similar molecular weight values are observed; however, charge heterogeneity has been shown due to carbohydrate modifications that occur on these enzymes. A number of the enzymes exist as homodimers under nondenaturing conditions [7]. NADases typically exhibit a broad pH-optimum with maxima at or near neutrality. The microbial enzymes have higher K_m values for NAD than the mammalian enzymes. For example, K_m values of 500 μM were estimated for *N. crassa* and *B. subtilis* enzymes [14], while the canine spleen enzyme has a K_m value of 9.9 μM [2], which is the lowest K_m value reported to date.

Many NADases utilize acceptors other than water as illustrated by their ability to catalyze a reaction in which the nicotinamide ring is exchanged for other pyridine derivatives, a reaction termed transglycosidation [7]. In addition, some NADases will transfer ADP-ribose to other nucleophiles such as alcohols and imidazole compounds [18, 19]. In contrast to many protein: mono-ADP-ribosyltransferases, guanidino compounds are ineffective as acceptors for NADases. The activity of NADases is often monitored by the release of nicotinamide from NAD. This led to the discovery of a novel enzyme from the ovotestis of *A. californica* that releases nicotinamide from NAD yet the other product is not ADP-ribose but cADPR [20]. This cyclic form of ADP-ribose was discovered by Lee and co-workers as a metabolite of NAD that mediates Ca^{2+} release from sea urchin microsomes [21]. The synthesis of cADPR from NAD involves cleavage of the nicotinamide-ribose bond and cyclization of the ribose to the adenine ring. Structural characterizations by Lee and co-workers demonstrated that cyclization involves linkage of the anomeric carbon of ribose to a nitrogen of adenine, but could not conclusively determine the position of the linkage. They proposed a structure for cADPR with cyclization at N^6 of the adenine ring [3]. Recent spectral studies from our laboratory support a structure in which cyclization is to position 1 of the adenine ring [22]. Fig. 2 shows the structure that we have proposed for this nucleotide.

Several lines of evidence support the postulation that

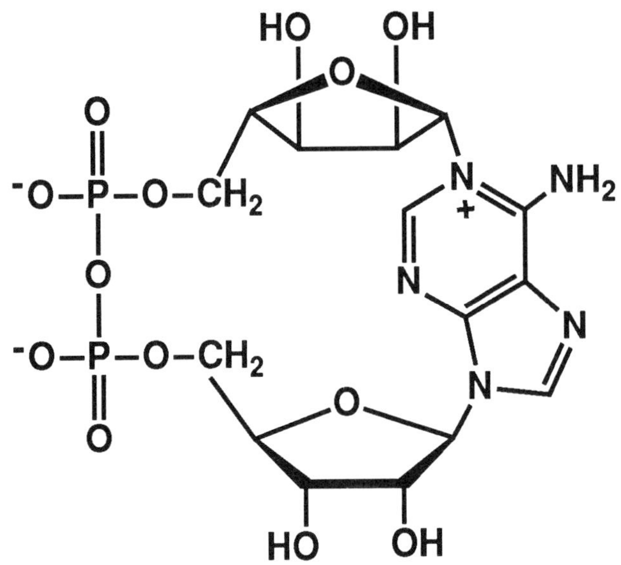

Fig. 2. Proposed structure of cyclic ADP-ribose.

cADPR is a second messenger involved in the regulation of Ca^{2+} channels is mammalian cells. The ubiquitous occurrence of low levels of cADPR in mammalian tissues has been demonstrated [6] and it has been shown to be a potent mediator of Ca^{2+} release in mammalian cell microsomes [23]. Other studies have implicated cADPR in Ca^{2+} induced Ca^{2+} release [24, 25] and insulin secretion [4]. If cADPR is a second messenger, it must be degraded soon after fulfilling its signalling function. We searched for cADPR metabolizing activity in canine tissues and detected activity in homogenates from several different tissues. We purified cADPR degrading enzymes from membrane fractions of canine liver and spleen. The purification of the enzyme from canine spleen to electrophoretic homogeneity and its characterization have been described recently [2]. The enzymes from both liver and spleen catalyze the hydrolysis of cADPR to ADP-ribose. They exhibit high affinity for cADPR with apparent K_m values of 42 μM (liver) and 46 μM (spleen). During the characterization of the enzymes, it was observed that NAD was a potent inhibitor of the cADPR hydrolase activity with apparent K_i values of 6.3 μM (liver) and 8.3 μM (spleen). The relationship of the cADPR hydrolases to NADases was established when it was demonstrated that the enzymes efficiently catalyzed the conversion of NAD to ADP-ribose with apparent K_m values for NAD of 7.3 μM (liver) and 9.9 μM (spleen). Thus, the enzymes are both cADPR hydrolases and NADases.

The mechanism of action of the NADases of calf spleen and snake venom have been extensively studied [13, 18, 26-31]. Chemical modification studies have suggested the presence of essential arginine and carboxylate residues [19, 32]. The arginine residues are putatively involved in binding the pyrophosphate moiety of NAD and the carboxyl residues possibly bind the pyridinium ring or stabilize an oxocarbonium ion intermediate. Steady-state kinetic analysis indicates that the kinetic mechanism for NADases is ping-pong bi-bi- when using acceptors other than water and ordered uni-bi when water is the acceptor [26]. Studies of isotope exchange, the transglycosidation reaction, and competition between water and methanol as acceptor suggest that the rate limiting step is the formation of an enzyme-ADP-ribosyl intermediate [26]. Studies of kinetic α-secondary deuterium isotope effects and linear free-energy relationships provided evidence that the reaction proceeds through a unimolecular decomposition of the substrate involving an enzyme-stabilized oxocarbonium ion intermediate rather than a covalent intermediate [27, 31]. A general reaction scheme for NADases that summarizes mechanism studies is shown in Fig. 3. The oxocarbonium ion intermediate [E·ADPR*] can react with nicotinamide to regenerate NAD, with other nucleophiles (X:) such as pyridines, imidazoles, and alcohols with retention of configuration, or with water to generate ADP-ribose.

The ability of the enzymes isolated from canine liver and spleen to hydrolize both NAD and cADPR has led us to propose a mechanism in which NAD and cADPR share the common reaction intermediate, [E·ADPR*] as shown in Fig. 3. We have examined the proposed mechanism with the enzymes from both canine liver and spleen, initially testing for cADPR conversion to NAD in the presence of nicotinamide. Fig. 4 shows that when the canine liver enzyme was incubated with cADPR and nicotinamide, NAD formation was observed that was dependent on time of incubation (panel A) and nicotinamide concentration (panel B). The conversion of cADPR to NAD in the presence of nicotinamide was also catalyzed by NADases from snake venom, porcine brain and calf spleen.

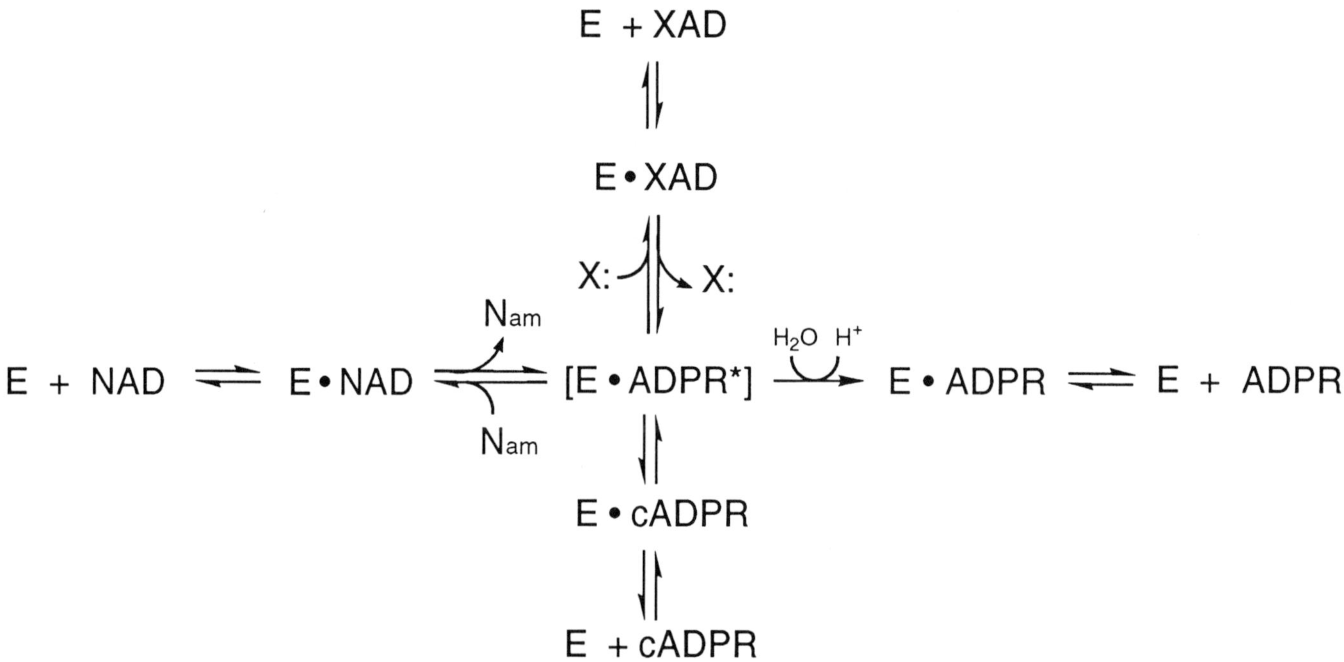

Fig. 3. Scheme depicting previously postulated mechanisms for reactions catalyzed by NADases and the mechanism for the synthesis and hydrolysis of cADPR. X: is used to represent nucleophiles such as pyridine derivatives, imidazole derivatives and alcohols.

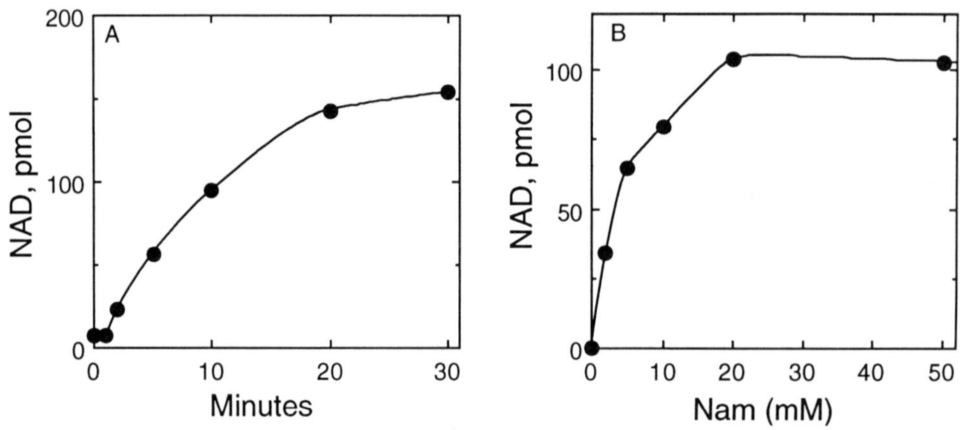

Fig. 4. Formation of NAD from cADPR and nicotinamide catalyzed by canine liver NADase. Panel A, Samples containing 200 mM cADPR and 20 mM nicotinamide in 50 µl total volume were incubated at 37°C for the indicated times in the presence of 1 milliunit of enzyme [2]. Reactions were terminated by placing samples on ice and adding 4 µl of ice cold 70% perchloric acid. Samples were neutralized with 25 µl of 2 M KOH, and 0.1 M K_2CO_3 before assay. NAD was quantified in 10 µl aliquots by an enzymic cycling assay [45] adapted to microtiter plates. Panel B, Incubations containing varying amounts of nicotinamide analyzed following incubation for 10 minutes.

From the mechanism proposed in Fig. 3, it can be predicted also that the enzyme would catalyze the synthesis of cADPR from NAD. The enzymes from both canine liver and spleen enzymes as well as the snake venom NADase converted NAD to cADPR [2]. Fig. 5 shows that the production of cADPR from NAD catalyzed by the liver enzyme increased progressively with time and reached a maximum at 20 min, representing a 2.3% net conversion of NAD to cADPR at that time. After 20 min, the amount of cADPR gradually declined as cADPR was degraded by the enzyme. The net conversion of NAD to cADPR reflects the relative rates of two competing events, namely release of cADPR from the enzyme or hydrolysis of cADPR to ADP-ribose. At present, our studies cannot distinguish between two possibilities for the conversion of NAD to cADPR. It is possible that, following conversion of NAD to the oxocarbonium ion intermediate, that a competition between water and adenine results in only a small fraction of con-

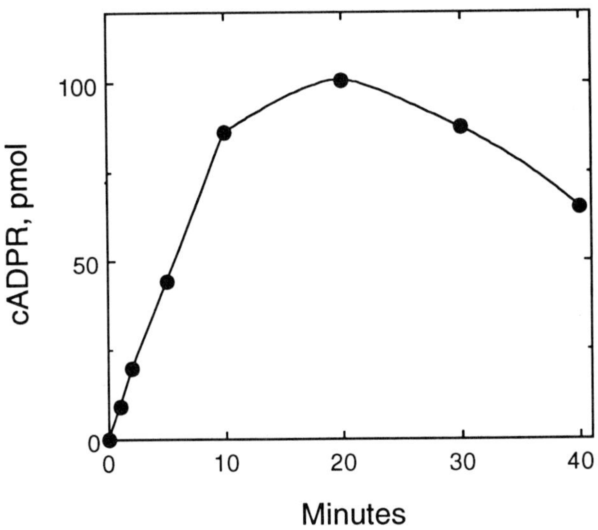

Fig. 5. Time course of formation of cADPR from NAD catalyzed by canine liver NADase. Each reaction mixture contained 100 μM [^{32}P]NAD (1.71 x 10^6 cpm), and 0.16 milliunit of enzyme in a total volume of 50 μl containing 20 mM HEPES, pH 7.2 and 0.1% Triton X-100 with incubation at 37°C for the times indicated. Samples were analyzed for cADPR as described elsewhere [2].

version to cADPR, the remainder being converted directly to ADP-ribose. However, it is also possible that, following the conversion of NAD to the oxocarbonium ion, water is not accessible to the active site. This would result in complete conversion of NAD to cADPR. Following a conformational change, cADPR would then be converted via the oxocarbonium ion intermediate to ADP-ribose. Further studies will be necessary to distinguish between these two possibilities. In either case, our studies show that NADases can function as cADPR synthetases.

The search for the physiological function of NADases has spanned more than five decades. A regulatory function for NADases was suggested by observations that the tissue activity of these enzymes changed in response to disease states [33–35] and drug treatments [36-38]. It has been proposed that metabolic processes could be regulated by NADases through an effect on cellular NAD levels [39]. However, correlations between cellular NAD levels and NADase activity are variable and contradictory. The ability of NADases to both synthesize and degrade cADPR raises the interesting possibility that NADases play a fundamental role in cADPR metabolism and Ca^{+2} homeostasis. It is interesting to note that alterations in both NADase activity and Ca^{+2} levels have been shown to accompany differentiation [40, 41] and fertilization [42, 43]. Also, the studies of Richter and co-workers have established a strong correlation between NAD hydrolysis and Ca^{+2} release in mitochondria [8].

The role of cytosolic free Ca^{+2} as a regulator of numerous physiological processes is now well established [44]. Alterations in intracellular Ca^{+2} are effected by the transient activation of membrane Ca^{+2} channels in response to specific signals. The role of inositol polyphosphates in the binding to specific components of Ca^{+2} channels causing channel activation has been well documented [44]. Lee and co-workers have provided evidence that cADPR also activates specific Ca^{+2} channels by binding to a cADPR receptor protein [5].

It is possible that the cADPR synthetase and cADPR hydrolase activities of the NADase function independently in cADPR metabolism. Under certain conditions, the hydrolysis function of the enzyme may be blocked in which

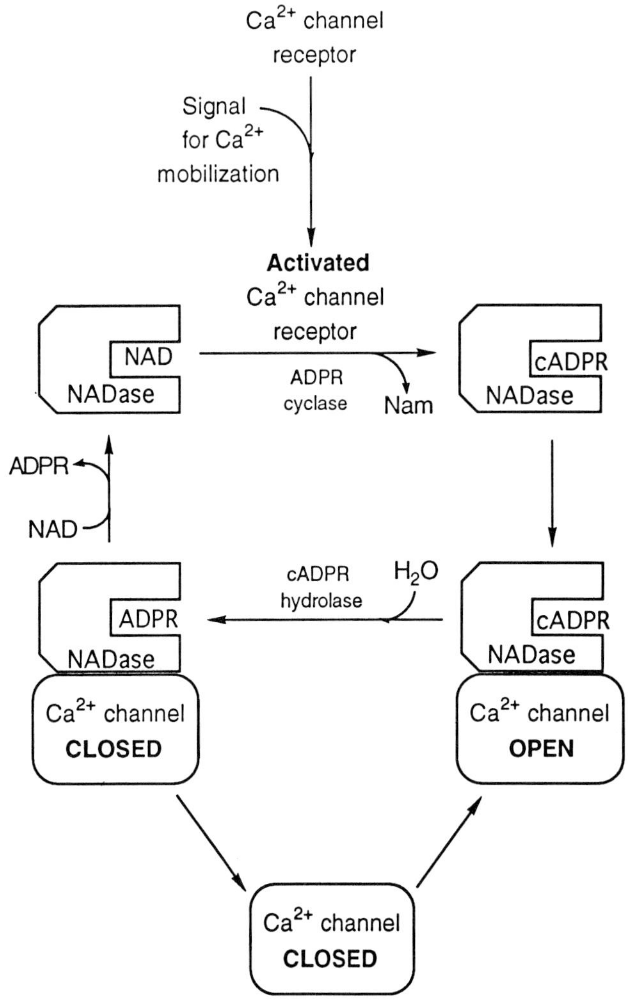

Fig. 6. Hypothetical scheme for the possible involvement of a bifunctional NADase in the transient activation of a calcium channel.

case the enzyme would function as a cADPR synthetase, releasing cADPR to bind to a specific Ca^{+2} channel protein resulting in activation. Under other conditions, the enzyme might function to degrade cADPR. However, the finding that NADases are bifunctional enzymes that can catalyze both the synthesis and hydrolysis of cADPR raises the interesting possibility that the NADase may be involved in the activation of the Ca^{+2} channel *per se*. Presented in Fig. 6 is a scheme depicting a hypothesis yet to be tested concerning this possibility. In this scheme, the NADase would serve as a link between the Ca^{+2} channel and the Ca^{+2} channel receptor(s) which receive signals for opening of the channel. Activation of the receptor would result in activation of the NADase, resulting in the conversion of enzyme bound NAD to enzyme bound cADPR. The enzyme bound cADPR could directly effect opening of the channel. In this model, the opening of the channel would be transient due to the intrinsic cADPR hydrolase activity, which would convert cADPR to ADP-ribose, resulting in closing of the channel. Exchange of ADP-ribose by NAD would be required before another round of activation could occur. It is possible that the ability of cADPR to directly activate Ca^{+2} channels *in vitro* [5] involves direct binding of cADPR to the NADase, eliminating the need for activation of the enzyme by the receptor. Studies to test this model are ongoing in our laboratory.

Further studies will be necessary to determine whether NADases function in cADPR metabolism *in vivo* and, if so, how they may relate to the regulation of Ca^{+2} homeostasis. However, their ability to both synthesize and degrade cADPR indicates that these fascinating enzymes should be investigated from a new perspective.

Acknowledgements

This work was supported by National Institutes of Health Grant CA 43894 and National Science Foundation Grant DMB 8917274.

References

1 Handler P, Klein JR: The inactivation of pyridine nucleotides by animal tissues *in vitro*. J Biol Chem 143: 49–57, 1942
2. Kim H, Jacobson EL, Jacobson MK: Synthesis and degradation of cyclic ADP-ribose by NAD glycohydrolases. Science 261: 1330–1333, 1993
3. Lee HC, Walseth TF, Bratt, GT, Hayes RN, Clapper DL: Structural determination of a cyclic metabolite of NAD^+ with intracellular Ca^{2+}. J Biol Chem 264: 1608–1615, 1989
4. Takasawa S, Nata K, Yonekura H, Okamoto, H: Cyclic ADP-ribose in insulin secretion from pancreatic B cells. Science 259: 370–373, 1993
5. Lee HC: Specific binding of cyclic ADP-ribose to calcium-storing microsomes from sea urchin eggs. J Biol Chem 266: 2276–2281, 1991
6. Walseth TF, Aaarhus R, Zeleznikar RJ Jr, Lee HC: Determination of endogenous levels of cyclic ADP-ribose in rat tissues. Biochem Biophys Acta 1094: 113–120, 1991
7. Price SR, Pekala PH: Pyridine nucleotide-linked glycohydrolases. I: D Dolphin, O Avramovic and R Poulson (eds). Pyridine Nucleotide Coenzymes: Chemical, Biochemical, and Medical Aspects. Wiley-Interscience, New York, 1987, pp 513–548
8. Richter C, Kass GEN: Oxidative stress in mitochondria: Its relationship to cellular Ca^{+2} homeostasis, cell death, proliferation, and differentiation. Chem–Biol Interactions 77: 1–23, 1991
9. Kim U, Rockwood SF, Kim H, Daynes RA: Membrane-associated NAD^+ glycohydrolase from rabbit erythrocytes is solubilized by phosphatidylinositol-specific phospholipase C. Biochem Biophys Acta: 965. 76–81, 1988
10. Swislocki NI, Kaplan NO: Purification and characterization of diphosphopyridine nucleotidase from pig brain. J Biol Chem 242: 1083–1088, 1967
11. Diaugustine RP, Abe T, Voytek P: Purification and comparative properties of the glycoprotein nicotinamide adenine dinucleotide glycohydrolase from rat liver microsomal and plasma membranes. Biochem Biophys Acta 526: 518–530, 1978
12. Schuber F, Travo P: Calf-spleen nicotinamide-adenine dinucleotide glycohydrolase: Solubilization, purification and properties of the enzyme. Eur J Biochem 65: 247–255, 1976
13. Yost DA, Anderson BM: Purification and properties of the soluble NAD glycohydrolase from *Bungarus fasciatus* venom. J Biol Chem 256: 3647–3653, 1981
14. Everse J, Everse KE, Kaplan NO: The pyridine nucleotidases from *Bacillus subtillis* and *Neurospora crassa*: Isolation and structural properties. Arch Biochem Biophys 169: 702–713, 1975
15. Grushoff PS, Shany S, Bernheimer AW: Purification and properties of streptococcal nicotinamide adenine dinucleotide glycohydrolase. J Bacteriol 122: 599–605, 1975
16. Yuan JH, Anderson BM: Bull semen nicotinamide adenine dinucleotide nucleotidase: I. Purification and properties of the enzyme. J Biol Chem 246: 2111–2115, 1971
17. Swislocki NI, Kalish MI, Chasalow FI, Kaplan NO: Solubilization and comparative properties of some mammalian diphosphopyridine nucleotidases. J Biol Chem 242: 1089–1094, 1967
18. Yost D, Anderson BM: Adenosine diphosphoribose transfer reactions catalyzed by *Bungarus fasciatus* venom NAD glycohydrolase. J Biol Chem 258: 3075–3080, 1987
19. Yost, DA, Anderson BM: Studies of *Bungarus fasciatus* venom NAD glycohydrolase: Inhibition, modification, and transglycosidation reactions. J Biol Chem 257: 767–772, 1982
20. Hellmich MR, Strumwasser F: Purification and characterization of a molluscan egg-specific NADase, a second messenger enzyme. Cell Regul 2: 193–202, 1991
21. Clapper DL, Walseth TF, Dargie PJ, Lee HC: Pyridine nucleotide metabolites stimulate calcium release from sea urchin egg microsomes desensitized to inositol trisphosphate. J Biol Chem 262: 9561–9568, 1987
22. Kim H, Jacobson EL, Jacobson MK: Position of cyclization in cyclic ADP-ribose. Biochem Biophys Res Commun 194: 1143–1147, 1993
23. Koshiyama H, Lee HC, Tashjian AH Jr: Novel mechanism of intracellular calcium release in pituitary cells. J Biol Chem 266: 16985–16988, 1991
24. Galione A, Lee HC, Busa, WB: Ca^{2+} induced Ca^{2+} release in sea urchin egg homogenates: Modulation by cyclic ADP-ribose. Science 253: 1143–1146, 1991
25. Lee HC: Potentiation of calcium-and caffeine-induced calcium release by cyclic ADP-ribose: J Biol Chem 268: 293–299, 1993
26. Schuber F, Travo P, Pascal M: Calf-spleen nicotinamide-adenine dinucleotide glycohydrolase: Kinetic mechanism. Eur J Biochem 69: 593–602, 1976
27. Bull HG, Ferraz JP, Cordez EH, Ribbi A, Apitz-Castro R: Concerning the mechanism of the enzymatic and nonenzymatic hydrolysis of nicotinamide nucleotide coenzymes. J Biol Chem 253: 5186–5192, 1978

28. Tarnus C, Muller H, Schuber F: Chemical evidence in favor of a stabilized oxocarbonium-ion intermediate in the NAD+ glycohydrolase-catalyzed reactions. Bioorg Chem 16: 38–51, 1988
29. Schuber F, Travo P, Pascal M: On the mechanism of action of calf spleen NAD+ glycohydrolase. Bioorg Chem 8: 83–90, 1979
30. Muller-Steffner H, Malver O, Hoise L, Oppenheimer NJ, Schuber F: Slow-binding inhibition of NAD+ glycohydrolase by arabino analogue of β-NAD+. J Biol Chem 267: 9606–9611, 1983
31. Tarnus C, Schuber F: Application of linear free-energy relationships to the mechanistic probing of nonenzymatic and NAD+-glycohydrolase-catalyzed hydrolysis of pyridine dinucleotides. Bioorg Chem 15: 31–42, 1987
32. Schuber F, Pascal M, Travo P: Calf spleen nicotinamide-adenine dinucleotide glycohydrolase: Properties of the active site. Eur J Biochem 83: 205–214, 1982
33. McKee RW, Dickey A, Parks ME: NAD-ase activity and glycolysis in Ehrlich-Lettre carcinoma and liver cell particles. Arch Biochem Biophys 126: 760–763, 1968
34. Skala H, Lenoir GM, Pichard AL, Vuillaume M, Dreyfus JC: Elevated NAD(P) glycohydrolase activity: A possible enzymatic marker for malignancy in Burkitt's lymphoma cells. Blood 60: 912–917, 1982
35. Lombard Y, Hartman D, Coupin G, Benezra R, Poindron P: Human B-like lymphoblastoid cell lines obtained by long-term culture of normal spleen leukocytes. Biol Cell 50: 295–298, 1984
36. Shah NS: Subcellular distribution of nicotinamide adenine dinucleotidase in the spleen of tuberculous guinea pigs and tuberculous guinea pigs treated with isoniazid. Amer Rev Res Dis 96: 11–16, 1967
37. Toida I, Ando F, Yamamoto S: Biochemical study on the host response to tuberculous infection IV: Influences of isoniazid, streptomycin, nicotinamide, and cortisone. Amer Rev Res Dis 98: 424–428, 1968
38. Tsukagoshi S, Rao MH, Goldin A: Effects of 1,3-bis(2-chloroethyl)-1-nitrosourea (NSC-409962) and cyclophosphamide (NSC-26271) on nicotinamide adenine dinucleotide glycohydrolase in normal and leukemic mouse tissues. Can Chemother Rep 52: 569–578, 1968
39. Kaplan NO: Current aspects of biochemical energetics. In: NO Kaplan, and EP Kennedy (eds). Curr Aspects Biochem Energ Academic Press, New York, 1966, pp. 447–458
40. Hemmi H, Breitman TR: Induction by retinoic acid of NAD+-glycohydrolase activity of myelomonocytic cell lines HL-60, THP-1 and U-937, and fresh human acute promyelocytic leukemia cells in primary culture. Biochem Biophys Res Commun 109: 669–674, 1982
41. Rovera G, Santoli D, Damsky C: Human promyelocytic leukemia cells in culture differentiate into macrophage-like cells when treated with a phorbol diester. Proc Natl Acad Sci USA 76: 2779–2783, 1979
42. Williams GT, Ford CC, Shall S: NAD-glycohydrolase activity in *Xenopus laevis* oocytes and early embryos. Biochem Biophys Res Commun 108: 36–41, 1982
43. Carafoli E: Intracellular Ca^{2+} homeostasis. Annu Rev Biochem 56: 395–433, 1987
44. Berridge ML, Irvine RF: Inositol phosphates and cell signalling. Nature 341: 197–205, 1989
45. Jacobson EL, Jacobson MK, Pyridine Nucleotide Levels as a Function of Growth in Normal and Transformed 3T3 Cells. Arch Biochem Biophys 175: 627–634, 1976

Molecular and Cellular Biochemistry **138:** 245–251, 1994.

NAD hydrolysis: Chemical and enzymatic mechanisms

Norman J. Oppenheimer
Department of Pharmaceutical Chemistry S-926, University of California, San Francisco, CA 94143-0446, USA

Abstract

The pyridine nucleotides have important non-redox activities as cellular effectors and metabolic regulators [1–3]. The enzyme-catalyzed cleavage of the nicotinamide–ribosyl bond of NAD^+ and the attendant delivery of the ADPRibosyl moiety to acceptors is central to these many diverse biological activities. Included are the medically important NAD-dependent toxins associated with cholera, diphtheria, pertussis, and related diseases [4]; the reversible ADPRibosylation-mediated biological regulatory systems [5,6]; the synthesis of poly(ADPRibose) in response to DNA damage or cellular division [7]; and the synthesis of cyclic ADPRibose as part of an independent, calcium-mediated regulatory system[8]. As will be presented in this chapter, all evidence points to both the chemical and enzyme-catalyzed cleavage of the nicotinamide–ribosyl bond being dissociative in character via an oxocarbenium intermediate. (Mol Cell Biochem **138:** 245–252, 1994)

Key words: ADP-ribosylation, NAD hydrolysis, chemical mechanisms, NADase

Chemical Hydrolysis of NAD^+

NAD^+ + :Acceptor ® ADPR-Acceptor + Nic

pH-dependence of NAD^+. NAD^+ has long been known to be unstable toward basic conditions, undergoing specific base-catalyzed hydrolysis with no evidence for buffer catalysis [9]. At concentrations above 1 M hydroxide, the primary reaction shifts to formation of 2-pyridone-3-carboxaldehyde, generated by nucleophilic attack of hydroxide on the ring and a subsequent ring opening and reclosure [10, 11]. The change in mechanism and the origins of specific base catalysis were explained by Johnson *et al.* [12]. They found that the pH profile for the non-enzymatic hydrolysis of NAD^+ paralleled the ionization of the ribose diol (pK_a = 11.9 at 37°C), establishing that the strongly electron-donating diol anion was responsible for the alkaline lability of NAD^+, with a rate approximately four orders of magnitude faster than for the neutral diol. In the pH-independent region above pH 12.5 for hydrolysis, increasing concentrations of hydroxide only serves to generate the pyridone [12]; there is no evidence for hydroxide-dependent displacement of nicotinamide. Underscoring the importance of diol ionization, $O^{2'}$,$O^{3'}$-isopropylidene nicotinamide riboside shows no pH-dependent cleavage of the nicotinamide–ribosyl bond and only formation of pyridone at high pH [12]. The rate of cleavage of the ribosyl linkage is little affected by solvent or the presence of external nucleophiles. The rate of cleavage in 50% aqueous methanol is essentially the same as in water, and incubation with 2 M azide accelerates the cleavage by <40%, when contributions from salt effects are excluded [13]. Finally, partitioning between water and methanol is on the basis of their molefraction and not their intrinsic nucleophilicity, a result consistent with the presence of a highly reactive oxocarbenium that is not solvent equilibrated [12, 14].

Secondary deuterium kinetic isotope effects on the chemical hydrolysis are consistent with extensive rehybridization at the anomeric carbon [15–17]. The values listed in Table 1 for k_H/k_D of 1.144 and 1.12 for the chemical hydrolysis of [^{2}H-N 1′]-nicotinamide riboside at pH 4.5 and 9.7, respectively, establish that both the diol and diol anion have approximately the same effects on the rehybridization of the activated complex for hydrolysis. The large values for the 2° KIE also argue against significant nucleophilic participation in the reaction.

Address for offprints: N.J. Oppenheimer, Department of Pharmaceutical Chemistry S-926, University of California, San Francisco, CA 94143-0446, USA

Table 1. α-Secondary Deuterium Isotope effects for Hydrolysis of [²H-N1′]-NAD, [²H-N1′]-Nicotinamide Mononucleotide, and [²H-N1′]-Nicotinamide Riboside

	Isotope effect (kH/kD)		
Reaction	NAD^+	NMN^+	NR^+
pH-independent hydrolysis(16) (pH 3.4 and 4.5)	1.101	1.135	1.144
Hydroxide-catalyzed(16) (pH 9.74)	1.15	–	1.12
Pig Brain NADase(17)	0.985	1.132	1.083
N. crassa NADase(17)	1.003	1.100	
E. coli NMNase(17)	–	1.00	1.02

Substituent effects on the rate of NAD^+ hydrolysis. The rate of hydrolysis of NAD^+ analogues containing 3- or 4-substituted pyridines correlates with their pK_a, with a β_{LG}=-1.11 for data extrapolated to 37°C [18]. A plot of the log of the rate of pH-independent hydrolysis of the nicotinamide glycosyl bond for a series of 2′-substituents β-nicotinamide arabinosides [19] depends linearly on the inductive Taft constant, σ_i, with a slope, ρ = -6.7 (see Fig. 1). The corresponding plot for a series of 2-substituted β-nicotinamide ribosides (Fig. 1) gives a slop of -7.1 [20]. The large negative values of both β_{LG} and ρ provide strong evidence for a dissociative reaction with extensive bond cleavage via a highly electron deficient activated complex.

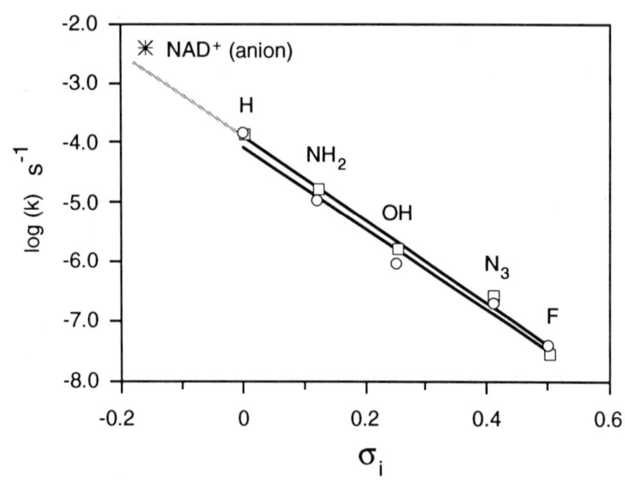

Fig. 1. Plot of the log of the rate of hydrolysis of 2′-substituted NAD analogues vs. the Taft σ_i. The results for 2′-substituted ribo analogues, designated by (❑), and for 2′-substituted arabino analogues with a (O). Data was obtained at 25°C and under conditions where the reaction was pH independent.

Anchimeric assistance with formation of 1,2-anhydrosugar intermediates has been previously invoked in the alkaline hydrolysis of glycosyl bonds and nucleosides [21]. The linearity of the σ_i plot, especially the points for the 2′-amino arabinoside and riboside, a substituent with ample precedent for anchimeric involvement in other systems, precludes significant anchimeric assistance in these cases. More importantly, the ribose diol anion of NAD^+ also falls on the line, thus there is no indication of the special rate acceleration anticipated for anchimeric assistance. Any observation, of a 1,2-anhydrosugar as a product would not even be evidence for this mechanism because it could arise from collapse of an oxocarbenium intermediate subsequent to bond cleavage. Finally, because nicotinamide nucleosides are intrinsically cationic, only the substituent effect on the oxocarbenium will be observed. In contrast, substituent effects for the acid-catalyzed hydrolysis of glycosides and nucleosides are a function of both the rate of cleavage and the pK_a of the conjugate acid of the leaving group [14, 17].

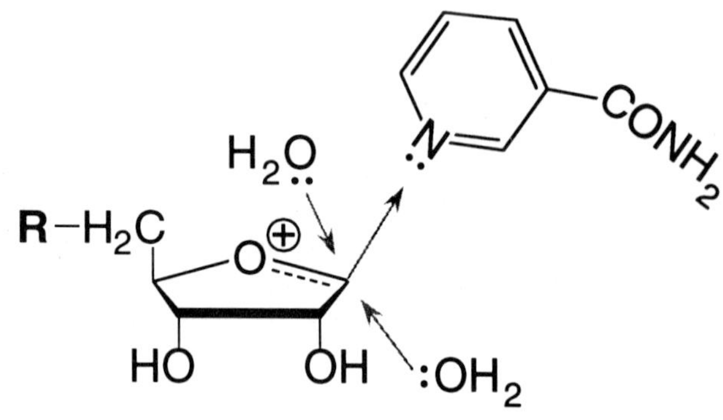

Fig. 2. Proposed exploded transition state based on Young and Jencks [22].

Another important issue is the anticipated stability/lifetime of the putative oxocarbenium ion intermediate. Young and Jencks [22] originally estimated lifetimes for pyranosyl oxocarbenium intermediates of $<10^{-15}$s, based on extrapolation from studies of sulfite addition to ketones and aldehydes. Recently this number has been revised to *ca.* 10^{-12}s [23, 24], although the more strained furanosyl cation should be less stable. As they discuss, if the lifetime of a putative intermediate is less than an atomic vibration, $<10^{-13}$s, then some form of stabilization is needed because under such circumstances the intermediate could not exist as a free entity, let alone as a solvent-equilibrated species.

Nucleophilic stabilization via a single "exploded" transition state, as shown in Fig. 2, has been invoked to explain the results for solvolysis of α and β-glucopyranosides. The reaction would then proceed with the weak, but *obligatory participation* of both the leaving group and the incoming nucleophiles, to stabilized the activated complex. The observed mixture of products would then result from the participation of the incoming nucleophile with either retention or inversion.

A different pictures emerges from PM3 calculations of the transition state and reaction coordinates for the hydrolysis of nicotinamide riboside [25]. The approach of the nucleophile to the anomeric carbon is energetically unfavorable until the bond order to the nicotinamide is10%, at which time the barrier collapses. Although this transition state appears 'exploded', there are two distinct activated complexes differing by about 1.5 kcals, one leading to retention and the other to inversion. Stabilization of the oxocarbenium by the incoming nucleophile is minimal. The observation that the gas-phase dissociation of 2′-substituted nicotinamide arabinosides correlates with the rates for the solution reaction indicates that the stabilization from the leaving group is comparable, with or without solvent, a result in accord with the calculations [25]. In either case the evidence points to an activated complex with extensive bond breaking and little nucleophilic participation to stabilize the oxocarbenium intermediate.

Enzyme-catalyzed Hydrolysis of NAD⁺

Stereochemistry of NAD cleavage and nucleophilic substitution. All known enzymes that cleave the nicotinamide-ribosyl bond act only on the β-anomer. The enzymes can be divided into three distinct categories with respect to the stereochemistry of the linkage they generate with the ultimate acceptor of the ADPRibosyl moiety: retention of configuration, inversion of configuration, and inability to form a configurationally stable product. Enzymes catalyzing formation of α-linkages include ADPRtransferases [26], toxins [27, 28] and polyADPRibose synthase [29]. These enzymes also show a remarkable substrate specificity, catalyzing the slow hydrolysis of NAD^+ without catalyzing methanolysis [26,27].

The calf spleen and pig brain NADases catalyze the exchange of the nicotinamide moiety with exogenous nucleophiles. They are able to conduct this transglycosidation because they retain the energy of the nicotinamide ribosyl bond in the form of a stabilized ENZ•ADPR intermediate. This intermediate behaves as if it were a stable, solvent-equilibrated oxocarbenium reacting with exogenous nucleophiles, including solvents, all with retention of configuration [30,31]. For example, methanol is 100 times more effective than water at trapping the enzyme-bound intermediate, a result in accord with their intrinsic nucleophilicity [32]. In contrast, the non-enzymic reaction in aqueous methanol partitions according to the molefraction of the reactants [12]. The nicotinamide insensitive NADases; e.g., the *N crassa* NADase, catalyzes a rapid and highly selective hydrolysis of NAD^+, but without any detectable methanolysis. The rapid isomerization of the 1′-hydroxyl of ribose prevents determination of the stereochemistry of the hydrolysis reaction.

The origins of these stereochemical differences is unclear. They may result from fundamentally distinct mechanisms for these enzymes. Alternatively, the mechanism could remain the same, but access of the acceptor to the intermediate in the active site could differ. Where amino acid sequences are available, there is no indication of any evolutionary relationships among the enzymes. Indeed, even for toxins with similar function, such as diphtheria toxin and *P. Aeruginosa* exotoxin A, the sequences are so different that only photolabeling of an active site glutamate led to the discovery of weak alignments [33].

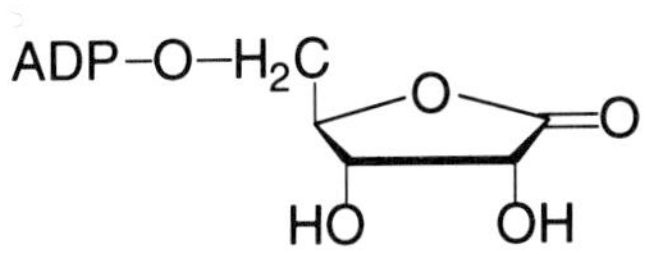

1

Structure 1

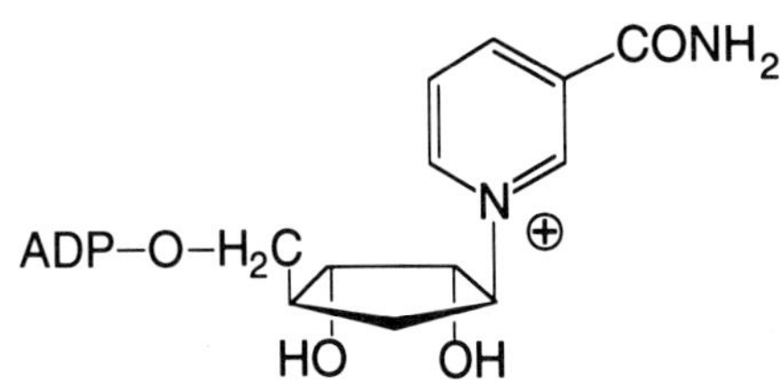

2

Structure 2

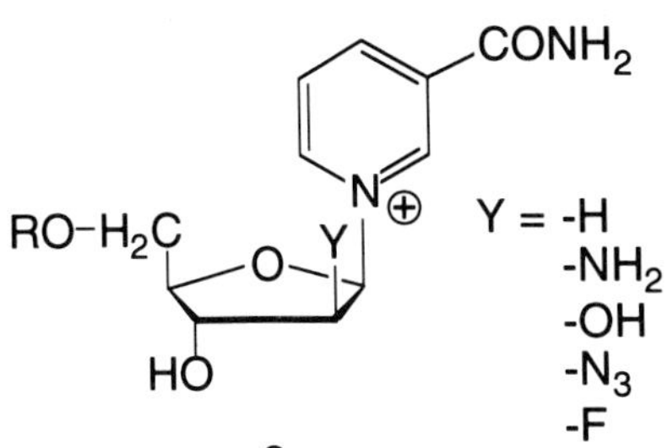

3

Structure 3

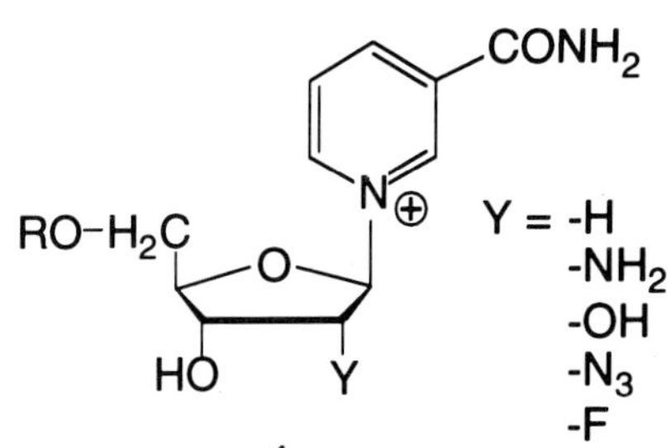

4

Structure 4

Inhibition of NAD-glycohydrolases. The design of inhibitors of NAD$^+$-glycohydrolases have focused on the sp^2 oxocarbenium intermediate. The first such compound was ADP-ribonolactone,1, with a K_i of 115 μM, only 9 times better than ADP-ribose (K_i = 1.0 mM) despite the planar atom at C1′ [34]. This strategy had been successfully used to inhibit lysozyme, another enzyme that cleaves glycosyl bonds and is thought to generate an oxocarbenium intermediate [35]. The carbocyclic analogue of NAD$^+$ containing a 1-N-nicotinamide-2,3-dihydroxycyclopentane adenine dinucleotide (carba-NAD$^+$) is a structural analogue of the substrate containing a C–N bond to the nicotinamide ring that should be resistant to hydrolysis [36]. As anticipated, it is not a substrate for NAD$^+$ glycohydrolase, but neither is it a particularly good inhibitor; its inhibition constant is comparable to that for ADP-ribonolactone [37]. Interestingly, the L-carbocyclic ribose analogue ψ-carbaNAD$^+$,2, is a better inhibitor with a K_i of 6.7 μM for the bovine brain NADase.

Although the calf spleen NAD-glycohydrolase is insensitive to the substitution of the pyridine ring, it is especially sensitive to modification of the configuration at the 2′-position of the sugar. Nicotinamide arabinoside adenine dinucleotide,3, (araNAD$^+$) and nicotinamide 2′-deoxy-2′-fluoroarabinoside adenine dinucleotide,4,(2′-F-araNAD$^+$) are not hydrolyzed at measurable rates (an upper bound of $<10^{-4}$ and 10^{-6} respectively) and are the first documented examples of reversible slow-binding inhibitors of this class of enzyme [38]. The kinetic data are consistent with both slow k_{on} and k_{off} rate constants in the formation of an enzyme • inhibitor complex; i.e., the association rate constants are about 10^6 and 10^4 slower than diffusion rates respectively for araNAD$^+$ and 2′-F-araNAD$^+$ and the half-life of the complex is about 3 m for both analogues. The 2′-F-araNAD$^+$ analogue, with a K_i = 30 nM, is also the most potent nucleoside-based inhibitor described for an NADase. In the absence of structural date on NAD$^+$-glycohydrolases, one can only speculate on the origins of the slow association rate constants, the tight binding, and the lack of catalysis. Inhibition is not limited to the NADases catalyzing transglycosylation. Other enzymes that catalyze hydrolysis of the nicotinamide–ribosyl bond, including diphtheria toxin, cholera toxin, and ADPRibose cyclase, are competitively inhibited by these 2′-substituted arabino analogues [39]. At the very least these inhibitors demonstrate that there is a heretofore unsuspected mechanistic importance of the ribosyl 2′-hydroxyl for the enzyme-catalyzed reaction.

Kinetic isotope effects. Secondary deuterium isotope effects have been measured for the hydrolysis of [^{2}H-N1′]-NAD$^+$,[^{2}H-N1′]-NMN$^+$, and [^{2}H-N1′]-nicotinamide riboside catalyzed by a series of NADases [16, 17] and the results are listed in Table 1. For enzymes showing a kinetic isotope effect, the values are similar to those for the chemical reaction, indicating that the enzyme-catalyzed cleavage of the nicotinamide–ribosyl bond also occurs with extensive rehybridization of the anomeric carbon, consistent with a dissociative mechanism.

Substrate specificity and leaving group effects. Linear free energy relationships for the calf spleen NADase-catalyzed hydrolysis of NAD$^+$ analogues containing a series of pyridines substituted at either C-3 or C-4 show that the rate of hydrolysis is inversely correlated with the pK_a of the departing pyridine [18]. A Brønsted plot of Log(V_{rel}) vs. pK_a gives a slope of β_{LG} = -0.9 over 4 pK_a units. The sign and magnitude of this slope indicates the development of considerable electron deficiency with a large degree of bond-breaking in the transition state. Interestingly, only NAD$^+$ and NMN$^+$ deviate from the otherwise linear plot and are hydrolyzed 14 and 2.8 times faster, respectively, than anticipated from their pK_as. These results suggest that interactions at the carboxamide, such as increased electron withdrawal by strong hydrogen bonding to a proton donor, might alter the pK_a of the nicotinamide, thus further labilizing the ribosyl bond, although the interactions involved are less than 1.5 kcals. The strong dependence of the rate of hydrolysis on the pK_a of the base demonstrates that the reaction is insensitive to the location of the substituent on the ring. This result is consistent with the active site residues having only a minimal interaction with the base.

Hydrolytic activity with the 2′-substituted ribo analogs. In contrast to the arabino analogues, substitutions at the 2′-position having a ribo configuration are well-tolerated by the enzyme and serve as substrates. A plot of the log of the relative rate for the hydrolysis of the etheno derivatives of

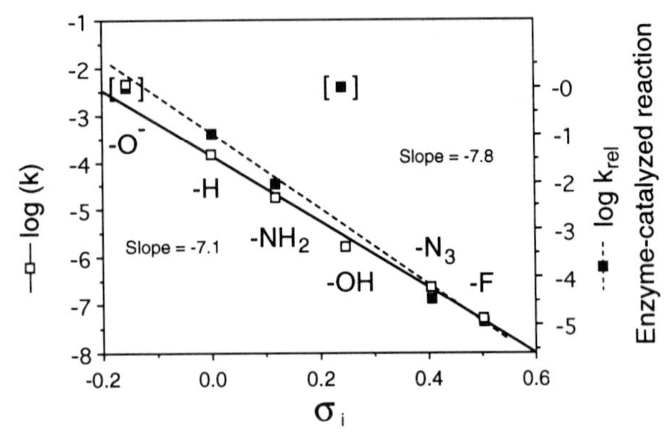

Fig. 3. Plot of the log of the rate of hydrolysis of 2′-substituted NAD analogues vs. the Taft σ_i. The axis on the left and the solid line plot (❑) refer to the log of the rate constants for the pH-independent chemical hydrolysis of the NAD analogues. The axis on the right and the dashed line plot (■) refer to the relative rates of calf-spleen NADase-catalyzed hydrolysis of the analogs. The two points in brackets designate the relative rate for the enzyme-catalyzed hydrolysis of NAD corresponding to σ_i values of an alkoxide and hydroxyl substituent respectively.

NAD$^+$, 2′deoxy-, 2′-amino, 2′-azido-, and 2′F-NAD$^+$ against their Taft σ_i values gives a slope, ρ = -7.8, a result nearly identical to the ρ for the non-enzymatic reaction of -7.1 (see Fig. 3). A crucial issue in this study is the appropriate value of σ_i to use for NAD$^+$ in the plot. If the σ_i value of a neutral hydroxyl is used, then the point lies far above the line defined by the other analogues. On the other hand, if the σ_i value for an alkoxide is used, then the point falls directly on that line. The results of these studies have two important consequences: first, they establish that enzyme catalysis has essentially the same sensitivity to substituent effects as the non-enzymic reaction, and second, the reactivity of NAD$^+$ toward enzyme-catalyzed hydrolysis is comparable to that of the *diol* anion.

The Mechanism of Cleavage of the Nicotinamide–ribosyl Bond. The central chemistry catalyzed by the ADPRTases and NADases is cleavage of the nicotinamide ribosyl bond. All evidence points to this reaction being a dissociative, Type IV S_N1 hydrolysis, but therein lies a mechanistic dilemma. How can enzymes catalyze an S_N1 reaction, which is unimolecular by definition?

$$:Y^{\circ} + X\text{-}R^{+} \rightarrow :X^{\circ} + Y\text{-}R^{+}$$

Type IV nucleophilic substitution reaction

The Cordes dilemma. Cordes and coworkers first elaborated the mechanistic dilemma regarding enzymes conducting cleavage of a nicotinamide–ribosyl bond via a dissociative mechanism [16]. General acid catalysis is rule out because the leaving group is already cationic. General base or nucleophilic catalysis is precluded because of the large secondary kinetic isotope effects reflect major rehybridization in the activated complex, consistent with formation of an oxocarbenium ion intermediate. Finally, because both the activated complex and substrate are positively charged, the importance of differential electrostatic stabilization is greatly diminished. The cleavage of the nicotinamide bond results in only a short range redistribution of charge from the ring nitrogen to the adjacent anomeric carbon.

In the absence of viable mechanistic alternatives, catalysis by distortion or strain was suggested to explain the enzyme-catalyzed rate acceleration of cleavage of the nicotinamide–ribosyl bond by up to ten orders of magnitude from that for the uncatalyzed reaction [16]. More importantly, *this dilemma is not limited to the NADases*. Once an aglycone is protonated, the mechanism of other enzymes catalyzing dissociative cleave of glycosyl bonds should operate in an analogous manner. Indeed, the ultimate mechanism of such enzymes as lysozyme or β-galactosidase comes down to debates over the relative energies of distorted intermediates; for example, see the review by Sinnott [21].

Implicit in mechanistic arguments based on strain is the concept that an enzyme can convert binding energy into decreased strength of the scissile bond, leading to rate enhancements comparable to those afforded by more direct chemical means [40]. For the calf spleen NADase there is little discrimination with respect to the pyridinium moiety based on shape or location of substituents. The reactivity primarily correlates with the basicity of the pyridine leaving group [18]. Likewise, the enzyme cleaves 2′-substituted ribo analogues with rates that reflect the relative stability for the oxocarbenium ions, and not the size or shape of the 2′-substituent. These properties are inconsistent with the predicted tight binding needed at both ends of the scissile bond to promote catalysis by strain. If catalysis by strain is thus precluded, then more fundamental aspects of the chemistry must be examined in order to explain the mechanism of enzyme catalysis.

Proposed mechanism for enzyme catalysis. The following discussion provides at least a partial explanation for the observed rate acceleration and properties of the enzyme (see Fig. 4). Approximately four orders of magnitude of enzymic catalysis can be accounted for if the ribose diol were effectively ionized to the functional equivalent of an anion. Note that a fully formed diol anion (pK_a~ 12) need not be invoked in the active site, but rather that the inductive polarization at the 2′ position approach that of the anion. The same semantics are used for discussions of the active site serine hydroxyl of proteases, that it carries 15% of a 'negative charge', but it is not 15% ionized [41].

The model shown in Fig. 4 accounts for the mechanistic involvement of the 2′-OH in bond scission, as suggested in previous studies [12], and the inability of NADases and ADPRTases to cleave araNAD$^+$ or 2′-F-araNAD$^+$. Second,

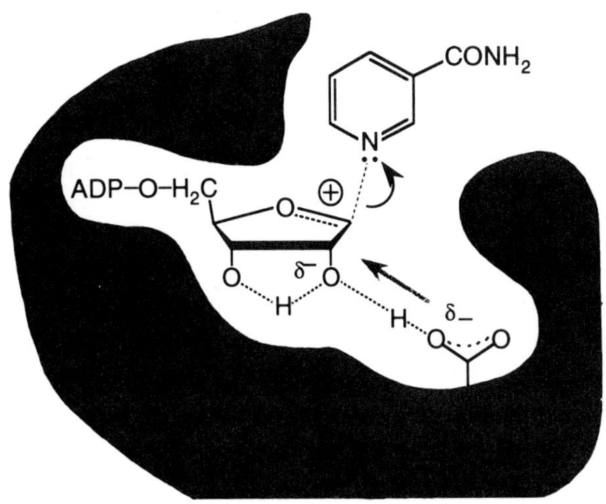

Fig. 4. Proposed mechanism for the inductive stabilization of an oxocarbenium ion intermediate via an interaction between the 2′-hydroxyl and an active site carboxylate.

it predicts the presence of a basic group in the active site, preferably anionic. Photoaffinity reactions of NAD^+ and site-directed mutagenesis have shown a glutamate in the active sites of diphtheria toxin and Pseudomonas exotoxin A [42,43].

There is an important literature precedent for the properties of the NAD analogues. In the AMPase-catalyzed hydrolysis of AMP and 2′dAMP, the V_m/K_m for AMP is 48 times greater, whereas in the acid-catalyzed hydrolysis, 2′dAMP is favored over AMP by ca. 1000 fold [42, 43]. Extensive investigations of the kinetic effects of heavy isotope replacement demonstrate that AMPase works by preprotonation of the purine at N-7 for both ribo- and deoxyribonucleotides. Therefore, the greater rate for AMP in the enzyme reaction reflects either a major suppression of the intrinsic reactivity of the protonated 2'dAMP, or that the enzyme strongly enhances the reactivity of AMP (see Table 2). Because 2′AMP is otherwise an excellent substrate, it is hard to imagine that its reactivity is being suppressed by such a large extent (ca. 3.5 x 10^4 times slower [12]). Rather, as observed

Table 2. Comparison of rates for the acid-catalyzed and AMP nucleosidase-catalyzed hydrolysis of 5′-AMP and 5′-dAMP

Reaction	5′-AMP	5′-dAMP	5′-AMP/5′-dAMP
Acid catalyzed(45) (k x 10^5 s^{-1})	16.5	12,500	0.00132
AMP-Nucleosidase(42) (V_m/K_m s^{-1} mg^{-1})	16.8	0.35	48
Acid catalyzed and diol anion[a] (k x 10^5 s^{-1})	165,000	12,500	13.2

[a]The anticipated rate acceleration of 10^4 for hydrolysis if the ribose diol were to be ionized for an AMP with a protonated purine ring, a situation impossible in solution but easily envisioned in an enzyme active site. No significant difference in rate is anticipated for dAMP because the 3′-hydroxyl is both further removed, hence less able to stabilize an oxocarbenium ion, and its pK_a is at least 2 pH units higher than the hydroxyls of the ribose diol.

for 2′-substituted NAD^+ analogues, it is the riboAMP that is accelerated by approximately the amount anticipated for the ionization of the ribose diol while maintaining the protonation at N-7. Note that the AMPase can hydrolyze nicotinamide mononucleotide [42], suggesting that in this example the protonated aglycone and the cationic pyridinium are mechanistically equivalent with respect to the cleavage of the glycosyl bond.

As shown in Fig. 1, 2′d-NAD^+ hydrolyzes more rapidly in solution than the ribonucleotide, yet it is hydrolyzed enzymatically at a slower rate than NAD^+. The results for these two different enzymes suggest that at least one component of catalysis is the participation of the equivalent of the ribose diol anion stabilizing the oxocarbenium intermediate. Clearly, from the results shown in Fig. 3, the calf spleen NADase does more; it promotes catalysis of all the 2′-substituted-NAD analogues, even the putative diol anion by an additional factor of 10^4 to 10^5.

Remaining questions

The chemistry catalyzed by the Calf Spleen NADase parallels the solution chemistry. It is subject to the same substituent effects with respect to the 2′-position of the sugar moiety (19) and to the same leaving group effects with respect to substituted pyridines [18]. The effective ionization of the diol relative to other 2′-substituted analogues provides about half the observed rate acceleration at neutral pH, explaining the efficiency for NAD^+ hydrolysis. Therefore a major component of enzyme catalysis has been identified for NAD^+ itself, but there still remains the question as to the mechanism by which hydrolysis is accelerated for all the 2′-substituted analogues, independent of the 2′-substituent or its ionization state. How does an enzyme catalyze a unimolecular dissociative reaction via a highly electron-deficient complex, when the standard modes of catalysis have been effectively dismissed?

A highly speculative, yet provocative possibility emerges from the preliminary investigations into the gas-phase stability of nicotinamide nucleosides [44]. The results suggests that solvation is a critical component of the stability of the ribosyl linkage in solution. It is the presence of the surrounding solvent molecules that impedes the departure of the base and promotes recombination with the oxocarbenium. When the solvent molecules are removed; e.g., in a monomolecular dispersion in vacuum, the nucleoside (< 5 microseconds) dissociates to the oxocarbenium (or related products resulting from internal attack) and nicotinamide, with essentially no possibility for their recombination. Likewise, desolvation of the nucleotide in the active site in the *absence* of tight binding can be envisioned as providing sufficient separation of the species to decrease the efficiency of their recombination, hence catalyzing the cleavage of the glycosyl bond. This suggestion clearly represents a dramatic departure from the classical paradigm for enzymic catalysis first proposed by Pauling in which transitions states are bound more tightly than the reactants. In the case of NADases this idea may have to be turned on its head. That in order to promote cleavage of the ribosyl linkage, catalysis might involve a transition state that is bound *less* tightly than the reactant.

Summary

The hydrolysis of nicotinamide–nucleosides appears simple. Nicotinamide is a good leaving group, the reaction is strongly favored thermodynamically, and the putative oxocarbenium intermediate is of low but accessible stability. Nonetheless, there are a number of puzzling features about the reaction that are not intuitively obvious. Why is the reaction dissociative? Why does direct solvent stabilization appear to be absent? Why are powerful nucleophiles like azide or hydroxide unable to displace nicotinamide? What is the nature of the barrier (e.g., dipolar, geometric, etc.) that prevents access of nucleophiles to C1′ until the bond to nitrogen is essentially broken? And finally, how can an enzyme catalyze the reaction? These are non-trivial questions that address fundamental issues of nucleophilic substitution reactions and enzyme catalysis.

References

1. Oppenheimer NJ, Handlon AL: Mechanism of NAD-dependent Enzymes. The Enzymes 20: 454–505, 1992
2. Jacobson M, Jacobson E: ADP-Ribose Transfer Reactions. Springer-Verlag. New York. 1989
3. Moss J, Vaughan M: ADP-Ribosylation and G Proteins. American Society for Microbiology. Washington, D.C. 1990
4. Moss J, Vaughan M: ADP-Ribosylation of Guanyl Nucleotide-Binding Regulatory Proteins by Bacterial Toxins. Adv Enzymol 61: 303–379, 1987
5. Lowery RG, Ludden PW: Purification and Properties of Dinitrogenase Reductase ADP-Ribosyltransferase from the Photosynthetic Bacterium *Rhodospirillum rubrum*. J Biol Chem 263: 16714–16719, 1988
6. Fu H, Burris RH, Roberts GP: Reversible ADP-ribosylation is Demonstrated to be a Regulatory Mechanism in Prokaryotes by Heterologous Expression. Proc Natl Acad Sci USA 87: 1720–1724, 1990
7. Althaus FR, Richter C: ADP-Ribosylation of Proteins. Springer-Verlag. Berlin. 1987
8. Lee HC, Aarhus R: ADP-Ribosyl Cyclase: an Enzyme that Cyclizes NAD^+ into a Calcium-Mobilizing Metabolite. Cell Regul 2: 203–209, 1991
9. Anderson BM, Anderson CD: The Effect of Buffers on Nicotinamide Adenine Dinucleotide Hydrolysis. J Biol Chem 238: 1475–1478, 1962
10. Johnson SL, Morrison DL: The Alkaline Reaction of Nicotinamide Adenine Dinucleotide, a New Transient Intermediate. J Biol Chem 245: 4519–4524, 1970
11. Guilbert CC, Johnson SL: Isolation and Characterization of the Fluorescent Alkali Product from Diphosphopyridine Nucleotide. Biochemistry 10: 2313–2316, 1971
12. Johnson RW, Marschner TM, Oppenheimer NJ: Pyridine Nucleotide Chemistry. A New Mechanism for the Hydroxide-Catalyzed Hydrolysis of the Nicotinamide-Glycosyl Bond. J Am Chem Soc 110: 2257–2263, 1988
13. Ta-Shma R, Oppenheimer NJ: Unpublished results.
14. Tarnus C, Muller HM, Schuber F: Chemical Evidence in Favor of a Stabilized Oxocarbonium-Ion Intermediate in the NAD+ Glycohydrolase-Catalyzed Reactions. Bioorg Chem 16: 38–51, 1988
15. Cordes EH, Bull HG: Mechanism and Catalysis for Hydrolysis of Acetals, Ketals, and Ortho Esters. Chem Rev 74: 581–603, 1974
16. Bull HG, Ferraz JP, Cordes EH, Ribbi A, Apitz-Castro R: Concerning the Mechanism of the Enzymatic and Nonenzymatic Hydrolysis of Nicotinamide Nucleotide Coenzymes. J Biol Chem 253: 5186–5192, 1978
17. Ferraz JP, Bull HG, Cordes EH: Kinetic α-deuterium Isotope Effects for Enzymatic and Nonenzymatic Hydrolysis of Nicotinamide Riboside. Arch Biochem Biophys 191: 431–436, 1978
18. Tarnus C, Schuber F: Application of Linear Free-Energy Relations to the Mechanistic Probing of Nonenzymatic and NAD+-Glyco-hydrolase-Catalyzed Hydrolysis of Pyridine Dinucleotides. Bioorg Chem 15: 31–42, 1987
19. Handlon AL, Oppenheimer NJ: Substituent Effects on the pH-Independent Hydrolysis of 2′-Substituted Nicotinamide Arabinosides. J Org Chem 56: 5009–5010, 1991
20. Handlon AL, Oppenheimer NJ: Unpublished results.
21. Sinnott ML: Glycosyl Group Transfer. In: M.I. Page and A. Williams (ed), Enzyme Mechanisms, Burlington House, London, 1987, pp 259–297
22. Young PR and Jencks WP: Trapping of the Oxocarbonium Ion Intermediate in the Hydrolysis of Acetophenone Dimethyl Ketals. J Am Chem Soc 99: 8238–8248, 1977
23. Banait NS and Jencks WP: General-Acid and General-Base Catalysis of the Cleavage of α-D-Glucopyranosyl Fluoride. J Am Chem Soc 113: 7958–7963, 1991
24. Banait NS, Jencks WP: Reactions of Anionic Nucleophiles with α-D-Glycopyranosyl Fluoride in Aqueous Solution through a Concerted $A_ND_N(S_N2)$ Mechanism. J Am Chem Soc 113: 7951–7958, 1991
25. Schroder S, Buckley N, Oppenheimer NJ, Kollman PA: A Quantum Chemical Study of the Type IV Nucleophilic Substitution Reaction and Dissociation of the β-Nicotinamide Glycosyl Bond in the Gas Phase Using Semiempirical PM3 Calculations. J Am Chem Soc 114: 8232–8238, 1992
26. Moss J, Garrison S, and Oppenheimer NJ: NAD-dependent ADP-ribosylation of Arginine and Proteins by Escherichia Coli Heat-labile Enterotoxin. J Biol Chem 254: 6270–6272, 1979
27. Oppenheimer NJ: Structural Determination and Stereospecificity of the Choleragen-catalyzed Reaction of NAD^+ with Guanidines. J Biol Chem 253: 4907–4910, 1978
28. Oppenheimer NJ and Bodley JW: Diphtheria Toxin. Site and Configuration of ADP-ribosylation of Diphthamide in Elongation Factor 2. J Biol Chem 256: 8579–8581, 1981
29. Ferro AM, Oppenheimer NJ: Structure of a Poly(ADP-ribose) monomer: 2′-(5″-phosphoribosyl)-5′-adenosine monophosphate. Proc Natl Acad Sci USA 75: 809–813, 1978
30. Schuber F, Travo P, Pascal M: On the Mechanism of Action of Calf Spleen NAD^+ Glycohydrolase. Bioorg Chem 8: 83–90, 1979
31. Pascal M, Schuber F: The Stereochemistry of Calf Spleen NAD^+-Glycohydrolase-Catalyzed NAD Methanolysis. FEBS Letters 66: 107–109, 1976
32. Sinnott ML, Viratelle OM: The Effect of Methanol and Dioxan on the Rates of the β-Galactosidase-Catalyzed Hydrolyses of some β-D-Galactopyranosides:Rate-Limiting Degalactosylation. Biochem J 133: 81–87, 1973
33. Carroll SF, Collier RJ: Active Site of Pseudomonas aeruginosa Exotoxin A. J Biol Chem 262: 8707–8711, 1987
34. Schuber F, Pascal M: ADP-ribonolactone: a Potential Activated Intermediate Analogue of NAD Glycohydrolase. FEBS Lett 73: 92–96, 1977
35. Secemski II, Lehrer SS, and Lienhard GE: A Transition State Analog for Lysozyme. J Biol Chem 247: 4740–4748, 1972
36. Slama JT, Simmons AM: Carbanicotinamide Adenine Dinucleotide: Synthesis and Enzymological Properties of a Carbocyclic Analogue of Oxidized Nicotinamide Adenine Dinucleotide. Biochemistry 27: 183–193, 1988
37. Slama JT, Simmons AM: Inhibition of NAD glycohyrdolase and ADP-ribosyltransferase by carbocyclic analogues of oxidized NAD^+. Biochemistry 28: 7688–7694, 1989
38. Muller-Steffner HM, Malver O, Hosie L, Oppenheimer NJ, Schuber F: Slow-Binding Inhibition of NAD^+ Glycohydrolase by Arabino-analogues of β-NAD^+. J Biol Chem 267: 9606–9611, 1992
39. Oppenheimer NJ: Unpublished results.

Molecular and Cellular Biochemistry **138**: 253–255, 1994.

Index